W9-BLE-191
3 1611 00160 1795

PLANT–ANIMAL INTERACTIONS

PLANT–ANIMAL INTERACTIONS

Edited by
WARREN G. ABRAHAMSON
Department of Biology
Bucknell University

MCGRAW-HILL BOOK COMPANY
New York St. Louis San Francisco Auckland
Bogotá Hamburg London Madrid Mexico
Milan Montreal New Delhi Panama
Paris São Paulo Singapore
Sydney Tokyo Toronto

Library of Congress Cataloging-in-Publication Data

Plant-animal interactions.

Bibliography: p.
Includes index.
1. Biotic communities. 2. Ecology. 3. Insect-plant relationships. I. Abrahamson, Warren G.
QH541.P58 1989 574.5'247 88-13190
ISBN 0-07-000179-0

1 2 3 4 5 6 7 8 9 0 DOCDOC 8921098

ISBN 0-07-000179-0

For more information about other McGraw-Hill materials, call 1-800-2-MCGRAW in the United States. In other countries, call your nearest McGraw-Hill office.

CONTENTS

PREFACE

Until recent decades, scientists often considered the biochemistry, physiology, morphology, or development of plants as being essentially independent of these subdisciplines for animals. This independence was reflected to some extent by the separation of biology into the disciplines of botany and zoology. More common today is the view that plants and animals should be considered in a holistic, interacting manner. The biochemistry of a plant species, for example, may have been markedly altered by the selective pressures created through long-term attack by various invertebrate and/or vertebrate animals. Likewise, the physiology of these herbivores may be appreciably different today from that of a few thousand years ago as adaptations for the detoxification or avoidance of plant secondary chemical compounds have arisen. This evolutionary, holistic approach to plants and animals has produced a renaissance within fields as diverse as biochemistry, physiology, morphology, behavior, and most significantly, ecology.

In ecology, this approach flourishes in studies that focus on plant–animal interactions. These studies, using a highly analytic approach to biology, are underpinned by theories founded in genetics. These theories assume that genetic shifts occur in the populations of interacting species as a result of selective pressures produced by the interactions themselves. This selection, by acting upon the heritable variation in the interacting organisms' traits, produces evolutionary responses in the form of adaptations.

This edited textbook attempts to synthesize the general evolutionary and ecological patterns of plant–animal interactions through the survey of numerous kinds of interactions (e.g., pollinator–plant, herbivore–plant). Further, the volume examines the conditions that may or may not favor coevolution between plants and animals. The contributing authors convey the dynamic state of plant–animal interactions by treating controversial concepts and by presenting plausible hypotheses. It is, however, likely that some of the hypotheses this book offers may be discarded with the information provided by future experiments. There are few irrefutable facts in any science.

The broad survey of interactions presented in this text demonstrates that the principles of the various plant–animal interactions are frequently the same. One of my goals for this volume was to identify the common perspectives and approaches utilized in a wide variety of plant–animal interactions. Each of the following chapters describes what is known about the specific kind of plant–animal interaction discussed. We hope that these syntheses will serve as guides for future research.

This textbook took form over the past two decades as I have watched the subdiscipline of plant–animal interactions grow and develop. Appreciable progress has been made in that time span and the rate of publication of new information on plant–animal interactions increases each year. Thus, this textbook is best seen as one of the initial steps towards synthesis. It is not *the* synthesis.

The book has been written for the advanced undergraduate and beginning graduate student. The volume provides a convenient entry into the theory, empirical data, interpretation, and literature of many kinds of plant–animal interactions. The authors have assumed that the textbook's readers have had thorough introductions to the basic principles of ecology, evolution, genetics, morphology, physiology, biochemistry, and natural history of plants and animals. However, since today biology curricula vary markedly in perspective and topics covered, the authors define many terms within the contexts of their chapters. The book should also be valuable to researchers in plant–animal interactions since it provides ready access and summary of much plant–animal interactions theory, data, and literature.

Depending on the format used for the course in which this book is to be used, instructors may wish to emphasize some topics in greater detail than other chapters. The plant–animal interactions course that spawned the concept for this textbook uses a lecture, discussion, and laboratory format. Although that course does cover all the topics included in this text, it does not give each topic equal weight. We leave it to individual instructors to determine their particular emphasis. The laboratory component of a plant–animal interactions course is not dealt with in this book, but plant–animal interactions courses are best taught with both field and laboratory exercises to supplement lecture and discussion materials. The laboratory/field experience is important to help hone students' understanding of plant–animal interactions.

I offer my deep appreciation to my friends, colleagues, and students, too numerous to list, who encouraged me to proceed with this task. I thank Otto Solbrig and Winslow Briggs, both of whom urged me to develop my initial plant–animal interactions course (1972–1973) while I was at Harvard University. Gregory Payne, Richard Root, Mark Rausher, and Jon Roughgarden provided the immediate encouragement necessary to set this project into motion. I thank the book's artist, Karen Prather, for her excellent illustrations and Karen Shrawder for help in typing parts of the manuscript. I appreciate the critical technical reviews provided by Ann Johnson and Doris Dysinger. My deep respect and hearty thanks are offered to Irene Kralick for her attention to detail throughout the course of the book's production and for her care in typing the bulk of the book and its related correspondence. I especially thank the authors of the following chapters for sharing their ideas, knowledge and synthesis. Finally, I thank my wife, Chris, who was willing to listen and discuss my ideas and goals for this edited textbook.

Warren Abrahamson

FOREWORD

When I was a graduate student in entomology in the early 1960s I was told not to waste my time on courses in the botany department. During 4 years of graduate school in insect ecology I never heard the words plant defensive chemical or secondary compound. Yes, plants were known to be food for animals, but people also asked, in all seriousness, why, since all the world was green, it was not eaten to the ground by herbivores. Pollination biology consisted of how to get more seed set from crop plants, and lists of bees caught at flowers. Ant–plants resided in yellowing dusty pages from long dead explorers in exotic places like Africa, Java, and the Amazon. Seed dispersal by animals consisted of lists of animals in fig trees and compendia of robin stomach contents. A mere 20 years later, Abrahamson and his chapter writers are able to orchestrate studies by at least 1000 researchers, 90 percent of whom are still living in the prime of their academic careers, into a complex statement with an inescapable implicit conclusion. What we call "plants" today were generated almost entirely by interactions with animals. And, the majority of what we call animals today (and yesterday) were generated by their interactions with plants and other animals that eat plants. Oh, what a dull world it would be if in fact plants were just green fodder for animals.

In the explosive development of a scientific subfield, a great wealth of examples, hypotheses, interpretations, errors, myths, traditions, etc., appear—a jungle of quick history, seeded on virtually bare ground. One academic generation later, new minds and personalities are still entering the jungle in search of new opportunities to add yet more facts, coherence, and understanding. But the task is now enormous, greatly impeded by the thickets of fast-growing early successional plants on the intellectual landscape. A book such as this one—and quite specifically this one—offers a new baseline to the new explorer. It distinguishes the bramble patches from shrub-free but resource-poor understory in the deep shade of large trees, and points to tree falls and landslides as resource-rich sites for growth. A couple of weeks spent thoroughly digesting this book is, in my opinion, a must for the newcomer to animal–plant interactions. Follow it with a month of library natural history, reading those papers that really caught your attention as you read of them. *Then* is the time to ask where to start your personal quest into this still very rich, untapped habitat.

There is a widespread feeling that (what remains of) the tropics are the epitome of habitats rich in animal–plant interactions, yet only 26 percent of the 1100+ references cited here deal directly with tropical organisms. Abrahamson's authors have generally brought us back to the reality that there is an enormous extratropical audience that needs to know about its own animal–plant interactions as well as those in Brazil. On the other hand, the cipher of 26 percent should also be viewed in the context that less than 10 percent of the biologists who might conceivably work on animal–plant interactions live in extratropical regions. Since well over 70 percent of the world's terrestrial species and more than 90 percent of the animal–plant interactions occur in the tropics, it is evident that only the tip of the tropical animal–plant iceberg has been viewed by the research community. While this book brings the new reader into the animal–plant arena, there is clearly much left to sift through and analyze.

What stands in the way of a second round of concentrated study, the one to be conducted by the students who read this book? The world's biggest pool of animal–plant interactions is ours to behold for only one or two decades more, at best. Genetic engineering is removing the single greatest conservation force that ever occurred in the tropics—agricultural inviability. Humans are the most coevolved of any organism; the coevolution has at its center the removal of threats to a multiplicity of wild and tame domesticates. Well, we have won. The biggest threat is that the animal–plant interactions will be destroyed before we can study them. There is a tropical–extratropical gradient in restorability; North America is the easiest, the rainforest the hardest. Why? For many decades after the first heavy wave of agricultural pioneers, the agrospace is often still bestrewn with living debris, yet the interactions with the animals—pollen and seed dispersers, and herbivores—are now almost all gone. There may be hundreds of species of orchids hanging on to fencerows as living dead, yet they do not form a viable population when restoration occurs because the pollinators disappeared with the forest. Isolated trees left standing in tropical pastures often have rotting piles of fruit beneath them, mute testimony to the extinct vertebrates that will not reappear even if the pasture trees are given back terrain on which to grow.

Wildland animal–plant interactions are what make the world go around, and wildland animal–plant interactions are the major avenue to understanding the form and function of the majority of plant traits that attract our attention. Interactions do not survive in zoos and botanic gardens, in fencerows and woodlots. The species without their interactions are about as useful as an alphabetized list of the words in a Shakespeare play. This book will go far to generate the second stage of study of animal–plant interactions, but its practitioners will have to have a widespread collaboration with many other areas of science if there is to be anything left to study.

Daniel H. Janzen
Department of Biology
University of Pennsylvania

CONTRIBUTING AUTHORS

Warren G. Abrahamson Department of Biology, Bucknell University, Lewisburg, PA 17837

May R. Berenbaum Department of Entomology, University of Illinois, Urbana, IL 61801

Robert I. Bertin Department of Biology, College of the Holy Cross, Worcester, MA 01610

William R. Bromer Department of Biological Sciences, Purdue University, West Lafayette, IN 47907

Thomas J. Givnish Department of Botany, University of Wisconsin, Madison, WI 53706

Kathleen H. Keeler School of Biological Sciences, University of Nebraska-Lincoln, Lincoln, NE 68588-0118

Richard L. Lindroth Department of Entomology, University of Wisconsin, 237 Russell Laboratories, 1630 Linden Drive, Madison, WI 53706

Kerry N. Rabenold Department of Biological Sciences, Purdue University, West Lafayette, IN 47907

Edmund W. Stiles Department of Biological Sciences and Bureau of Biological Research, Rutgers University, Piscataway, NJ 08854

Benjamin R. Stinner Department of Entomology, The Ohio State University, Ohio Agriculture Research and Development Center, Wooster OH 44691

Deborah H. Stinner Department of Entomology, The Ohio State University, Ohio Agricultural Research and Development Center, Wooster, OH 44691

Arthur E. Weis Department of Biological Sciences, Northern Illinois University, DeKalb, IL 60115

PLANT–ANIMAL INTERACTIONS

1

Plant–Animal Interactions: An Overview

WARREN G. ABRAHAMSON

Department of Biology
Bucknell University
Lewisburg, PA 17837

Introduction

Interactions among organisms can strongly influence the morphology, behavior, and ecology of those organisms. Why else, for instance, would many angiosperms produce elaborate and brightly colored petals, copious amounts of floral and extrafloral nectar, or diverse types of compounds toxic to herbivores? Natural history has recorded countless examples of adaptations that are clearly the result of past selective episodes created by species interactions. Although these interactions are more ephemeral and less tangible than a species' morphological traits, interactions are just as fundamental in a species' evolution. Not only can interactions produce the selective forces necessary to cause evolutionary changes within organisms, but these interactions linking the life histories

I thank C. R. Abrahamson, S. S. Anderson, R. I. Bertin, T. J. Givnish, A. F. Johnson, K. H. Keeler, R. L. Lindroth, K. D. McCrea, and B. R. Stinner for critical reading of this chapter.

of the Earth's inhabitants are themselves the products of evolution. As is argued later in this chapter, interactions can create new resources (e.g., nectar) that in turn allow the development of additional interactions. Species interactions contribute to the complex networks that link species into communities and ecosystems. Interactions should be viewed as one of the major organizers of community and ecosystem functions.

Until the 20th century the interactions of species, particularly those among plants and animals, received limited investigation. Ecologists now realize, for instance, that the bizarre synchronous fruiting of many bamboo species is related to seed predation and that the extinction of the common dodo more than 300 years ago may be responsible for the loss of *Calvaria* forests on the Indian Ocean island of Mauritius. In order to understand the evolutionary basis of the ecology of various species, it is necessary to critically evaluate the ways in which they interact with other species. Studying organisms without considering their interactions is akin to thinking that species exist only as collections in zoological parks or botanical gardens rather than in natural communities. Interactions are fundamental to all living organisms.

This book is about plant–animal interactions and how they have evolved. By surveying a variety of interactions and the natural history of the species involved, some general patterns will become obvious; the patterns for other, more subtle interactions have yet to clearly emerge. The following chapters focus on specific examples of interactions that represent a wide range of species relationships. These chapters do more than merely describe; they also provide a conceptual framework for understanding the evolution of interactions. Continued research and reflection on organismal interactions will allow advances in our understanding of the ecology and evolution of the biosphere.

Types of Plant–Animal Interactions

Plants and animals evolve in response to a background of selective pressures created by their physical environment and the other organisms with which they interact. This setting provides for numerous kinds of interactions, some of which are mutualistic, while others are antagonistic (e.g., parasitism, predation) or commensalistic. These types are defined on the basis of whether the effects of the interaction are beneficial, harmful, or neutral for each interacting species (Table 1-1). It is important to realize, however, that sharp boundaries often do not occur among these types of interactions and that interactions, whether beneficial, harmful, or neutral, can change in time and space. Thus, these distinctions are unclear in many cases.

Table 1–1. Distinctions of several interaction types based on whether the effects are beneficial (+), harmful (−), or neutral (0) for each species of the interaction.*

Interaction	Effect on Species A	Effect on Species B
Mutualism	+	+
Commensalism	+	0
Antagonism	+	−
(Neutralism)	0	0
(Amensalism)	0	−
Competition	−	−

*Interaction types shown in parentheses are beyond the scope of this text.

Mutualism

The mention of plant–animal interactions commonly invokes thoughts of mutualism, a relationship beneficial to both participating species. There are numerous examples of mutualism involving both plants and animals. The usual relationship between a pollinator and a flower, for instance, includes a plant-produced reward (e.g., nectar, brood-place) that benefits the pollinator and an animal activity (e.g., carrying pollen to another blossom) that benefits the plant. As described in the next chapter, however, not all pollinator–plant interactions are mutually beneficial. In some cases (e.g., carrion flowers, *Stapelia*), plants deceive floral visitors and gain pollen transfer but without providing any benefit to the pollinator.

Mutualism is very common, and Thompson (1982) has argued that the high frequency of mutualism in communities depends evolutionarily on the richness of antagonistic interactions. There is good evidence, for example, that the majority of the early insect interactions with plant reproductive structures were detrimental to the plants (Crepet, 1979). Mulcahy (1979) has argued that the highly successful closed carpels of angiosperms were likely to have evolved as a defense against antagonistic floral visitors. Whatever the precise evolutionary pathway might have been, the point remains that the interaction itself can evolve from an antagonistic to mutualistic relationship. Antagonistic interactions are usually less specialized than mutualistic ones, creating the opportunity for selection to produce mutual benefits. In addition, Thompson (1982) theorized that because mutualisms often create new resources (e.g., fruits, nectar), they can become foci for the development of new interactions. Thus, mutualisms contribute to the complex network of interactions and the organization of communities.

Antagonism

Antagonistic interactions occur between populations because organisms are concentrated parcels of limited resources such as energy and nutrients (Thompson, 1982). Plants, for example, fix carbon dioxide, producing carbohydrates and other energy-rich products. These plant products are essential to the survival of almost all animals, many of whom feed directly on plants, stealing materials crucial to the plant's survival. Over evolutionary time, this process results in selection pressure favoring plant genotypes with adaptations that reduce the loss to herbivores. Simultaneously, these processes will favor herbivores that can counter plant modifications with their own set of adaptations enabling them to detoxify plant toxins, digest plant parts that are high in fiber, or circumvent a plant's physical barriers such as spines or thorns.

One form of antagonism is parasitism, an interaction in which an organism (parasite) spends much of its life either attached to or within a single host organism taking nourishment from the host. This relationship typically results in reduced reproduction and/or survival of the host. Traditionally, the term parasite is used in reference to microorganisms or small animals that attack larger animals. But many other organisms, including invertebrate herbivores such as gall-makers or aphids that spend the immature portion of their life on a single host, are clearly parasitic.

Another form of antagonism is predation, an interaction in which one organism, the predator, obtains energy or nutrients by consuming, usually killing, another organism, the prey. Typically, a predator is thought of as an animal that catches, kills, and consumes its prey. However, predation can also include the capture and digestion of insect prey by carnivorous plants or the consumption of seeds by insects.

Predation is similar to parasitism in that it is a means of feeding that brings harm to the consumed organism. The distinction between predation and parasitism is often not clear cut, especially for insect populations. There is actually a continuous gradation of interactions from predation to parasitism.

Just as not all pollinator–plant interactions are mutually beneficial, not all predator–prey relationships are entirely antagonistic. For example, Givnish (Chap. 7) suggests that certain pitcher plants may actually be involved in a mutualism with their social insect prey.

Commensalism

Commensalisms include those interactions in which one species, the commensal, benefits from another unaffected species. There are numerous examples of commensalism, such as the use of trees as nesting places by birds. Animals utilize the physical habitat provided by plants in a variety of commensalistic ways.

Some commensalisms may involve plants only indirectly. In the Serengeti Plains of East Africa, for example, large herds of mammalian herbivores migrate in temporal succession (Bell, 1971). Zebra, feeding on tall, protein-poor stems and grass sheaths (the sheath is the split tubular portion of a grass leaf that surrounds the stem), migrate first and are followed by wildebeest, which consume grass sheaths and the more nutritious grass leaves. Next come the Thompson's gazelles feeding on the nutritious regrowth of grasses and the protein-rich herbaceous vegetation. In this example, the wildebeest is the commensal of the zebra and the Thompson's gazelle is the commensal of the wildebeest. While the relationships among these animals are commensalistic, the relationship between each of these herbivores and the plant communities on which they feed is predominantly antagonistic.

Competition

In simple terms, competition is an interaction in which one organism consumes a resource that would have been available to, and might have been consumed by, another (Begon et al., 1986). Although competition is not often directly considered within the context of plant–animal interactions, it virtually always effects these interactions in many indirect ways. Rabenold and Bromer (Chap. 8) show the important role competition plays in determining animal distribution. For example, slight variations in adaptations to feeding in a particular plant community inevitably lead to differences in efficiency. These differences result in either a segregation of similar species to different resource bases within a habitat or extirpation of one of a pair of very similar species from a habitat. Thus, the variety of resources provided by plants and their attendant fauna determines the likelihood of coexistence of similar species by controlling the degree to which niche separation is possible.

It should be apparent from the foregoing discussion of mutualism, antagonism, commensalism, and competition that the division of species interactions into such categories is somewhat arbitrary, but nonetheless useful. The fundamental issue is not which category is to be applied to a given example; rather, it is the recognition that there is a continuum of interactions ranging from antagonistic to mutualistic. This recognition is crucial to our understanding of how and why such interactions exist at all.

Multitrophic-Level Interactions

Much of the ecological theory of plant–animal interactions addresses only two trophic-level systems. For instance, most discussions of herbivore–plant interactions exclude the effects of the herbivore's natural enemies (Abrahamson et al., 1983; Weis et al., 1985). Studies have shown that parasitoids (parasites that kill their hosts as a consequence of their own development) that could potentially control herbivore populations can be adversely affected by the host plant's defenses against its herbivores (van de Merendonk and van Lenteren, 1978). This suggests the necessity of evaluating not just the plant–herbivore interaction but also the plant–herbivore–natural enemies interactions if the evolution of a plant and its herbivores is to be understood.

Price et al. (1980) pointed out that all terrestrial communities based on living plants are composed of at least three interacting trophic levels, not merely two (Fig. 1–1). Some communities may have four or more interacting trophic levels. A further complication is that each trophic level typically consists of a number of species that not only interact with other trophic levels but also interact with the species within the same trophic level. These guilds of species generate selective pressures as a

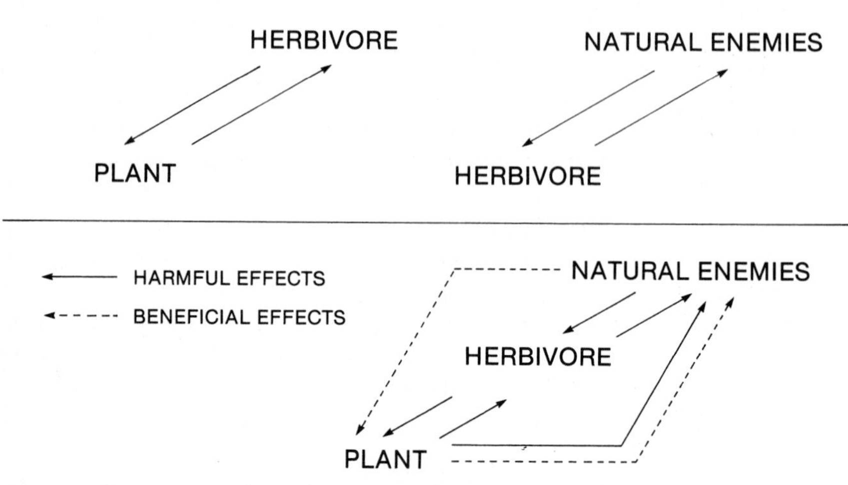

Figure 1–1. Many ecological studies isolate trophic-level interactions, as shown in the upper portion of the figure, by considering only the reciprocal effects of a plant population and its herbivores or the reciprocal effects of a herbivore and its enemies. When all three trophic levels are considered simultaneously, as illustrated in the lower portion of the figure, the additional effects of interactions involving the first with the third trophic levels become apparent. Modified from A. E. Weis (unpublished).

group (e.g., insect herbivores attacking a host plant population) that can provide opportunity for the evolution of interactions. Thus, theory on plant–animal interactions will not progress realistically without consideration of all trophic levels that can influence a given system of interactions. Many studies have begun this holistic approach to plant–animal interactions, but as the succeeding chapters show, much more needs to be learned about multitrophic-level interactions. Studies of this type have become an exciting bridge between the traditionally separate fields of population and community ecology.

Evolution of Interactions

Interactions can occur as a result of a number of different ecological or evolutionary processes. These processes include reciprocal coevolution, diffuse coevolution, mutual congruence, and evolutionary *tracking*. Although plant–animal interactions can be studied without reference to the particular process that created the interaction, investigation of the evolution of interactions is currently an active field within biology.

Reciprocal Coevolution

Various definitions for coevolution have been proposed including Janzen's (1980a) restrictive definition that defines coevolution as

> . . . an evolutionary change in a trait of the individuals in one population in response to a trait of the individuals of a second population, followed by an evolutionary response by the second population to the change in the first.

Although this is one of the better working definitions for coevolution and one adopted by many, it does have a drawback: its overemphasis of specificity and reciprocity between the interacting populations. The definition requires specificity and reciprocity since the traits of each population are due to the traits of the other population and both sets of traits must evolve (Futuyma and Slatkin, 1983). Coevolution could, for example, occur either simultaneously or sequentially among interacting taxa (Schemske, 1983).

Some ecologists have approached the problem of creating an adequate definition for coevolution by relaxing restrictions. Roughgarden's (1979) definition, for instance, simply states that coevolution is "the simultaneous evolution of interacting populations." However, Schemske (1983) points out that this definition is inadequate because it does not sufficiently emphasize that coevolution results from selective pressures on heritable traits that are created by the interactions between taxa. He further argues that the definition should emphasize that coevolution requires genetic variation in the characters relevant to the interaction as well as in those characters that are genetically correlated with selected characters. Schemske (1983) has defined coevolution as the joint selective effects on characters of interacting taxa, based on heritable variation in these characters.

The term coevolution was popularized by Ehrlich and Raven (1964) in their classic paper describing the interactions of butterflies and plants. Unfortunately, Ehrlich and Raven's paper did not define coevolution, and the example they chose to illustrate it was inappropriate. In their example, the butterflies were neither stated nor implied to have been the single populations or array of herbivores that generated the selective pressures on plant traits that produced the observed butterfly distributions on host plants (Janzen, 1980a).

The following hypothetical example illustrates the concept of reciprocal coevolution. Since all organisms are potentially food resources for other organisms, it is expected that individuals should have adaptations (e.g., chemical, physical, behavioral, morphological) to defend themselves. Envision, for example, a plant population that varies in its chemical defense against an herbivore population that in turn varies in its ability to deal with the plants' defenses. Those herbivores that are best able to circumvent the plants' defenses are likely to leave the most descendants, and through many generations their traits are likely to spread among the herbivore population. But the interaction should not stop there. Because of its variable resistance, the plant population should also change evolutionarily by acquiring either greater quantities of the defensive chemical or a novel chemical. A dynamic evolutionary interaction should develop that would continue as long as there is close interaction of the plant and herbivore populations. In this coevolutionary situation, the evolution of each interacting taxon is partially dependent on the evolution of the other. This concept is logical, unifying, and aesthetically pleasing; however, the role of reciprocal coevolution as a widespread and important evolutionary force, at least in Janzen's restrictive sense, is not yet established (Begon et al., 1986). Futuyma (1983) has suggested that in most cases it may be difficult to show that reciprocal coevolution has ever occurred since the reasons for the evolution of a plant's chemical defense or an herbivore's detoxification adaptations are lost in unknowable past evolutionary events.

But, reciprocal coevolution is still a tenable concept. Ecology does offer a number of examples that are likely the result of coevolutionary events. For example, many ant–plant interactions (e.g., bullthorn acacia–ant relations, see Chap. 6) appear to be the result of reciprocal coevolutionary interactions as do the limpet–corralline alga association (Steneck, 1982) and some pollinator–plant associations (e.g., pronuba moth–*Yucca,* see Chap. 2).

The term coevolution has often been misused, probably because of the lack of a rigorous working definition for the concept (Janzen, 1980a; Schemske, 1983). In some literature, coevolution has been incorrectly used synonymously with *interaction*, *symbiosis*, *mutualism*, or even *animal–plant interactions*, processes that do not necessarily involve coevolved traits (Janzen, 1980a).

The following examples from Janzen (1980a) illustrate the interpretative dangers inherent in the misuse of the term coevolution:

1. It is commonly assumed that a pair of species whose traits are mutually congruent (i.e., coinciding traits, such as long floral tube and long pollinator mouthparts) have coevolved. Janzen (1980a) offers the example that fruit traits of a mammal-dispersed seed could have coevolved with the mammal's dietary needs. However, it is also possible that the mammal entered the plant's habitat with its dietary preferences well established and subsequently began feeding on the fruits of those species that fulfilled them. The mammal whose dietary requirements are most congruent with the plant's fruit characteristics may appear the most coevolved, yet as Janzen points out, is likely to be the least coevolved.
2. Invertebrate herbivores attacking a plant are typically assumed to be coevolved with the plant's chemistry or morphology. In many situations, however, the herbivore may arrive in a new habitat and simply begin feeding on those plant species whose defensive traits it can circumvent. Unfortunately, such cases cannot be distinguished from ones in which, during the course of feeding on the plant, the herbivore evolved the ability to circumvent the defenses.
3. Janzen provides an additional example in which coevolution is assumed because a parasite has traits that circumvent the defenses of its host. However, there is no reason to conclude that the defense overcome was the result of selective pressures created by that parasite. The plant's defense could be the product of its earlier coevolution with a parasite no longer present in its habitat or no longer parasitizing that particular plant species.

These examples should make it clear that biologists must be careful when invoking coevolution. Further, the point remains that the study of plant–animal interactions will yield important results regardless of whether reciprocal coevolution is or is not involved in the development of a particular interaction.

Diffuse Coevolution

Some authors (Janzen, 1980a; Fox, 1981; Feeny, 1982; Futuyma, 1983) have suggested that although reciprocal coevolution may be rare, diffuse coevolution may be relatively common. Diffuse coevolution occurs when either or both interacting populations are represented in an array of species that generates selective pressures as a group. Diffuse coevolution, in contrast to reciprocal or pairwise coevolution, may consist of events widely separated in evolutionary time or involve selection pressures created by a guild of species. For instance, most plants are simultaneously attacked by a wide variety of insect herbivores. Selection should therefore be most effective for plant traits that provide a broad-spectrum defense. Indeed, most plant chemical defenses are active against a wide variety of insects, vertebrates, and pathogenic microorganisms (Futuyma, 1983). Such broad-spectrum defenses are most likely the result of diffuse coevolution, not reciprocal coevolution.

Mutual Congruence and Tracking

It should now be clear that interactions can evolve through a number of different evolutionary processes. Some interactions, for instance, are simply the result of the mutual congruence (i.e., coincidental co-occurrence) of traits possessed by the interacting organisms (e.g., an introduced pollinator able to forage on the flowers of native plant species). Others may occur because one of the interacting species responds to (tracks) the evolutionary changes of another population, while still others, as just seen, may be due to reciprocal or diffuse coevolution.

Figure 1-2 illustrates the difference between an animal that simply tracks the evolutionary modifications of the plant with which it interacts and reciprocal coevolution. The traits of the plant population are shown as notches, while the traits of the animal population are indicated as cogs. Also illustrated are the selection intensities created by pressures on the plant population by the animal population, s_a, and those selection effects produced by the plant population on the animal population, s_p. In the example labeled "tracking," the plant population is unaffected by the animal population. This is indicated by the unchanging zero value for the selection intensity s_a. However, as the plant population's traits evolve through time in response to other selective pressures in the environment, the animal population responds to these plant modifications with changes of its own and thus improves the congruence of characters important to the interaction. During the times of nonconformance of traits, the selection intensity created by the plant population on the animal population is greater than zero. This process favors those animal genotypes with traits more congruent to those of the plants.

In the case of reciprocal adaptation or coevolution (Fig. 1-2), both the plant and the animal population create selective pressure on each other. An alteration of animal traits, for example, produces selection (s_a >0) on the plant population. Those plant genotypes most congruent with the modified animal traits should, on average, leave more offspring and eventually restore a degree of congruence in the interaction. Later, a change in plant traits is countered by modification in animal traits.

It should now be evident that the term interaction is not synony-

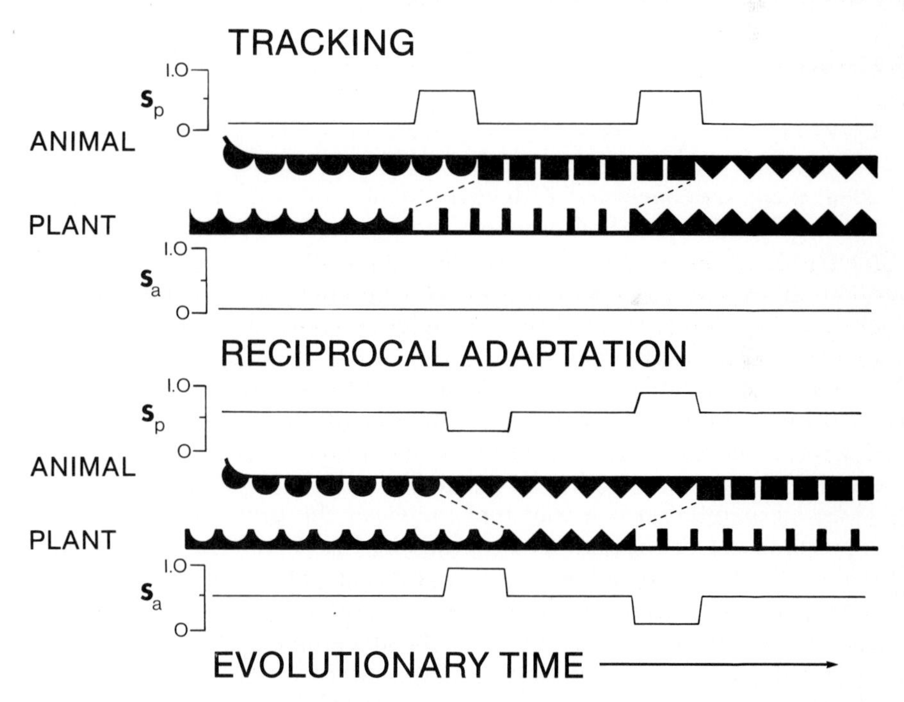

Figure 1–2. Differences between (a) evolutionary "tracking" and (b) reciprocal adaptation or coevolution. Plant traits are shown as notches, while animal traits are indicated as cogs. Selection intensities created by pressures on the plant population from the animal population are illustrated as s_a, while the selection effects produced by the plant population on the animal population are indicated as s_p. From A. E. Weis (unpublished).

mous with coevolution, and that it is a fallacy that interactions are necessarily coevolved. Schemske (1983) in discussing flower specialization in North American bees suggests that there is virtually no evidence that the tremendous diversification and specialization of these bees is related to coevolution between plants and bees. One must proceed with caution in studies of plant–animal interaction. Without a quantitative assessment of selection intensities on the characters involved in an interaction between taxa, coevolution cannot be assumed. Further, plant–animal interactions can be highly asymmetrical, both in the selective pressures that the interactions create and in the evolutionary responses to those pressures. Thus, in many cases there is no reason to expect coevolution at all.

Certainly, coevolution is a process that could potentially influence the patterns of variation and adaptation in plants and animals. But progress toward understanding coevolution will be markedly slowed if the concept is misapplied.

Fitness

Every organism possesses an immense number of adaptations as a result of natural selection acting on an organism's traits. Many of these adaptations are concerned with gas exchange, internal transport, regulation of fluids, and other processes not necessarily influenced by any plant–animal interaction. But others, such as the leaf morphologies of carnivorous plants or the extrafloral nectaries of an ant-tended plant, are at least partially the product of selective pressures influenced by plant–animal interactions.

These adaptations are genetically controlled traits that increase an organism's fitness. In evolutionary biology, fitness refers to an individual's (or deme's or allele's) genetic contribution to future generations, relative to the contribution by other individuals (or demes or alleles). Thus, an adaptation is a trait that increases the probability of genetic representation in succeeding generations. It is important to recognize that adaptations do not necessarily enhance the individual's survival probability at all stages of life. Since an adaptation must improve the chance of leaving descendants to future generations, it would be expected to increase prereproductive survival, but not necessarily postreproductive survival. Indeed, in some species (e.g., century plant, *Agave)*, evolution has resulted in monocarpy (i.e., the "big-bang" reproductive pattern of reproducing only once and dying in the process).

Fitness is generally based on the reproductive performance of genotypes (or alleles) relative to one another and in a particular habitat. Genotypes, for example, vary in their reproductive success (relative fitnesses) in a given environment and the relative fitness of a given

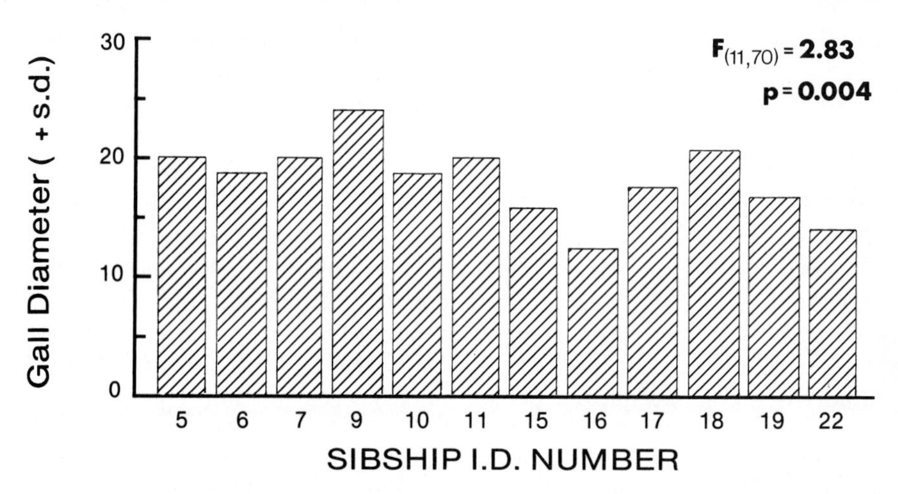

Figure 1–3. The abilities of the offspring of 12 goldenrod ball gallmakers, *Eurosta solidaginis,* to stimulate gall growth vary on one genotype of tall goldenrod *(Solidago altissima).* Offspring of fly 16, for instance, were less successful in stimulating gall growth on this plant genotype than offspring of fly 9. If tested on another plant genotype, the relative abilities of these flies to stimulate gall growth would likely change. From A. E. Weis and W. G. Abrahamson (unpublished data).

individual can change according to the environment in which fitness is measured. Although the offspring of certain gall-maker genotypes perform poorly on one host plant genotype (Fig. 1-3), those offspring might perform well on other host plant genotypes (Weis and Abrahamson, 1986). As a result, the alleles responsible for the gall-maker's ability to manipulate its hosts should change over evolutionary time in response to alterations of the host plant. A given allele or set of alleles may increase in frequency in some localities while declining in others. Recognition of the genetic basis of plant–animal interactions is critical to the development of theory within the field.

Principle of Allocation

Each individual organism has a limited amount of resources available to devote to such primary activities as growth, maintenance, and reproduction. Because of this limit, the resources spent on one activity are unavailable for allocation to other activities. For example, increased reproduction during one growing season will reduce current growth. A large commitment to reproduction in the early years of a long-lived organism's life will decrease fecundity (an organism's reproductive capacity) in future years. In long-lived clonal plants (e.g., aspens, goldenrods, mayapples), for example, clonal growth should be favored over fecundity, especially when space for clonal expansion is available. This strategy increases lifetime reproductive output (Abrahamson, 1980; McCrea and Abrahamson, in preparation). Alternatively, in species with short life-spans the strategy of allocating more resources to reproduction rather than growth is expected.

This principle of allocation is applicable to many evolutionary alternatives faced by organisms. There are trade-offs between seed size (i.e., provisioning of the embryo) and seed number, between fruit size and number, and between floral secondary attractants (e.g., petals or colorful sepals) and primary attractants (e.g., nectar, pollen). In discussions of plant–animal interactions, it is useful to keep this principle in mind when considering the evolutionary alternatives implemented by a given species in a particular situation.

It must also be emphasized that organisms are not perfect in their design, even though there is close correspondence between an organism's morphology, physiology, and behavior, and its environment. As Gould (1980) has argued, design is not inherent to the process of natural selection. The environment itself is the template for the pattern seen in organisms. Selection merely acts as the process that creates the pattern. Gould (1980) has illustrated this point with the imperfection of the panda's thumb. The panda's thumb is functionally a sixth "digit" that is derived from a wrist bone. This unusual development converts what would otherwise be a paw into a manipulative hand able to grasp stems of bamboo, the panda's primary food. The panda's thumb is not an ideal design in the sense of engineering perfection, but it works. Natural selection can only act on the heritable variation available to bring about evolutionary change. Organisms are constrained by their evolutionary past.

In discussing the evolution of plant and animal traits important to an interaction, it is tempting to perceive organisms as completely malleable by selection pressures. However, the form that an adaptation may take is strongly limited by the species' prior evolution (Gould, 1980). Natural selection can only work with existing genetic variability. This evolutionary history along with the randomness of mutation and recombination constrain the possibilities and directions of new evolutionary modifications. As a consequence, perfect correlations between theory, or what a designer would produce, and what nature (or human selection of domesticated plants and animals) creates are not realized.

Ecological vs Evolutionary Time

A considerable amount of ecological theory deals with competition among species. Virtually every student taking an introductory course in ecology is introduced to the competitive exclusion principle (the hypothesis that two or more species cannot coexist on a single resource that is limited relative to the demand for it). This principle provides various theoretical outcomes of interspecific competition depending on numbers of individuals initially present and the competitive ability of the interacting populations. But these outcomes consider only the immediate, or ecological, time frame. More important to the discussions of the subsequent chapters are the implications of competition (see especially Chap. 8) and other types of interactions in an evolutionary time frame. Thus, one can view competition for floral visitors among flowering plants as a selective force that can alter the representation of plant genotypes in future generations. The nectar rewards or floral morphology of some genotypes may be such that these genotypes set more seeds than other genotypes resulting in higher fitness. The study of plant–animal interactions has an evolutionary time perspective. Although interactions may be ephemeral in a paleontological sense, they are the evolutionary outcome of ecological processes.

Plant–Animal Interactions as a Subject of Study

The study of plant–animal interactions encompasses much of the disciplines of evolution and ecology and therefore cannot claim any distinction as a separate field of study. However, as Futuyma and Slatkin (1983) suggest, progress is frequently made by drawing attention to particular definable subjects within a field. Focus on plant–animal interactions facilitates an evolutionary approach to classical ecological studies. Studies addressing various plant–animal interactions have initiated a highly analytical view of ecology. The theories are based on the fact that genetic shifts can occur in the populations of interacting species as a result of selective pressures. These pressures are created by the interactions themselves and act on the heritable variation in the organism's traits. These genetic shifts can be driven both by the immediate interactions and by feedback through the rest of the community as a result of diffuse coevolution (Roughgarden, 1983).

An important attribute of many studies on plant–animal interactions is their perspective of evolutionary time. These interactions are dynamic and change as selective pressures are altered. While discussing coevolution, Futuyma and Slatkin (1983) note that if one species is considered alone, it might be expected to evolve until it has resolved whatever challenges it has met and then stop. But if two or more species respond evolutionarily to one another, then continued evolutionary change should be expected. A plant species could continually elaborate new chemical defenses while its herbivores evolve adaptations to over-

A
B
C
L
D
I
J
K
M
O
N
E
G
F
P
H
© KJPrather 1987

come those compounds. Recent studies of plant–animal interactions, regardless of whether they involve coevolution, have utilized this dynamic, evolutionary time perspective. It is now clear that this approach is fruitful and should provide important insights into the evolutionary history of many groups of organisms.

There are practical spin-offs to these studies in fields such as medicine and agriculture. For example, a number of plant chemicals, presumably evolved in response to herbivores, have useful therapeutic effects on humans. Cardiac glycosides, defensive chemicals common in foxglove *(Digitalis)*, possess cardiotonic (having a tonic effect on the heart) activity which is of value in the treatment of congestive heart failure. Digitalin increases the force of heart muscle contraction without a concomitant increase in oxygen consumption, making the heart a more efficient pump (Lewis and Elvin-Lewis, 1977). Indeed, most useful drugs originated as plant products that serve ecological functions in nature. There is an abundance of species yet to be discovered and screened for possible human uses.

Other applications of studies of plant–animal interactions will be particularly obvious in this book's final chapter on plant–animal interactions in agricultural systems. The fundamental processes of ecology and evolution operating in natural ecosystems also operate in managed ecosystems. Understanding the biology of plant–herbivore–natural enemy interactions in natural ecosystems, for example, can provide

Figure 1–4. *Calathea ovandensis* (Marantaceae) hosts a variety of mutualistic and antagonistic plant–animal interactions. These interactions include relationships with pollinators, seed dispersers, ant guards, and herbivores. A = *Euglossa heterosticta* (Hymenoptera: a common pollinator), B = *Drosophila* sp. (Diptera: adults oviposit in older *Calathea* inflorescences and the resulting larvae feed on reproductive tissues), C = *Solenopsis geminata* (Hymenoptera: ant tending the specialist herbivore larva, *Eurybia elvina*, see also L; *Solenopsis geminata* also serves as a *Calathea* seed disperser), D = *Crematogaster sumichrasti* (Hymenoptera: feeds from extrafloral nectaries while serving as an ant guard against generalist herbivores; this ant, like other ants visiting *Calathea,* feeds on the nectarlike secretions on *Eurybia elvina),* E = *Atta* sp. (Hymenoptera: leaf-cutting ant and its damage), F = *Pachycondyla harpax* (Hymenoptera: seed disperser), G = *Calathea ovandensis* seed with its lipid-rich elaiosome, H = *Calathea ovandensis* (Marantaceae), I = Hispine beetle (Coleoptera) damage to *Calathea* leaf, J = *Heliconius ismenius* (Lepidoptera: an occasional pollinator), K = *Eurybia elvina* (Lepidoptera: specialist herbivore as a larva but the adult visits *Calathea* flowers), L = *Eurybia elvina* larva (Lepidoptera: specialist herbivore that secretes a nectarlike fluid harvested by ants, such as *Solenopsis geminata,* while feeding on flowers and developing fruits); M = various members of the Hesperiidae (Lepidoptera: common flower visitors but poor pollinators), N = *Saliana* sp. (Lepidoptera: herbivorous larva and its leaf damage), O = *Podalia* sp. (Lepidoptera: herbivorous larva), P = Orthopteran leaf damage (herbivore). Illustration created from materials provided by C. C. Horvitz and D. W. Schemske.

crucial understanding of such interactions in systems based on crop plants.

Progress made in the field of molecular genetics involving the direct transfer of genes from one organism to another raises expectations for the improvement of modern crops using these techniques. Pest resistance could be obtained through the transfer of genes responsible for resistance to a particular set of herbivores from a wild plant species to a domesticated crop species. Knowledge of natural plant defenses against herbivores will be important to these attempts.

Breadth of Plant–Animal Survey

A comprehensive survey of plant–animal interactions is beyond the scope of this textbook or any single volume. The approach taken here is one of providing description and theory for a set of important, representative types of plant–animal interactions. There are obvious omissions (e.g., marine plant–animal interactions), but this approach does provide a theoretical framework useful in considering virtually any interaction. This textbook offers a way of thinking about organisms rather than detailing as many types of interactions as could possibly be described.

There are many well-studied plant–animal interactions including the New and Old World *Acacia* tree systems, goldenrod–herbivore interactions, and many pollinator–plant systems. A series of studies by Carol Horvitz and Douglas Schemske and their coworkers nicely illustrates the complexity of such interactions within one plant species, *Calathea ovandensis* (Marantaceae, arrowroot family). This neotropical herb hosts a variety of mutualistic and antagonistic plant–animal interactions, each of which affects several stages of the plant's life history (Fig. 1-4, Table 1-2). These interactions include pollinators, seed dispersers, ant guards, and herbivores of both reproductive organs and leaves (Horvitz and Beattie, 1980; Horvitz and Schemske, 1984, 1986; Schemske and Horvitz, 1984).

In *Calathea* there are at least three mutualistic interactions. There is a pollination system with bees as primary pollinators, but also other Hymenoptera (bees and wasps) and Lepidoptera (butterflies and moths), each of which vary in pollination efficiency (Schemske and Horvitz, 1984). It has a seed dispersal system with ants that are attracted by the *Calathea* seed's lipid-rich elaiosome (edible oil body attached to the outside of some plant species' seeds). Mature fruits dehisce (split open) so that seeds are scattered onto the forest floor near the plant. Here they attract several ant species that vary in their ability to disperse seeds (Horvitz and Beattie, 1980; Horvitz, 1980, 1981). These ant dispersal

Table 1–2. Plant–animal interactions in *Calathea ovadensis*.*

Stage	Effect on Animal					
	Pollinators	*Ants at Extrafloral Nectaries*	*Hebivores of Repro. Tissues*	*Post-dispersal Seed Predators*	*Dispersal Agents*	*Leaf Herbivores*
Ovules	+	+	–	0	0	0
Seeds (on plant)	0	+	–	0	0	0
Seeds (in soil)	0	0	0	–	+	0
Seedlings	0	0	0	0	+	– (Survival)
Larger Plants	0	0	0	0	+	– (Growth)

*Effects are indicated as beneficial (+), harmful (–), or of no appreciable benefit or harm (0) for each guild of animals that interact with *Calathea*.

From: D. W Schemske and C. C. Horvitz, personal communication.

agents probably have important effects on seedling establishment and long-term effects on seedling success because the ants determine where seedlings will grow (Horvitz and Schemske, 1986). Finally, there is an ant guard defense system based on the plants extrafloral nectaries (Horvitz and Schemske, 1984). Variation in the magnitude of beneficial effects by ant mutualists is likely. Factors such as the distance of plants from ant nests as well as the spatial and temporal variation in the particular ant species involved can markedly influence the ant's benefit to the plant (Keeler, 1977; Schemske, 1980b).

Calathea has antagonistic interactions with (1) an herbivorous butterfly larva, *Eurybia elvina* (family: Riodinidae) which, while feeding on the flowers and developing fruits, secretes a nectarlike fluid harvested by ants (Horvitz and Schemske 1984), and with (2) herbivores feeding on vegetative tissues, particularly leaves. The numbers of interactions involving this one plant species themselves create complexity. Moreover, the relationships are further complicated because none of these interactions are isolated from one another. For example, the inflorescences of *Calathea* possess extrafloral nectaries that provide sugar-rich food to several ant species. But, in addition to feeding on nectar, these ants prey on most but not all invertebrate herbivores encountered while patrolling *Calathea*. These same ants that offer *Calathea* defense against generalist herbivores also tend the larvae of the specialized herbivore *E. elvina* which in turn produces a nectarlike fluid used as food by the ants (Schemske, personal communication).

The *Calathea* system demonstrates the complexity of interactions among species on several trophic levels. But this example is not unique. Most systems are equally complex. In the chapters that follow, these types of interactions and many others will be explored. The examples will be drawn from a diverse selection of plant and animal taxa. The organization of chapters moves from population-level to community-level interactions and from interactions involving plant reproductive organs to vegetative organs.

Chapter Topics

The second chapter, written by Robert Bertin, discusses the interactions between flowers and their pollinators. Many of these interactions are mutualistic. The chapter examines plant fitness, floral mechanisms for pollinator attraction, plant breeding systems, pollinator sensory abilities, and pollinator behavior and energetics.

A discussion of fruits, seeds, and dispersal agents, follows the pollination chapter. Here Edmund Stiles explores fruit and seed characteristics (e.g., structure, chemistry, phenology, allocation) in relation to the characteristics of frugivores (fruit-eating animals) and seed predators. Throughout, there is an emphasis on the evolutionary responses to the potential selective pressures resulting from the interaction.

Next, Arthur Weis and May Berenbaum shift attention to antagonistic interactions involving insect herbivores with plants. These authors consider the ecological impacts of insect herbivores on plants and the resultant evolutionary responses. Also outlined are the evolutionary responses of insects to plants. Weis and Berenbaum deal with natural enemies of herbivores providing a holistic look at multitrophic-level interactions.

Insects are not the only herbivores of plants. Richard Lindroth guides an examination of herbivorous mammals and their relationships to plants. This chapter describes these interactions from both the plant's perspective (e.g., plant adaptations, impacts of mammalian herbivores on plant populations and communities) and the mammal's perspective (e.g., mammalian adaptations to herbivory, effects of plants on herbivore population dynamics and community organization).

Kathleen Keeler's chapter on ant–plant interactions describes the many relationships that ants and plants have evolved. Ants are seen as fungus gardeners, seed harvesters, plant defenders, and herders. These interactions, illustrating both mutualistic and antagonistic interactions, provide some of the best examples of coevolution.

Some plants, the carnivorous ones, seem to have turned the tables on insects. Thomas Givnish details much of what is known about the ecology and evolution of carnivorous plant species. This encounter with predaceous plants includes, among other topics, the mechanisms of prey attraction, capture, and digestion as well as the nutritional benefits, possible mutualism with social insects, and a cost–benefit analysis of plant carnivory. The chapter concludes with a discussion of the evolutionary pathway to carnivory.

The remaining two chapters emphasize community and ecosystem patterns. Kerry Rabenold and William Bromer look at plant communities as animal habitats. The theme running throughout their chapter is that plants form much of the physical and biological environment for animals and thereby greatly affect animal population dynamics (e.g., competitive interactions) as well as the patterns of animal distribution and abundance. The approach of this chapter is, of necessity, more one-sided than that of all but the preceding chapter, but in this case considers primarily animals rather than plants.

In the final chapter, Benjamin Stinner and Deborah Stinner examine plant–animal interactions in human-created agricultural ecosystems. They draw heavily from the theories developed for natural ecosystems and the theories discussed in the earlier chapters to provide an understanding of current problems in agriculture. The problem of limited genetic variation in crop species, for instance, is outlined relative to attempts to enhance a crop's genetically based resistance to pests. This chapter offers an opportunity to consider the application of basic ecological and evolutionary theory to an economically relevant situation.

These topics provide a wide variety of natural history descriptions of interactions but also offer a solid theoretical framework in which to consider the evolution of these interactions. The topics included represent the biases of the editor rather than an attempt to comprehensively examine all possible types of plant–animal interactions. The goal was to initiate a reader's discovery of the complexity and wonder of ecological interactions.

Selected References

Boucher, D. H., ed. 1985. The biology of mutualism. Oxford University Press, NY.
Futuyma, D. J., and M. Slatkin, eds. 1983. Coevolution. Sinauer Associates, Sunderland, MA.
Janzen, D. H. 1980. When is it coevolution? *Evolution* 34:611–612.
Nitecki, M. H., ed. 1983. Coevolution. University of Chicago Press, Chicago, IL.
Thompson, J. N. 1982. Interaction and coevolution. John Wiley and Sons, NY.

2

Pollination Biology

ROBERT I. BERTIN
Department of Biology
College of the Holy Cross
Worcester, MA 01610

Introduction

Pollination is the transfer of pollen from male to female reproductive structures of plants. In this process lies strategy and competition, intrigue, deceit, rare beauty, and economic livelihood. It is the purpose of this chapter to introduce the natural history, essential terminology, and theoretical developments of pollination biology from the perspective of both plant and animal. The emphasis will be ecological and evolutionary, and the treatment of each topic will necessarily be brief.

To understand the whys of pollination it is essential to keep in mind the benefits derived from this interaction by both plant and animal. A plant benefits from the movement of pollen from male to female parts, a necessary prerequisite to sexual reproduction. The interest of animals in pollination comes not from any sympathy for a plant's needs, but to meet animal needs, which are often nutritional. Animal traits promoting feeding may not be those that promote plant reproduction, or vice versa, injecting an element of antagonism into many plant–pollinator interac-

I thank W. G. Abrahamson, A. F. Johnson, and J. Lanza for their extensive comments on earlier versions of this chapter, and S. Galvin for preparing Figs. 2-3, 2-4, and 2-7.

tions. A major theme in this chapter is how various plant and animal traits provide adaptive benefits to the organisms involved.

Biotic and Abiotic Pollination

Pollination can be biotic (accomplished by animals) or abiotic, resulting from physical factors such as wind and water currents. It is useful first to consider briefly abiotic pollination, particularly in contrast to biotic pollination.

Pollination in or on water (hydrophily) has been recorded from 25 genera in 11 families (Cox, 1983; Fig. 2-1). However, most genera of aquatic plants do not exhibit hydrophily, but rather bear flowers on or above the water's surface, and are pollinated by wind or animals. On theoretical grounds, Cox (1983) predicted that a pollen grain's likelihood of finding a stigma (female receptive surface, see the following) is greater if it moves in two dimensions rather than three, and if it has one markedly elongate axis. In fact, most hydrophilous species disperse pollen on the water surface (a two-dimensional environment), and their pollen grains (or units in which pollen is dispersed) tend to be elongate. In several genera (e.g., *Zostera, Thalassodendron, Amphibolis),* the filamentous grains are up to several millimeters long. The relative rarity of hydrophily among aquatic plants suggests that it is inefficient, and self-pollination appears to be common (Philbrick, 1986).

Wind pollination (anemophily) is much more common than hydrophily, being especially common among several groups of monocots (grasses, sedges, rushes) and woody plants (gymnosperms, willows, birches, oaks; Fig. 2-1).

Several features are usually associated with wind pollination. Petals and other attractive flower parts are absent or greatly reduced. They are unnecessary for pollinator attraction and could interfere with pollen release and capture (Faegri and Pijl, 1979). Nectar is usually absent. Stigmas are large and feathery, and the pollen-containing anthers are often large and pendulous. Pollen grains are small, dry, and occur singly, rather than in clumps, as is true in some animal-pollinated species. Immense numbers of pollen grains are often present. A single ragweed *(Ambrosia)* plant can release 8 billion pollen grains in 1 day. Pollen capture may be aided by patterns of airflow created by the positioning of stigmas, other floral parts, and even nearby leaves (Niklas and Buchmann, 1985).

What are the relative merits of biotic and abiotic pollination? Because hydrophily is relatively uncommon, this question is often restricted to a comparison of wind- and animal-pollinated species. Whitehead (1983) concluded that the evolution and maintenance of

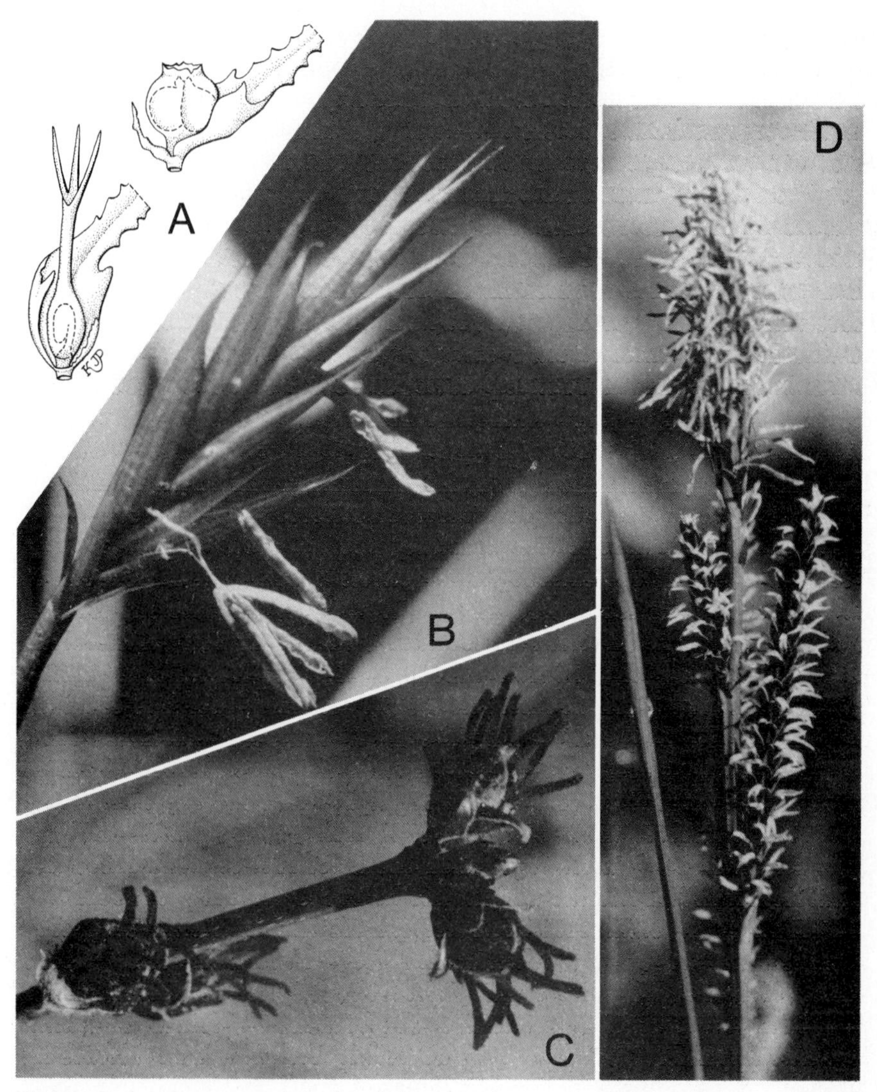

Figure 2–1. Flowers of hydrophilous and anemophilous species. A. Male (upper) and female flowers of *Najas marina,* a hydrophilous species; note the extreme reduction. B. Spike of *Arundinaria gigantea,* a grass, illustrating the pendulous anthers typical of many wind-pollinated species. C. Female flowers of silver maple *(Acer saccharinum),* showing the conspicuous stigmas. D. Sedge (*Carex* sp.) showing the distinct male (upper) and female spikelets. Figure 2–1A from Sculthorpe (1967), *The Biology of Aquatic Vascular Plants,* Edward Arnold Ltd., London.

wind pollination is favored by (1) close spacing of conspecifics (individuals of the same species), (2) low probability of interception of pollen by intervening vegetation, (3) low frequency of rainfall, (4) unambiguous environmental cues to coordinate flowering, and (5) sites with short growing seasons in which early flowering is favored, at a time when animal vectors may be unreliable. The opposites of the above conditions would therefore favor biotic pollination. Beyond these factors, biotic pollination is believed to be more efficient than wind pollination (Faegri and Pijl, 1979) because it increases the likelihood of a pollen grain reaching a conspecific stigma, and is more likely to result in outcrossing (Regal, 1976). Apparently no studies have specifically examined these points, and systematically collected comparative data would be useful.

The Plant's Perspective

Some Botany Basics

Floral Structure

The diversity of flower shapes, sizes, colors, and structures is truly one of nature's greatest splendors. Although it is impossible to adequately summarize these features in a few paragraphs, some knowledge of floral structure and terminology is essential for what follows.

A generalized flower such as a buttercup or lily consists of sepals, petals, stamens, and pistils (Figs. 2-2, 2-3A). The sepals are often green, protect the flower in bud, and are collectively referred to as the calyx. The petals, typically brightly colored, lie inside the sepals and are collectively termed the corolla. The term perianth refers to the calyx and corolla jointly. The stamens are collectively the androecium, or male

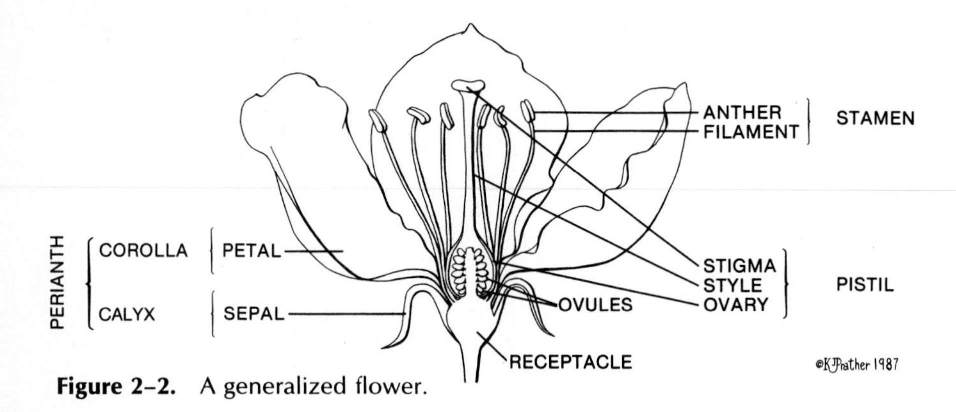

Figure 2–2. A generalized flower.

structures. A stamen typically consists of a slender filament bearing the pollen-containing anther. The pistils collectively are the gynoecium. Each pistil consists of a stigma (the surface receptive to pollen), an often elongate style, and an ovary containing one or more ovules. Following successful fertilization, ovules and their surrounding integuments develop into seeds. Ovaries, sometimes together with nearby structures, develop into fruits.

Many modifications of this basic structure exist, including the presence or absence of structures and variations in their number, size, shape, aggregation, degree of fusion, color, and function. Some flowers, for example, possess functional organs of only one sex, in which case they are unisexual (and termed either staminate or pistillate). Some flowers on

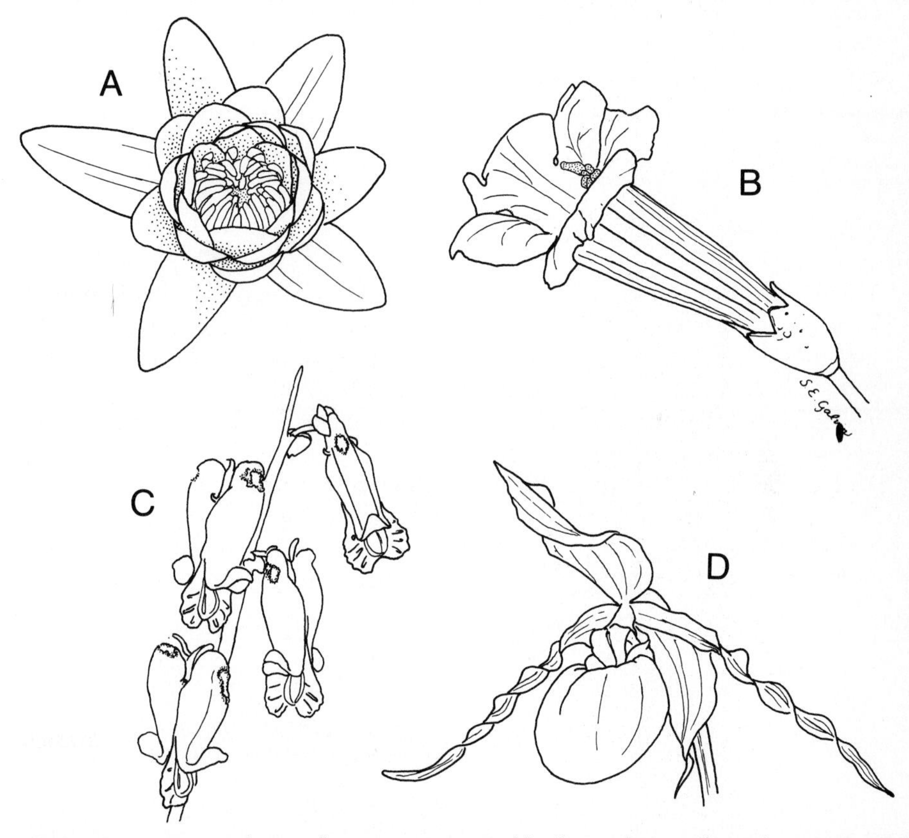

Figure 2–3. Some showy flowers. A. Water lily *(Nymphaea odorata),* an open, relatively unspecialized flower likely to attract a variety of floral visitors. B. Trumpet creeper *(Campsis radicans),* visited by hummingbirds and bees. C. Squirrel corn *(Dicentra canadensis),* visited by bees; note the holes in the perianth made by nectar robbers. D. Yellow lady's slipper *(Cypripedium calceolus),* presumably pollinated by large bees.

certain plants may lack functional sexual parts entirely and are termed sterile. They probably have an attractive function. Partial or complete fusion of petals and sepals is common in many plant families (Fig. 2-3). Such fusion restricts nectar access to a subset of potential flower visitors, potentially affecting pollen transfer. Stamens and pistils are sometimes found in unusual arrangements, as in *Hibiscus* spp. in which stamens are fused with the pistil over most of their length.

In many angiosperms, flowers are aggregated in inflorescences. This has been carried to an extreme in the Compositae, which includes asters, daisies, and goldenrods. What appears to be a single flower (Fig. 2-4) is actually a dense aggregation of tiny flowers (a capitulum). In many members of this family, two flower types are present within each capitulum: the outer ray flowers each bear one showy petal, while the inner disk flowers have very reduced petals. One consequence of the capitulum is a fine division of nectar and pollen resources, perhaps

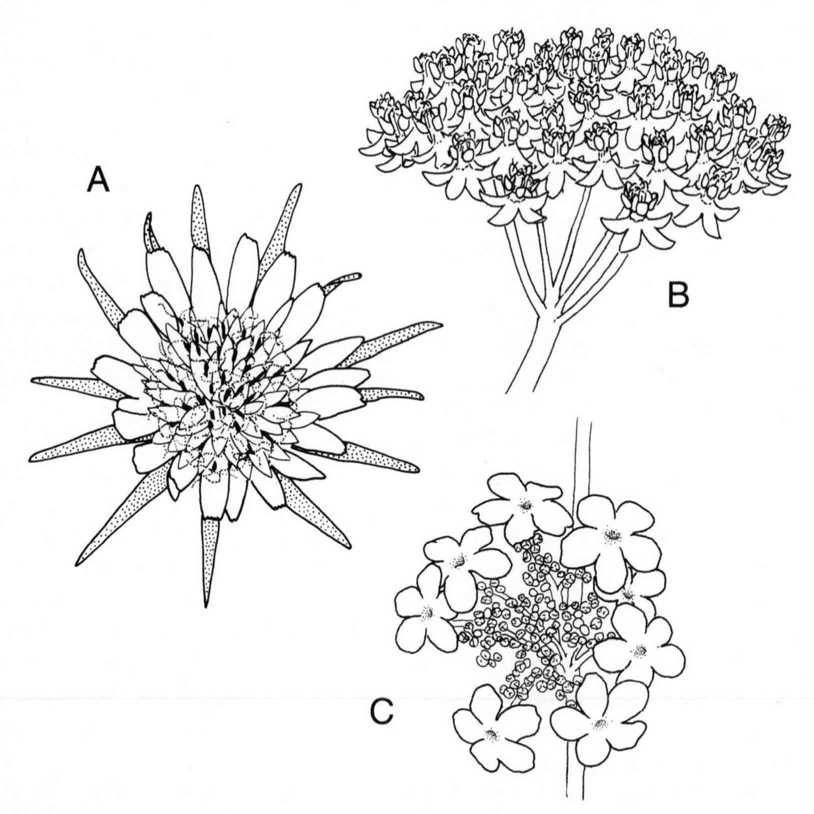

Figure 2–4. Inflorescences. A. Salsify *(Tragopogon porrifolius),* a composite, illustrating the peripheral ray flowers and central disc flowers. B. Swamp milkweed *(Asclepias incarnata).* C. Hobblebush *(Viburnum alnifolium);* the showy peripheral flowers lack sexual parts; the smaller fertile flowers are still in bud.

requiring pollinators to spend more time in such a capitulum and transfer more pollen than they would in a single large flower. The arum family (Araceae) also exhibits much aggregation and reduction of flowers (Fig. 2-5). Here a modified leaf, the spathe, subtends and surrounds a thick spike, the spadix, bearing numerous tiny flowers.

Pollination in a Plant's Life Cycle

In many groups of plants there is an alternation between a haploid, or gametophytic, generation and a diploid, or sporophytic, generation. In some lower plants these generations are physically independent. In ferns, for example, the conspicuous fronds together with their underground parts are sporophytes. These diploid plants produce spores which undergo meiosis, eventually giving rise to tiny green haploid gametophytes that live independently on the substratum, and give rise to gametes, which fuse to form the next generation of sporophytes.

The life cycle of a seed plant is similar except that the gametophytic generation is further reduced and physically dependent on the sporophyte. The life cycle of an angiosperm can be generalized as follows, although the exact pattern varies among taxa. The sporophyte (what is recognized as the plant) bears flowers. In the flowers' ovaries, diploid megaspore mother cells divide meiotically to produce haploid megaspores, while in the anthers, diploid microspore mother cells give rise to microspores. The megaspores and microspores are comparable to the spores produced by the sporophytes of ferns. Of the four megaspores derived from one megaspore mother cell, three degenerate (in some species two or none degenerate), leaving one functional megaspore. This megaspore enlarges and divides three times to produce eight nuclei. Of these, one is an egg and two are polar nuclei (the other five will not concern us here). The eight-nucleate structure is the mature female gametophyte, and it is entirely enclosed within ovarian tissue. Meanwhile, the microspore has become surrounded by a two-layer coat. At about the time the anther opens, the microspore nucleus divides to produce a generative and a tube nucleus. The transfer of such a microspore, now called a pollen grain, to a stigma is pollination, the subject of the bulk of this chapter. Before or shortly after germination of the pollen grain, the generative nucleus divides to produce two sperm nuclei. The germinated pollen grain with its three haploid nuclei is the mature male gametophyte.

Pollen germination involves emergence of a pollen tube. Under appropriate circumstances this tube enters the stigma, traverses the style, and reaches the female gametophyte. Pollen tube growth initially involves mobilization of reserves from the pollen grain, although later growth may require substances produced by the pistil (Labarca and

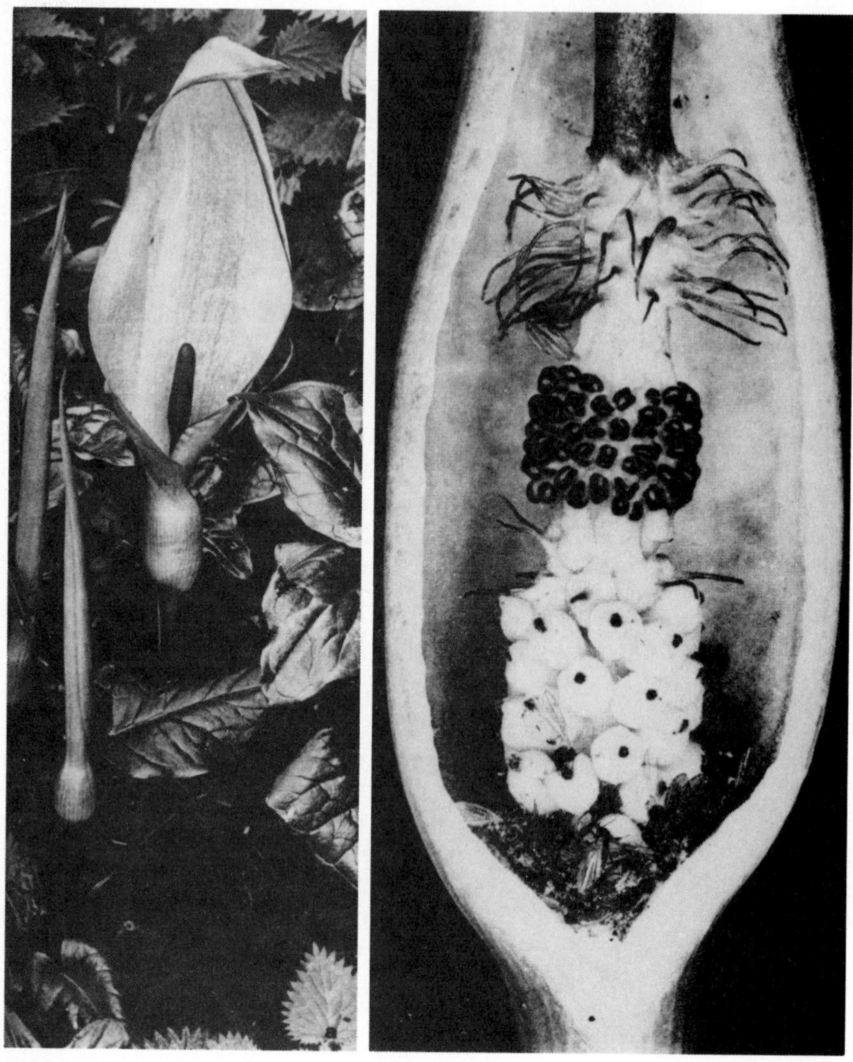

Figure 2–5. *Arum maculatum*. Left: Inflorescence showing the top of the spadix (dark) and the surrounding spathe. Right: Magnified view of the lower spadix with the spathe cut away. A zone of bristles lies above the dark male flowers, which lie above the female flowers. Several trapped insects have accumulated at the bottom. Plate 39 of Proctor and Yeo (1972); used with permission of Taplinger Publishing Co., Inc.

Loewus, 1973). Upon reaching the female gametophyte, one of the sperm nuclei fuses with the egg, producing a zygote. The other sperm nucleus fuses with the two polar nuclei, creating a triploid endosperm nucleus that gives rise to a nutritive tissue called endosperm. The time elapsed between pollination and fertilization varies from hours to a year or more in different plant taxa. One pollen grain is required for each egg, and it is typical to have many pollen tubes in each pistil.

The embryo, endosperm, and surrounding maternal tissues are referred to as ovules, each of which can develop into a seed. The ovary enlarges along with the developing ovules and ripens into a fruit. Released seeds then germinate and grow, producing a new sporophyte.

Evolution of Flower Form and Function

Potentially important selection pressures acting on floral biology include those related to pollinator behavior, recombination, sexual selection, and pollen/stigma interference. The importance of the first two has long been recognized. Pollinator behavior and its consequences for floral biology are examined at some length later in this chapter, and the other factors will be considered below. Of these three, recombination has long been accorded great importance in the evolution of floral traits, and some maintain that it has been overemphasized. Interest in the remaining two topics has developed only recently.

Sexual Selection

Natural selection favors traits that increase fitness, or the number of offspring surviving to reproductive age. Both the number and the quality of offspring are important in determining an individual's fitness.

For many years botanists considered a plant's reproductive performance only in terms of seed production (i.e., its maternal performance). But it is obvious that every sexually produced seed contains one maternal and one paternal genome, and therefore within a population the total (or average) male contribution must equal the total (or average) female contribution. Hence, an evolutionary interpretation of reproductive characteristics must consider their effects on both male and female performance.

Male and female reproductive costs differ in fundamental ways. The costs per female gamete are generally much higher than the costs per male gamete when the female's postfertilization investments in seeds are included. The timing of allocation to male and female functions also differs, with female allocation extending later than male allocation, which ceases with pollen release (Stephenson and Bertin, 1983). These and other observations have led to the expectation that in most species, female reproductive success is often limited by resources, while male success is more likely to depend on access to female gametes (Willson and Burley, 1983). Some cases in which pollen availability rather than resources limit female success are known (e.g., jack-in-the-pulpit, *Arisaema triphyllum;* Bierzychudek, 1981), but this condition appears less common than resource limitation of fruit production (Willson and Burley, 1983).

If resources are limiting, selection pressures acting on male and female reproductive functions may be very different. Specifically, females with access to many male gametes, but few resources, can enhance their fitness if they can discriminate among male gametes or zygotes of different qualities and incorporate only the best into their seed crop. It is known that gynoecia reject pollen grains and/or seeds of certain genotypes (Willson and Burley, 1983), and selective fruit abortion can enhance the quality of a plant's seed crop (Stephenson and Winsor, 1986), although the mechanisms are poorly known. It is also likely that plant traits influence what pollen is received (Janzen, 1977a), although there are few empirical data on this point.

If male fitness is enhanced only by access to more female gametes, traits that increase the likelihood of pollen removal and its delivery to stigmas will be favored. Such traits might include large inflorescences, large showy flowers, and sequential rather than simultaneous pollen presentation within a flower (to spread pollen among more floral visitors; see, for example, Willson and Rathcke, 1974; Stephenson and Bertin, 1983). Some of these factors (e.g., flower color and size) can also affect female reproductive output, but careful experiments indicate that such traits sometimes have their greatest effect on male success (Bell, 1985; Stanton et al., 1986).

The processes described above (discrimination by females, and competition between males) have parallels among animals. They are often referred to as intersexual selection and intrasexual selection, respectively. Together these processes are thought to be the major agents of sexual selection, which can be thought of as the differential production of offspring as a result of mating patterns. Animal traits have been interpreted in terms of sexual selection since Darwin's time, but only recently have biologists examined plants from the same perspective (Charnov, 1979; Willson, 1979). This viewpoint is useful in that it focuses attention on both male and female reproductive success and the different selection pressures acting on them. To date, however, most work on sexual selection in plants has been theoretical and there remain differences of opinion as to its applicability and importance to plants, and even as to the precise meaning of certain fundamental terms.

Recombination

During the reproduction of most organisms a balance is struck between the conservation and change of genetic material. Two processes contributing importantly to genetic change are meiosis, which can rearrange alleles between a pair of homologous chromosomes, and fertilization, which joins genetic material of two gametes into a novel combination. Both self-fertilization (wherein both gametes come from the same plant) and cross-fertilization (or outcrossing, involving gametes from different plants) may occur in many species. Seeds can also be produced asexually. Such offspring are genetically identical to one another and to their single parent, barring mutations. The most genetically conservative route is asexual reproduction, followed by self-fertilization, followed by cross-fertilization. Cross-fertilizations themselves can vary in the degree of heterozygosity that they yield, depending on the degree to which the two parents involved in the cross are related.

How much recombination is best for individuals of a particular species? If it is assumed that current plant populations are near evolutionary equilibrium, the answer varies widely among species. Populations range from obligate outcrossing to nearly total asexuality in their mode of reproduction. A more meaningful question is, "What are the advantages of each mode of recombination and under what circumstances is each favored?" Lower recombination levels may be advantageous in reducing the cost of reproductive structures (since pollinator attraction may be less important), lowering dependence on (sometimes unreliable) pollinators, decreasing alteration of already successful genotypes, and improving ability to colonize a new area (because conspecific individuals and pollinators are not required). Furthermore, a selfer passes on two copies of its genes to each offspring, while an outcrosser passes along only one. However, in many species, self-fertilization yields offspring of lower vigor and survivorship than cross-fertilization, presumably as a result of lower levels of heterozygosity.

The value of recombination is likely to vary among plants with different habits and living in different environments. For example, *Antennaria parlinii* exists in both sexual and asexual forms. Asexual individuals produce more seed than sexuals, but their seed is smaller and they yield offspring with lower survivorship (Michaels and Bazzaz, 1986). Thus the asexuals are likely to do better as colonizers, while sexuals do better in competitive environments. In general, asexual species are most common in disturbed habitats and those at high altitudes and latitudes (Levin, 1975). Certain plants, such as *Impatiens capensis* (jewelweed), bear two flower types: chasmogamous (open flowers), that are sometimes cross-pollinated, and cleistogamous (which self-pollinate in bud and never open to receive pollinators). Cleistogamous flowers are more frequent on jewelweed plants growing in shady, rather than sunny conditions. Energy expenditure per cleistogam seed is only one-half to one-third that of a chasmogam seed because of the greater investment in nectar and other floral tissues in the latter (Schemske, 1978). Thus the cheaper cleistogamous route is more common in habitats where light availability and, therefore, photosynthate production, is limited. Seedlings derived from chasmogamous flowers outperform seedlings from cleistogamous flowers, however, presumably reflecting heterozygote vigor (Waller, 1984). Thus chasmogamous reproduction can be advantageous if adequate light is available. These studies illustrate how outcrossing and selfing can each be advantageous under specific environmental circumstances.

Outcrossing between conspecifics growing at different distances from each other can affect reproductive output. In *Phlox drummondii,* seed abortion was inversely related to interparent distance, presumably because interparent distance was correlated with genetic distance (Levin, 1984). In two montane wildflowers, *Delphinium nelsonii* and *Ipomopsis aggregata,* both inbreeding and outbreeding depression appeared to occur, resulting in an intermediate optimal outcrossing distance of between 1 and 100 m (Waser and Price, 1983). The notion of outbreeding depression on this spatial scale is controversial, but if it exists, it presumably reflects sufficient genetic differences between parents that some problems occur in the resultant zygotes. Whatever the causes of these patterns, they show that outcrossing success can be influenced by the distance of pollen movement. This sets the stage for selection on any plant traits (e.g., floral display, nectar rewards) that can influence the distance of pollen movement.

Pollen–Stigma Interference

Male and female functions can interfere with each other when they occur on a single plant and particularly within a single flower. Movement of pollen from anthers to stigma within a flower may not result in successful fertilization because of self-incompatibility (discussed later) or may yield offspring of poor quality because of inbreeding depression. In either case, pollen is wasted and ovules may also be wasted if deposition of self pollen decreases the likelihood of successful cross-pollination. The potential importance of such interference in the evolution of breeding systems and floral biology has recently been emphasized (Bawa, 1980a; Lloyd and Webb, 1986; Webb and Lloyd, 1986). However, there have been few empirical studies of the extent and effects of pollen–stigma interference.

Phylogenetic Constraints

In discussing the evolution of plant reproductive traits (or any other traits), it is tempting to view organisms as totally malleable by selection pressures. They of course are not, as has frequently been emphasized (Jacob, 1977; Gould, 1980; Kochmer and Handel, 1986). Every species comes with the imprint of its prior evolution, and this history, together with the vagaries of mutation and recombination, places constraints on the likelihood and direction of new evolutionary changes. As a consequence, perfect correlations between theory, or what a designer would produce, and what nature creates are not expected. Natural selection can only work with existing genes, which constrains the course of evolution.

Intrafloral Ecology

While the gross patterns of arrangement of floral parts have already been described, there remain many less conspicuous details of much importance in producing efficient pollination. A variety of cues direct a visitor's movement appropriately to effect pollination. Prominent among these cues are guide marks. These are sometimes called nectar guides, but the former term is preferable because they sometimes occur in flowers that do not produce nectar. A guide mark is simply a pattern in a flower that allows visitors to orient correctly and/or find the floral rewards. Guide marks often involve a contrasting color at the center of a flower, or lines of color radiating from the center. Because the patterns are sometimes in the ultraviolet part of the spectrum, they are not always visible to humans. The function of guide marks has been studied experimentally (Waser, 1983a, and references therein). Artificial flowers with guides stimulate feeding responses more often than those without such guides. Alteration of naturally occurring guide marks either causes pollinators to avoid landing on the flowers or interferes with behavior after landing. Therefore, what to us is simply part of the beauty and distinctiveness of a flower, is a matter of considerable importance to a pollinator. As will be seen in the following, guides may change color, sometimes disappearing with age, and this seems to communicate additional information to pollinators. Also of potential importance in flower identification and the orientation of insects within flowers are olfactory and tactile cues, although these have received less study than visual guides (Faegri and Pijl, 1979; Kevan and Lane, 1985).

The spatial arrangement of stigma and anthers is critically important in ensuring effective pollen pickup and delivery, particularly in specialized flowers. Flowers of many species exhibit herkogamy, the spatial separation of anthers and stigma. Presumably the advantage lies in the reduced interference of male and female functions within a flower (Webb and Lloyd, 1986). In a self-incompatible species, for example, deposition of self pollen yields no seeds and therefore will decrease male fitness. Given that anthers and stigma are in specific positions, it is also necessary that the pollinator's body is positioned in such a way that it makes effective contact with either part. This is generally accomplished by the relative positioning of rewards, anthers and stigma, the presence of guide marks, and the shape and size of the perianth.

Visitors to flowers of many orchids are more actively encouraged to pick up or deposit pollen. *Calopogon pulchellus*, a nectarless orchid of the eastern United States and Canada, has an uppermost petal bearing a group of antherlike hairs that bees presumably mistake for anthers. When a bee lands on the upper petal, it bends forward under the bee's weight. The bee falls on its back onto a lower petal called the column, which deposits a sticky substance onto the bee's back. The bee, still on its back, slides down the column, passing the stigma (and perhaps depositing pollen) as it does so. The bee then contacts the anther, and the orchid's four pollinia (pollen sacs) attach themselves to the sticky patch on the bee's back (Meeuse and Morris, 1984). Similarly complex mechanisms occur in other orchids, many of which were described by Darwin (1890).

The flowers of many species change during their lifetime in ways that are critical to successful pollination. Flowers of many plants are dichogamous, that is, pollen release and stigma receptivity occur at different times. In protandrous flowers the pollen is released first while in protogynous flowers the stigma is receptive first. Dichogamy may be advantageous in avoiding pollen–stigma interference and minimizing self-fertilization (Lloyd and Webb, 1986). In some protogynous species, the male and female functions overlap, which, in self-compatible species may permit "backup" self-pollination to occur if the flower has not already been cross-pollinated. Dichogamy can also be of great significance in allowing considerable independence of male and female functions in a bisexual flower. Thus the duration of male and female phases can be different, as can the levels of nectar production in each phase, as well as certain aspects of floral morphology. In flowers of *Lobelia cardinalis*, for example, the mature staminate phase lasts 5.5 days and the mature pistillate phase only 2.8 days (Devlin and Stephenson, 1985). The male function is further emphasized because nectar production per day is nearly twice as great in male phase flowers as in female phase flowers.

In some mallows (Malvaceae) the style can play an active role in achieving self-fertilization (Buttrose et al., 1977). The long style of *Hibiscus trionum* has five branches. Early in the period of stigma receptivity, the five branches are pressed together and erect. If no pollination occurs, they subsequently bend outward, eventually reflexing so that the terminal stigmas are pressed into the anthers, resulting in self-fertilization. If pollination occurs during stigma descent, the descent stops and may reverse. This mechanism gives first preference to outcrossing, but permits selfing if the former fails.

Changes in the structure of the stigma itself may also regulate the degree of self-pollination. In *Campsis radicans*, for example, the stigma is bilabiate (two-lobed) and receptive only on the inner surface of each lobe (Bertin, 1982a). The lobes are pressed together during the male phase, which lasts about 12 hours. Subsequently they separate, and eventually become reflexed. The stigmas of *Campsis* and many other plants are also sensitive, tending to close after being touched or receiving pollen, a pattern that is not well understood (Bertin, 1982a).

Changes other than those involving the sexual parts also occur in the flowers of many species. Prominent among these are color changes, but others involve nectar and odor production, flower orientation, and shape. These changes may be facultative, and induced, for example, by successful pollination, or they may be fixed, occurring in all flowers of a given age. Gori (1983) proposes three hypotheses to explain these changes. First, "extra" pollinator visits to an adequately pollinated flower may reduce reproductive success by damaging floral parts, dislodging pollen from the stigma, or depositing self pollen. Second, the changes may mark unrewarding flowers for pollinators, increasing the foraging efficiency of pollinators, and encouraging them to spend more time on the plant. Third, by restricting pollinator visits to receptive flowers, removal of inviable pollen (if such pollen is present in the unvisited flowers) and deposition of pollen on unreceptive stigmas is minimized. In eight Australian shrubs studied by Lamont (1985), color changes in all species coincided with loss of stigma receptivity. The above hypotheses, however, do not address the question of why the changed flowers should be retained at all. Suggested benefits are that: (1) retention permits recapture of some nutrients in the flowers by the plants, and (2) the changed flowers may enhance the plant as a visual target for pollinators (Gori, 1983). The latter hypothesis was supported by data from *Lupinus* (Gori, 1983), but not by data from *Cryptantha humilis* (Casper and LaPine, 1984). The two hypotheses are not, of course, mutually exclusive.

Self-Incompatibility

In many species, self-pollinations do not lead to production of viable seed. This phenomenon is known as self-incompatibility (SI). Some disagreement exists as to the scope of phenomena subsumed by this term (Nettancourt, 1977; Richards, 1986). In particular, some researchers include only processes occurring before fertilization (Nettancourt, 1977), while others include events (e.g., seed abortion) occurring after zygote formation. Historically, SI has been considered largely a stigmatic or stylar phenomenon, but Seavey and Bawa (1985) suggest that late-acting (ovarian, sometimes post-fertilization) SI may be more widespread than previously thought.

Two major categories of SI systems are recognized. In the sporophytic system the incompatibility phenotype of pollen is determined by the genotype of the pollen-producing plant. Thus, each pollen grain produced by a plant with this system has an identical SI reaction. Other species have gametophytic SI, wherein the incompatibility phenotype of a pollen grain is determined by its own (haploid) genetic constitution. A plant with this system of SI may produce pollen grains with more than one incompatibility phenotype.

In either system, a pollen grain's SI phenotype is conferred by one or more SI genes, whose alleles are termed S alleles. One or more genes may be involved, and the number of alleles at a particular locus ranges from two to dozens or even hundreds. The genetics of SI is further complicated by the possibility of dominance/recessiveness and epistatic interactions among alleles and genes, respectively. Because different individuals in a population sometimes share one or more S alleles, certain cross-pollinations are unsuccessful, or at least less successful than others.

Self-incompatibility is often not an all-or-nothing phenomenon (Nettancourt, 1977). Self-pollination of immature and very old stigmas, or flowers at the end of the blooming season sometimes results in fertilization in normally SI species. The simultaneous or prior application of compatible pollen may also permit self-fertilization. The term cryptic SI has been applied to plants wherein self pollen can effect fertilization, but because of the slower growth rate of self pollen tubes, few self-fertilizations occur following pollination with a mixture of self and cross pollen (Weller and Ornduff, 1977). Considerable variability in the degree of SI may occur among individuals of a population and populations of a species, and many instances of the evolutionary breakdown of SI have been reported (Barrett, 1987).

Sexual Systems

Plant populations differ greatly in the disposition of male and female sexual functions among plants and among flowers on a plant. The particular advantages of each sexual system are not completely understood, although the subject has received much recent attention. The reader is referred to Bawa and Beach (1981), Willson (1983), and Wyatt (1983) for more information. The following is largely a synopsis of their treatments.

Hermaphroditism

This term can refer either to flowers containing functional male and female organs, or to plants that bear only such bisexual flowers. In this chapter, the term is used in the latter sense. Most plant species (perhaps 70 percent) have this sexual system. Its advantages include the need for only one perianth and perhaps one dose of nectar for both male and female functions, rather than one for each. Pollinators can both deposit and pick up pollen in a single visit. Where pollen is the food reward for pollinators, bisexual flowers are required to ensure visits to gynoecia. Finally, in bisexual flowers the opportunity for self-pollination exists unless other factors prevent it.

Heterostyly

Some hermaphroditic populations include plants of two or three floral morphs differing in the positions of anthers and stigma, a condition referred to as heterostyly. In distylous species, individuals have either flowers with a long style and short stamens (pin flowers) or flowers with short style and long stamens (thrum flowers). A few species have three floral morphs and are termed tristylous (Fig. 2-6). These morphs are referred to as short-, mid-, and long-styled, reflecting the relative position of the stigma. In many species, plants of a given morph are incompatible with themselves as well as with other individuals of their morph, a condition referred to as sporophytic heteromorphic SI.

A traditional adaptive explanation of heterostyly is that the complementary placement of anthers and stigmas in different morphs facilitates insect-mediated cross-pollination between morphs (legitimate pollination; Fig. 2-6; Darwin, 1877). Specifically, pollen from the anthers of one morph would adhere to the pollinator's body at a place from where it would be likely to be deposited on a stigma of the other morph. Studies have shown, however, that legitimate pollen does not necessarily make up the bulk of stigmatic pollen loads (Barrett and Glover, 1985), suggesting that facilitation of legitimate pollinations cannot be a general explanation for the evolution and maintenance of heterostyly. It has also

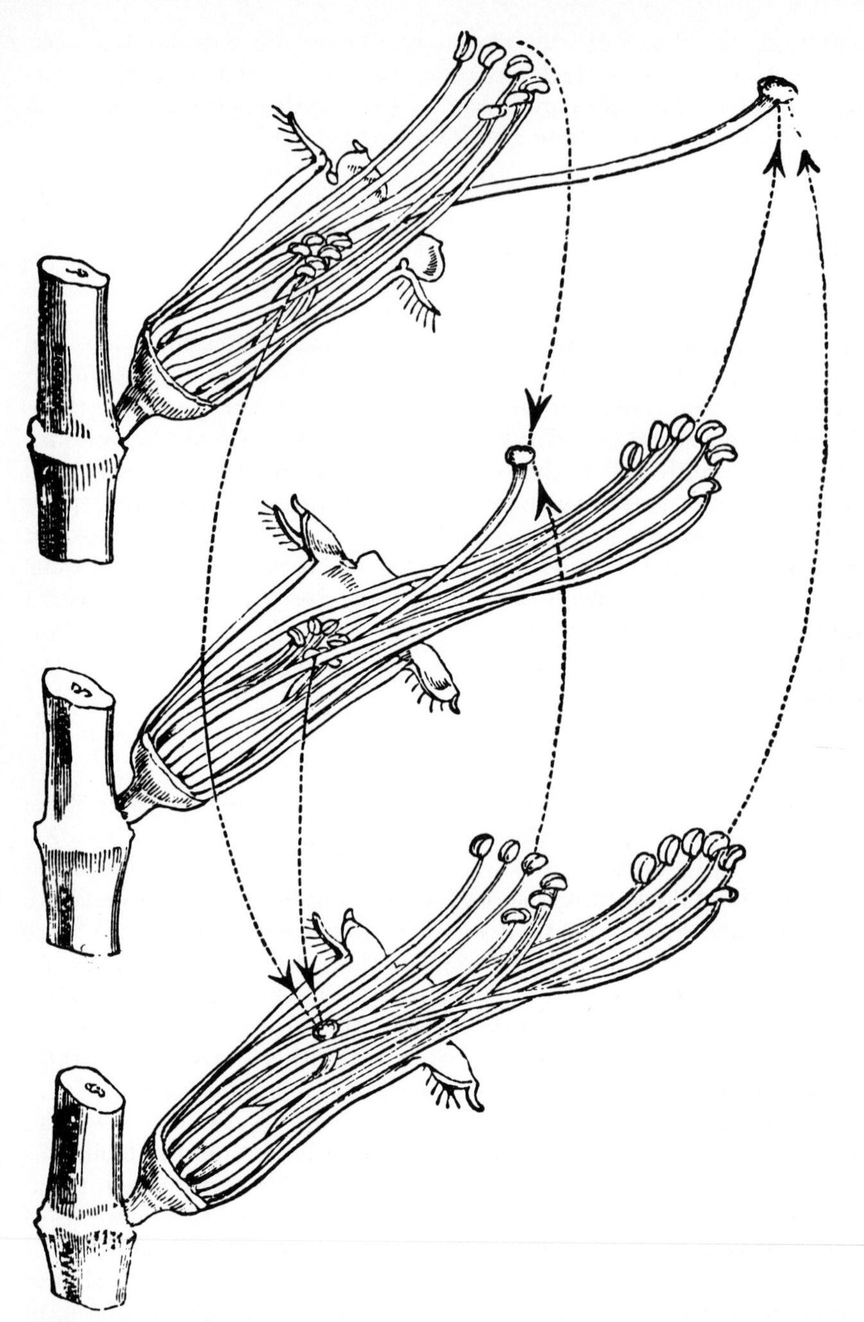

Figure 2–6. The three flower types of a tristylous species, *Lythrum salicaria:* long-styled (top), midstyled (middle), short-styled (bottom). Arrows indicate patterns of "legitimate" pollen movement. Quantitative studies show that movement of pollen between other anther–stigma combinations can also be common. From Darwin (1877).

been suggested that heterostyly reduces interference between male and female functions (Lloyd and Yates, 1982), and that it is associated with sexual specialization of the different morphs, with the pin morph being more female (Bawa and Beach, 1981).

Andromonoecy

Andromonoecious individuals have both male and bisexual flowers (Fig. 2-7). The ratio of the two flower types may be relatively constant among individuals, as in umbellifers, or variable and influenced by environmental factors such as light availability. Despite some claims to the contrary, andromonoecy probably does not promote outcrossing, because each pistil is still in a flower containing functional anthers. Some andromonoecious species are large-fruited (e.g., *Aesculus* spp.) and this sexual system may reflect the diversion of resources away from excess pistils which the plant has inadequate resources to mature into fruit. The retention of male flowers might reflect their importance in pollinator attraction or in some fitness gain through pollen donation (Bertin, 1982b).

Gynomonoecy

Plants with female and bisexual flowers occur in only a few families and are most common in the Compositae. In this family, flowers are in densely packed heads, and in many species the outer ray flowers are female (Fig. 2-4). These ray flowers bear elongate petals that make a composite inflorescence attractive to pollinators. If stamens and ray flowers are mutually exclusive evolutionary and developmental options, the advantage of male sterility is evident. In fact, flowers of the composite *Senecio squalidus* do show a negative correlation between ray length and degree of another development (Ingram and Taylor, 1982). In chicory *(Cichorium)* and its relatives, however, all flowers are bisexual and also have rays. Lloyd (1979) notes that, given the uniovulate nature of all composite flowers, gynomonoecism is the only way to add ovules without the cost of adding pollen. This argument assumes that deletion of stamens is an easier evolutionary route than increasing the number of ovules per flower. Our knowledge would benefit from detailed examinations of interactions between flowers and pollinators, the degree of

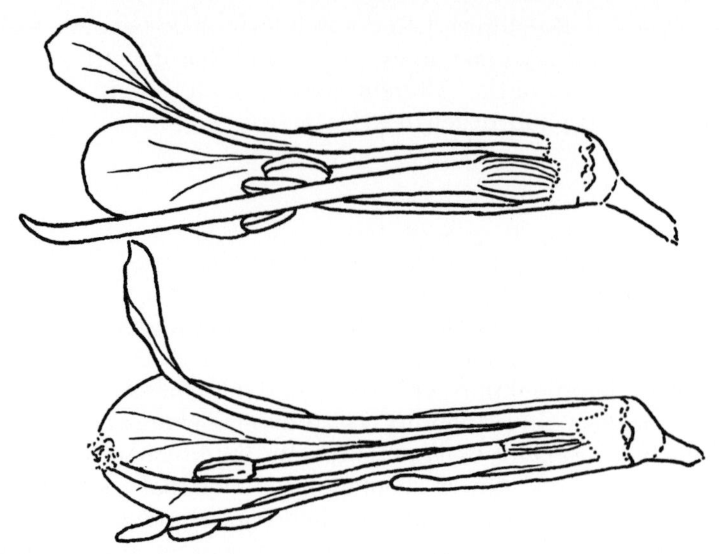

Figure 2–7. Flowers of *Aesculus pavia,* an andromoecious shrub. Part of each perianth has been removed to reveal the sexual parts. Bisexual flowers (top) are protogynous, here illustrated during stigma receptivity, before elongation of stamens and dehiscence of anthers. Note the vestigial pistil in the male flower (bottom).

variability in ratios of flower types, and the factors causing this variability.

Monoecy

A monoecious plant possesses functional male and female organs in separate flowers. Several advantages of this system have been proposed. It may reduce the likelihood of self-pollination and it permits a greater degree of specialization of the male and female functions, which are not always compatible. For example, the presence of stamens or the deposition of self pollen may interfere with the deposition of incoming pollen. Furthermore, the optimum number and positioning of male and female flowers may differ, as may the best timing of anthesis and stigma receptivity. In wind-pollinated species, a monoecious condition with male flowers above female flowers may enhance both pollen receipt and the effectiveness of pollen donation. Self-pollination could be avoided by different times of opening of male and female flowers. Many wind-pollinated trees in the temperate zone are monoecious (e.g., *Acer,*

Castanea, Fagus, Quercus; Fig. 2-1), but in the tropics many insect-pollinated species also have this system (Bawa and Beach, 1981).

Dioecy

A dioecious population includes individuals bearing only male flowers and other individuals bearing only female flowers. The traditional advantage assigned to this sexual system is that of ensured outcrossing. However, a large number of ecological correlates of dioecy have also been suggested or reported, and the advantage of this system probably reflects factors other than outcrossing in some species (Bawa, 1980a). Dioecy could, for example, permit individuals of the two sexes to inhabit different microsites. This has been suggested for several wind-pollinated species in arid regions of North America (Freeman et al., 1976). Females

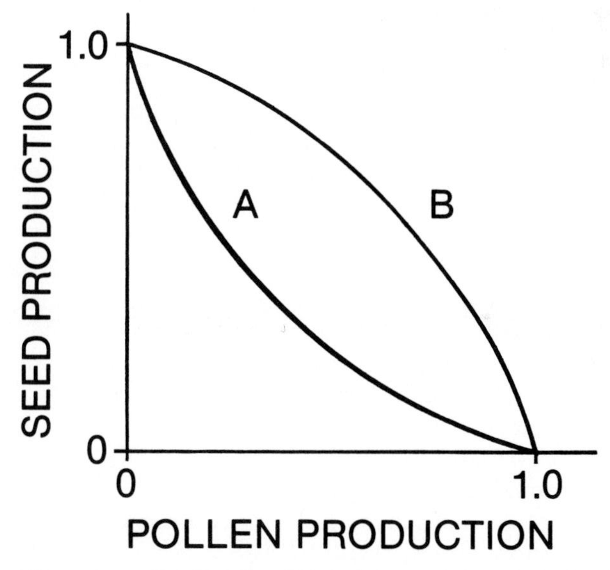

Figure 2–8. Fitness sets for the allocation of resources to reproductive function. Along the x-axis is pollen production of an hermaphrodite relative to that of a male. Along the y-axis is seed production of a hermaphrodite relative to that of a female. Each curve represents the evolutionary options available to a hypothetical population. Given a concave fitness set (A), the most productive options are maleness ($x = 1$) or femaleness ($y = 1$), and dioecy is favored. Given a convex fitness set (B), hermaphroditism should be maintained because the sum $x + y$ is greater when both pollen and seeds are produced ($x < 1$, $y < 1$). The use of fitness sets has been largely theoretical because of the difficulty of gathering empirical data to construct the curves. Redrawn by permission from *Nature* 263:125–126, Copyright © 1976, MacMillan Journals Ltd.

were more common than males in moister microsites. This microhabitat differentiation was suggested to increase the reproductive output of both sexes. Wind may be more likely to disperse pollen from males on the sparsely vegetated xeric sites, while soil moisture may be more important to female reproduction, which takes place over a long time and requires resource inputs even after pollen production ceases.

Correlations have also been reported between dioecy and small, inconspicuous flowers, pollination by small, unspecialized insects, and fleshy fruits, particularly with bird-dispersed seeds. Various explanations for these correlations have been suggested (Bawa and Beach, 1981; Lloyd, 1982), but no consensus has yet been reached. A frequent theoretical approach to assessing the relative merits of dioecy and hermaphroditism has been to use fitness sets, an example of which is given in Fig. 2-8. Dioecy has also been suggested as a possible result of intense intrasexual selection (Willson, 1979).

Gynodioecy

In some populations each individual has either female flowers or bisexual flowers. Such populations are thought to have descended from populations of hermaphrodites by the invasion of male-sterile (female) individuals. Because male steriles do not obtain any fitness via the male function, they generally must possess compensatory advantages if they are to invade a population of hermaphrodites. One advantage that females may have over hermaphrodites is that the lack of male parts prevents (1) self-pollination, (2) any accompanying inbreeding depression (and therefore lower maternal reproductive success), and (3) pollen-stigma interference within flowers if individuals are self-incompatible. In some species maternal fecundity is lower in hermaphrodites than in females (Willson, 1983). A complicating factor is that male sterility may be inherited through either nuclear or cytoplasmic factors or both. The mode of inheritance influences the ability of male-steriles to invade hermaphroditic populations (Willson, 1983).

Androdioecy

Few if any species exhibit this sexual system, characterized by male and hermaphroditic individuals in the same population. The rarity of androdioecy has been attributed to the necessity for male individuals to have at least twice the male success of hermaphrodites if they are to invade a population of hermaphrodites. Androdioecy has been reported from a few species of *Solanum* in which pollen is the only reward for pollinators, but these reports are controversial and require further study.

Further Considerations

The above descriptions of sexual systems were made as though each is distinct and recognizable, which frequently is not true. For example, *Aesculus pavia* is generally described as andromonoecious because most individuals have some male flowers and some bisexual flowers (Bertin, 1982c). The proportion of male flowers is often low, however, and some plants lack them entirely. Furthermore, individuals that are exclusively male in one year sometimes bear bisexual flowers in another year. Finally, rare individuals have functionally female flowers. Thus there are elements of andromonoecy, androdioecy, dioecy, and gynodioecy in a single population. These terms are in part labels of convenience, and a continuous rather than discontinuous approach to the classification of sexual systems may be superior.

Lloyd (1980) proposed a quantitative approach to the representation of breeding systems. The functional gender of a plant can be represented by G and A, the proportion of genes likely to be transmitted via female and male functions, respectively. These are estimated as $g/(g + a)$ and $a/(g + a)$, respectively, where g and a are the numbers of offspring from female and male gametes, respectively. Because g and particularly a are difficult to quantify, Lloyd suggests that counts of other gynoecial and androecial units (e.g., ovules and anthers) may be used instead. When such units are used, the measure is more appropriately referred to as phenotypic gender rather than functional gender, since the ultimate success of pollen and ovules is unknown. Gender calculations for several individuals can be combined to provide a graphic description of the distribution of gender within a population (Fig. 2-9).

Floral Syndromes

The kinds of floral visitors that a plant receives is related to its floral morphology. Indeed, a regular part of many volumes on pollination biology (e.g., Faegri and Pijl, 1979; Proctor and Yeo, 1972) is an enumeration of the floral characteristics associated with pollination by various animal groups. A knowledge of such floral syndromes is important in allowing predictions of expected pollinators for plants whose flowers are known but whose pollination has not been studied, and it emphasizes the differing abilities and needs of different pollinator groups. It should be kept in mind, however, that most plants are pollinated by animals belonging to more than one major group (e.g., Waser, 1983a), and that the only way to be sure of a plant's pollinators is to make detailed observations of its floral visitors, their pollen loads, and their ability to pick up and deposit pollen. The following treatment is drawn largely from Faegri and Pijl (1979).

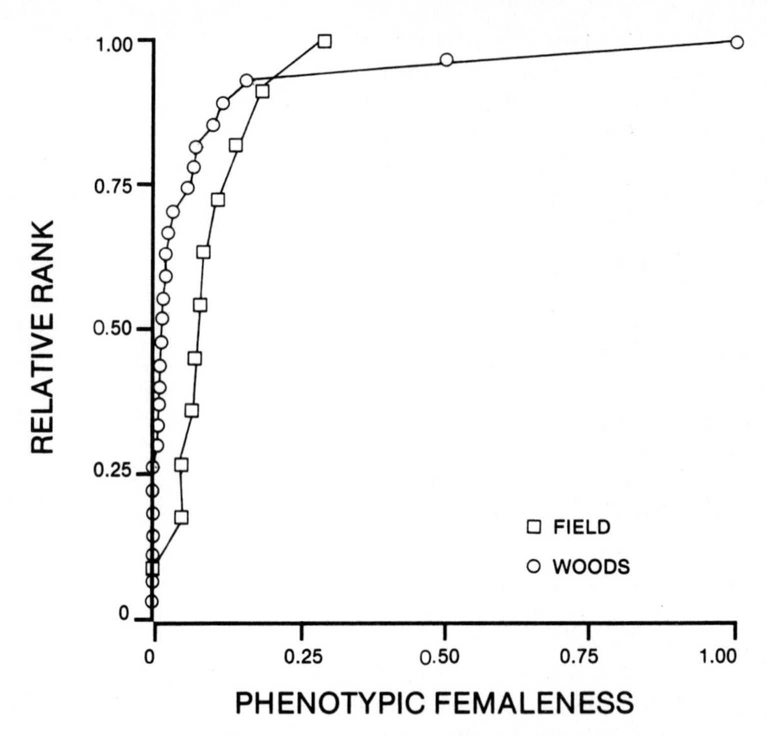

Figure 2–9. Graphic depictions of population gender in *Aesculus pavia*. In the woods population, most individuals have all or nearly all male flowers, although one individual has only bisexual flowers (femaleness = 0.5) and one has only female flowers (femaleness = 1.0). Most individuals in the field population have 10- to 30-percent bisexual flowers. This difference was correlated with differences in light intensity, with the field plants exposed to more light and presumably having more resources for fruit production.

Insect Flowers

Beetles (Coleoptera) are considered to be relatively inefficient pollinators due to their smooth, hard exteriors unsuited to adhesion of pollen, their chewing (rather than, for example, sucking) mouthparts, and their ungainly movements. The limited effectiveness and imprecise movements of these visitors has led to the phrase "mess and soil pollination" to describe their floral activities. Nevertheless, a number of flowers seem to be visited primarily by beetles. These flowers tend to be either large and solitary, like *Magnolia* and *Calycanthus*, or small and aggregated into dense inflorescences (e.g., some *Viburnum*, *Sorbus*). Visual attractants are unspectacular, colors are whitish or dull, and the flowers are open, with easy access to sexual organs and rewards. Odors are often strong and, to a human nose, generally unpleasant.

Two types of flowers pollinated by flies (Diptera) are generally recognized. One group is characterized by small, open, unspecialized flowers, usually dull or light colored, without odor, but sometimes with guide marks. Flowers of the other group, the specialized fly flowers, are typically green, purple, or brown. Visitors to the latter flowers receive no rewards, and are lured by carrion-like odors and are often trapped or have their patterns of movement manipulated by strategically located transparent windows, angled bristles, slipways, and the like.

Proctor and Yeo (1972) describe pollination of a specialized fly flower, *Arum nigrum*, as earlier reported by Knoll. Inflorescences of this species, like those of the American jack-in-the-pulpit, consist of a column (spadix) bearing many small unisexual flowers, and surrounded and overtopped by a modified leaf, the spathe. The spadix produces a fecal odor during the first night that the inflorescence is open. Visitors, particularly dung-frequenting flies and beetles, land in the upper part of the inflorescence, but lose their grip on the smooth surface with downward-projecting papillae. In their descent or subsequent attempts to clamber out they brush against stigmas of the female flowers, potentially effecting pollination. The insects remain trapped in the base of the spathe, and on the second night the male flowers release their pollen, dusting the trapped insects. On the second day, odor production ceases and the papillae that bar the insects' escape shrink, allowing the latter to crawl out of the inflorescence laden with their host's pollen.

The Hymenoptera (bees, wasps, ants, and relatives) probably pollinate more plant species than do any other animal group. From an economic viewpoint, bee pollination is certainly more important than all other forms of biotic pollination combined. The complex social interactions found in certain bees (discussed later) allow particularly efficient exploitation of floral resources.

A few plant species (e.g., *Polygonum cascadense*, Hickman, 1974; *Diamorpha smallii*, Wyatt, 1981) are pollinated by ants, although the hard exoskeletons, small size, and limited mobility of these insects reduce their effectiveness as pollinators. Although studies of ant pollination are too few to be confident of generalizations, an expected syndrome for ant pollination might include high plant density, small conspicuous flowers borne near the ground, low ovule number and nectar production, small amounts of sticky pollen, and a warm, dry habitat. Most of these traits complement the small size and low energy demands of these visitors (Hickman, 1974; Faegri and Pijl, 1979; Wyatt, 1981). Beattie et al. (1984) showed that exposure of some pollen to ants for brief periods reduces performance of the pollen. They suggest that this reduction may be caused by antibiotic secretions of the ants required by their nest-building and brood-rearing activities. The existence of such secretions may help explain the rarity of ant pollinators.

Most wasps include floral resources as just one component in their

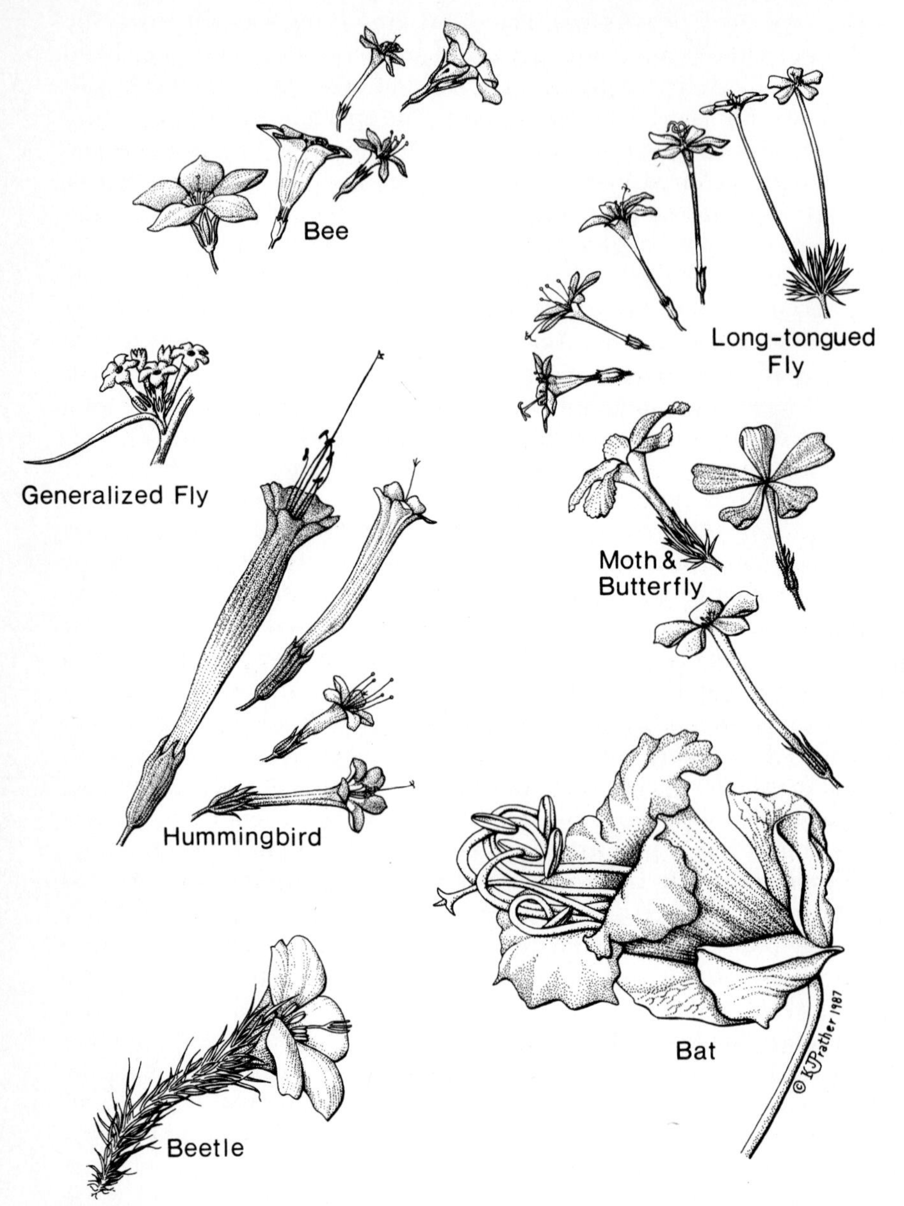

Figure 2–10. Flowers of the Polemoniaceae, grouped by putative pollinator type. Redrawn from Grant and Grant (1965).

mixed diet. Because of their typically short mouthparts, they usually visit shallow blossoms having exposed nectar or pollen. Some of the solitary bees also visit flowers of this type. Indeed "unspecialized" flowers with accessible rewards may receive visits from bees, wasps, flies, beetles, and even Lepidoptera (moths and butterflies). Some Australian orchids have flowers that mimic female wasps and they are pollinated by male wasps exhibiting copulatory behavior (Stoutamire, 1974; see the following).

Social bees include the familiar honeybee *(Apis mellifera)* and bumblebees *(Bombus* spp.; Fig. 2-11), which can be found on a wide variety of flowers, including both less specialized ones (e.g., *Solidago, Pyrus)* and more elaborate zygomorphic flowers (those exhibiting bilateral symmetry), such as members of the pea and orchid families. These latter blossoms are often exclusively pollinated by bees. Flowers of this group are zygomorphic, often semi-closed, and possess a landing surface, bright colors such as yellow and blue (the latter a relatively uncommon color among flowers), guide marks, fresh scents, and hidden nectar (Faegri and Pijl, 1979; Fig. 2-3).

Social bees are often considered more intelligent than other pollinators due to their ability to learn how to tap the rewards in flowers such as *Aconitum,* which are partly or completely closed, and which require a specific pattern of manipulation for obtaining nectar. The large size of many bees is undoubtedly an added advantage in manipulating these blossoms.

Butterflies are day-flying Lepidoptera, and most species have a long, slender proboscis. The classic butterfly flower has a long, very narrow tube and a flat rim, and is represented by such genera as *Phlox, Lantana,* and *Buddleia* (Fig. 2-10). Cruden and Hermann-Parker (1979) note that several neotropical butterfly-pollinated flowers have a yellow target on a red background, with the target often fading to red on nonreceptive flowers. Interestingly, some butterflies see further into the red end of the visual spectrum than do other insects, and therefore the above color patterns may increase the proportion of visitors that are butterflies. Flowers of many other colors are also visited by butterflies, however, including ones at the other (blue) end of the visual spectrum. Except for the most primitive taxa, chewing mouthparts are lacking and therefore pollen is not consumed. However, some *Heliconius* butterflies obtain nutrients from pollen that leach into nectar when the two are mixed. Nutrients obtained in this manner are of considerable importance in reproduction (Gilbert, 1972).

Moths include mostly nocturnal or crepuscular forms, and many of the important flower visitors hover rather than land on their flowers. Moth blossoms tend to be large, tubular, light-colored, horizontal or pendant, open in the evening or at night, heavily scented, and lack a landing platform (Fig. 2-10). Some of the most spectacular moth-pollinated species are African orchids with long, narrow, nectar-containing spurs up to 30 cm long. They are pollinated by sphingid moths with similarly long probosces (Nilsson et al., 1985). Some North American genera with moth-pollinated representatives include *Datura, Silene,* and *Oenothera.* A specialized relationship exists between certain moths and *Yucca* plants, and this is described later.

Bird Flowers

Major bird pollinators include the New World hummingbirds (Trochilidae), the Old World sunbirds (Nectariniidae) and honeyeaters (Meliphagidae), and representatives of a few other taxa (Fig. 2-11). Bird flowers are brightly colored, often red or yellow, and lack odor and a landing platform. These features correspond to the visual orientation of these pollinators and the fact that they do not generally perch on the flowers. Nectar is abundant and deeply held, accommodating the high energy requirements and long bills and/or tongues of birds. Red coloration has been suggested to be especially appropriate for attracting birds exclusively because it is conspicuous to them but indistinguishable from a gray background to many insects (Raven, 1972). North American hummingbird-visited species include *Lobelia cardinalis, Campsis radicans, Fouquieria splendens,* and *Ipomopsis aggregata* (Fig. 2-3B).

Mammal Flowers

Bat pollination is common in the tropics and rare or absent elsewhere. It is associated with a distinctive suite of characteristics including flowers that are often borne in unusual positions: on long, hanging stalks, outside the foliage, or springing directly from large stems (cauliflory). These arrangements facilitate visitation by large, awkward visitors. Flowers open in the evening or at night and frequently last for just one night. They are whitish or dull-colored, but emit a strong odor and are large and open, providing easy access. Pollen and nectar are produced in large quantities, the latter sometimes amounting to several milliliters per flower. Some tropical genera exhibiting bat pollination include *Bauhinia, Ceiba, Adansonia,* and *Mucuna* (Fig. 2-10). Bat-pollinated agaves *(Agave)* are found in the southwestern United States.

In the past few years, evidence has accumulated that some plants are pollinated by mammals other than bats, including lemurs, rodents, and marsupials, such as the honey possum *(Tarsipes spencerae)* and various phalangers. Plants visited by these mammals are similar to bat

Figure 2–11. Pollinator types. A. Honeybee *(Apis mellifera).* B. Bumblebee *(Bombus americanorum.)* C. Euglossine bee *(Euglossa* sp.). D. Hawkmoth *(Celerio lineata* (redrawn from Grant and Grant 1965). E. Beefly *(Eulonchus smaragdinus);* flies in this family have much longer tongues than most other flower-visiting flies and often visit flowers with deep nectar. F. Anna's hummingbird *(Calypte anna);* a generalist species. G. Longnose bat *(Leptonycteris* sp.).

flowers in having crepuscular or nocturnal opening of flowers, a strong odor, copious nectar, and inconspicuous coloration (Carpenter, 1978; Sussman and Raven, 1978). They may also have leathery or fleshy petals, and these or other flower parts are frequently eaten by mammalian visitors. In some species, flowers are borne near the ground, which facilitates visitation by rodents.

Limitations to the Syndrome Approach

Most species of plants are visited by pollinators belonging to more than one species and often more than one major taxon. Two examples will illustrate this point.

Milkweeds *(Asclepias* spp.) have their pollen aggregated into pollinia (pollen sacs, Fig. 2-12). These pollinia are readily visible on floral visitors, and therefore one can readily enumerate species able to carry pollen. At four sites in Illinois, honeybees carried the most pollinia, and at three of these, a noctuid moth ranked second. A beetle, *Chauliognathus pennsylvanicus,* and another noctuid were also frequent carriers of pollen (Willson and Bertin, 1979). At a study area in Maine, however, the major pollinators seemed to be bumblebees, honeybees, and vespid wasps (Fritz and Morse, 1981). Nocturnal noctuid and geometrid moths were frequent visitors but contributed little to pollination (Morse and Fritz, 1983). Thus, milkweed receives frequent visits and some pollination from insects of three orders, and the relative contribution of the two major orders (Hymenoptera and Lepidoptera) appears to differ between regions. Such differences are undoubtedly caused in part by habitat differences which make each area more or less suitable to particular pollinators. Such differences illustrate a potential cost to a plant relying exclusively on a single species or small group of similar pollinators. This may restrict the range of habitats and/or geographical areas that can be occupied by the plant. A final consideration is that honeybees are not native to North America. Therefore, it is unlikely that much evolution has occurred between any native plant and this bee. One can only wonder to what extent this bee has displaced native pollinators and obscured ecological relationships between the plant and its original pollinator fauna.

Another example of a plant species visited by members of many taxa is *Nelumbo pentapetala,* which has large, open, actinomorphic (dish-shaped) flowers with yellow petals (Sohmer and Sefton, 1978). Representatives of 70 insect species representing 53 families and 10 orders were collected from flowers of *Nelumbo.* Some of these species were

Figure 2–12. Pollination in milkweed *(Asclepias).* When an insect draws its leg upward through one of five vertical slits on the flower margin (A), it removes the pollinarium, consisting of a corpusculum, which clamps onto the leg, and two translator arms, each bearing a pollinium. The translator arms twist through 90 degrees within minutes of removal and the pollinia can then be inserted if the insect draws its leg upwards through the stigmatic slit (B). From Meeuse and Morris (1984),,used with permission.

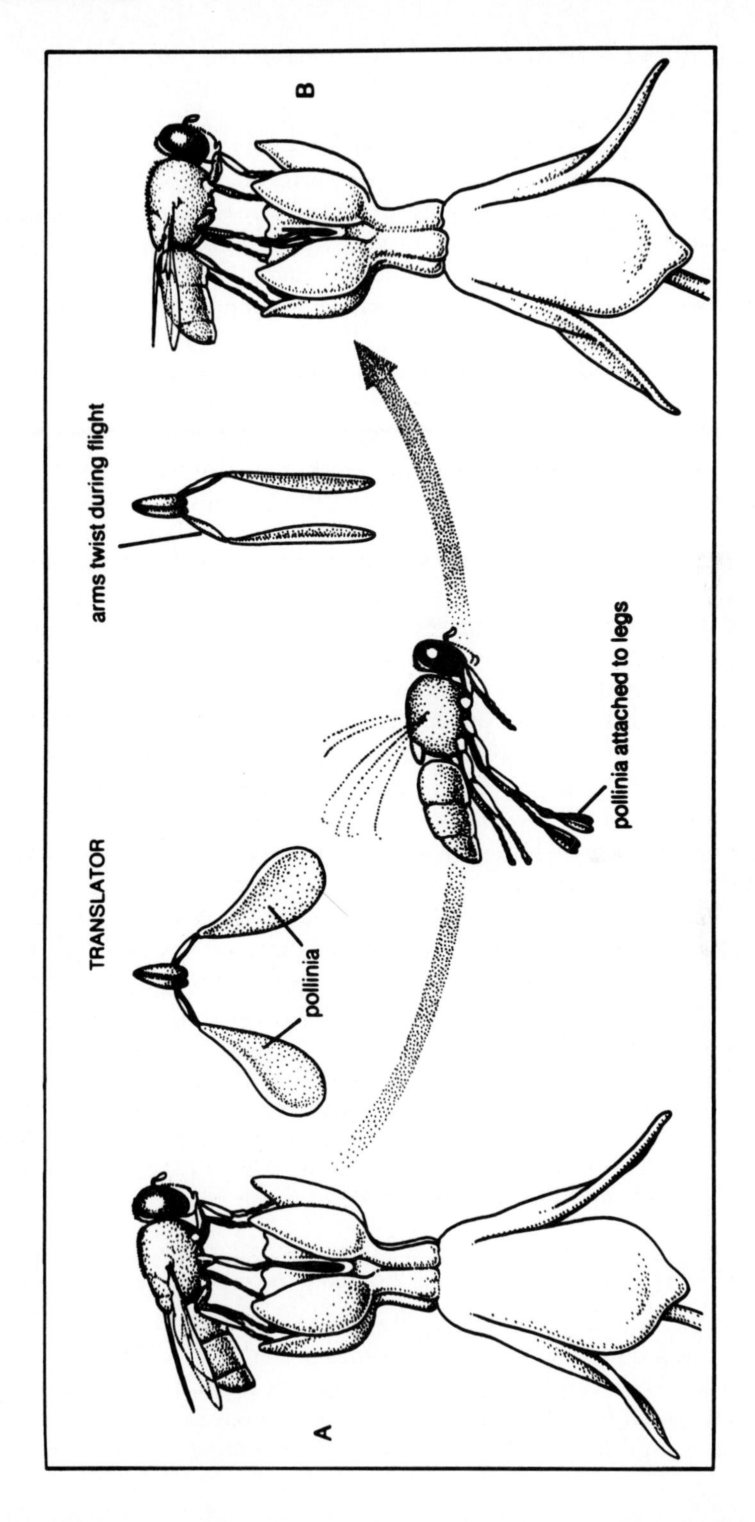
A
TRANSLATOR
pollinia
arms twist during flight
pollinia attached to legs
B

uncommon, and many probably effected little or no pollination, but several members each of the Diptera, Hymenoptera, and Coleoptera appeared to be important pollinators.

Different species of floral visitors are usually not equally effective as pollinators. To determine the relative performance of floral visitors, a variety of approaches has been used. A time-honored method is detailed observation of the behavior of insects in flowers. Frequently one can determine which visitors contact anthers and stigma, and thus obtain a qualitative measure of their performance. A quantitative approach is to expose virgin stigmas to a single pollinator visit and then count the deposited pollen grains under a microscope. Using such an approach, Bertin (1982a) found that hummingbirds deposited about 10 times as many pollen grains on stigmas of *Campsis radicans* as did honeybees and bumblebees, and these in turn deposited several times as many as did halictid bees. Such figures must be weighted by the visitation frequencies of the different visitors. If pollinators differ greatly in size, it may be possible to estimate the effects of the larger visitors by excluding them with screening of appropriate mesh size. For example, exclusion of hummingbirds from flowering ocotillo *(Fouquieria splendens)* reduced seed set by about 30 percent, with carpenter bees *(Xylocopa)* being the major remaining pollinators (Waser, 1979). While it cannot be assumed that hummingbirds cause 30 percent of seed set when both birds and bees are present, this figure does provide a rough idea of the bird's importance.

Pollen

Pollen is relevant to pollination in several ways. From the plant's perspective, a pollen grain is a male gametophyte that gives rise to the male gametes essential to sexual reproduction. Pollen movement is one of two main contributors to gene flow, and therefore influences the genetic structure of plant populations. From the viewpoint of the pollinator, pollen is sometimes irrelevant, sometimes an encumbrance, and sometimes food.

Pollen Structure and Composition

Pollen grains consist of two or three nuclei surrounded by two walls. The intine, or inner wall, is composed principally of cellulose and pectin. Around this is the exine, consisting of the acid- and enzyme-resistant lipid sporopollenin, which is sometimes mixed with other substances. The exine is occasionally smooth, but often sculptured in various patterns, which have both taxonomic and ecological significance.

The composition of pollen varies greatly among species in terms of starch, sugar, lipid, protein, and various minerals (Stanley and Linskens, 1974). Herbert and Irene Baker (1983b) extensively investigated the distribution of starchy and starchless pollen grains and concluded that small pollen grains are likely to be starchless and presumably lipid-rich. Lipids are, of course, a more compact means of energy storage. Autogamous (self-fertilizing) species typically have starchy grains, which the Bakers suggest could reflect the lower metabolic cost of producing starch than lipid. Starchy grains may be a cheap alternative for plants that do not need to satisfy other nutritional needs of pollinators, and in which there is no premium on pollen compactness. Starchless grains are common in species whose regular visitors feed on pollen, and such grains may contain higher levels of other substances such as protein or lipid, that have nutritional importance. Clearly much more work is needed to understand the significance of pollen constituents in the plant–pollinator interaction. A few studies have examined the functional significance of the exine. Pollen grains of anemophilous species generally have smoother surfaces than those habitually carried by animal vectors. This reduces the likelihood of grains sticking together, which could interfere with effective wind dispersal. The spiny surfaces of insect-borne pollen may allow the grains to stick together or stick to floral visitors (Faegri and Pijl, 1979).

Pollen of some species is packaged in units (polyads) of two to many grains. The smaller units contain two (dyads) or four (tetrads) pollen grains, and the large pollinia of some milkweeds and orchids contain hundreds or thousands of pollen grains. There is no general agreement as to the adaptive significance of these features. Pollinia result in simultaneous pickup or deposition of many pollen grains from a single individual. This may be advantageous for male fitness, especially in populations serviced by infrequent but species-constant visitors. In certain polyad-bearing Australian *Acacia*, most stigmas that receive any pollen obtain only one polyad (Knox and Kenrick, 1983). If pollen were loose and only one pollen grain were deposited, this would probably be insufficient for fruit production and would therefore make no contribution to male fitness. Aggregation of pollen increases the likelihood of success from donation of one pollen unit. Pollen donors may also benefit by having pollinia if insertion of a pollinium into a stigmatic chamber (a specialized receiving structure) precludes subsequent insertions, minimizing pollen competition (Willson, 1979).

Pollen Grain Number

The number of pollen grains per flower or per plant varies widely among species, among varieties of cultivated species, and among years for a single plant. Pollen grain numbers are often expressed relative to numbers of ovules. The ratio of number of pollen grains produced by a plant to the number of ovules produced (*P/O*) ranges from a few to over a million, and is correlated with mode of pollination and breeding system (Cruden, 1977). *P/O*s increase with degree of outbreeding. The mean *P/O*s for samples of species exhibiting obligate autogamy, facultative outcrossing, and obligate outcrossing were 28, 797, and 5859, respectively (Cruden, 1977). Some wind-pollinated species have *P/O*s above a million. The explanation for this pattern may be in the increased opportunities for pollen to reach more stigmas in the systems with the higher *P/O*s (Bertin, 1988). For example, in an obligate selfer a plant's ovules are saturated with pollen even at relatively low levels of pollen production, meaning no benefits will come from producing more grains. In contrast, in an obligate outcrosser the available stigmas will not be saturated at such low levels, and substantial fitness benefits may accrue from increased pollen production.

Nectar

Nectar is the most common floral reward for pollinators, and there is little doubt that the composition, amount, and concentration of floral nectar, and the timing of nectar production, have evolved primarily in response to their effects on pollinator visitation.

Amount and Periodicity

Flowers produce nectar in amounts ranging from zero to several milliliters. There is a general correlation between a flower's nectar production and the energetic demands of its habitual pollinators. Based on small samples of plant species, Cruden et al. (1983) report the following ranking of nectar amount per flower for several pollinator groups: bat ($\simeq$140 μL), hawkmoth, sunbird, hummingbird, bee, butterfly ($\simeq$2 μL). Much variability also existed among plants pollinated by one type of pollinator.

The timing of nectar secretion varies among species and is generally thought to be related to the times of pollinator visitation. For example, Cruden et al. (1983) report that many hawkmoth-pollinated flowers begin secreting nectar between 1600 and 1800 hours, 2 to 4 hours prior to hawkmoth activity. In many species the pattern is less simple. Nectar production of *Asclepias syriaca* in Illinois peaked at 1400 to 2200 hours (Willson and Bertin, 1979), whereas pollinator activity, pollinia removal and pollinia insertion were highest during the day. However, the fewer nocturnal pollinators outperformed the diurnal ones in effecting seed set (Bertin and Willson, 1980). The duration of nectar production varies and is frequently shorter than the life of the flower. In some species nectar production ceases when the flower is effectively pollinated, and some species resorb at least the sugar moiety of unused nectar in older flowers (Cruden et al., 1983).

Composition

Nectar is an aqueous solution of sugars, and it often contains small amounts of other substances, such as amino acids, proteins, alkaloids, phenolics, antioxidants, lipids, saponins, dextrins, and vitamins (Baker and Baker, 1983a). The significance of many of these substances is poorly known.

A survey of the sugar components of nectar was made for 765 species (Baker and Baker, 1983c). Of these, 649 had detectable amounts of glucose, sucrose, and fructose, and most of the rest had detectable levels of two of these sugars. Other sugars (e.g., maltose, melezitose, raffinose, melibiose) occurred less frequently.

Several ecological correlations of sugar types are known. Percival (1961) showed that flowers with deep tubular corollas tend to have sucrose-rich nectar. Baker and Baker (1983c) reported that sucrose-rich nectars occur in flowers assumed to be pollinated by hummingbirds, hawkmoths, butterflies, and long-tongued bees, while nectars rich in hexose (glucose and fructose) are more typical of flowers assumed to be pollinated by perching birds, short-tongued bees, flies, and bats (at least New World species). Hummingbirds prefer sucrose over hexoses (Hainsworth and Wolf, 1976), but honeybee preferences are less clearly defined.

Sugar concentrations are not usually reported separately for each sugar, but rather as a combined figure, expressed as percentage of sucrose (generally sugar weight per weight of solution). High concentrations are often found in bee flowers and lower concentrations in hummingbird, butterfly, and bat flowers. The low concentrations of nectar in hummingbird flowers may reflect the hummingbird's use of capillarity to fill grooves on the tongue with nectar. Because dilute solutions exhibit higher capillarity, a hummingbird might feed on them more rapidly than on concentrated solutions. Likewise, the lower viscosity of dilute nectars might be easier for butterflies, which suck nectar through a very narrow proboscis. In contrast, some nectar feeders, such as flies and bees, can ingest crystallized nectar by regurgitating liquid onto the crystals and taking up the resultant solution.

Another possible reason for correlations between sugar concentrations and major pollinators is the relative importance of feeding costs and transit costs in different pollinator groups. Heyneman (1983) suggests, for example, that feeding costs of hummingbirds and hawkmoths are high because they hover while they feed and use energetically cheaper forward flight between flowers. Therefore, they are under a premium to extract sugar rapidly, which occurs at concentrations of about 26 percent. Bees, in contrast, use little energy while on flowers, but may expend much energy in long-distance travel to get to feeding areas. Consequently, concentrated nectars are most rewarding.

The total amount and concentration of sugar per flower vary greatly and are influenced by several environmental factors. It has been suggested that arctic and alpine species tend to have concentrated nectar, perhaps because large amounts of energy must be provided to attract any visitors in such cold climates. Nectar of hummingbird flowers seems to decrease in concentration with increasing elevation (Cruden et al., 1983), perhaps because more concentrated nectar would be too viscous for rapid feeding at the cooler temperatures characteristic of higher elevations (Baker, 1975). Temperature, humidity, soil conditions, genotype, flower age, and other factors have all been shown to influence nectar concentrations in one or more species.

Amino acids are frequent, minor constituents of nectar, and the amount and kinds of these nitrogenous compounds have been examined by Baker and Baker (1975). They report correlations between amino acid levels and certain ecologically relevant parameters. For example, higher amino acid levels occurred in flowers of herbaceous than woody species, in flowers of plants with concealed nectar than plants with exposed nectar, and in flowers of species with fused petals than species with distinct petals. The evolutionary reasons for these differences generally are still unclear, and some may in fact be due to other, unmeasured, variables with which the recorded variables happen to be correlated.

Of greater interest are correlations between histidine scores and presumed pollinators. The greatest amino acid concentrations were in flowers pollinated by carrion-frequenting flies, followed in order by flowers visited by beetles, butterflies, long-tongued bees, short-tongued bees, generalized flies, and hummingbirds. Several explanations for these results can be proposed. For example, the higher value for long-tongued than short-tongued bees might reflect the former's inability to digest much pollen. The latter substance is an important nitrogen source for short-tongued bees (e.g., honeybees) and they therefore might not benefit from feeding on nectar rich in amino acids. The high score for butterflies could reflect the fact that nectar is one of the few sources of nitrogen available to adult butterflies. The high scores for beetle and specialized fly flowers might result from the nitrogen-rich food sources (e.g., carrion) with which these flowers compete for their pollinators. These ideas are hypotheses rather than conclusions, as there is no universal agreement that pollination systems and levels of amino acids in nectar are related (Gottsberger et al., 1984). No convincing evidence has yet been presented indicating that pollinator species can detect the levels of amino acids typically found in floral nectars (Hainsworth and Wolf, 1976; Inouye and Waller, 1984).

It has long been known that the nectar of certain species is toxic or unattractive to certain floral visitors. Much work has been done with honeybees (Pryce-Jones, 1944), although other visitors are also affected. Potassium, glycosides, nonprotein amino acids, phenolics, and various alkaloids may function in this regard (Baker and Baker, 1975). The effect of these substances varies among floral visitors. For example, the nectar of *Catalpa speciosa* contains iridoid glycosides. These do not affect the feeding preferences or behavior of the bees *Bombus* and *Xylocopa*, the legitimate pollinators, but cause pronounced behavioral abnormalities in and are avoided by an ant and a skipper, which are nectar thieves (Stephenson, 1981a, 1982a). Thus a major consequence of at least some toxic substances is to limit nectar thievery (Rhoades and Bergdahl, 1981).

Other Attractants and Rewards

To attract pollinators consistently, a plant must provide attractants (color, odor) that signal its location and suggest to pollinators that a visit will be rewarding. In most cases the plant must also provide a reward so that pollinators associate the attractant with some benefit. Exceptions to the latter are plants providing no reward and relying instead on deceit to ensure pollination. The previous two sections deal with the two most important rewards. In the following, some less common rewards and attractants are discussed; these are involved in some of the more bizarre and exotic examples of plant-pollinator relationships.

Miscellaneous Nutritive Rewards

In several species, flower or inflorescence parts that usually have an attractive or structural function are consumed by pollinators. Examples include fleshy bracts in bat-pollinated *Freycinetia*, parts of the spadix in certain beetle-pollinated arums, and specially modified regions of the tepals (undifferentiated units of the perianth) in *Calycanthus* (Simpson and Neff, 1981). In other species, specially modified trichomes (plant hairs) appear to have a nutritional function and some of these may mimic pollen.

Flowers of species in several plant families produce oils from small structures termed elaiophores. These oils are collected by female bees in the family Anthophoridae, which have hairs on their legs used in foraging. The oils apparently are fed to larvae. Resins, gums, and stigmatic exudates are rewards in a few species (Simpson and Neff,

1981). Other resins and oils have nonnutritive functions, being used in nest construction by insects (Armbruster, 1984) and to aid adhesion of pollen to pollinators (Steiner, 1985), respectively.

Flowers and inflorescences are sometimes used as brood places for pollinator larvae, in which case the insects obtain both shelter and larval food from the plant. Frequently, larval food is obtained at some reproductive cost to the plant (e.g., loss of ovules), and it is not always clear that the benefits provided in pollination outweigh these costs. The line between pollinator–plant mutualism and herbivory therefore is sometimes thin. Two dramatic examples of plant–pollinator relationships wherein pollinator offspring develop in floral structures involve figs and fig wasps, and yuccas and yucca moths. These are presented as case studies later in this chapter.

Aromatic Substances

Aromatic droplets are produced in the flowers of various neotropical orchids. These droplets contain mixtures of compounds, such as 1,8-cineole, methyl salicylate, and benzyl acetate, with each species of orchid apparently having a distinct mixture (Dodson et al., 1969). These flowers are visited almost exclusively by male euglossine bees. The euglossines are a subfamily of bees related to the bumblebees. They are brightly metallic in appearance, with blue, green, or golden coloration. Male euglossines have greatly enlarged tibiae on their hind legs, and brushes of hair on their forelegs. The forelegs are used to scrape or mop aromatic substances from the flowers, and the droplets are then transferred to the hind tibiae. The collected compounds were originally thought to be used as sexual attractants. Current thinking, however, is that the compounds are used as olfactory markers to attract other males in mating condition, with females then being attracted for mating by the males' group activities (Dodson, 1975). A further interesting feature of this aromatic fluid is that it appears to have a slightly intoxicating effect, causing visiting male euglossines to slip and fall from the upper to lower parts of the flower in such a manner that their subsequent departure requires them to move past the stigma and pollinia, increasing the chances of pollination.

Mate Mimics

European orchids of the genus *Ophrys* use a different approach for attracting pollinators (Kullenberg et al., 1984). The lower lip, or labellum, of the orchid flower is inflated and has a velvety surface (Fig. 2-13). A strong smell is produced from glands on the anterior edge, and these secretions, like those of the orchids previously described, contain a mixture of compounds, of which the terpene γ-cadinene is often prominent. The odors of these compounds attract males of various species of Hymenoptera, including ants, bees, and wasps. Each species of *Ophrys* attracts only a single species of visitor, although some insect species visit more than one orchid. The male insects land on the flower and by all appearances attempt to mate with the labellum. Experiments have shown that flower size, shape, color, and scent as well as tactile cues all stimulate the male's copulatory movements. During this activity the male picks up the orchid's pollinia, or deposits pollinia that it is carrying. Thus these orchids provide no reward to the male hymenopterans, but are good enough mimics of female bees that frequent deception is possible, allowing pollinia to be transported among plants. The insect behavior that these orchids depend on for pollination is called pseudocopulation. It has also been demonstrated in many Australian orchids visited by wasps (Stoutamire, 1974).

Prey Mimics

The orchid *Epipactis consimilis* bears structures resembling aphids and is visited by several species of hoverfly, all of which have larvae that eat aphids. During their visits the female hoverflies lay an egg on the labellum, presumably because they perceive the orchid as an aphid-infested plant suitable for the development of young. In the course of their activities they pollinate the flower. The orchid also produces nectar, on which both male and female hoverflies feed during pollination activities. However, the importance of the prey-mimicking structures in the pollination system is suggested by the absence of visits by local species of hoverflies that do not feed on aphids (Meeuse and Morris, 1984).

Aggressive Mimics

Male bees of several genera defend mating territories against other males. In tropical America, orchids of the genus *Oncidium* sometimes occur in territories of male *Centris* bees (Dodson and Frymire, 1961). *Oncidium* flowers are borne on long, flexible stalks that sway in even a light breeze. *Centris* males apparently mistake the flowers for other males and attack them. Pollinia are attached to the bee's forehead in the process, and these are subsequently deposited during attacks on other flowers. Because *Centris* males apparently have great site fidelity, however, it seems unlikely that distant pollen movement is accomplished, and self-pollination may be frequent.

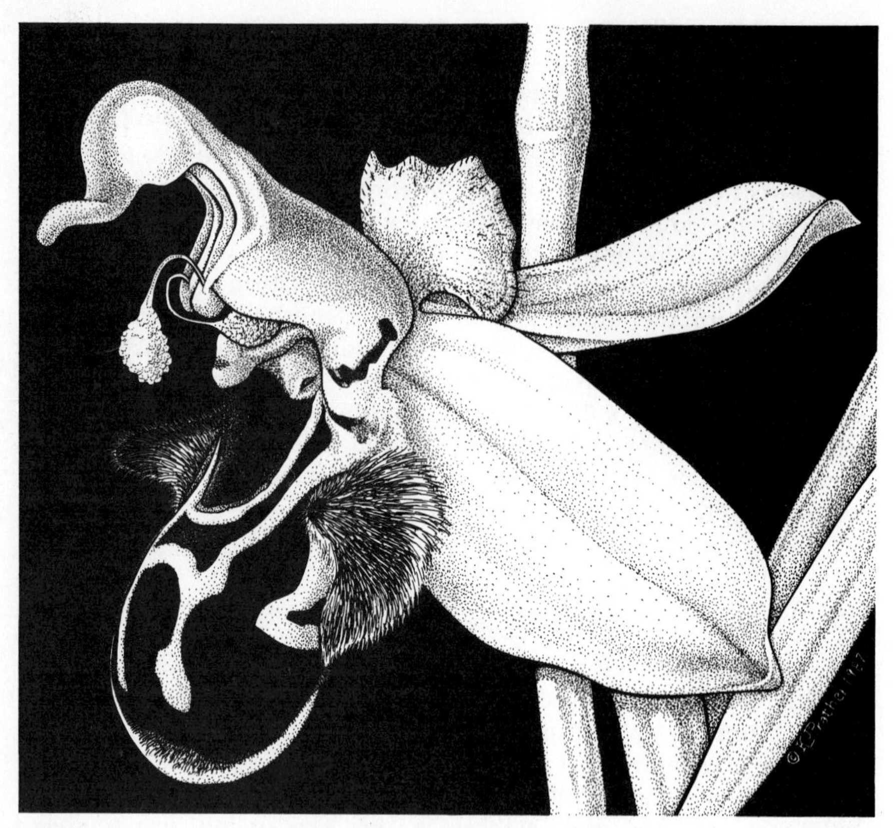

Figure 2–13. Bee orchid *(Ophrys apifera),* a species pollinated by pseudocopulation. Note the pattern and texture of the labellum (the inflated lower petal). One anther has swung up into contact with the stigma. Redrawn from Plate IIC of Proctor and Yeo (1972), used with permission of Taplinger Publishing Co., Inc.

Other Deceits

Several plants use attractants other than those described above to lure pollinators and then send them away without rewards. Such attractants may be visual, olfactory, or both. Brown and Kodric-Brown (1979) describe a population of *Lobelia cardinalis* in Arizona where individuals secrete no nectar and depend for pollination on hummingbirds habituated to visiting red tubular flowers of other local species. *Calypso bulbosa* is a nectarless orchid that relies on newly emerged and presumably naive bumblebee queens for pollination (Boyden, 1982). Individual bees apparently learn to avoid these flowers over time. *Cephalanthera rubra* is a nectarless pink-flowered orchid that is visited by solitary bees that also frequent nectariferous blue-flowered bellflowers *(Campanula* spp.). While these flowers are very different to the human eye, Nilsson (1983) showed that the two colors are very similar in the bee's visual spectrum, and suggests that the orchid mimics the bellflowers. In several species with unisexual flowers, the female flowers produce no nectar and are apparently visited by mistake by pollinators accustomed to finding rewards in male flowers (Baker, 1976; Bawa, 1980b).

Many arums (Araceae) and some plants in other families lure insects with odors of decaying protein and feces, but provide no rewards. Beetles and flies are the typical insects attracted to such flowers, in search of food, a suitable place to lay eggs, or both. Some arums (e.g., skunk cabbage, *Symplocarpus foetidus)* generate much heat during their early spring flowering (to 15°C above ambient), which may be a means of volatilizing the odorous attractants. Many flowers or inflorescences trap their visitors and hold them for varying lengths of time before releasing them.

Basking Place

Certain flowers in the Arctic and perhaps elsewhere have light-colored parabolic corollas that track the sun. Kevan (1975) showed that temperatures in such flowers could rise up to 10°C above ambient if wind speeds were low and the sun unobscured. Insects frequently bask in these flowers and in doing so pollinate them. Thus the corolla shape, color, and heliotropism may have evolved because of the pollination benefits they provide, although other explanations are also possible.

Floral Larceny

That not all flower visitors effect pollination has long been known (Kerner von Marilaun, 1878). Inouye (1980b) suggests that the term "robbing" should be used when some violence accompanies removal of nectar or pollen by nonpollinating visitors, and "thieving" when it does not. For example, a bird slashing the perianth of a flower to obtain nectar is a nectar robber. Major nectar robbers include birds, some bumblebees and ants that use powerful biting mandibles, and carpenter bees that use their probosces for slashing (Fig. 2-3C). One can also distinguish primary and secondary nectar robbers. The animals described above that actually make the holes are primary nectar robbers. Visitors that subsequently use these holes would be secondary nectar robbers. In a southern Illinois stand of red buckeye, carpenter bees *(Xylocopa)* were primary robbers, and honeybees later fed from these holes as secondary robbers. Nectar thievery could result when a very small visitor, such as an ant, crawls into a flower and obtains nectar without contacting the sexual parts. It could also occur if a visitor is able to obtain nectar by probing between the calyx and corolla or between petals. Pollen robbing and thievery can also occur, although the former is uncommon because pollen is frequently displayed openly, and therefore force is unnecessary. Small bees, especially in the Halictidae, are often pollen thieves because they can land on anthers, collect pollen, and depart without ever contacting the stigma.

Not surprisingly, illegitimate visitors can reduce a plant's reproductive performance. Butterfly-weed *(Asclepias curassavica)* is frequented by ants that sip nectar from the staminal hoods, but which perform little or no pollination. Wyatt (1980) noticed that the butterfly pollinators of butterfly-weed visited more flowers but spent less time on each when ants were present. Plants visited by ants received fewer pollinia than those from which ants were experimentally excluded, presumably depressing their fitness. In an experiment with the small tree *Quassia amara* (Simaroubaceae), Roubik et al. (1985) found that exclusion of nectar robbers raised seed production by factors of four to twelve. These researchers observed reduced nectar production in flowers from *Quassia* populations on certain Caribbean islands and speculated that the absence of robbers may have selected for reduced nectar production.

Nectar larceny may increase the reproductive success of plants under some circumstances. Increased success could result if larceny forced legitimate pollinators to visit more flowers or to make more frequent trips between plants (Heinrich and Raven, 1972; Gentry, 1978). Red clover sets more seed when exposed to nectar robbers than when they are excluded (Hawkins, 1961). Nevertheless, McDade and Kinsman (1980) conclude that floral larceny generally reduces a plant's reproductive success.

Phenology and Species Interactions

The phenology, or temporal pattern, of flowering influences plant fitness and has consequences at the community level. The question "Why does species *x* flower when it does?" can be answered in either a proximate or ultimate sense. The proximate cues for flowering usually involve temperature, moisture, or photoperiod (Rathcke and Lacey, 1985). For many species the cues are unknown. Providing an ultimate explanation for patterns of flowering phenology is often difficult. Certain limits are undoubtedly imposed by climate, e.g., cold weather during the winter in temperate and arctic climates. Of more interest, however, is how phenologies have evolved in the more hospitable seasons. Certain species bloom during the time when their major pollinators are available, and it might seem reasonable to explain the plant's flowering time as having evolved in response to the pollinator's phenology. In such cases it is rarely clear, however, which member evolved in response to the other (Rathcke and Lacey, 1985). The nature of selection pressures on flowering phenology of the tropical shrub *Hybanthus prunifolius* were revealed in an elegant study by Augspurger (1981). Plants experimentally induced to flower outside the normal blooming peak had lower reproductive success than those blooming during the peak due to lower pollination success and greater seed predation.

Flowering of two species at similar times in the same community theoretically could result in either divergence or convergence of flowering times. The former might reflect competition for pollinators, the latter facilitation (i.e., if close blooming times increase overall pollinator visitation rates and plant reproductive output; Rathcke, 1983).

Mosquin (1971) examined flowering phenologies of insect-pollinated plants in the Canadian Rockies. He argued that competition among pollinators had been important in the evolution of flowering times, based on visual inspection of flowering phenologies and unquantified observations of pollinator abundance on different plant species. In more recent work, attempts have been made to quantify and statistically examine flowering phenologies in particular communities (Poole and Rathcke, 1979). An overdispersed temporal blooming pattern might be expected if competition for pollinators is critical, but most studies have shown random or underdispersed patterns. A serious difficulty with this statistical approach is in choosing a null model for purposes of comparison. The use of different models can lead to different interpretations of the same community pattern (Thomson, 1978; Stiles, 1979; Rathcke, 1983).

More convincing data come from Waser's (1983b) studies of *Ipomopsis aggregata* and *Delphinium nelsonii*. These species flower sequentially, but in different local populations the relative order of flowering differs. That their relative timing varies, yet is never superimposed, suggests that their flowering times may be evolving in response to each other. Furthermore, when artificially created populations were induced to flower at the same time, the seed set of *Delphinium* was significantly reduced, possibly due to clogging of stigmas with *Ipomopsis* pollen (Waser, 1978a, 1978b). Campbell (1985) demonstrated competition between *Claytonia virginica* and *Stellaria pubera* by removing *Claytonia* from the vicinity of *Stellaria* and observing increased seed set in the latter. In both of the above examples, problems associated with interspecific pollen transfer appeared to be more significant than competition for pollinator visits.

Evidence for convergence of flowering time (or other floral traits) as a result of facilitation is poor (Rathcke and Lacey, 1985). Flowering times of different species may be clumped during the season (Poole and

Rathcke, 1979; Anderson and Schelfhout, 1980), but many selective forces other than facilitation could cause such patterns. It has been suggested that species blooming at the same time and having flowers of similar appearance may mimic one another. Two wildflowers in Colorado, a *Delphinium* and an *Ipomopsis,* may benefit from their pattern of sequential flowering. When few earlier flowering *Delphinium* were present, migrating hummingbirds did not stay long enough to pollinate the later-flowering *Ipomopsis* (Waser and Real, 1979). In another study of Rocky Mountain wildflowers, Thomson (1981) demonstrated a positive correlation between pollinator visitation rates to one plant species and the local abundance of plants of other species that used the same pollinators. However, the potential negative effects of interspecific pollen transfer must be considered in any complete assessment of the effects of such interspecific interactions.

Interpretation of flowering pattern in communities requires making a distinction between present and past competition. Competition among plant species for pollinators may have been intense enough in the past to produce divergent blooming times resulting in no present competition. Thus the present absence of competition cannot be used as evidence that patterns seen today were not due to competition between plants. Conversely, the demonstration of present competition does not mean this force has been important in producing characteristics of plants.

Community Studies

A few studies have been made of the reproductive biologies (phenology as well as other factors) of entire plant communities. Pojar (1974) compared a subalpine meadow, a salt marsh, and two bogs in British Columbia. He observed an inverse relationship between number of plant species in the community and average duration of a species' bloom. Autogamous species had a shorter than average blooming period. He suggested that interspecific competition for pollinators had been important in the evolution of blooming times and various floral traits.

Ostler and Harper (1978) examined the floral ecology of 25 communities in the Wasatch Mountains of Idaho and Utah. They observed that increased plant species diversity was accompanied by an increase in the proportion of animal-pollinated species, an increase in the diversity of flower colors present in open communities but not in forested communities, and an increase in the frequency of complex flower types (zygomorphic and/or with restricted access to nectar). They suggest that these trends reflect greater selection for precision of pollen transfer in the more diverse communities. The lack of an increase in color diversity with species diversity in forests might be due to the low visibility of heavily pigmented flowers against shaded forest floors. Indeed most species flowering in the forests were either white or light-colored. It has also been argued that the predominance of white flowers in forested habitats may be due to the abundance of moths and scarcity of bees in such habitats (Baker and Hurd, 1968).

The Animal's Perspective

Sensory Abilities of Pollinators

A knowledge of the stimuli to which pollinators can respond is of obvious importance in understanding the plant–pollinator relationship. Such information tells us how a plant might attract pollinators and how pollinators can orient themselves. Very few pollinator species have been examined in detail, and extrapolations to other species must be made with caution. Important senses include vision, olfaction, taste, touch, and ability to detect polarized light and perhaps electrical charges.

The visual spectrum of birds is thought to resemble that of humans, though possibly enhanced in the yellow to red wavelengths. A few bird species, including hummingbirds, have also been shown to be sensitive to near ultraviolet wavelengths (at least down to 370 nm) (Goldsmith, 1980). Bats do not see color, and this is probably also true for other mammal pollinators such as rodents and marsupials. These pollinators are usually crepuscular or nocturnal, and color vision may be less advantageous than great visual sensitivity, a characteristic of the rods associated with noncolor vision. Color vision is, however, widespread in the major orders of flower-visiting insects: Lepidoptera, Hymenoptera, and Diptera (Proctor and Yeo, 1972). The visual spectrum of many insects does not extend as far into the long wavelengths as it does in a human being or hummingbird. Such red-blindness has been demonstrated in some members of each of the three orders mentioned above. This does not mean that structures appearing red to us are invisible to these insects, but that red objects may not stand out from backgrounds of other colors. The visual spectrum of some insects, including several bee species, extends to ultraviolet wavelengths. The honeybee's visual spectrum, for example, ranges from 325 to 650 nm, compared to 400 to 800 nm for humans. Like humans, bees have three types of color-sensitive cells in their eyes, but in bees one is sensitive to ultraviolet wavelengths. This cell type is the most common and, indeed, bees can detect and respond to very small amounts of ultraviolet light. As a consequence, bees do not perceive colors as do humans, and many flowers that appear uniform in color to us have distinct patterns visible to bees, caused by differential reflection and absorption of ultraviolet light (Fig. 2-14). Furthermore, two colors that appear identical to us can appear quite different to a bee because they reflect differing amounts of ultraviolet light, undetectable to us but not to the bee. Bees can also detect polarized light, which is considered in the next section.

Olfactory sensitivities are less well known than visual systems. Birds apparently have a poor sense of smell, and many flowers frequented by birds lack odors detectable by humans. Mammal pollinators probably have much better olfactory abilities and the plants they visit typically are highly aromatic. Among the insects, the use of olfactory cues has been shown in beetles, bees, flies, butterflies, and moths. In a few cases odor is responsible for long-distance attraction of pollinators, particularly flies

Figure 2–14. Inflorescences of *Rudbeckia fulgida* (top) and *Rudbeckia triloba* (bottom) as seen in daylight (left) and ultraviolet light (right). Because many insects see ultraviolet light they can distinguish some features that humans cannot. Photos courtesy of Ken McCrea.

and beetles (Pellmyr and Patt, 1986). For many insects, however, sight is more important for long-distance attraction, and odors are used to a greater extent at close range. Scents sometimes exert a strong influence on pollinator behavior within a flower, and experiments show that bees are quite capable of distinguishing different parts of a flower by scent. Insects are much more sensitive than humans to certain fragrances. Honeybees, for example, detect scents at one-tenth to one-hundredth of the minimum strength detectable by humans. They also excel at discriminating between slightly different mixtures of fragrances (Proctor and Yeo, 1972).

Taste has been evaluated largely in terms of simple sugars commonly found in nectar. Both honeybees and hummingbirds can discriminate among glucose, fructose, and sucrose, and various concentrations thereof. Some flies can detect sucrose at concentrations of less than 1 percent of the minimum detected by humans. In taste tests with hummingbirds, sucrose was the most preferred single sugar and glucose was least preferred, although some mixtures were preferred to any of the single sugars (Hainsworth and Wolf, 1976). Both bees and hummingbirds were able to distinguish differences in sugar concentration, although this might partly reflect differences in viscosity rather than taste.

The tactile sense of honeybees and its potential importance in foraging behavior was demonstrated by experiments of Kevan and Lane (1985). This sense involves mechanoreceptors on various parts of the body, including the tips of the antennae. The value of this sense may lie in flower recognition or in orientation within flowers.

Erickson and Buchmann (1983) reviewed evidence that electric charges may be important in the pollination process. Plants have negative surface charges under some conditions, and bees, at least, develop positive charges during flight. When a bee visits a flower presumably these charges will tend to dissipate. Thus, an insect able to detect plant surface charges may be able to avoid visiting flowers that were recently visited. Bees and other insects are sensitive to electrical stimuli, but it is not yet clear whether this sensitivity is sufficient for bees to take advantage of electrical information that plants provide.

Honeybees, at least, can tell time. The experiments of Frisch (1967) have shown that bees fed at a particular place at a certain time of day will return to that site at the same time on subsequent days. Such behavior is of obvious value because the food rewards of many flowers are available exclusively or predominantly at certain times of day. For example, flowers of *Pyrrhopappus carolinianus* are open only for about 2 hours in the early morning (Estes and Thorp, 1975). In other species, nectar availability may exhibit marked periodicity even if flowers are open all day. A time sense is also important in orientation using the sun (see the following).

Honeybee Orientation

An ability to orient is essential for a pollinator to return to rewarding patches of flowers or to find its nest. Almost all work on pollinator orientation has been carried out with honeybees (summarized by Frisch, 1967), and I will restrict my discussion to them.

Honeybees use conspicuous visual landmarks, when available, to find their way between hive and food sources. Bees also orient using the sun and, remarkably, they are able to compensate for its change in position during the day. Suppose, for example, they find rich food west of the hive, toward the setting sun, on the evening of one day. Overnight the hive is moved to a new location so that landmark navigation cannot be used. On the morning of the second day the workers know to fly away from the direction of the rising sun to look for the food source west of the hive. This clearly demonstrates the presence of an ability to detect the sun's position and to determine the time of day.

Bees can also find their way when the sun is obscured by clouds. A patch of blue sky whose diameter corresponds to a visual angle of only 10–15 degrees provides all the information they need. A series of experiments summarized in Frisch (1967) indicates that this information comes from polarized light. The light coming from a patch of blue sky is polarized, that is, it vibrates in a particular plane, and this plane bears a specific relationship to the position of the sun. Even if the sky is completely cloudy, bees apparently can locate the sun because the clouds in the direction of the sun radiate different proportions of ultraviolet and long wavelengths than clouds elsewhere in the sky (Barth, 1985).

Pollinator Behavior and Energetics

Optimal Foraging

Food is the most common reward obtained by pollinators from the plants they visit. It is commonly supposed that natural selection will act on pollinator behavior so as to maximize the pollinator's net return (i.e., the excess of benefits over costs). Pollinators demonstrably foraging so as to maximize their net returns are sometimes called optimal foragers, and the predictive body of theory that surrounds this notion is called optimal foraging theory.

To evaluate pollinator behavior in terms of optimal foraging theory, the appropriate currency for measuring costs and benefits must be identified. Typically this is assumed to be energy, but other factors cannot always be ruled out. For some pollinators, nutrients might be more important than energy. In arid environments, dilute nectars might be favored over concentrated nectars to supply water as well as energy. Under other circumstances, minimization of water load may be more important than maximization of energy intake (Bertsch, 1984). Added complexity occurs if species feed on pollen instead of, or in addition to, nectar. The quality of different pollens is not as easily compared as is the quality of different nectars. A further problem arises in pollinators that do other things while they forage. Such activities include searching for mates or nonfloral food, and avoiding predation or the wrath of other, territorial, foragers. Their behaviors then would probably not be ones that provide the greatest net returns in terms of feeding alone. Many of these variables are difficult to measure, and therefore an element of uncertainty remains in predictions using an optimality approach.

These sorts of problems are probably of least concern (but certainly not eliminated) for pollinators feeding on nectar. Sugars are the only major constituents of most nectars, they are easily measured, and they are assumed to be completely assimilated to supply energy. Therefore much work on ecological energetics has been done with nectar-feeding animals.

What sorts of behaviors maximize the rate of energy intake for a nectar-feeder? One can consider behaviors at several levels, and the following categories follow in part those of Heinrich (1983).

Choice of Flower Species

Pollinators may be generalists (visiting flowers of many species) or specialists (visiting flowers of few species). It is often thought that generalists benefit by having low dependence on any one resource but at the cost of lower efficiency in harvesting each resource. Specialists benefit by foraging with greater efficiency, but at the potential cost of greater dependence on a few resources. Some evidence supports this notion. Bees take time to learn how to handle some complicated flowers (e.g., *Aconitum)* and to learn which plant species are most rewarding (Heinrich, 1979a). If such learning is required for the most efficient handling of many flower types, generalist visitors that handled each type infrequently might indeed be at a disadvantage in terms of efficiency. To illustrate, *Echium vulgare* is visited by several pollen-feeding bees, including the specialist *Hoplitis anthocopoides* and four generalists, which also collected pollen from other plants (Strickler, 1979). The specialist bee required much less time to collect the pollen needed to provision one offspring than did the generalists (8 vs 13 to 39 minutes). While specialization may confer benefits in terms of efficiency to the animals involved, it is not necessarily also advantageous to the plant. Motten et al. (1981) demonstrated that visits to *Claytonia virginica* by a specialist bee and a generalist fly had equal effects on seed set.

Bumblebees use a mixed approach to foraging. Individual bees may make most of their visits to one (major) species and make a few visits to one or a few other (minor) species (Heinrich, 1979a). Such behavior presumably provides many of the advantages of specialization, while allowing a bee to keep track of additional resources. When a minor species was enriched by additions of sugar syrup, this species became the bee's major species, sometimes within the duration of a single foraging trip. Presumably the same shift would have occurred if this minor species had naturally become more rewarding than the original major.

Perhaps the ultimate means of tracking complex floral resources is provided by the sophisticated social interactions of honeybees. In bumblebees, there is apparently no means of communicating information about floral resources among the members of a colony. Thus, each individual must rely on its own perceptions of the relative worth of different resources. Honeybees, however, use dances to share information about the location and richness of food sources with their hivemates (Fig. 2-15). This dance language was first worked out by the Nobel laureate Karl von Frisch, and much of his work is summarized in Frisch (1967).

If a high-quality food source is less than about 80 m distant, the

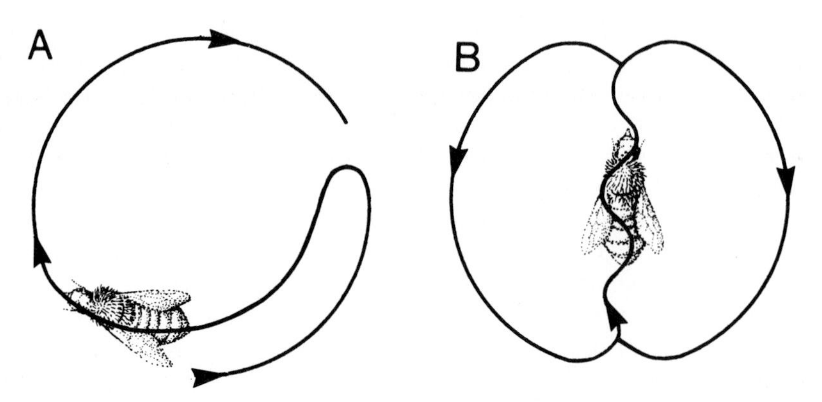

Figure 2–15. Honeybee dances. A. Round dance, used to indicate nearby food sources: the richer the source, the faster the worker makes her circles. B. Waggle dance, used to indicate more distant food sources: the more distant the food, the slower the circuits. If the dance is performed on a horizontal surface, the bees move towards the food source in the straight run (cutting across the circle). On a vertical surface, the upwards direction symbolizes the direction of the sun, and the direction of the straight run is oriented appropriately to indicate the direction of the food source.

worker bee performs the round dance upon returning to the hive. She runs rapidly in a circle, frequently changing the direction between clockwise and counterclockwise. The vigor of her dance indicates the quality of the food source, and additional information is provided by floral odors remaining on the dancer. Information on more distant sources is provided by the tail-wagging dance. The worker runs in a straight line wagging her abdomen, and then returns to the starting point in an arc on either the left or the right side. The distance of the source is communicated by the speed with which the cycles are completed, greater speed indicating closer food. The direction of the food source is indicated by the straight-line part of the dance. When a bee is dancing on a horizontal surface, she simply does the straight run in the direction of the food source. More often, however, the dance is done on a vertical surface, in which case the vertical direction symbolizes the position of the sun. When the straight run points upward, this means "fly toward the sun"; when pointed down, it means "fly away from the sun," etc. Thus worker bees are constantly appraised of the quality of a variety of food sources around the colony, and the colony's foragers can be mobilized accordingly.

Because the rewards in different types of flowers vary in quality, it might be expected that only the most rewarding species would be visited. This is not usually the case, for at least two reasons. First, different species of pollinators are likely to differ greatly in the ease with which they extract rewards from different plant species. Clearly a hummingbird and a fly are unlikely to extract food with equal efficiency from a flower with a long, narrow corolla. Within a particular taxonomic group of pollinators great differences may also exist. For example, bumblebee species differ greatly in proboscis length, and this influences the range of flowers that can be profitably foraged. Second, the foraging of pollinators might influence the availability of food to other individuals and other species. Heinrich (1983) suggests that any pollinator starting to forage in a particular area will usually concentrate its foraging on the species with the richest reward. In doing so, however, it decreases the richness of that species by its own consumption. Another forager may find that some other plant species is now the most rewarding. This kind of behavior exhibited by numerous foragers will tend to equalize (and depress) the rewards of all plants. It can also be predicted that the number of pollinator individuals specializing on flowers of a particular species should be proportional to the rewards they produce. These predictions were borne out in an experiment in which bumblebees were introduced singly into an enclosure containing several plant species varying in the amount of their nectar rewards (Heinrich, 1979a).

Several factors other than average nectar amount will also affect flower choice by pollinators. Real (1981) demonstrated that variance in nectar levels among flowers affects foraging preferences. Bees and wasps were allowed to forage in an array of artificial flowers of two colors, with each color having the same mean nectar content, but a different variance. Both insects preferred the color having the least variance in nectar. Color itself can often play a role in feeding preferences, independent of reward levels. In the above study, for example, wasps preferentially visited yellow flowers even when blue flowers contained equal rewards. Other important variables influencing feeding preferences include handling time of flowers, and plant and flower density (Waddington and Holden, 1979; Harder, 1983).

Flower Choice and Handling

Not all flowers of a species are equally rewarding, reflecting environment, plant genotype, flower age, and time elapsed since the most recent pollinator visit. Some cues by which flower age can be determined were reviewed above, but it has also been suggested that flower visitors may be able to remotely sense the nectar content of a flower. For example, *Bombus* workers often rejected individual flowers of *Apocynum androsaemifolium* after contacting them lightly with the antennae, or simply hovering in front of them (Marden, 1984). These rejected flowers contained significantly less nectar than randomly chosen flowers on the same plant. The nature of the cues used by the bumblebees is unknown. Foraging efficiency could also be enhanced if individuals do not revisit flowers that they have recently visited. Experiments with several species of hummingbirds suggest that they learn to visit a flower in a new location more easily than they learn to return to a flower in a position recently visited (Cole et al., 1982). Such a predisposition would presumably increase foraging rewards under natural settings.

Individual foragers must use different techniques to feed with greatest efficiency from different kinds of flowers. An example is provided by the varied approaches that bumblebees must use to efficiently extract pollen and nectar from some common plants (Heinrich, 1979a). To collect pollen from the flat-topped inflorescences of wild carrot *(Daucus carota)*, a bee must walk rapidly over the surface, pressing its body down to pick up pollen from the many tiny florets. Nightshade *(Solanum dulcamara)* has tubular anthers and the bee must grasp each anther with its mandible and vibrate its wings to shake the pollen loose. Timothy *(Phleum pratense)* has loose, normally wind-dispersed pollen, and a bee has only to clamber up the inflorescence to strip the pollen free. Collected pollen of each species must then be transferred from wherever it collects on the bee's body to the corbiculae, or pollen baskets, on the hind legs. To obtain nectar from a monkshood *(Aconitum)* flower, a bumblebee must enter at the bottom, pass over the anthers, and probe into the tips of two modified petals lying under the hood formed by modified sepals. The petals of a turtlehead *(Chelone)* flower must be pried apart for the forager to reach nectar inside the flower. Clearly a generalist pollinator uses varied behaviors when foraging.

It is not surprising that handling time influences foraging preferences. For example, the flowers of cow vetch *(Vicia cracca)* vary in the length of the corolla. Morse (1978) found a positive correlation between the tongue length of individual bumblebees of one species, and the length of the corollas that they were probing. In the same vein, Inouye (1980a) compared the tongue lengths of *Bombus flavifrons* workers feeding on *Delphinium,* which has long corollas, and *Mertensia,* with shallower corollas. The former group had the longer tongues. Other studies show that long-tongued bees forage more rapidly on species with long corollas, and short-tongued bees feed more rapidly on flowers with short corollas. These foraging patterns presumably reflect flower choices on the part of the bees that minimize handling time and increase their net return.

Movement among Flowers

Within inflorescences certain visitors seem to have stereotyped patterns of movement. In vertically elongate inflorescences, for example, bees typically begin foraging near the bottom and move upward. In many of these plants, flowers open sequentially from the bottom to the top of an inflorescence, the older (lower) flowers have more nectar, and the flowers are protandrous. Therefore male-phase flowers occur above female-phase flowers in the inflorescence. It has been suggested that this pattern of dichogamy and nectar production may be an adaptation to enhance cross-pollination. Visitors would tend to first visit the rich lower flowers and deposit pollen they are carrying on these stigmas. After depleting nectar in the lower flowers they move up to the less rewarding male flowers. Eventually they leave the plant, carrying pollen from the male flowers to the female flowers on another plant. Self-pollination is thus minimized. The upward movement of insects could, however, be due to factors other than the distribution of nectar, such as the direction in which visitors must face while foraging from the flowers. In *Linaria vulgaris,* for example, *Bombus hortorum* workers face upward and also tend to move upward when foraging, even though upper flowers contain more nectar than lower flowers. In another bumblebee species *(B. terrestris),* workers face downward while foraging (robbing) and have a slight tendency to move down the inflorescence while foraging (Corbet et al., 1981). Other experiments with artificial inflorescences showed that bumblebees foraged upward even if all flowers were equally rewarding or if the top two were the most rewarding (Waddington and

Heinrich, 1979). They also visited upward in inflorescences of *Scrophularia aquatica* when nectar content was independent of flower position (Corbet et al., 1981). These results show that this behavioral pattern is strongly held, and it may simply be a genetically programmed behavior that minimizes repeat visits to flowers (Heinrich, 1975). Pollen receipt and/or export could be enhanced incidentally by such behavior.

Another decision facing a pollinator is how many flowers to visit on one plant. This is likely to be determined by the nature of the rewards and the relative costs of staying on one plant versus visiting other plants. Specific variables that have been shown to increase the number of flowers visited by some pollinators include large nectar volumes, low variance in nectar volumes among flowers, and low plant density, which increases interplant traveling costs (Gill and Wolf, 1977; Hartling and Plowright, 1979; Cibula and Zimmerman, 1984).

Because pollinators respond to such variables, plants have some ability to manipulate the visitation patterns of their pollinators through control of energetic benefits derived from their flowers (Heinrich and Raven, 1972; Heinrich, 1975). Is there likely to be an optimum number of visits to flowers on the same plant from the plant's perspective? Probably so. In a highly self-fertile plant this optimum number may be very high. For a self-incompatible plant or one that suffers from inbreeding depression, however, the optimum number of flower visits may be much lower. After visiting a certain number of flowers, a pollinator will deposit mostly self pollen, potentially reducing both male and female reproductive success. Evaluating the optimum number of flower visits for a plant requires knowledge of patterns of pollen pickup and delivery. The few available data suggest that deposition of non-self pollen tapers off rapidly after the first 5 to 15 flowers visited (Thomson and Plowright, 1980; Galen and Plowright, 1985).

Flight distances between consecutively visited flowers are generally leptokurtically distributed, with the majority of distances being very short, including many to the nearest neighbors. Increases in floral density decrease the mean flight distance because of the greater proximity of nearest neighbors (Waddington, 1983). Average flight distance also tends to be shorter when average nectar volume is greater and longer when variance in nectar volume is greater (Waddington, 1983; Ott et al., 1985).

A second aspect of pollinator movement is the direction of movement between successively visited flowers. Several species of insects tend to move predominantly in one direction when they forage. Bumblebees and bird pollinators sometimes show random choice of directions in successive flights. This behavior is potentially disadvantageous to the pollinator in that it increases the chance of revisiting an already-visited, and therefore depleted, flower. Zimmerman (1979) suggests that random movements are most likely to be found where they yield only a low probability of revisiting a flower, as when inflorescences are large and few flowers are visited on each inflorescence. The richness of floral rewards may influence turning patterns just as it influences the distances of visits. Pyke (1978) and Heinrich (1979b) found that bumblebees turned more from the previous flight direction if rewards were richer. This behavior may be advantageous to the bees in keeping them in the local area when they encounter rich rewards. If successive flights were most nearly in a straight line, the pollinators would tend to move through a local area more rapidly. However, not all studies show this behavior.

Foraging Range

The spacing systems of pollinators and the maximum spatial area over which they forage vary greatly among species and these differences have important consequences for gene flow in the plant species they visit.

Pollinators that restrict their visits to very small areas for extended periods can be called tenants (Schemske, 1980a). An example is a single bee observed by Heinrich (1976) that foraged for several days in an area less than 7 m long in a hay field. Territorial pollinators not only stay in one place for extended periods, but also defend the area against conspecifics and sometimes against individuals of other species as well. The movement of pollen from such defended plants has been compared with the movement of pollen from undefended plants by using colored dyes (Linhart, 1973). Dye is placed on the flowers of interest, and flowers at varying distances from the dyed flowers are later examined for dye particles. The pattern of dye movement is assumed to parallel the pattern of pollen movement. Dye placed on flowers within hummingbird territories moved a shorter distance than dye placed on flowers outside territories. In instances where nearby deposition of pollen is disadvantageous (as when nearby individuals are relatives and inbreeding depression occurs), being in a pollinator territory is disadvantageous for a plant.

Territorial behavior in many pollinators is highly facultative, and territory size also varies considerably. Optimal foraging considerations lead us to expect a territory to be maintained if this behavior yields greater net rewards than not maintaining a territory. The net rewards are determined by the richness of floral resources and the cost of their defense. Golden-winged sunbirds *(Nectarinia reichenowi)* are nectar feeders that adjust the size of their territories as the density of flowers changes through time (Gill and Wolf, 1975). A sunbird requires an estimated 13 to 14 kcal of energy per day, which could be provided by 1500 to 2500 flowers of the food plant *Leonotis nepetifolia* secreting at the observed rate of 3.4–4.4 μL/day. In fact, sunbird territories contained about this many flowers (1000 to 2500). The average nectar volume in flowers within sunbird territories was consistently greater than that in flowers outside sunbird territories, presumably because the latter flowers received more pollinator visits. Defense of a territory increases the daily energy costs of a sunbird by some 900 cal, but under the observed conditions of territoriality, this expenditure was more than offset by the greater availability of nectar within the defended area. The effect of defense costs was illustrated by the effects of intruders. When short-term intruder frequencies (and presumably also defense costs) were very high, the territories were sometimes abandoned. Similar observations have been made on territorial hummingbirds (Stiles and Wolf, 1970). Furthermore, on a day when undefended nectar levels were very high, the sunbirds eschewed territoriality until later in the day when the nectar volumes were lower and competition intensified.

Most pollinators are not territorial, and move around to a greater degree. Honeybees sometimes fly several kilometers from their hives to feed, although shorter excursions are preferred. More distant food sources must, of course, be successively richer to justify the increased costs of getting to them. It has been suggested that some flower visitors exhibit a form of behavior that combines long-distance travel and site specificity (Janzen, 1971a). This behavior, termed traplining, involves visiting many different foraging sites in a specific sequence, much as a trapper runs a trapline. Unprofitable food sources will be omitted from the trapline, and new ones added periodically, but the frequency of such changes is low. This approach could be an efficient way for a forager to harvest scattered but relatively rich rewards that are predictable in space and time. Evidence exists for some features of traplining (Ackerman et al., 1982), but it has not yet been documented in the precise manner in which it was originally described.

Community Interactions

Most studies of pollinator "communities" or guilds attempt to elucidate the way in which a group of pollinator species divides up the available resources. A common and unsurprising observation is that the available resources, rather than being widely shared, are partitioned. In its broadest sense, this partitioning reflects the floral syndromes described earlier. Bats and bees generally are not found feeding from the same flowers, for example, because of their different sensory abilities, behaviors, and energy requirements.

Of greater interest is how a group of similar pollinators (e.g., bumblebees or hummingbirds) partition a resource. In successional habitats in Monteverde, Costa Rica, Feinsinger (1976) identified 14 species of hummingbirds. One of these was a medium-sized hummingbird, *Amazilia saucerottei,* that dominated most other species. Some individuals were territorial, and defended the most nectar-rich patches of flowers. The smallest hummingbird was *Chlorostilbon canivetii,* the fork-tailed emerald. This species was a transient, or trapliner, and foraged mostly at dispersed flowers which could not support territorial individuals. Magenta-throated wood stars *(Philodice bryantae)* were intraspecifically territorial, but their territories overlapped with those of other species. They were usually not displaced, apparently because they were mistaken for bees. The green violet-ear *(Colibri thalassinus)* was an opportunistic forager. The remaining 10 species were either less common or were found predominantly in different habitats and/or at different elevations. Feinsinger (1976) identified several dimensions in which nectar was partitioned to some degree. These included habitat, reflecting, in part, elevational differences. The density of floral rewards was clearly important, reflecting the territorial/transient dichotomy. Spatial partitioning was sometimes important, with different species feeding at different heights, or on the inside as opposed to the outside of certain shrubs. Some diurnal partitioning also occurred, which could reflect the change in average nectar volumes during the day, altering the profitability of particular plants for each hummingbird species. Finally, there was a seasonal dimension, reflecting changes in the intensity of flowering of particular plant species and seasonal movement of birds with respect to habitat, elevation, or latitude. Thus what on superficial examination may seem like a group of similar pollinators obtaining their food in a similar manner from the same place, usually turns out to be a group that divides up a resource in ways that correspond to important morphological and behavioral differences among members of the group.

How are such differences in feeding preferences or pollinator morphology brought about? Coevolution between particular plants and pollinators (or small groups of either) has often been suggested to explain the above patterns. Indeed, because of their great specificity, some pollination relationships seem unlikely to have arisen in any other way. Examples include fig and yucca pollination described in the next section, and instances where a conspicuous morphological match exists between pollinator and plant [e.g., pollination of some passion flowers with greatly elongate, curved corollas by sword-billed hummingbirds *(Ensifera ensifera)* with bills of nearly equal length and curvature]. In many cases, however, coevolution may have little or nothing to do with pollinator relationships, and even where it does, very little is known about the degree of heritable variation and response to selection in relevant traits (Feinsinger, 1983). Janzen (1985) makes a related point. Most species, he argues, have acquired many of their traits in a small subset of their present range, which has subsequently expanded. For such species, evolution occurring within many parts of their current range may have been relatively minor. Their presence in such areas simply reflects their ability to survive given the available resources and the other organisms already there. Thus it cannot be concluded from an observed "fit" between plants and their pollinators that coevolution has necessarily been the cause.

A different approach to the analysis of pollinator communities was taken by Ranta (1984). He examined bumblebees from various communities in North America and Europe, noting the abundance of each species, and ranking them by proboscis length. He then observed differences in proboscis lengths between species with adjacent rankings and related these spacings to those expected if the community composition was organized largely with regard to competition. In most communities the species were more similar than expected, reflecting either the unimportance of competition as an organizing force, or as Ranta (1984) believes, the uneven distribution of floral resources with regard to corolla depth. This illustrates the difficulty of finding appropriate null models with which existing communities should be compared, a problem also encountered in deciphering phenological patterns in plants.

Current patterns of pollinator visitation may reflect present interactions (e.g., current competition) as well as past events. Competition between pollinators is important in maintaining dietary differences between pollinators in several communities. Inouye (1978) demonstrated that the feeding preferences of two species of *Bombus* in a Colorado alpine meadow were maintained by competition. Each species visited primarily a different species of plant, but if one bumblebee species was removed, the other bee expanded its foraging to include the species usually visited by the removed bee. Another example of pollinator competition involves two bumblebees and the ruby-throated hummingbird feeding primarily on jewelweed (Laverty and Plowright, 1985). Resource partitioning existed, with the hummingbirds visiting the outermost flowers and lowering their nectar below the level where bumblebees could reach it. The bumblebees visited mostly the inner flowers which hummingbirds did not visit because of difficulties in maneuvering. At times when hummingbirds were rare, the number of foragers of *B. fervidus* increased, and they visited more outer flowers than when hummingbirds were common. *Bombus vagans* did not show such a response, suggesting they did not compete greatly with hummingbirds. *B. vagans* foraged mostly in the early morning, while *B. fervidus* was more active later in the day. When *B. fervidus* was removed, however, the numbers of *B. vagans* increased and they foraged over a greater part of the day.

Experiments such as these demonstrate that competition can be an important force influencing pollinator visitation patterns. If it remained consistent over a long period of time, one would expect such competition to lead to genetically based divergences in behavior and morphology and therefore to fixed rather than facultative partitioning of resources.

Pollination and Evolution

Evolution of Biotic Pollination

The first pollinators were probably mandibulate insects that visited flowers to consume pollen, ovules, or other floral parts (Crepet, 1979). If some pollen transfer was accomplished incidentally during such activities, selection pressures would have favored floral traits that enhanced pollen transfer and minimized damage to critical flower parts. Selection would also have acted on plants visited by a variety of insects to favor traits encouraging visits by those taxa most effective in pollination and to discourage visits by the poorer pollinators.

Of the modern groups of pollinators, the beetles may have been the first to visit plant reproductive structures, in the Permian [280 to 230 million years before present (BP)]. Their mouthparts were well-suited to feeding on pollen grains and ovules, and the diets of some species undoubtedly included microspores, ovules, young seeds, and perhaps other reproductive parts of seed ferns, cycads, and other plant taxa. Recent studies of primitive angiosperms in the South Pacific *(Drimys, Zygogynum, Degeneria,* and *Belliolum)* show that flies are important pollinators, and on this basis, Thien (1980) suggests that they too may have been among the early pollinators. The Diptera do not, however, appear in the fossil record until the Triassic (230 to 180 million years BP), along with the Hymenoptera, the most important modern order of insect pollinators. These orders as well as the beetles underwent major radiations in the Jurassic (180 to 130 million years BP). The remaining important order of insect pollinators, the Lepidoptera, appeared and underwent a major radiation in the Cretaceous (130 to 60 million years BP). Bat and bird pollinators were probably not important until much more recently.

It is unclear when pollination or spore movement was first carried out by insects on a regular basis. Crepet (1979) observes that some seed ferns from the Upper Carboniferous had pollen grains so large (600 nm diameter) that wind dispersal was unlikely. Furthermore, the characteristics of pollen-bearing structures in these plants are more easily interpreted in terms of insect pollination than wind pollination. A variety of features suggests increasing insect visitation to cycad cones in the Jurassic and Cretaceous. Such evidence includes crowding of cones (perhaps to decrease access for herbivorous insects) and fossilized plant parts showing insect damage (Crepet, 1979). The first unequivocal angiosperms are Cretaceous, although suggestions for an ancestry extending back to the Permian have been made. In any event, by the Cretaceous there existed bona fide angiosperms and their four major orders of insect pollinators.

Pollination and Angiosperm Success

As a group, angiosperms are much more successful than their closest living relatives, the gymnosperms. The closed carpel of angiosperms may be one reason for this success, because of the opportunities that it affords for self-incompatibility, and intense pollen tube competition (Mulcahy, 1979; Zavada and Taylor, 1986), both of which potentially enhance offspring quality.

Angiosperm success may also reflect the predominance of animal pollination in most angiosperm groups compared to the ubiquity of wind pollination in gymnosperms. There is certainly more room for differentiation in the pollination niche dimensions if pollination is biotic than if it is only by wind. Regal (1976) further suggests that animal pollination allows more outcrossing among widely dispersed individuals than does wind pollination, permitting angiosperms to be successful when growing at low densities.

The potential importance of pollination in the radiation of a particular angiosperm group is illustrated by a detailed examination of the phlox family (Polemoniaceae; Fig. 2-10; Grant and Grant, 1965). According to the authors' reconstruction, bee-pollinated ancestors have given rise to species pollinated by beetles, hummingbirds, and several kinds of Lepidoptera and Diptera, as well as to autogamous species. Such evolutionary shifts have apparently taken place many times during the evolution of the family. Several species currently possess two or more races visited predominantly by different types of pollinators. This could be an early stage in the speciation process. Given adequate time and genetic isolation, modifications could accumulate in each race making the flowers more suited to the local pollinators, and less likely to be visited by members of other pollinator groups. This could eventually produce complete reproductive isolation of these plants.

Several studies have examined the importance of pollinators in the reproductive isolation of plant populations in ecological time (Straw, 1956; Whitten, 1981). In an experimental study of two interfertile species of *Cercidium* (Leguminosae), for example, Jones (1978) showed that bee pollinators rarely moved from one species to the other. This pollinator specificity seemed to result especially from the different ultraviolet floral patterns of the two species. Hybrids were uncommon in nature, despite the physical proximity and flowering synchrony of individuals of the two species.

Appearances are sometimes deceiving, however, and pollinators do not always discriminate flowers that are quite different to the human eye.

The columbines *Aquilegia formosa* and *A. pubescens* have red and white flowers, respectively, and are interfertile. Chase and Raven (1975) tested and rejected an earlier suggestion that pollinator specificity was important in their reproductive isolation. Despite the species' very different flower colors and morphologies, they were not differentially visited by pollinators, and their isolation is likely to be caused instead by somewhat divergent flowering times and the rarity of appropriate intermediate habitats for hybrids. Nevertheless, if the Grants' (1965) conclusions about pollinator-induced reproductive isolation in the Polemoniaceae apply to other families, pollinators have played a substantial part in generating the present diversity of angiosperms.

Case Histories

Figs and Fig Wasps

The edible fig is the product of a specialized inflorescence called a syconium. The flowers of fig syconia are pollinated by tiny wasps of the family Agaonidae in a complicated sequence of events (Fig. 2-16). Here the focus is on *Ficus sycomorus,* and its most common pollinators, *Ceratosolen arabicus,* which have been studied in detail by Galil and coworkers (Galil and Eiskovitch, 1968, 1974).

The fig syconium is a spherical inflorescence with a distal scale-covered opening. Lining the inside of the inflorescence are several hundred minute flowers. Most flowers are female, and these vary considerably in the length of their styles. A smaller number of male flowers are concentrated near the distal opening.

Winged female wasps are chemically attracted to young figs and enter a syconium through the distal opening, losing their wings in the process. The females carry pollen in pockets on either side of the thorax. They spread this pollen over the receptive stigmas of the female flowers, a few grains at a time. The wasps also oviposit through the styles and into the ovaries of some of the female flowers, particularly those with shorter styles. The insects then die, remaining inside the syconium. Meanwhile, fig embryos develop in the unscathed ovaries, and wasp larvae develop in the remainder. The latter ovaries enlarge greatly as galls, protecting the larvae which feed on the plant tissues from within. Male wasps hatch before the females, puncture the galls containing the females, and impregnate them. At this time the males of many species of fig wasp die and the scales covering the opening to the syconium loosen, permitting the females to escape. In *F. sycomorus,* however, the scales do not separate and the male wasps tunnel around them. The female wasps emerge and gather pollen by scraping it into the pollen pockets with a row of bristles on each foreleg. The females depart via the tunnel made by the males, seek new syconia, and the cycle is repeated.

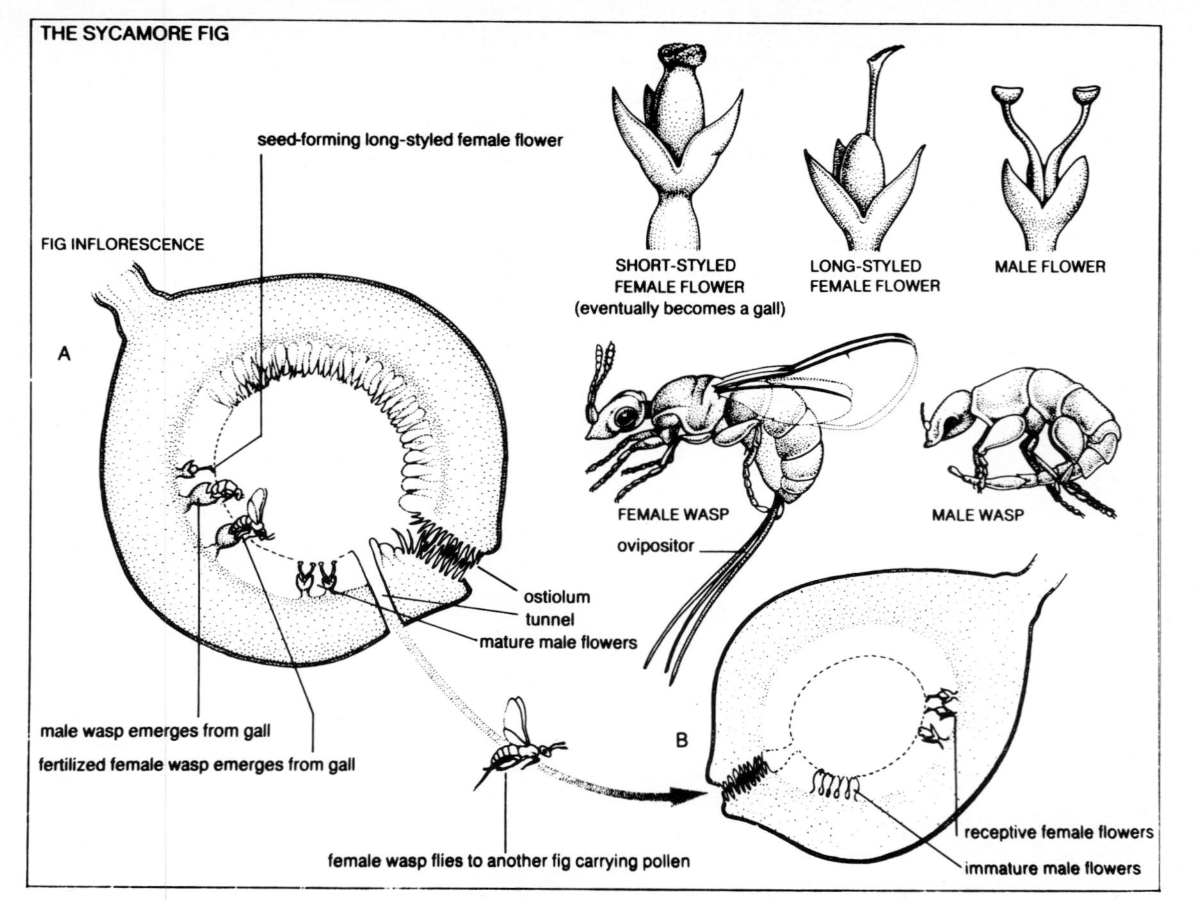

Figure 2–16. *Ficus sycomorus* and its wasp pollinators. From Meeuse and Morris (1984), used with permission of Rainbird.

While *F. sycomorus* and many other figs are monoecious, *F. carica* (the edible fig) and about 125 other species are dioecious, further complicating the plant–pollinator relationship (Valdeyron and Lloyd, 1979). In *F. carica* the male plants are termed caprifigs. Two or three crops of syconia are borne on the caprifigs, staggered so that wasps emerging from one generation enter the next. In late summer the syconia on the female plants are receptive and are invaded by pollen-bearing female *Blastophaga* wasps freshly emerged from male syconia. These wasps pollinate the female flowers therein, but cannot oviposit because all the styles are too long for their ovipositors to reach the ovary. Thus the attractiveness of the female syconia depends on their similarity to the caprifig syconia, wherein oviposition can occur. For the wasps, this deception is lethal because they cannot fly to other syconia. The continuity of this cycle depends on some wasps finding and ovipositing in caprifig syconia at this time.

Yuccas and Yucca Moths

The genus *Yucca* comprises 40 species of plants, and all are dependent for pollination on female moths of the genus *Tegeticula*. The three species of moth in this genus depend exclusively on *Yucca* to provide a rearing place and food for their larvae (Keeley et al., 1984). Early work on this system was done in the late 1800s, and additional observations and/or recent summaries are given by Proctor and Yeo (1972) and Aker and Udovic (1981).

Yuccas are robust plants, producing large inflorescences of large white flowers. These have an odor, which is most apparent at night. Nectar may be produced, but it is apparently not consumed by *Tegeticula* and it may serve to keep other insects away from the stigma. Yucca moths emerge from overwintering sites in the soil during the blooming period of yucca, which may last a month or so. Timing of emergence is critical because the moths are probably active for only a few days. A fertilized female moth enters a flower and climbs up a stamen to the anther. Here she uses her long maxillary palps to gather pollen into a ball, which she holds under her head with her maxillae and forelimbs. After collecting pollen from one or more stamens, she flies to another inflorescence. This behavior is important because yuccas exhibit some self-incompatibility, and deposition of self pollen is less beneficial than cross-pollination for both moth and plant.

Female moths are often very selective in choosing an oviposition site. Ovipositing moths seem to prefer young flowers and they may also avoid flowers in which oviposition has already occurred. The ovary of a yucca flower contains three locules (compartments), and the short styles and stigmas of these are united. The moth pushes her ovipositor into the ovary and lays an egg, a process that takes about half an hour. A small droplet of fluid exudes from the flower ovary at the point where the ovipositor was removed, and it has been suggested that some chemical marker is left by the ovipositing female to alert subsequent visitors to previous oviposition (Aker and Udovic, 1981). Such behavior would probably benefit each female moth by reducing competition among their larvae for food. A female may oviposit several times into a single ovary, but generally places each egg in a different locule, thereby minimizing competition among the larvae for food. In some ovaries, however, up to 17 larvae have been found (Keeley et al., 1984), and it is not clear whether this number reflects the efforts of one or more females. Periodically the moth moves to the stigma and deposits pollen, which is essential for the survival of her offspring because unpollinated flowers soon die, killing the larvae within. The moth larvae develop in the yucca ovary, consuming some seeds in the process. In a study of a few individuals of each of nine species of yucca, Keeley et al. (1984) found that 3 to 45 percent of the seeds were destroyed. As the fruits are ripening the larvae emerge and pupate in the ground. Their emergence as adults occurs in the following three blooming seasons, which protects the moths against the yucca's occasional failure to flower in a particular year (Proctor and Yeo, 1972).

As in figs, the yucca pollination system exhibits complete interdependence of plant and pollinator species. Behavioral patterns of the moths also exhibit many aspects which increase the likelihood of successful reproduction of both themselves and the plant, the two of course being related. These include the tendency to move to other plants after acquiring pollen, the limited oviposition in any one ovary, and the active pollination. Here is one extreme in the specialization and coevolution between plants and pollinators. The many very generalized pollination systems represent another extreme. The success of a system is judged not by its degree of specialization, but by its persistence in evolutionary time.

Selected References

Barth, F. G. 1985. Insects and flowers. Princeton University Press, Princeton, NJ.
Faegri, K., and L. van der Pijl. 1979. The principles of pollination ecology. Pergamon Press, Oxford.
Jones, C. E., and R. J. Little. 1983. Handbook of experimental pollination biology. Van Nostrand Reinhold, NY.
Real, L. 1983. Pollination biology. Academic Press, NY.
Richards, A. J. 1986. Plant breeding systems. Allen and Unwin, London.
Willson, M. F. 1983. Plant reproductive ecology. John Wiley and Sons, NY.

3

Fruits, Seeds, and Dispersal Agents

EDMUND W. STILES

Department of Biological Sciences
and
Bureau of Biological Research
Rutgers University
Piscataway, NJ 08854

Introduction

Vascular plants are primarily sessile organisms. The selective pressures associated with being fixed in space will become apparent in the subsequent three chapters that examine the influences of mobile herbivores. This chapter, however, asks questions about the mobile stage in vascular plants' life histories, that phase when plant seeds are moved away from the parent plant. The plant part distributed by dispersal is called a diaspore, regardless of its developmental and morphological origins (Sernander, 1927; Pijl, 1982).

After pollination is accomplished and a seed develops, a plant gains a selective advantage if it successfully moves its seeds to safe sites that yield high probabilities of successful germination and offspring survival

I wish to thank Alexander Brash, Margaret Garguillo, and Douglas White for critical reading of the manuscript and the editor, Warren Abrahamson, for significant improvements in readability.

to reproductive maturity. Over evolutionary time, genotypes yielding poor solutions to this problem are eliminated from the flora.

To complete this mobile phase of the life cycle, some plants interact with abiotic factors such as wind and are carried aloft by wings or plumes. Others are distributed by fresh or salt water, floating with the water's currents. Some plants can eject seeds ballistically with explosive capsules that throw the seeds from 1 to as much as 10 m. Many plants interact with animals that disperse seeds unharmed from the parent plant (i.e., seed dispersers). But virtually all plants interact with animals that eat seeds (i.e., seed predators). There are selective pressures on plants to attract and reward dispersers, but at the same time to repel and discourage seed predators. These different selective pressures have led to many interesting adaptations in fruits and seeds.

This chapter begins by discussing the extent of coevolution in seed dispersal and seed predation interactions. Next, the fruit and seed characteristics important to plant–animal interactions are described, and finally, the interactions of seed dispersers with fruits and of seed predators with seeds are examined.

Importance of History

Phylogenetic Considerations

It is important to keep in mind that selective pressures acting on one phase of the life history of a plant or animal are only part of the selective environment in which an organism exists. To consider seed dispersal and seed predation, it is necessary to understand that organisms are not isolated in time and space. The conservatism in the transmission of genetic material dictates that a large component of any organism is a product of past selective pressures (i.e., organisms are adapted to past environments).

The degree of conservatism among related species is important when we consider isolated aspects of the life history of an organism. For example, the fruits of *Taxus* species (yew) are virtually identical to those of the extinct *Palaeotaxus,* a close relative that lived 175 million years ago in the Upper Triassic (Emberger, 1968). Certainly, fruits of this taxonomic line have interacted with a wide variety of dispersal agents and seed predators but have changed little since the time of dinosaurs (Herrera, 1985a). However, members of the dogwood genus *Cornus* show extreme variation in fruit characteristics. This variation is ascribed to selective pressures exerted by the interactions of dogwood fruits with specific groups of dispersers in different parts of the world. Big-bracted dogwoods have a common evolutionary origin (Bate-Smith et al., 1975;

Brunner and Fairbrothers, 1978), yet compound fruits represented by Kousa dogwood *(Cornus kousa)* of Asia (Fig. 3-1) appear to have evolved from earlier simple fruits as a result of contact with primates 5 million years ago. This contact yielded large fruits rich in sugars. These are markedly different from smaller, lipid-rich fruits of the flowering dogwood *(Cornus florida)* that are attractive primarily to birds (Stiles, 1980). Flowering dogwood has evolved in the eastern United States where nonhuman primates were never found (Eyde, 1985). The radiations of birds and mammals in the late Cretaceous and early Tertiary (approximately 65 million years ago) are implicated as selective forces leading to larger diaspore size in the angiosperms and the rapid appearance of modern taxa (Tiffney, 1984).

Changing Distributions of Plants and Animals

Plants and animals must be coincident in time and space if animals are to exert selective pressures on the evolution of plant characteristics and if plants are to exert selective pressures on the morphology, physiology, and behavior of animals. However, absolute coincidence over either time or space or both is certainly a rare or nonexistent event. Plant and animal species, with the diverse array of selective pressures impacting them, evolve at different rates and persist for different lengths of geological time (Tiffney, 1984; Herrera, 1985a). In addition, geographic ranges of

Figure 3–1. Big-bracted dogwoods, such as *Cornus florida* (upper left) of North America, are adapted for bird dispersal, while *Cornus kousa* (lower right) of Asia are adapted for monkey dispersal.

plants and animals change over time, placing potentially interacting pairs together in some parts of their ranges and apart in others. In some instances, limits to ranges may be determined by absences of one member of a pair. For example, the introduction of multiflora rose *(Rosa multiflora)* in the northeastern United States from Asia has been accompanied by a subsequent northward expansion (more than 800 km) of the winter range of the northern mockingbird *(Mimus polyglottus)* (Stiles, 1982a). The mockingbird, however, feeds not just on overwintering rose hips, but also on other native fruits such as sumacs *(Rhus* spp.). This extensive range change has occurred over the past 30 years and has created selective pressures acting on the mockingbird for endurance of severe northern winters. The selective pressures acting on sumacs have also changed since they are now receiving different dispersal service than before the mockingbird's range expansion. This human-induced change is illustrative of the kind of changes that can take place in ecological, and especially, geological time periods.

In seed dispersal and seed predation interactions, the degree to which a plant or animal has changed over evolutionary time is dependent upon the genetic background, the intensity of selective pressures exerted, and the time over which selection has been operating. As described both here and in the subsequent chapters, the degree of specificity depends on the short-term and long-term fluctuations in the environment important to the interacting pair (Levins, 1968). For example, a change in the characteristics of a plant's seeds may result in a large number of selective deaths for an insect seed predator that depends exclusively on those seeds. The result would be a subsequent change in the characteristics of the insect or extinction (Rhoades, 1979). However, for a bird eating a variety of fruits, a change in the characteristics of the fruits of one species may simply shift the bird's dependence on that fruit. Unless dependence on that fruit is high, the number of selective deaths will be small, and the degree of change in the disperser population will be minimal.

Fruits and Seeds of Plants

The Movement Phase of Plant Life Histories

Seed Movement Vectors

Plants utilize a diverse array of mechanisms to get their seeds moved (Ridley, 1930; Pijl, 1982). Whatever the mechanism is, however, the result is a distribution of seeds in the environment, usually with the highest density closest to the parent plant (Janzen, 1970). This distribution is called a seed shadow and is measured by determining the densities of seeds at increasing distance from the parent plant. The characteristics of a seed shadow are influenced by the mechanism or vector used by the plant to disperse its seeds. Vectors may be abiotic or biotic with the primary abiotic vectors being wind, water, and gravity. Although these dispersal mechanisms do not fall directly under the realm of this book, the seed shadows they generate are important in understanding seed predator interactions (Janzen, 1970).

Abiotic Vectors

Water and gravity are uncommon as sole dispersal vectors for seeds. Gravity alone generates a seed shadow primarily under the parent which, for reasons discussed in the following, is probably disadvantageous. Water dispersal creates a unidirectional seed shadow in moving freshwater and one that is similar to gravity-generated seed shadows in lakes. Water becomes an effective vector only in oceans where tidal floods and ebbs and major currents move seeds (Ridley, 1930). However, under certain circumstances, such as tropical rivers that predictably flood, seeds may be well distributed at least within the floodplain (Goulding, 1980) by multidirectional currents. Likewise, wind may interact with water in lakes by blowing seeds along the surface to create a broad seed shadow.

Wind is an effective vector for seed dispersal in those regions and habitats where wind is predictably present. Wind is predictably present in the temperate zones and in the canopies of many plant communities. Wind is less predictable in the tropics and in the understories of most plant communities. Wind is also markedly seasonal, and wind-disseminated plants typically fruit when wind is most available and predictable (Pijl, 1982). The maples of eastern North America, for example, flower in early spring and either disperse seeds immediately before the leaves have fully expanded or in the autumn following the abscission of leaves.

Seeds dispersed by the wind include dustlike seeds (orchids), plumed diaspores (dandelions), and winged diaspores (maples), and seeds that are thrown ballistically (evening primroses, mulleins) through the action of elastic stalks bent by the wind (Ridley, 1930). Seeds dispersed by wind are strongly influenced by gravity, and their adaptations reduce the rate of fall, increasing the length of time they may be pushed by the wind. The interaction of wind and gravity produces a decay curve of seed deposition with the majority of seeds deposited under the canopy of the parent and a rapidly declining number of seeds deposited at increasing distances from the plant. There is a marked decline in the numbers of seeds reaching increasing distances from the source, and an even greater decrease in the density of seeds since concentric circles of increasing radius from the parent plant enclose a greatly expanding area for each increment in radial distance.

Biotic Vectors

Animals are major dispersers of seeds. Plants have capitalized on animal mobility to generate seed shadows that presumably increase the probability of seed survival. The great array of plant adaptations associated with animal dispersal includes fruit colors, odors, size, shape, presentation, chemistry, and ripening times, to mention a few. The primary animal dispersers of seeds are ants and vertebrates, except the amphibians. Ants are considered in Chap. 6 and are not discussed here. Numerically, birds and then mammals interact most strongly with plants as seed dispersers, but fishes and reptiles also have important disperser roles for certain species of plants.

Advantages and Disadvantages to Seed Movement by Vertebrates

There are trade-offs associated with attracting vertebrates to move seeds. Plants profit from the interaction because by moving offspring away from the parent plant, competition for spatially restricted resources (e.g., sunlight, water, or nutrients) may be reduced (Harper et al., 1970). Placing seeds in a variety of remote locations may increase the probability of landing on a favorable site for germination and survival. Plants of early successional stages are more successful if they germinate in an earlier seral stage than that occupied by their parent. Environmental disturbance is common, but in most circumstances it is not predictable in space (Pickett and White, 1985). Therefore, a wide dissemination of seeds away from a parent plant should increase the probability that a seed will encounter one of these disturbances (i.e., an earlier seral stage). In addition, seeds moved away from the parent may be less likely to be eaten by predators (Janzen, 1970).

The disadvantages associated with using animal dispersal agents involve the costs required to attract and reward animals. Animals are attracted to fruits in ways similar to those discussed for flowers in the previous chapter. Plants offer various signals to advertise the potential presence of an available source of food. The word potential is important, as food may not be present if colorful bracts or leaves are the attracting signal. This signal is, however, typically reinforced by the presence of a food reward (part of the fruit). The amount of resource provided as a reward is a cost that must be weighed, in an evolutionary sense, against the benefits accrued from seed dispersal.

Not all animal-dispersed seeds have associated attractants and rewards (Ridley, 1930; Pijl, 1982). Some seeds are disseminated by attaching to the feathers of birds or the fur of mammals (Agnew and Flux, 1970; Pijl, 1982; Sorensen, 1986). Seeds can also be eaten with other foods and passed in viable condition (Janzen, 1982b) or seeds may be transported in mud on the feet of animals (Darwin, 1859; Pijl, 1982). Seeds in these latter situations, however, have no obvious adaptations to these dispersal modes.

Seed movement may be disadvantageous if a disperser leaves seeds in inhospitable locations. Dispersal of seeds into inappropriate habitats or over very long distances (where environmental conditions are not appropriate for success of a particular plant species) reduces the fitness of a plant.

Fruit Structure

Botanical Definitions

In angiosperms, the gynoecium is the collective term for the carpels (probably evolved from leaves) that produce one or more ovules (see also Chap. 2). The ovary is the lower part of the carpel or gynoecium that contains the ovules. The fertilized ovules develop into seeds. Technically, the fruit is a matured ovary, but in common usage the fruit is considered a derivative of the gynoecium and sometimes extracarpellary parts (Esau, 1977). A fruit may include the receptacle, as in strawberries; the calyces, as in mulberries; or bracts, as in pineapples. Simple fruits consist of a single ovary with one or more carpels. Aggregate fruits, such as blackberries, are derived from groups of separate (apocarpous) ovaries, and multiple fruits are the gynoecia of many flowers combined, as in mulberries. Any one of the three types may be combined with extracarpellary parts, which is then termed an accessory fruit. Thus, a mulberry is a multiple accessory fruit.

The pericarp is the matured ovary wall and may be divided into outer, middle, and inner layers: the exocarp, mesocarp, and endocarp, respectively. When extracarpellary structures are included in the matured fruit, as is often the case, the fruit wall is used as the inclusive term to describe the layers surrounding the seed.

Simple fruits may also be divided into dry fruits and fleshy fruits. Dry fruits may be either dehiscent, with the fruit wall splitting along one or both sutures, or indehiscent. Dehiscent, dry fruits include specific types known as follicles, legumes, capsules, and siliques, whereas indehiscent, dry fruits include caryopses, achenes, schizocarps, nuts, grains, and samaras. Fleshy fruits may be divided into those derived from monocarpellate or multicarpellate gynoecia. The two major types of fleshy fruits are drupes and berries, but pomes, hesperidia, and pepos are also found. In drupes, the endocarp forms a hard, stony covering surrounding the seed or seeds, whereas in berries the entire fruit wall is fleshy. This can be confusing as many fruits with the common name berry are actually drupes, including blackberries (multiple drupelets), blueberries, and strawberries.

Although terminology of angiosperm fruit types can be somewhat complex, and is different yet for gymnosperms, animals utilizing fruit or seed parts for food resources are not concerned with either their terminology or embryological origins. Of concern to animals are the amounts and nutritional characteristics of the fleshy pulp, the endosperm of the seed, and/or the hard indigestible portions of the fruit.

For dispersers, a major component of the hard indigestible portion of a fruit is the seed. Seeds, fertilized matured ovules, consist of the embryo (i.e., developing sporophyte), the nutrients stored for its development (i.e., endosperm), and the seed coat or testa. The seed coat is formed by different numbers and thicknesses of integuments and by different patterns of vascularization. The seed coat serves as a protective barrier for seeds ingested by seed dispersers and is scarified (cut or softened) to varying degrees during seed passage through a disperser's gut. For seed predators, the seed coat is a deterrent, decreasing the efficiency with which they can consume the embryo and endosperm. The seed may have associated appendages that serve as food for seed dispersers in a manner similar to the fruit wall of fleshy fruited species. Seed appendages include arils, caruncles, and elaiosomes. Elaiosomes, for example, are oil-containing appendages that are utilized by ants and are discussed in Chap. 6.

Fruit Chemistry

Chemical Compartmentalization: Pulp and Seed

For seed dispersers, fruits represent a resource of highly variable quality. Through differential fruit selection, animals influence transportation of seeds of fruits of different phenotypes, and probably different genotypes, altering plant survival and subsequent reproduction (Janzen, 1983a).

The nutritional value or profitability of a fruit for a disperser depends on the nutritional characteristics of the fruit pulp (the fleshy portion of the fruit) and the ease with which nutrients are extracted. The profitability of a fruit for a seed predator is associated with the amount and nutritional characteristics of the embryo, the stored nutrients of the seed, and the relative ease of accessing them through the seed coat. Seed predators may or may not use pulp as food.

Nutritional Aspects of Pulp

The nutritional value of a fruit to an animal involves both the characteristics of the fruit and characteristics of the animal. The amount of pulp available to animals on each fruit is only the initial consideration. Other factors that must be considered are the relative amounts of digestible and indigestible fruit parts. Seed dispersers can extract nutrients from pulp, but indigestible materials take up gut space that could be occupied by nutritious food (McKey, 1975; Howe and Estabrook, 1977; Herrera, 1981; White and Stiles, unpublished). For many fruits the majority of the indigestible parts (particularly seeds) can be separated from pulp either in the mouth or in the upper digestive tract and either spit out or regurgitated, respectively (Ridley, 1930; McKey, 1975; Sorensen, 1984). These processes reduce the amount of indigestible materials occupying space in the gut. Thus, fruits that have easily separated seeds may be more valuable to animals than those that do not. In an evaluation of fruits from the eastern North American deciduous forest, seeds from fruits with high seed loads (low pulp/fruit ratio) were regurgitated more often than seeds of fruits with low seed loads. Instances where seeds from fruits with high seed loads are defecated are very rare (White and Stiles, unpublished).

Water contents of fruits can be highly variable. Pulps of fruits that are consumed by animals for the pulp vary from less than 10 percent water in the dry pulps of *Rhus* (sumac) and *Toxicodendron* (poison ivy) to more than 90 percent water in juicy fruits of some *Rubus* (raspberry), *Fragaria* (strawberry), *Cornus* (dogwood), *Clintonia*, and *Streptopus* (twisted stalk) (White and Stiles, unpublished). Water may be an important component of fruits. If fruit pulp is dry, an animal may need to drink additional water to provide an intestinal water content that maximizes digestive efficiency. This is an added expense in foraging time and energy. Water may also serve as an attractant in very dry habitats (Herrera, 1984).

Carbohydrates, lipids, and proteins are the primary nutrients in fruit pulp, although the proportions of these constituents vary widely. Hot-water-soluble carbohydrates range from less than 5 percent in some *Myrica* (bayberry), *Toxicodendron*, *Aronia* (chokeberry), *Rhus*, and *Sambucus* (elderberry) to more than 90 percent in some *Vaccinium* (blueberry) and *Fragaria* when measured as a percentage of the dry weight of pulp. Protein values are usually estimated by determining nitrogen content and multiplying by 6.25, as proteins contain approximately 16 percent nitrogen. Protein values in temperate fleshy fruits are generally low, ranging from less than 1 percent in many species to more than 25 percent pulp on a dry-weight basis. Most values, however, fall between 2 and 6 percent. Insects and other animal food eaten by most frugivores (fruit-eating animals) contain about 70 percent protein (Morton, 1973), making fruit a poor source of amino acids for these animals. Lipids yield approximately twice the amount of energy on catabolism as either carbohydrates or proteins. Pulp provides less than 1 percent lipid for many species but is as high as 50 percent in *Sassafras* and *Lindera* (spicebush) and is more than 70 percent in some *Magnolia*.

Fruit pulp contains a variety of minerals, with the most common ones being potassium, calcium, phosphorus, magnesium, and sodium. However, little information has been collected on the significance of these minerals to frugivores (Johnson et al., 1985). The physiological significance of these nutrients depends on synergistic and competitive interactions with other minerals in their bodies. For example, the potassium/sodium ratio is important in nervous system function in vertebrates. Fruit pulp, as well as other plant material, is very high in potassium but is low in sodium. As the kidney eliminates potassium, sodium is also lost. All herbivores, including frugivores, face the problem of too much potassium and not enough sodium. Sodium deficiencies in animals can cause reduced growth, softening of bones, corneal keratinization, gonadal inactivity, loss of appetite, weakness and incoordination, shock, and death (Robbins, 1983). Calcium, phosphorus and magnesium deficiencies result in equally severe problems.

Any of these conditions could reduce fitness, so feeding behaviors that reduce these deficiencies should be favored by selection. Although no obvious correlations between fruit minerals and frugivore selection have been documented, few fruits have been evaluated (Johnson et al., 1985). Even less is known about the importance of micronutrients (e.g., iron, zinc, manganese, copper, iodine, selenium, cobalt, molybdenum, fluoride, and chromium) and vitamins (Janzen, 1983a). The adaptive significance of different fruit compositions awaits a more complete catalog of both constituents and diets of frugivores.

Palatability Inhibitors and Toxins

After a flower is pollinated and the fruit begins to develop, plants consolidate nutrients in seeds and pulp (Esau, 1977). Nutrients are appreciably more concentrated in reproductive organs than in other parts of the plant, and are vulnerable to seed and fruit predators. Timing of pollination and fruit maturation are under different selective pressures, and the time between these two events may be as short as a few weeks or as long as a year or more. During this interval, plants minimize the availability of pulp and seed to predators. Unripe fruits often contain palatability-inhibiting compounds (e.g., tannins) or toxins (e.g., alkaloids) (Goldstein and Swain, 1963; Janzen, 1971b; Herrera, 1982a). These compounds reduce attack by fungi and insects that kill seeds or reduce their dispersability. During this preripening time, fruit may chemically deter predatory animals in a fashion similar to that of a leaf, and concepts surrounding this are discussed at length in the following chapter. Some seeds [e.g., the maples *Acer negundo* (box elder) and *A. saccharum* (sugar maple)] expand the fruit (samara) wall immediately after pollination but do not put appreciable energy into the developing seed until almost 3 months later. During the intervening time, the winged samara is green and photosynthesizes, functioning much like a leaf (Bazzaz et al., 1979), but the undeveloped seed is not vulnerable to predation. During this period the fruit is cryptic and difficult to distinguish among the green leaves.

The fruit-ripening process involves a complex set of chemical modifications that change an unpalatable, cryptic fruit into a nutritious, conspicuous fruit (Hulme, 1971). Chloroplasts in the fruit wall change to chromoplasts (Esau, 1977) to yield the colors that contrast with the green, leafy background (Ridley, 1930; Stiles, 1982b). Ripening also may involve fruit softening and the associated reduction in astringent tannins that polymerize into high molecular weight polyphenolics (Goldstein and Swain, 1963). Toxins often are destroyed or created at this time. For example, in papaya, one proteolytic enzyme, the alkaloid papain, is destroyed, while in pineapple, the proteolytic enzyme, bromelin, first appears during the ripening process (Czyhrinciw, 1969; Janzen, 1983a). Changes such as these create different profitabilities for both rotting agents and dispersers.

Ripening increases the nutritional value of fruits for dispersal agents, but it also appreciably decreases defense mechanisms that deter nondispersing fungi, bacteria, and insects (Janzen, 1977b; Thompson and Willson, 1979; Stiles, 1980; Herrera, 1982a). There is a selective premium on having the appropriate dispersers eat fruit before the defenses have been eliminated, but after the seed is mature (Foster, 1977). Likewise, selection should favor adaptations that retain defense against nondispersers while attracting dispersers. This problem is often overcome by compartmentalization of the nondisperser defenses in fruit skin. The fruit skin is often voided with the seeds by disperser species. Under these selection pressures, timing of ripening should be closely associated with availability of dispersers (Thompson and Willson, 1979; Stiles, 1980; Stiles and White, 1982).

Seed Morphology and Chemistry

Seeds contain both the embryo and rich nutrient materials that provide the germinating embryo with needed resources for early growth. Thus, seeds are compact food packets for animals because they are often rich in lipids and proteins and they store well. Large seed size may often be a disadvantage if dispersal vectors are wind or animals. Therefore, there are few inert fillers such as cellulose or silica in seeds, but rather compact, rich sources of food for the embryo. Soybeans contain 20 percent oil and 40 percent protein, mustard seeds contain 40 percent oil and 30 percent protein, while cereal grain seeds contain from 70 to 80 percent starch (on dry-weight bases) (Esau, 1977). Thus, seed predators can place strong selective pressures on the protection of these resources.

Seed Coat

The seed coat or stony endocarp of drupes decreases the ease and speed of access by seed predators to the embryo and stored resources of seeds. Generalist predators may be deterred by the physical difficulty of cutting through the hard seed coat, but in most cases specialists have evolved mechanisms to penetrate even the hardest seed coat (Janzen, 1971b). Still, even for specialists, the time involved in breaking through the seed coat is an important factor in their foraging time/energy budgets (Pyke et al., 1977). Turkeys can crush pecans *(Carya illinoensis;* a 30- to 36-kg force is required to crush a nut) in their gizzard in about 1 hour, but it takes more than 30 hours for them to crush shagbark hickory nuts *(Carya ovata;* 76–117 kg required to crush a nut; Schorger, 1960). Seed predators may distinguish toxins in seed coats, as found with mourning doves and the polymorphic seeds of dove weed *(Eremocarpus setigerus;* Cook et al., 1971), but these toxins may or may not be important deterrents depending on the method of removing the seed coat. In an experiment with squirrels *(Sciurus)* and cotton rats *(Sigmodon)*, acorns coated with the poison endrin poisoned rats but not squirrels, because rats gnawed through the seed coat ingesting part of it, whereas squirrels broke open the acorns discarding the seed coat (Johnson et al., 1964). Seed coats of some conifer species [e.g., *Abies* (firs) and *Cedrus* (true cedars)] have numerous resin glands that may decrease the ease of access by mice and squirrels (Smith, 1970).

The seed coat also functions to protect the embryo during passage through the gut of the dispersing animal. Occasionally seeds are crushed or germinate during gut passage and are killed (Roessler, 1936; Krefting and Roe, 1949; Janzen, 1981a, 1981b, 1981c, 1982a). Although increased seed coat strength may decrease seed mortality in animal guts, it could delay or even inhibit germination for seeds that are not passed through the acidic, grinding environment of a disperser's gut (Krefting and Roe, 1949; Temple, 1977).

The advantages of seed dispersal by animals have selected for a more resistant seed coat, and are great enough to incur the cost of delayed germination in undispersed seeds. Temple (1977) has proposed that an extreme case occurs on the island of Mauritius in the Indian Ocean. He found that a species of tree *(Calvaria major)* had had no recruitment of young trees for more than 300 years, even though there is a vigorous, seed-producing population of large, old trees. The fruits of this species contain a large seed with a thick, hard, seed coat. Temple hypothesized that the dodo, a large terrestrial pigeon, had been the major consumer of the fallen fruits of *Calvaria*, but the dodo became extinct in the late 1600s due to human persecution and the introduction of exotic predators. The grinding of seeds in the dodo's acidic gut broke down the heavy seed coat, but without this scarification, germination is entirely prevented. Temple was able to mimic the passage of *Calvaria* seeds through dodos by feeding the fruits to turkeys. These scarified seeds germinated successfully. Alternatives have been proposed to explain the loss of recruitment in *Calvaria*, but this example serves to illustrate the potential strength of advantages gained from animal dispersal.

Seed Toxins

Toxins in the seed's endosperm may be a greater deterrent to more specialized seed predators than a thick or toxic seed coat. Seeds contain from 1 to 10 percent of such toxins in the form of alkaloids, saponins, cyanogenic glycosides, rare amino acids, gossypol, and selenium (Trelease and Trelease, 1937; Whiting, 1963; Aronow and Kerdel-Vegas, 1965; Hardee and Davich, 1966; Jones, 1966; Lukefahr et al., 1966; Fowden et al., 1967; Janzen, 1969a, 1971b; Bell and Janzen, 1971). Endosperm toxins may be effective deterrents against the majority of seed predators. But as with the seed coat, specialists that survive the large number of selective deaths that accompany the evolution of a new toxic compound will proliferate and spread genes for a detoxification mechanism. Weevils of the families Bruchidae and Curculionidae have detoxification mechanisms for toxic alkaloids, toxic amino acids, and gossypol (Lukefahr et al., 1966; Bridwell, 1968; Janzen, 1971b).

Phenology (Timing)

The earth rotates around the sun during the course of a year, and rotates once daily on its axis that is tilted 23 degrees off perpendicular to the plane of annual rotation. As a result, the differential warming of the Earth's surface produces the warm/cold seasonality of the temperate zones and the wet/dry seasonality of the tropics. Seasonality places physiological constraints on plants and animals by influencing a variety of life-history characteristics. Most fruit production in plants is seasonal, even under conditions with relatively little environmental fluctuation (Foster, 1982). One of the best examples of the potential impact of seasonality on fruiting phenology is the eastern deciduous forest of North America because here seasonality has a pronounced amplitude.

In the temperate zones, cold winters restrict timing of flowering. This is especially true for insect-pollinated species because the timing of flowering is restricted to warmer periods when insects are active. Timing of fruiting is constrained only by the speed of fruit maturation following pollination. However, the temporal relationship between flowering and fruiting is also influenced strongly by the availability of dispersal vectors and seed predators. In the north temperate zone, most fleshy fruits ripen almost synchronously in autumn (e.g., late August to October; Thompson and Willson, 1979; Baird, 1980; Stiles, 1980; Herrera, 1982b). These species may have flowered as early as April or as late as August. During autumn the primary dispersal agents for these seeds are birds that are primarily insectivorous in summer (Martin et al., 1951).

As the majority of plants become dormant, the availability of insects markedly decreases and many bird species migrate to more southern latitudes. This migration requires large amounts of energy that is supplied from subcutaneous fat. Since fleshy fruits provide an excellent source of lipids and carbohydrates, they are a major food source for migrating birds. It is no coincidence that the timing of ripening of these fruits is correlated with the demand for energy by migrating birds. The specificity of this timing produces ripe fleshy fruits just when birds are present. Migration follows the reproductive period of birds but occurs before major juvenile mortality. Thus, fruits are quickly consumed, reducing the time that competing nondispersers (e.g., fungi and insects) have to use fruit resources. This is valuable since fruit maturation coincides with the loss of the fruit's defense system (Stiles, 1980). The migratory period is relatively short and by the end of October most migrants have departed (Fig. 3-2). The weather at this time is usually not severe, so the remaining resident birds continue to consume insects, since their requirements for energy are not as great as migrating individuals. Many fruits that ripen at the onset of bird migration, but that are not taken during the migratory period, will remain on plants until the weather becomes quite severe and insects are less available.

This pattern of frugivore availability may select against fruits of intermediate nutritional quality ripening at the onset of migration. Highly preferred fruits will be taken quickly by long-distance migrants or by short-distance migrants, or vagrant flocking frugivores (e.g., American robins, cedar waxwings). These flocking species are responsive to remaining high-quality fruits or concentrations of other fruits. Once high-quality fruits are depleted, birds will remove fruits that are lower in their preference ranking. If high-quality fruits remain uneaten when migrants depart, competing nondispersers are likely to consume or rot fruits. Fruits with high moisture content will freeze, rupturing the fruit skin and resulting in rapid desiccation. Lower-quality fruits will not be consumed as rapidly by dispersal agents, but they are more likely to persist in good condition on plants until severe winter weather induces

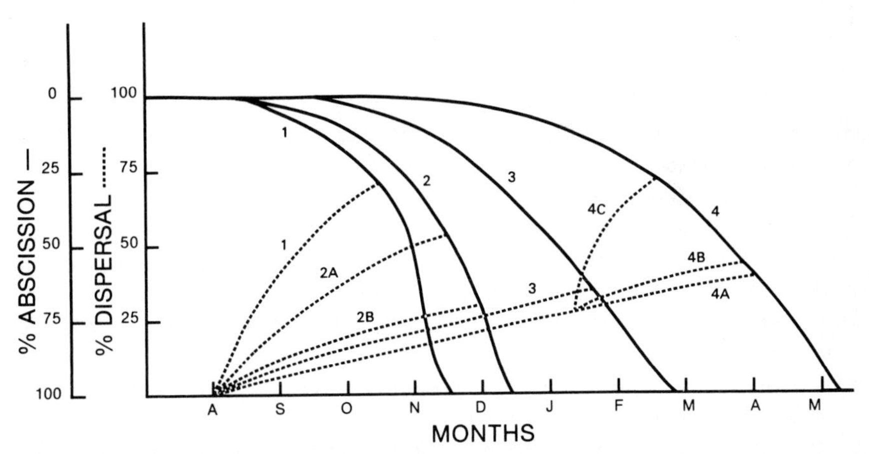

Figure 3–2. Abcission rates (solid lines) and dispersal rates (dashed lines) for four hypothetical bird-disseminated plants. Points on the curves represent the points in time when no fruits remain on the plants. 1 = fall high-quality fruit; 2A = fall high-quality fruit (lower quality than 1) when fruit species 1 is scarce; 2B = fall high-quality fruit when fruit from species 1 is abundant; 3 = intermediate quality fruit; 4A = fall low-quality fruit during mild winter conditions; 4B = fall low-quality fruit with a short period of time when other resources for frugivores are unavailable; 4C = fall low-quality fruit during very severe winter weather conditions, when other resources for frugivores are unavailable.

resident birds to consume more fruits. Some species of fruits persist even longer, and are consumed by early migrants the following spring (White and Stiles, unpublished). Intermediate quality fruits are not likely to be eaten during the migratory period, but are likely to rot before consumption by resident birds. Of autumn-ripening species in the eastern deciduous forest, many have very low lipid content, while others have high lipid content, but very few species have intermediate lipid content. This suggests disruptive selection against fruits with intermediate lipid levels.

During summer, birds produce eggs, feed offspring, and grow new feathers. These activities require a substantial investment of protein, and fruits are generally much poorer sources of protein than insects (Morton, 1973). Most temperate bird species consume primarily insects in summer (Martin et al., 1951). Thus, plants bearing fleshy fruits at this time encounter different frugivore feeding priorities from those fruiting in autumn. Species ripening during summer are also subject to higher temperatures and faster growth rates for those microorganisms that rot fruit. The fruits of these species are typically high in sugars, and most ripen asynchronously (Thompson and Willson, 1979; Stiles, 1980). They attract not only birds but also mammals and some reptiles (e.g., box turtles). Species that fruit during summer are either those with low growth form or have fruits that fall soon after ripening, making them available to terrestrial vertebrates. These plant species include almost all native species used as cultivars, including blueberries, strawberries, mulberries, blackberries, and cherries. Attraction of multiple classes of dispersal agents is uncommon in autumn when birds are more reliable as dispersal agents.

Using mammals as dispersal agents exposes seeds to potentially heavy predation by the eastern deciduous forest's most voracious seed predator, the white-footed mouse *(Peromyscus leucopus)*. Seeds of summer fruiting species are either very small or very large. Both small and large seeds may reduce the speed with which mice can extract nutrients from seeds. For example, a mouse must open each small seed of a blackberry for a small amount of food in each seed, and it must spend a long time gnawing through the heavy seed coat of a large seed like a cherry. Both situations reduce the energy per time gained by the mouse compared with an intermediate-sized seed. Species with intermediate seed size may be selected against if they fruit at this time and employ bird and mammal dispersers because they are subjected to heavier predation by mice.

As is evident from the above example, changing food requirements and numbers of frugivores have strong influences on the phenology of fruit ripening and fruit chemistry. In tropical regions, where timing of fruit ripening is not as restricted by a severe season, phenology is still influenced by availability of dispersal agents. Howe and De Steven (1979) suggested that timing of fruiting in *Guarea glabra* was influenced by northward movements of migrants, and Greenberg (1981) found that timing of fruiting in *Lindackeria laurina* was associated with the premigratory feeding of wood warblers. Morton (1971) has suggested a similar selective influence on timing of fruiting of *Didymopanax morototoni* in Panama, created by the migration of eastern kingbirds. However, more studies will be required to appreciate the full influence of frugivore availability on tropical fruiting phenology.

A further consideration for tropical climates is that competition among fruit species for service by frugivores may select for different fruit-ripening times. Snow (1965) found a temporal progression of fruit ripening in melostomes (fleshy-fruited shrubs and small trees) on Trinidad that were eaten by manakins (small frugivorous birds). This suggests a phenological shift in fruiting to reduce competition among fruiting plants for frugivores.

Signals, Fruit Size, and Fruit Production

Signals

Earlier production of more offspring yields a numeric advantage to a reproductive plant if each offspring's probability of dying does not increase concomitantly. This should select for large fruit crops. Plants with larger fruit crops may be visible at greater distances and may attract greater numbers of foraging frugivores (Snow, 1971; Howe and Estabrook, 1977). Initial experiments to test this hypothesis have produced conflicting results (Howe and Smallwood, 1982; Davidar and Morton, 1986). Howe and his coworkers (Howe, 1977; Howe and Estabrook, 1977; Howe and De Steven, 1979; Howe and Vande Kerckhove, 1979) have studied species of *Virola* in Panama that have a range of fruit and seed sizes both within (see also Jordano, 1984) and between species. They found that for trees with small fruits and small seeds, increasing the size of the fruit crop increased the numbers of birds visiting the plants. However, for trees with large fruits, increasing fruit crop size attracted additional birds only up to a point. Further, the small-fruited trees attracted a broad range of bird species, whereas large-fruited species attracted only large resident frugivores that have limited abundance (Fig. 3-3).

This relationship has also been studied by Wheelwright (1985) in the cloud forests of Costa Rica. He found that large-fruited species can only be handled by frugivores with large gape widths, but that smaller fruits attract and are eaten by both large and small frugivores. Whether

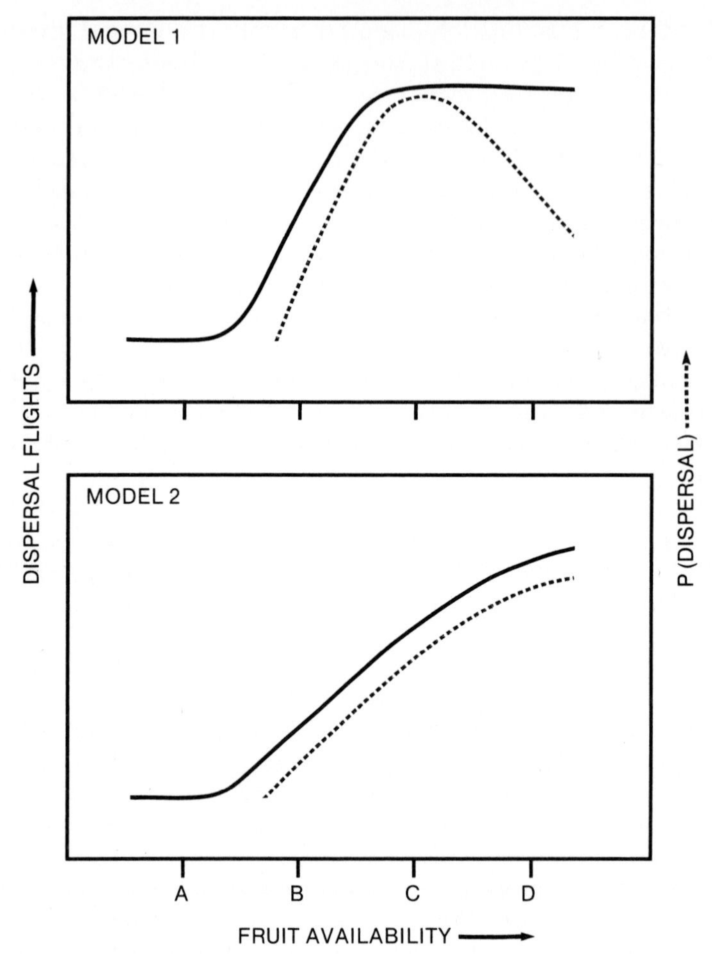

Figure 3–3. Expected relationships between visitation, fruit availability per unit time, and seed dispersal from an individual tree. The solid line represents the number of bird flights with potential for dispersing seeds away from the parent tree. The dashed line represents the probability of effective escape of a seed to a site suitable for growth. Model 1 describes the expected relationships for trees dependent on obligate specialists; model 2 describes the expected relationships for trees dependent on opportunists.

reduction in types of frugivores dispersing seeds of large-fruited species is an advantage to plant species (because of specialized treatment of seeds in a limited number of gut types) (McKey, 1975) or a disadvantage (because of other selective pressures for large seed size) is as yet undetermined (Wheelwright, 1985). There is evidence that larger bird-dispersed seeds often have reduced but highly nutritious flesh.

This effectively decreases fruit diameter without substantially reducing profitability of the fruit (Howe and Vande Kerckhove, 1980; Herrera, 1981).

There is evidence that signals are quite important for seed dispersal. Fruits are similar to flowers in this regard because they produce signals to alert dispersers. Signals may be odors that attract the olfactorily sophisticated mammals or reptiles, or they may be colors attracting visually oriented mammals or birds. Although attempts have been made to evaluate the importance of color in fruits (Turcek, 1963), selective pressures that yield the diversity of colors are poorly understood. The majority of fruits eaten by birds are black or red (Wheelwright and Janson, 1985), whereas fruits eaten by mammals are commonly yellow, green, orange, or brown (Janson, 1983; Knight and Siegfried, 1983). This, however, does not explain the wide range of fruit colors from purple and pink to white.

Preripening fruit flags (Stiles, 1982b) or bicolored displays (Willson and Thompson, 1982) involve a contrasting color signal that is presented before the fruit is ripe. It may be in the form of an unripe fruit color (often red prior to black or blue) or a brightly colored pedicel or bract. Preripening fruit flags are commonly found in the temperate zone for summer-ripening fruits and in the tropics. In both situations, resident birds are able to assess the imminent availability of ripe fruit. Preripening signals may lead to more rapid removal of fruit upon ripening, thereby reducing the chance for fungal or insect attack. These displays are usually associated with asynchronously ripening fruits and, thus, may also increase the size of display in the absence of ripe fruits (Willson and Thompson, 1982).

In the North American eastern deciduous forest, a nonrandom set of plants change leaf color before the majority of plants and thus create foliar fruit flags. These species are bird-disseminated plants whose color change is correlated with the peak of autumn frugivore migration (Stiles, 1982b). Unlike resident birds, migrants are rarely in one locality for a long period and are unaware of food locations. This large signal may identify the potential presence of fruit for spatially naive frugivores (Fig. 3-4). Reduced photosynthesis resulting from early leaf color change may be offset by gains from increased dispersal and decreased fruit loss to microorganisms and insects (Janzen, 1977b; Thompson and Willson, 1979; Stiles, 1980, 1982b). Many species that display foliar fruit flags are vines that are difficult to identify from a distance because of their varied form [*Vitis* (grape), *Parthenocissus* (Virginia creeper), *Toxicodendron*], species with fruits rich in lipids that would rot quickly if not dispersed by birds (*Lindera, Cornus, Sassafras, Magnolia*), or species bearing fruits singly or in small, somewhat isolated, clusters [*Lindera, Sassafras, Nyssa* (black gum)]. Without a large splash of color, fruits are difficult to locate from a distance.

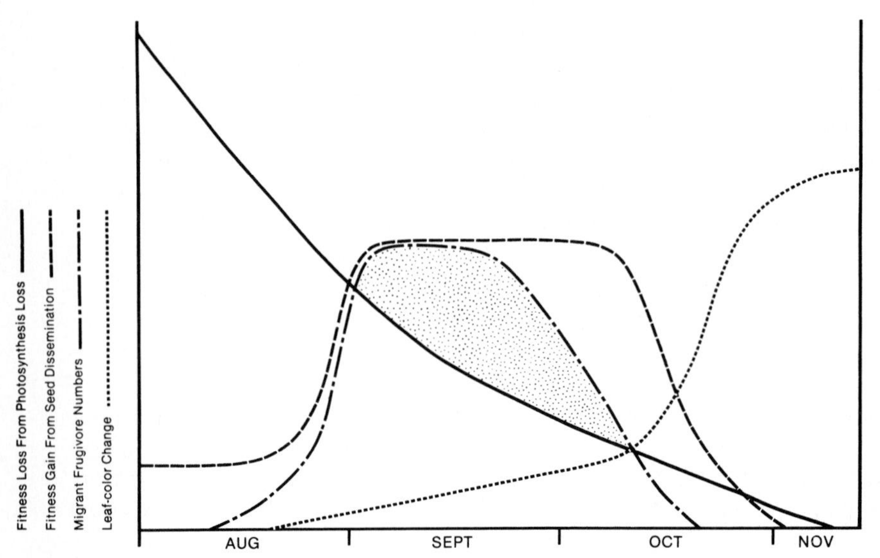

Figure 3–4. Foliar fruit flags have positive fitness value when the gain from increased seed dissemination exceeds the photosynthesis loss from leaf-color change (hatched area). This occurs only in early fall during fall bird migration, but before most plants change leaf color.

Fruit Size and Fruit Production

The number of fruits available on any plant in a given year is influenced by many parameters including plant size, pollination success, preripening predation, stored resources, predator avoidance, and fruit abortion due to nutrient limitation. While many factors influence resource allocation to each fruit, still other parameters affect the quality and quantity of pulp and the size and number of seeds.

Larger-fruited species have a reduced range of frugivores able to ingest their fruits largely because of bird gape-size limitations. Smaller-fruited species, however, are available to a broader range of frugivores. Larger, soft fruits, such as the aggregate fruits of strawberries and blackberries, may be eaten piecemeal by birds, mammals, and fish that have cutting mouthparts. Effective dispersal is still accomplished because either single-seeded fruits are carried off before the pulp is removed or small seeds are consumed along with the fleshy portions of the larger fruits.

Fleshy-fruited species have the evolutionary alternatives of making more fruits with nutrient-poor pulp or few fruits with nutrient-rich pulp, with the same resource budget. There is substantial variation in resource commitment to fruit pulp and profitability to birds among species with fruits of the same size (White, 1974; Stiles, 1980; Herrera, 1981; Johnson et al., 1985), and among individuals of the same species (Foster and McDiarmid, 1983; Stiles, unpublished). The effect of the variability among species may be the creation of a preference ranking of fruit consumption (Stiles and White, 1986; White and Stiles, unpublished; but see Frost, 1980; Sorensen, 1981, for exceptions). However, there is appreciable difficulty in assigning profitabilities against which preferences may be evaluated. Differential seed-handling abilities, gut passage times, and nutrient extraction abilities in frugivores all contribute to profitability assignments. The influence of within-species variation in profitability of fruit choice by frugivores awaits appropriate testing.

Dispersers

Range of Seed Dispersers

Major Vertebrate Groups

The two major groups of animal seed dispersers are vertebrates and ants. Ants are considered in Chap. 6 and are not considered here. Vertebrates are by far the major animal seed-dispersal agents. Of the classes of vertebrates, seed dispersers include, in order of importance, birds, mammals, bony fish, and reptiles.

Very few reptiles are important seed dispersers. Iguanas *(Iguana* spp.) of Central and South America are primarily herbivorous as adults and consume a variety of fruits and pass the seeds. The more terrestrial turtles including box turtles *(Terrapene)* (Martin et al., 1951; Klimstra and Newsome, 1960), gopher tortoises *(Gopherus)* (Woodbury and Hardy, 1948), and the Galapagos tortoise *(Testudo elephantropus)* (Rick and Bowman, 1961) are seed dispersers. Turtles may pass seeds quickly or may retain them in the gut for long periods [up to 21 days for Galapagos tortoises (Rick and Bowman, 1961) or 70 days for box turtles (Stiles and Offenloch, unpublished)]. Long retention time in part compensates for slow movement, but also increases the likelihood of germination and subsequent death in the gut (Janzen, 1981b, 1981c, 1982a).

Fish become important dispersal agents in areas with extensive, predictable flooding. This has been most extensively studied in the basin of the Amazon River (Gottsberger, 1978; Goulding, 1980; Smith, 1981) where in the Brazilian Amazon alone there are an estimated 70,000 km^2 of annually flooded forest. In large inundated areas, fish, primarily members of the family Characidae (e.g., *Colossoma* and *Brycon*), consume fallen fruits and act primarily as seed predators (see the following) but also as seed dispersers. Surviving seeds germinate after flood waters recede. Goulding lists 26 species of seeds that are dispersed or probably dispersed by members of the Characidae and three families of catfish. The importance of fish as dispersers is as yet poorly resolved.

The importance of mammals as dispersal agents has been recognized and documented by many authors in most parts of the world (Ridley, 1930; Pijl, 1982). Mammals may be functionally divided into four groups: bats, primates, rodents, and others. Frugivorous bats are confined to tropical regions of the world, but their importance in seed dispersal in these areas rivals that of birds. Flying vertebrates have easy access to ripening fruits, minimizing the time ripe fruits remain undispersed and maximizing the movement of seeds away from the canopy of the parent. Even though flying vertebrates can move seeds for long distances, the majority of seeds are not carried far. The leaf-nosed bats in the New World (Phyllostomidae) and fruit-eating bats of the Old World (Pteropidae) move seeds between fruiting trees and roosting locations, often distances not exceeding 200 to 300 m (Janzen et al., 1976; Pijl, 1982). Bats either ingest seeds and pulp, defecating the seeds (e.g., the leaf-nosed bat *Carollia* feeding on *Piper*, or *Artebius* feeding on *Cecropia;* Morrison, 1978; Fleming, 1981; Charles-Dominique, 1986), or they juice fruits by squeezing the fruit against the roof of their mouth with their tongue and spit out the fiberous pulp and seeds (e.g., many pteropids; Pijl, 1957; Ayensu, 1974).

Primates rival bats in their importance as dispersal agents in tropical areas. Both New and Old World monkeys are primarily gregarious. Family or larger-sized groups exploit the ripening fruits of a variety of species (Hladik et al., 1971; Lieberman et al., 1979; Milton, 1981; Leighton and Leighton, 1982; Janson, 1985). Prosimians (e.g., Madagascar lemurs) and some of the lorisids of Africa and Asia are also frugivorous (Charles-Dominique, 1977). Within their home ranges, individuals or groups of primates forage in fruiting trees consuming ripe fruits and subsequently disseminating seeds. Monkeys are among the few mammals with color vision similar to that of humans. Fruit colors of yellow, brown, or orange are exploited by New World primates, while red and black are used primarily by birds (Janson, 1983; Knight and Siegfried, 1983).

Rodents act as dispersal agents for a variety of nonfleshy-fruit seeds.

Their actions are viewed primarily as predatory, although they are important dispersal agents. Rodents will be considered under seed predators.

Other mammals that consume fruit and disperse seeds are primarily terrestrial and often eat large numbers of fruits of a given species at one feeding. In the process, they leave piles of defecated seeds that may create increased competition among plant siblings on germination. Ungulates (e.g., antelopes, elephants, zebras) consume a variety of vegetation and often fruits (Sikes, 1971; Dubost, 1979; Janzen, 1981b, 1981c). Small forest antelope in forested Gabon (especially duikers, *Cephalophus)* are highly frugivorous (Dubost, 1979). The nutritious pods of legumes are excellent food for ungulates. The pods are digested, but a high proportion of seeds pass through the ungulate's gut unharmed (Burtt, 1929; Gwynne, 1969; Pijl, 1982; Janzen, 1981a, 1981b, 1981c, 1982a; Janzen and Martin, 1982). Deposition of seeds in large aggregations may reduce the probability of successful seed establishment unless seeds are subsequently moved by seed-feeding rodents (Janzen, 1982b, 1982c).

Carnivores such as bears, raccoons, foxes, civet cats, some mustelids and some marsupials (especially opossums) are omnivorous and include fruits in their diets to varying degrees depending on availability and hunger levels. Black bears *(Ursus americanus)* consume large quantities of fruits such as blueberries *(Vaccinium)*, huckleberries *(Gaylussacia)*, and other members of the Ericaceae (Martin et al., 1951). Even the pine martin *(Martes americana)* feeds on fleshy fruits of *Aralia nudicaulis* (sarsparilla), *Rubus, Vaccinium,* and other herbs that fruit in late summer in Ontario (Francis and Stephenson, 1972). Large numbers of seeds are defecated in mammal dung, but the fates of these seeds are poorly known (but see Abbott and Quink, 1970; Van der Wall and Balda, 1977; Janzen 1978, 1982b, 1982c; Stapanian and Smith, 1984).

Birds are by far the most important animal dispersal vectors for plant seeds. Major fruit-eating orders of birds include cassowaries and emu (Casuariiformes), pheasants and relatives (Galliformes), pigeons (Columbiformes), turacos (Musophagiformes), colies or mousebirds (Coliiformes), trogons (Trogoniformes), hornbills, motmots, and relatives (Coraciiformes), woodpeckers, toucans, and relatives (Piciformes), and a large number of species of perching birds (Passeriformes). Major frugivore groups in the perching birds include cotingas, manakins, tyrant flycatchers, crows and jays, birds of paradise, bulbuls, mockingbirds, thrushes, Old World flycatchers, starlings, honey-eaters, flowerpeckers, white-eyes, troupials, tanagers, and finches. Many other groups in the Passeriformes have either smaller numbers of species or fewer members that are frugivorous. In short, a great many species of birds eat some fruit at some time in their lives.

Reliance on Fruits as Food

Very few, if any, vertebrates are totally frugivorous. This lack of extreme specialization stems from two characteristics of fruits. First, fruit production and availability is seasonal and, second, fruits may not provide a nutritionally balanced diet for vertebrates. In temperate zones, warm–cold seasonality yields periods when fruits are not common. Even though some fruits remain through winter, their presence is a function of the low nutritional gain received by animals consuming them (Stiles, 1980). Resident frugivores must diversify their diets during winter when only low-quality fruit remains and there is a general reduction in the numbers of available fruits, but they continue to consume what fruits are available.

Waxwings (Bombycillidae) are probably the most specialized frugivores in the temperate region (Berthold, 1976a; Holthuijzen and Adkisson, 1984). The efficiency of their digestion allows them to gain weight on diets of only fruit when other frugivorous species cannot. Waxwings, however, also feed on insects when they are available during spring and summer.

Temperate frugivores that migrate to the tropics use fruit in a predictable fashion. Their greatest reliance on fruits occurs during migratory periods, both when leaving the temperate zone in autumn and when leaving the tropics in spring. The nutrients obtained from these fruits are used for the formation of the subcutaneous fat utilized during migratory flights. Temperate mammals, as well, use fruits to build up fat reserves. Mammals use fat deposits for weathering cold periods with low food availability.

In the tropics, seasonal limitations on fruit production are not as severe and several groups of birds appear to be almost entirely frugivorous (Snow, 1962; McKey, 1975; Foster, 1977, 1978). Not only do the adults of these species feed primarily on fruit, but adults feed their developing young an almost exclusive fruit diet. One of these species, the oilbird (Steatornithidae), lives in caves by day but feeds on a variety of palm (Arecaceae), laurel (Lauraceae), and other fruits by night (Snow, 1962). Some of the cotingas and manakins (Passeriformes) are also almost entirely frugivorous, but the nutritional characteristics of the diets of these species and the efficiency of extraction of fruit nutrients by their digestive systems has yet to be evaluated (Snow, 1962, 1970; Wheelwright, 1983).

There are some disadvantages to frugivory. Young birds fed on exclusive fruit diets develop at slower rates than young of the same species or close relatives fed a varied diet (Kuroda, 1962; Morton, 1973; Berthold, 1976b, 1977). Development rate is important since the nestling period is a time of intense predation. Longer periods in the nest increase the probability of being discovered by a predator. Some temperate species such as American robins and northern mockingbirds also feed some fruit to nestlings (Breitwisch et al., 1984). But for temperate and most tropical frugivores, animal material constitutes a portion of their diet. The protein required by frugivores for tissue construction is found in much higher percentage in animal material. Nitrogen is often a growth-limiting plant resource and would be costly to use as a reward for frugivores.

Frugivores as Foragers

Visual, Olfactory, and Other Cues

Fruit is relatively easy to find. Unlike animals or seeds where crypsis may have selective advantages, conspicuousness is an advantage for fleshy fruits (Ridley, 1930; Pijl, 1982). However, this conspicuousness may carry with it the disadvantage of attracting seed predators (Herrera, 1982a; Grieg-Smith, 1986). Conspicuousness is in the eye, nose, or even ear of the beholder as it is associated with the sensory discrimination of dispersers. Bird- and primate-dispersed fruits are generally bright or contrastingly colored (Wheelwright and Janson, 1985). Bat fruits, although drably colored, often have musty, sour, or rancid odors (Pijl, 1982) and the sound of falling fruit may attract fish or rodents to fruiting trees (Janzen, 1971b; Goulding, 1980).

A number of factors influence the ease of locating fruit crops (e.g., the size of the signal whether created by the size of the fruit or fruit cluster, the presence of a preripening fruit flag or foliar fruit flag). The spatial memories of dispersers play an important role in return visits to a fruiting plant. If all the fruits of an individual plant are not consumed on a single visit, the plant gains an advantage if the disperser returns. During return visits, frugivores select for differential quality of fruits since they return first to the plants providing them with the most profit.

The behavior of frugivores at fruiting plants is determined in part by persistence of the food, and the size and profitability of the patch. Resident birds have been observed to defend fruiting plants (Pratt, 1984; Snow and Snow, 1984) which reduces the loss from a profitable patch. Small patches may not be worth defending or even revisiting, whereas very large patches may not be cost-effective to defend. However, defended individual plants have seldom been observed. One possible explanation for this may be the risk associated with remaining in the relatively stationary, exposed position necessary to defend a fruit resource (Howe, 1979). Further, most frugivores eat a full meal and then leave the fruiting tree. Predation risk from remaining while fruit are being digested may be great enough to favor short visits (Herrera and Jordano, 1981). Longer visits are found only in situations where frugivores are cryptic, such as the green fruit pigeons of Papua New Guinea (Pratt and Stiles, 1983).

Food Value and Fruit Selection

Fruit Size and Position and Disperser Morphology. Larger, soft fruits, such as the aggregate fruits of strawberries and blackberries, may be eaten piecemeal by birds, mammals, and fish that have cutting mouthparts. Effective dispersal is still accomplished because either single-seeded fruits are carried off before the pulp is removed or small seeds are consumed along with the fleshy portions of the larger fruits.

The position of fruit on a plant in relation to available perch sites may also be an important element in fruit selection (Best, 1981; Denslow and Moermond, 1982; Moermond and Denslow, 1983, 1985; Levey et al., 1984). In experiments with captive birds, fruit preference rankings were altered by varying the position of fruits relative to perch sites. For a given fruit type, fruits that were more difficult to reach were ranked lower than those reached more easily. Within and between species, individual fruits that are presented higher on a plant are often taken first (Best, 1981). Morphological characteristics of birds such as bill length, tarsus length, wing length, and hovering ability influence the ease with which fruits are procured and may alter preference ranking both within and between species (Moermond and Denslow, 1985) (Fig. 3-5).

Ripeness and Selectivity. Unripe fruits are appreciably more difficult to digest than ripe fruits and are avoided under most circumstances by frugivores (Ridley, 1930). However, under situations of resource limitation, frugivores may eat unripe fruit. Foster (1977) calculated that manakins eating unripe fruits of *Ardisia revoluta* would need to eat 6.6 times as many unripe fruits to equal the metabolizable energy of ripe fruits. Unripe fruits are characteristically high in phenolic compounds and other toxins which decrease their nutritional value (Janzen, 1977b; Herrera, 1982a). Plants, however, may gain by enticing frugivores to eat fruits before they are completely ripe (Foster, 1977; Stiles, 1982b). The seeds of many fleshy fruits are fully developed before fruits are ripened. Frugivores that consume unripe fruits remove fruit from potential degradation by insects, fungi, or other nondispersers (Thompson and Willson, 1979; Stiles, 1980).

At the opposite end of the ripeness scale, fruits that have been colonized by microorganisms often contain toxin, such as ethanol, which is excreted during the rotting process (Janzen, 1977b). Birds consuming too many alcohol-filled fruits become easy prey for carnivores or automobiles (Janzen, 1979a). Choice of ripe fruits is therefore a critical decision. Frugivores apparently use the dramatic changes in fruit color, softness, and chemistry at ripening as cues (McDonnell et al., 1984).

Fruit Mimicry. Bright red, or red and black seeds with no fleshy parts are common, especially in the Leguminosae (Ridley, 1930). Studies by McKey (1975) and Pijl (1982) have shown that these seeds are mistaken by frugivores for fleshy fruits or arilate seeds, and ingested. The hard seeds pass intact with birds acting as dispersal agents. McKey demonstrated that the ranges of those seeds looking like berries and those that appear like arilate seeds (i.e., part red, part black) approached the geographic ranges of fruits of those types. This supports the hypothesis that these seeds are mimics of locally available fleshy fruit models.

These mimic seeds apparently have no nutritional value. However, fruit mimicry need not be restricted to nonnutritional seeds mimicking fleshy fruits. Fruits that are high in nutritional quality or preference level may act as models for low-quality mimic fruits that gain elevated preference and dispersal (Howe, 1980; Manzur and Courtney, 1984; Stiles and DeVito, unpublished).

Seed Shadows

Seed Handling

Not all seeds handled by frugivores arrive at a germination site alive. Frugivores test fruits for ripeness, often rejecting unripe or rotting fruits, and fruits may be dropped intact under the parent. Seed coats may be broken by teeth or beaks, seeds may be ground up in the gizzard, or seeds may germinate in the intestine (Krefting and Roe, 1949; Janzen, 1971b, 1981a, 1981b, 1982a, 1983a, 1983b). The guts of highly frugivorous birds are not highly muscular and are relatively short (McKey, 1975). These animals treat seeds gently and reduce the probability of internal germination by short passage times (Janzen, 1983a). However, some frugivorous birds have muscular gizzards (e.g., dodo and many galliform birds) and turtles have very long gut passage times which expose seeds to false germination cues (Rick and Bowman, 1961).

Seeds may be deposited individually, as is typically the case with regurgitated seeds, or they may end up in a dung pile with hundreds of other seeds. Seed deposition with dung may either discourage predators or serve as a visual or olfactory cue to seed predators, depending on conditions.

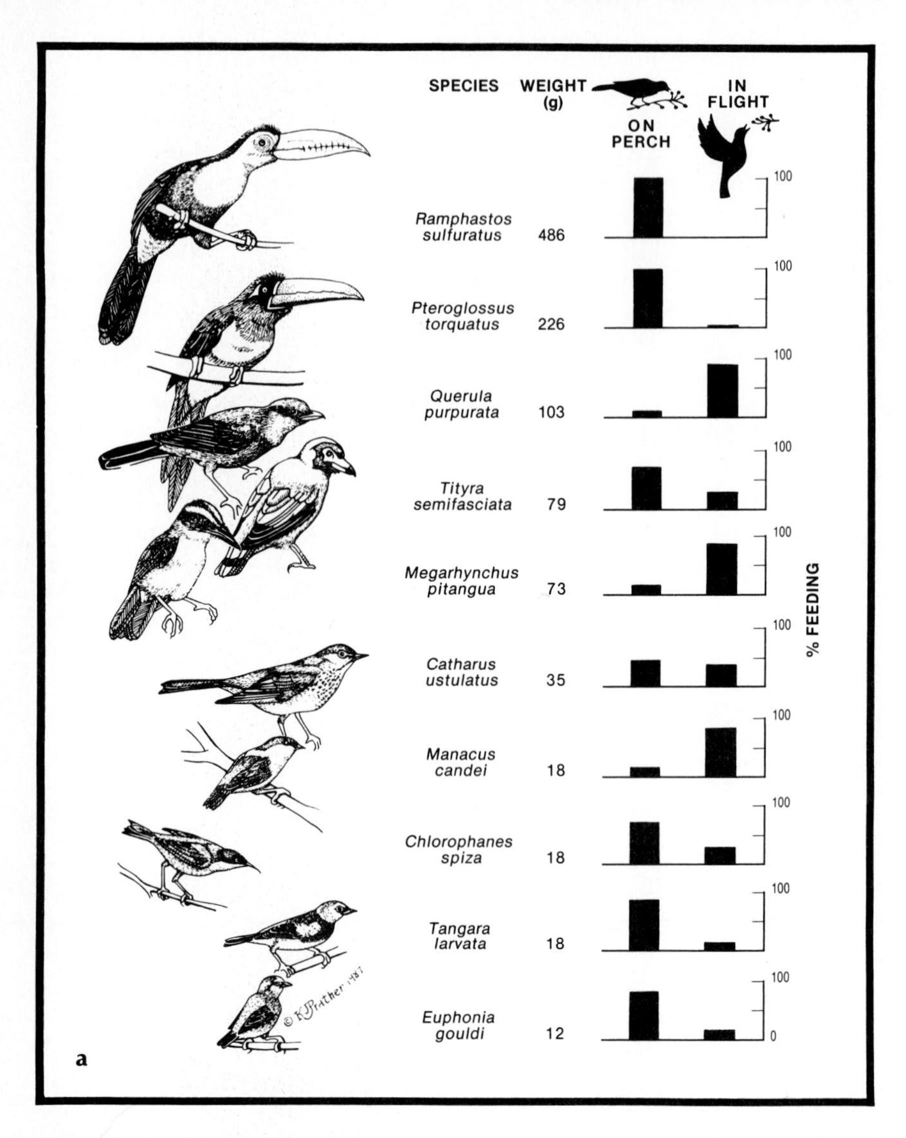

Figure 3–5. (a) Feeding behavior of the most common visitors to *Miconia multispicata* (Melastomataceae). Histograms show percentages of total fruit removals. (b) Distances of maxium reach in three directions from a sturdy perch for three small frugivores. *Manacus candei,* an aerial feeder, can reach only as far below the perch as the much smaller *Euphonia gouldi,* a perching feeder. Larger species have even more pronounced differences.

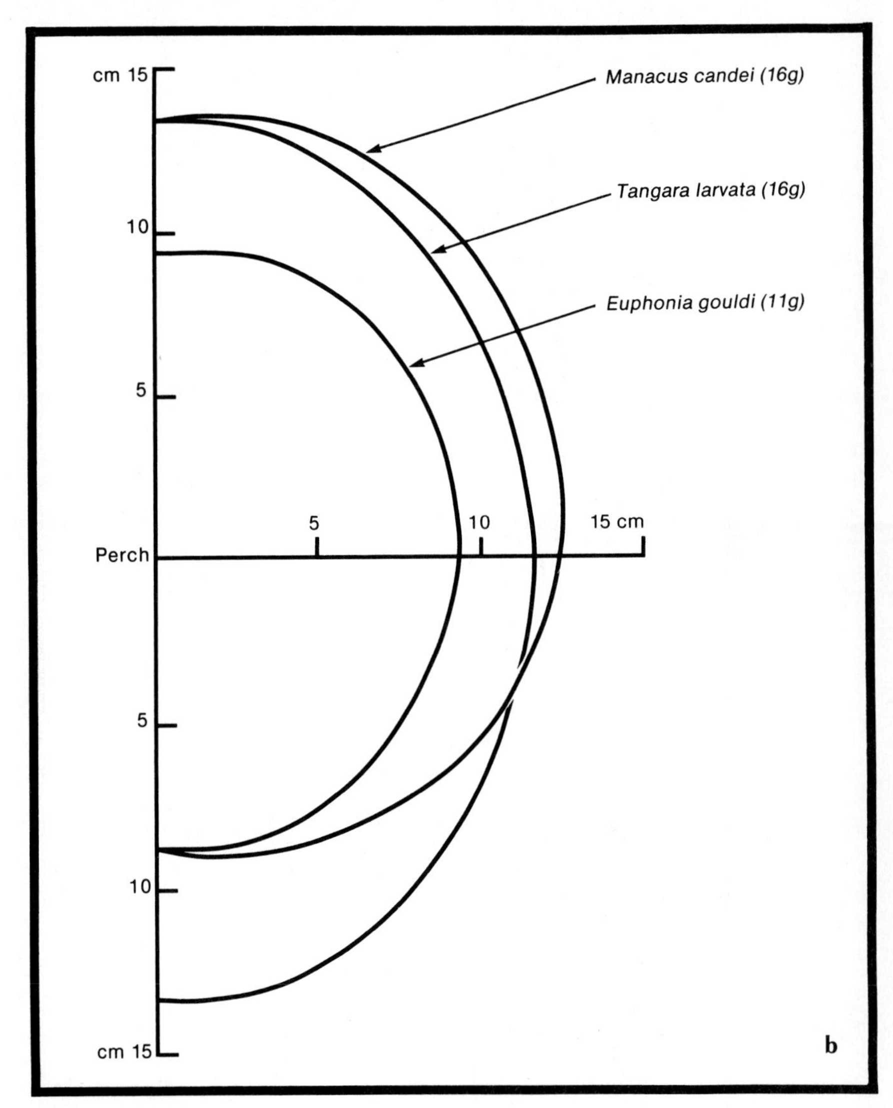

Figure 3–5. *Continued.*

Animal Movement Patterns

The length of time frugivores spend in fruiting plants influences the proportion of seeds deposited under the parent. Most frugivores spend only short periods in a fruiting tree (Leck, 1971; Herrera and Jordano, 1981; Pratt and Stiles, 1983), but frugivore movement back to the same plant, if it occurs, would increase the deposition of conspecific seeds under the tree's canopy. Information on seed contents of bird stomachs and bird droppings suggests that birds do not forage on the same individual for long periods (Herrera, personal communication; Stiles and

White, 1986), but instead feed on an array of different species of fruits over short periods. The seed contents in bird guts or in droppings represents a range of 20 minutes to several hours of feeding. Analysis of seed contents usually finds one to five species represented, with an average of more than two for frugivorous birds from Spain and eastern North America.

Seeds dispersed by animals do not follow the simple decay curves of wind-disseminated seeds. Both the direction of seed movement and the local density of deposition are influenced by the behavior of the frugivore. At the simplest level, local seed deposition density is determined by the direction and speed of movement of the dispersing animal and the speed that the frugivore processes fruits. For example, birds regurgitate seeds more quickly than they pass them through their digestive tract (Sorensen, 1984). Further, different bird species and different individuals of the same species process fruits at different rates. Although most seeds are regurgitated or defecated within a few hours, some may be retained in the gut for long periods (e.g., 340 hours for *Rhus glabra* in a killdeer; Proctor, 1968). The structure of the habitat also influences the patterns of seed deposition just as habitat structure influences animal movement (see Chap. 7; DeAngelis et al., 1977; Wegner and Merriam, 1979; Forman and Godron, 1986). Hedgerows and forest edges, for example, receive greater seed input than the surrounding fields, and elevated perches within fields act as recruitment foci for seeds (McDonnell and Stiles, 1983) (Fig. 3-6). Fruiting plants serve as foci attracting frugivores that have been feeding on the fruits of other species. This can yield contagious (i.e., clumped) distributions of species with similar preference ranking (Stiles and White, 1986; Stiles, unpublished). For example, roosting sites of both bats (Fleming and Heithaus, 1981) and birds (Snow, 1962) receive large numbers of seeds. In forests, the vertical position of perch sites relative to the height of feeding sites may influence where birds land and where seeds are deposited (Stiles, unpublished).

It is difficult to determine whether there is selection created by frugivores for directed dispersal (i.e., movement of seeds to specific sites favorable for germination). Plants attract different groups of dispersers through evolutionary manipulations of timing of fruiting and characteristics of the fruits. True reciprocal coevolution (Janzen, 1980a) between frugivores and fruits is unlikely, first because rewards are obtained by frugivores when the fruit is eaten and there are no specific rewards associated with movement to a specific seed deposition site (Wheelwright and Orians, 1982), and second, the pairs of co-occurring fruits and frugivores change over time (Herrera, 1985b). For example, at any given point in time and space, frugivores favor certain fruit traits while fruits select for certain frugivore traits. The extent of evolutionary

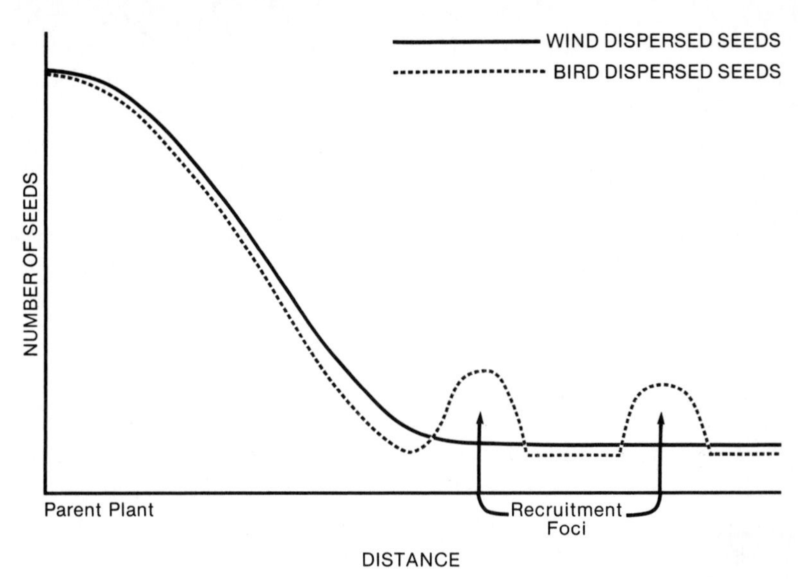

Figure 3–6. Animal dispersal creates patches of seed deposition (recruitment foci) at centers of animal activity.

change realized depends on the intensity of selective pressures, the time over which the association exists, and the available genetic variance. These evolutionary responses could alter the frugivores attracted simply by changing the ripening time or fruit chemistry. Alternatively, the responses could change the gut passage rates which would influence where seeds are deposited. Given all the interacting variables, true reciprocal coevolution may not be expected; however, diffuse coevolution may occur (see Chaps. 1 and 2).

Seed Predators

Types of Seed Predators

Seed predators are found among a diverse group of animals including insects, fish, birds, and mammals. Common insect seed predators include many beetles (Coleoptera) and especially weevils, ants (Hymenoptera, see Chap. 6), moths (Lepidoptera), true bugs (Hemiptera), and flies (Diptera). Adults of many insect seed predators (especially beetles, moths, and flies) lay their eggs on developing seeds. The insect larvae develop within the seed using the seed's stored nutrients for food. Others probe or chew into developing or mature seeds destroying the embryo and the seed's stored reserves.

Fish seed predators are from the same groups (characins and catfish) as fish seed dispersers. The strong jaws and crushing molariform teeth of these species are destructive to softer seeds. Goulding (1980) lists 6 species of plants in the Euphorbiaceae (spurges) and 14 other species from nine families whose seeds are consumed and destroyed by Amazonian fish.

Bird seed predators come primarily from the tinamous (Tinamiformes), pigeons (Columbiformes), grouse (Galliformes), parrots (Psittaciformes), woodpeckers (Piciformes), and several groups within the Passeriformes (especially finches, crows, jays, and weaver finches). Birds either shuck the hard seed coats with their strong beaks or grind seeds in their muscular gizzards.

Mammalian seed predators include some shrews (Insectivora), a large variety of rodents (Rodentia), and even- and odd-toed ungulates (Artiodactyla and Perissodactyla). Seeds are chewed, broken open with sharp teeth, ground between heavy molars, or digested during long passage through the digestive tract.

Seed Predators as Dispersers

The relationship between seed predation and seed dispersal must be evaluated carefully for each plant–animal association. For example, insect larvae developing in seeds are not dispersers but they may not be predators either. Insect larvae may remove part of the seed's endosperm without damaging the embryo; only larger numbers of larvae actually kill the seed (Janzen, 1971b, 1975; Leroi and Jarry, 1981). In still other cases a seed predator may be beneficial to the plant, as with yuccas and their pollinator, the yucca moth (see Chap. 2).

Birds that deposit seeds under the parent plant (Howe and Primack, 1975) or those that eat only the fleshy fruit parts leaving the seeds (called fruit thieves) reduce the fitness of the fruiting plant. Other birds that crush seeds with their bills or gizzards may still pass a proportion of seeds through their digestive tract unharmed. These seeds may be effectively dispersed. For some frugivores, virtually all seeds are viable after gut passage, while for other frugivores, none are. In a study on California linnets (house finches), Roessler (1936) found that only 7 of 40,025 seeds of 20 plant species successfully passed through the gut and germinated.

Mammals also vary in their effectiveness in dispersal versus predation. A large proportion (44 to 83 percent) of Guanacaste tree *(Enterolobium cyclocarpum)* seeds that passed through the guts of horses were destroyed and a smaller but substantial proportion (14 to 21 percent) were destroyed during passage through range cattle (Janzen, 1982a). Mammals such as opossums and civets may drop partially eaten fruits beneath the parent tree where they are subject to rotting or competition with their parent.

Caching and scatter-hoarding birds or mammals use seeds as the food rewards. Caching species, often called larder hoarders, store large groups of seeds together, often in locations where germination is unlikely. This behavior is found in wood rats *(Neotoma)*, kangaroo rats *(Dipodomys)*, and red squirrels *(Tamiasciurus)*. Scatter hoarders store seeds singly or in small groups by burying them very close to the surface. This may in fact enhance germination potential. Scatter hoarders include squirrels *(Sciurus)*, agoutis *(Dasyprocta)*, jays *(Cyanocitta* and *Perisoreus)*, and nutcrackers *(Nucifraga)*. Scatter-hoarding may be more effective in dispersal than larder hoarding, but even scatter hoarders are extremely adept at relocating and consuming hidden seeds. Cahalane (1942) found that 415 of 419 hickory nuts *(Carya)* buried by gray and fox squirrels were recovered and eaten during the following winter. Although few studies have been done, agoutis, mice, and kangaroo rats recover and eat the vast majority (usually more than 85 percent) of the seeds they bury (Abbott and Quink, 1970; Janzen, 1971b). However, few seeds in any dispersal system survive, and the loss of seeds to larder hoarders and scatter hoarders may be viewed as a cost of dispersal similar to the fleshy portion of fruits.

Seed Availability for Predators

Timing of Seed Ripening

Unlike fleshy fruits, selection pressures on seeds yield adaptations that reduce consumption by animals. As discussed above, the morphology and chemistry of seeds may diminish predation. Phenology (timing) of seed availability may also influence seed survival (Janzen, 1971b) since selective pressures from seed predators can alter the timing of seed ripening or seed dissemination. Chinese bamboo *(Phyllostachys bambusoides)* flowers and produces seeds once every 120 years. Janzen (1976) has hypothesized that this extreme reproductive cycle and similar (but less extreme) cycles in other species of bamboo are adaptations to increase seed survival through predator satiation (see the following).

Seed predation may occur either before seeds are dispersed (predispersal predation) or after seeds are dispersed (postdispersal predation) (Janzen, 1969a, 1971b). Predispersal predation for animal-dispersed seeds may be further divided into that occurring when the fruit is still attached to the parent and that occurring after the fruit falls from the parent but prior to dispersal by an animal. While the result of predation is seed death regardless of when it occurs, the adaptive responses to selective pressures are different.

Many seeds are attacked by predators while still on the plant. The percentage of seed crops destroyed at this time range from less than 1 to as high as 100 percent, but usual values range between 10 and 90 percent (Janzen, 1971b). The soft tissues of developing fruits are particularly vulnerable, since early colonization of seeds provides insects with a protected food source during their development. Insect damage on green fruits may allow colonization by microorganisms, especially fungi. While fruits are still attached to the parent, the plant may save energy by selectively aborting damaged fruits (Stephenson, 1981b). This is especially useful with immature fruit because little resource is lost through abortion. While abortion produces a reduction in fruit crop size, the saved resources can be stored for future reproduction. Some fruits delay development of their seeds (e.g., *Pinus* and *Acer*) for long periods following pollination. When pollination is separated from dispersal by an interval greater than that required for seed development, delayed seed development will reduce the period of vulnerability to predispersal seed predators.

Seed dispersal for wind- or water-carried seeds occurs upon separation from the parent. However, for seeds dispersed by terrestrial animals, fallen seeds are subject to additional predispersal predation. Since the parent plant can no longer recover resources through selective abortion, selective pressures, resulting in increased parental fitness from seed dispersal to a safe site for germination and establishment, act indirectly through survival of different genotypes.

Seed predators may be divided into two conceptual categories: distance responsive and density responsive (Janzen, 1970, 1971b). Distance-response predators identify the parent plant as the focus for feeding or resting and forage radially from this central location (Covich, 1976; Orians and Pearson, 1979). Density-responsive predators select a foraging location based on prey (seed) encounter rate regardless of parent plant location (Janzen, 1971b; Pyke et al., 1977). The constraints on distance-responsive predators are primarily time and possible risk of moving away from the central plant. Thus, selection for greater dispersal distance should be favored in plants attacked by distance-responsive seed predators. However, seed mortality due to density-responsive seed predators should favor plants that generate lower densities of seeds regardless of distance. Favored plants should have smaller seed crops, adaptations to attract dispersers that broadly disseminate seeds, or dispersal over a long time interval to reduce densities at any location in time and space (Janzen, 1970; Stiles, unpublished data) (Fig. 3-7).

Another method of escape involves predator satiation (Janzen, 1974a, 1976; Silvertown, 1980). Seed predators probably suffer high mortality in years when seed crops are small or absent. By producing large numbers of seeds at irregular intervals (mast fruiting), a plant can

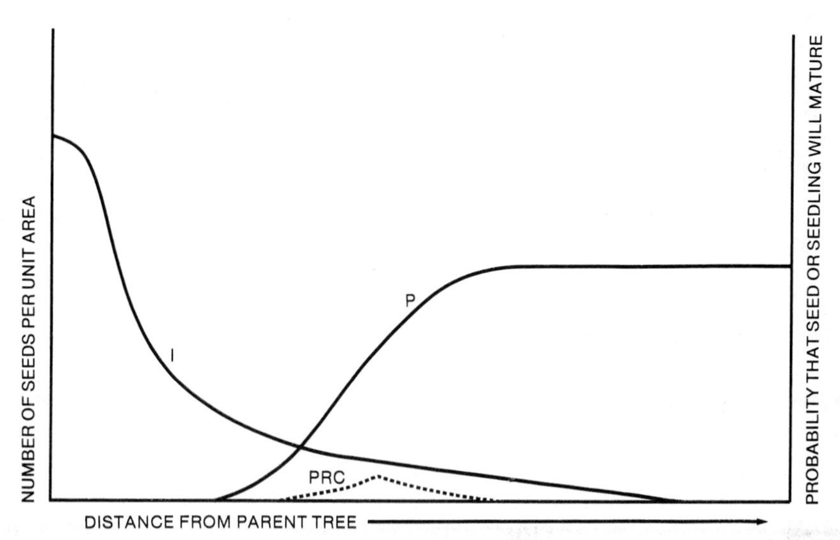

Figure 3–7. A model showing the probability of maturation of a seed or seedling at a point as a function of (1) seed crop size, (2) type of dispersal agents, (3) distance from parent tree, and (4) the activity of seed and seedling predators. With increasing distance from the parent, the number of seeds per unit area (I) declines rapidly, but the probability (P) that a dispersed seed or seedling will be missed by the host-specific seed and seedling predators, before maturing, increases. The product of the I and P curves yields a population recruitment curve (PRC) with a peak at the distance from the parent where a new adult is most likely to appear; the area under this curve represents the likelihood that the adult will reproduce at all, when summed over all seed crops in the life of the adult tree. In most habitats, P will never approach 1, due to a nonspecific predation and competition by other plants independent of distance from the parent. The curves in this figure are not precise quantifications of empirical observations or theoretical considerations, but are intended to illustrate general relationships only.

increase the number of seeds that escape predation. Seed predator populations that have survived years of low seed availability are unlikely to produce offspring fast enough to consume all seeds in a mast year. Individual plants or species that are out of phase with the mast year encounter many more predators per seed produced and incur heavy seed mortality. This mortality selects for fruiting synchrony among individuals and closely related species. Masting species are most likely to be found where large numbers of individuals of the same or closely related species are found in the same location (e.g., the temperate zone or low-nutrient tropical areas) (Janzen, 1974a; Silvertown, 1980). In these areas cold temperatures or low productivity depresses seed predators during some portion of the year. Plants with large, synchronous, unpredictable seed crops have the greatest possibility of satiating the predator population. Masting may be an important alternative under the conflicting pressures of increasing the number of surviving seeds and avoiding density- or distance-responsive predators in low-diversity plant communities.

Safe Sites

The successful transitions from seed to seedling and from seedling to reproductive adult are related to where the seed is deposited in the environment. Physical factors, competition, and predation play roles in the successful establishment of the next generation. But the degree to which the interaction of seed dispersers and seed predators with fruits influences arrival and persistence at safe sites is poorly known. Seed shadows generated by wind, water, and animals are highly variable from plant to plant and year to year. Seed shadows depend on seed crop size, variation in weather conditions, annual variability of the density of dispersal agents and predators in addition to a variety of community effects. Understanding of the relationship between a particular seed shadow and subsequent plant success will offer appreciable insights into the importance of seed dispersers and seed predators.

Selected References

Herrera, C. M. 1982b. Seasonal variation in the quality of fruits and diffuse coevolution between plants and avian dispersers. *Ecology* 63:773–785.

Howe, H. F., and J. Smallwood. 1982. Ecology of seed dispersal. *Annu. Rev. Ecol. Syst.* 13:201–228.

Janzen, D. H. 1971b. Seed predation by animals. *Annu. Rev. Ecol. Syst.* 2:465–492.

Janzen, D. H. 1983a. Dispersal of seeds by vertebrate guts. In: D. J. Futuyma, and M. Slatkin, eds. Coevolution. Sinauer Associateds, Sunderland, MA.

Pijl, L. van der. 1982. Principles of dispersal in higher plants, 3rd ed. Springer-Verlag, NY.

Ridley, H. N. 1930. The dispersal of plants throughout the world. L. Reeve, Ashford, Kent, UK.

Stiles, E. W. 1980. Patterns of fruit presentation and seed dispersal in bird-disseminated woody plants in the eastern deciduous forest. *Am. Natur.* 116:670–688.

Stiles, E. W. 1982b. Fruit flags: two hypotheses. *Am. Natur.* 120:500–509.

Herbivorous Insects and Green Plants

ARTHUR E. WEIS

Department of Biological Sciences
Northern Illinois University
DeKalb, IL 60115

MAY R. BERENBAUM

Department of Entomology
University of Illinois
Urbana, IL 61801

Introduction

Wherever terrestrial plants are found, insects are there, too. The success of the adaptive radiation of insects and other terrestrial arthropods into plant-feeding niches is a matter of record (Strong et al., 1984a). As shown in Fig. 4-1, there are approximately 308,000 species of terrestrial plants which constitute 21 percent of all species. Herbivorous insects account for more than 361,000 species or 26 percent of all species. Most

We would like to thank Cathy Crego, Ellen Heinninger, Jonathan Neal, James Nitao, Sherri Sandberg, and the Plant–Animal Interactions class of Northern Illinois University for their helpful comments and criticisms of earlier drafts of this chapter.

of these are distributed among 8 of the 29 insect orders: the Orthoptera (grasshoppers), Phasmida (stick insects), Thysanoptera (thrips), Hemiptera (true bugs), Coleoptera (beetles), Diptera (flies), Lepidoptera (moths and butterflies), and Hymenoptera (sawflies, ants, bees, and wasps) (see Fig. 4-2). In addition to insects, the mites and Collembola, or springtails, are also important phytophagous terrestrial arthropods.

The degree of specialization for herbivory varies among insect orders (Fig. 4-2). Some, such as the Lepidoptera and Phasmida, are essentially all plant feeders. In others, entire suborders are phytophagous, such as the Homoptera (cicadas, leafhoppers, and aphids) in the Hemiptera, or the Symphyta (sawflies) in the Hymenoptera. Specialization also occurs at lower taxonomic levels, such as the family. For example, all members of the Agromyzidae (leaf-mining flies) and Tephritidae (true fruit flies) in the Diptera are plant feeders.

Specificity to one or a few closely related food plants is a characteristic of many insect species. For instance, the larvae of the monarch

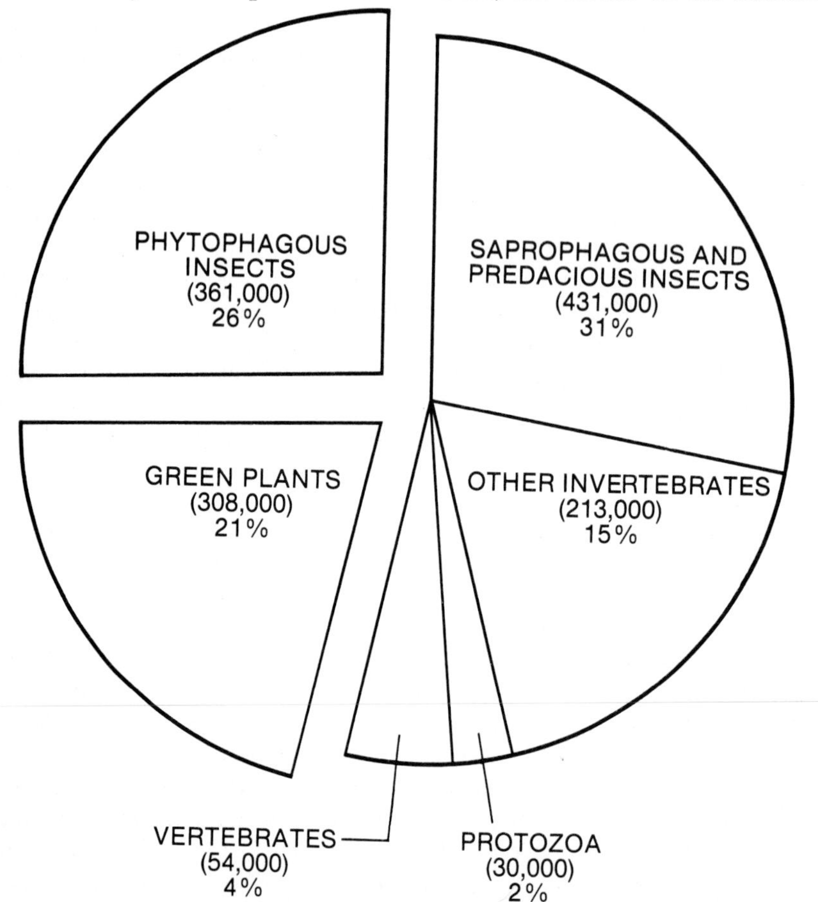

Figure 4–1. The number of plant and phytophagous insect species, as a proportion of all species (except algae, fungi, and microbes, modified from Strong et al., 1984a).

butterfly *(Danaus plexippus)* feed only on milkweeds, members of the plant family Asclepiadaceae. Of the North American butterfly species, 81 percent have diets restricted to plants belonging to one family (Futuyma, 1976). Insects with restricted plant preferences are termed monophagous. Other insects, such as many Orthopera, are polyphagous (i.e., they eat a taxonomically diverse array of plants). However, some so-called polyphagous insect species consist of a number of subpopulations that are locally monophagous (Fox and Morrow, 1981). The tiger swallowtail butterfly *(Papillio glaucus)* is reported to feed as a caterpillar on at least 13 plant families throughout its range. However, in any one geographic area, this species is restricted to only a subset of those plants (Scriber, 1984).

The Evolutionary Diversification of Plant and Insects

The association between plants and insects goes back to the time of their evolutionary origins. Insects are thought to have first colonized dry land approximately 320 to 280 million years BP during the Carboniferous. By this time, the early plants had already become established in terrestrial environments. The earliest known fossils of winged insects, representing a now extinct order, show structures for feeding on plant tissues (Wooten, 1981). The evolutionary pathway to herbivory may have started with feeding on pollen and spores that fell to the ground (Malyshev, 1968). The next step may have been feeding on spores and pollen at their source, plant reproductive structures. From this point, adaptation to eat other plant parts could have followed. Several insect groups that arose during the Permian (280 to 240 million years BP) have many surviving species that are still specialists on so-called primitive plants. For instance, many sawflies, members of the primitive suborder Symphyta of the Hymenoptera, feed on ferns and assorted gymnosperms (Smith, 1979).

One of the most important events in the evolution of the insects was the origin of the angiosperms, plants that produce seeds enclosed in fruits. These plants evolved during the Cretaceous [100 million years BP (Doyle, 1978)], as the dinosaurs reached their zenith. Entirely new phytophagous insect families made their first appearance at this time, including the aphids (Hemiptera: Aphididae) and the gall wasps (Cynipidae: Hymenoptera). In addition, new orders arose that are almost exclusively phytophagous, such as the Lepidoptera (Powell, 1980). In this order, both adults and larvae rely on plants for sustenance. Moths and butterflies collect nectar and pollen, and caterpillars are known to feed on almost any plant part. The evolutionary diversification of the angiosperms was mirrored by the diversification of the moths and butterflies (Ehrlich and Raven, 1964).

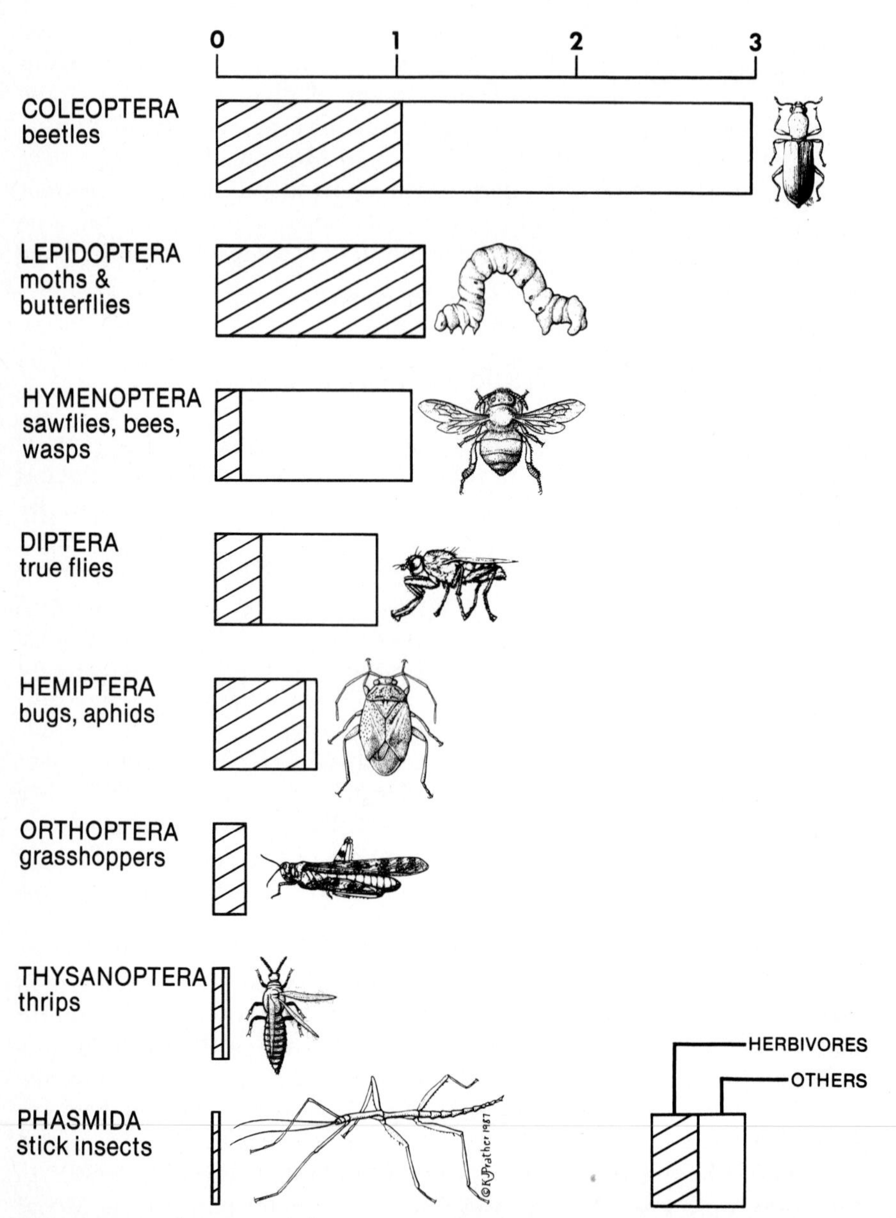

Figure 4–2. The size of the insect orders containing significant numbers of herbivores. The size of the order taken from Borrer et al. (1976) and Richard and Davies (1977). Proportion of herbivores in each order taken from Price (1977) except for Thysanoptera, which was estimated from Lewis (1973).

Nutritional Considerations

How to Eat a Plant: Insect Feeding Guilds

Most mature plants are orders of magnitude larger than most mature insects. This disparity in size makes it possible for many individuals of an insect species to feed on a single plant. Another consequence of large plant size relative to insect size is that insect species can evolve to feed on a single plant structure in a particular fashion (Strong et al., 1984a). Thus, an oak tree may have some caterpillars eating entire leaves, other caterpillars mining through leaf mesophyll, aphids and leaf hoppers sucking sap from twigs and petioles, cynipid wasps inducing and then eating galls, and weevils chewing out the embryo in the developing acorns.

The complexity of "plant architecture" is one determinant of the number of insect species that can feed on a plant species. Forbs, with the simplest type of architecture, support fewer insect species than shrubs, which in turn support fewer insects than trees (Lawton and Schroder, 1977). The greater architectural complexity of trees, in terms of the variety of structures and their persistence, make a greater number of feeding niches available (Strong et al., 1984a). The feeding styles of insects can be classified into several guilds, each with its own specializations.

Chewers

Most phytophagous insects process their food by chewing. Members of the orders Coleoptera, Hymenoptera, Orthoptera, and Phasmida have mandibulate, or chewing mouthparts, in both the juvenile and adult stages. Chewing mouthparts are present in larval lepidopterans and also in larval dipterans, in modified form.

Chewing mouthparts are ancestral in the insects and are derived from three pairs of segmental appendages. The segments bearing the mouthparts have fused with the ancestral head during the course of evolution to form a head capsule. The mandibles are the most anterior pair and do the actual chewing. Equipped with toothlike ridges (Fig. 4-3), the mandibles are constructed of highly sclerotized cuticle. Behind the mandibles are the maxillae, which manipulate the food as it is being chewed. The labium is derived from a pair of fused maxillalike appendages and closes the posterior of the oral cavity. Both the maxillae and the labium have antenna-like structures known as palps, which contain many taste receptors.

As strong as the mandibles may be, they are subject to wear and are replaced at each molt. Recently, Bernays (1986) discovered that the mandibles and the head muscles that control them can grow to accommodate to feeding on tough plant tissues. She demonstrated this phenomenon by rearing the armyworm *(Pseudaletia unipunctata)* on an agar-based artificial diet (very soft), grass seedlings (moderately soft), and mature grasses (hard).

Head-capsule size in mature larvae increased with increasing hardness of the food. These size differences are not simply due to the nutritional quality of the food because the head size trend remained after adjustment for overall body size.

The insect digestive tract consists of a foregut, midgut, and hindgut (Fig. 4-3). The foregut is a holding area for ingested food, where salivary enzymes swallowed with the food begin the digestive process. The midgut is the principal site of digestion and absorption, while the hindgut resorbs water from the feces (Wigglesworth, 1972). Because the indigestible cellulose and lignins of the cell walls account for a great part of plant tissue, herbivores eat a diet of low nutrient density. To compensate, the digestive tracts of chewing insects occupy a large proportion of the body cavity to handle their bulky diet.

Miners and Borers

A variety of insects are endophytophagous, that is, they are internal plant parasites. Miners and borers, a subset of chewing insects, have special morphological modifications and produce distinctive damage. Species that live in plant tissue are found in the orders Lepidoptera, Hymenoptera, Diptera, and Coleoptera. However, with the exception of some beetles, endophytophagy is restricted to the larval stage. Leaves, stems, roots, and fruits all harbor mining or boring insects.

Leaf miners feed in the leaf lamina without breaking the upper or lower leaf cuticle (Herring, 1951), and frequently only one of the leaf tissue layers is eaten. As the larva makes its way through the leaf, it produces a characteristic mine (Fig. 4-4). In some cases the morphology

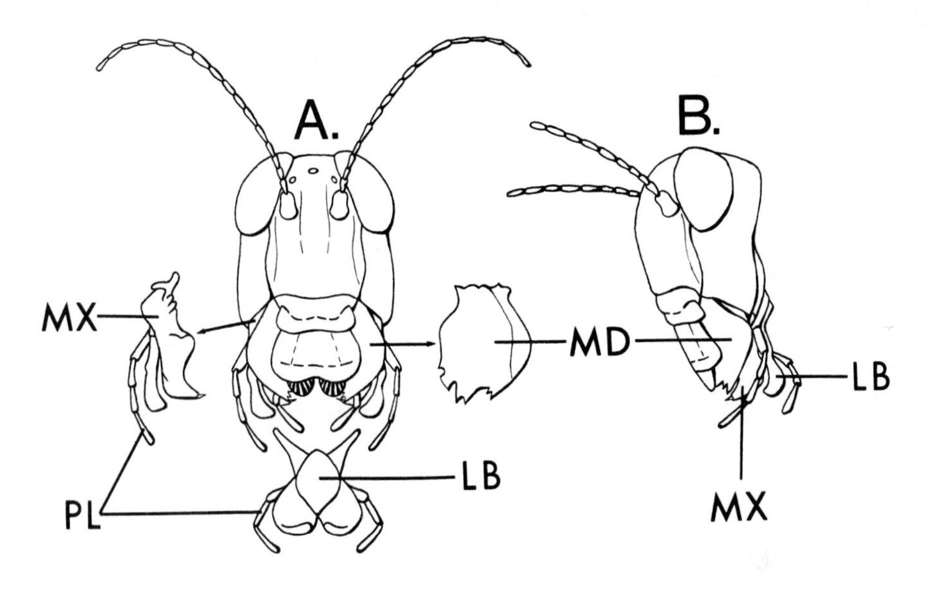

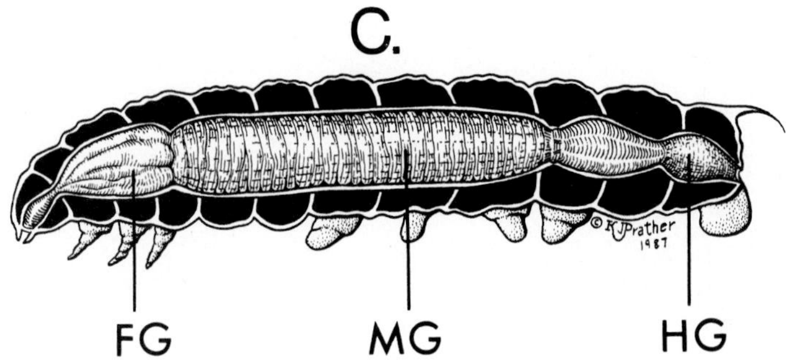

Figure 4–3. Mouthparts and digestive tract of typical chewing insect herbivores. (A) and (B) mouthparts of a typical grasshopper; MD, mandible; MX, maxilla; LB, labium, PL, palps (modified from Ross, 1965). (C) Digestive tract of a typical caterpillar; FG, foregut; MG, midgut, HG, hindgut (modified from Snodgrass, 1935).

of the mine can be reliably used to identify the insect that made it. For instance, some agromyzid flies produce distinctive spiral mines early in development that widen into serpentine tunnels.

The morphology of leaf miners reflects the constraints of living in tight spaces; most miners are small relative to external feeders. Blotch-

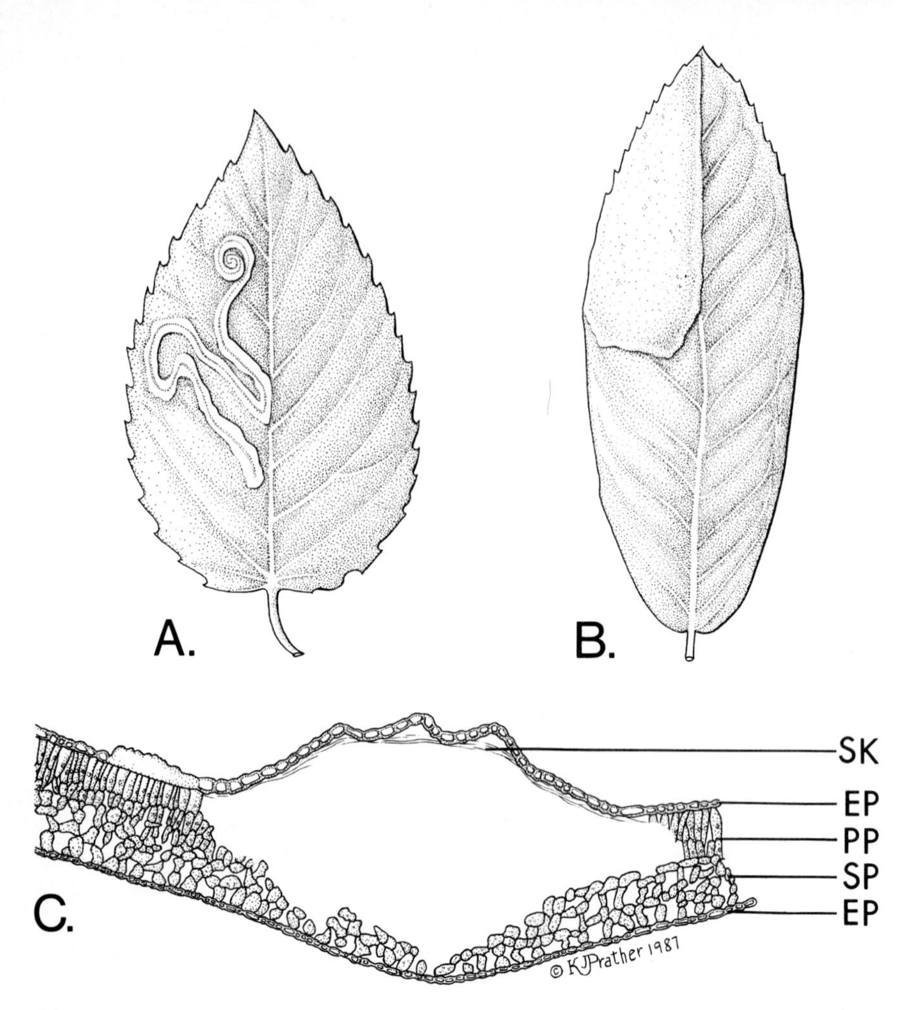

Figure 4–4. (A) Leaf mine of the dipteran *Liriomyza eupatorii* on *Galeopsis*. The newly hatched larva started the mine at the center of the spiral. As the larva grows, so does the width of the mine. (B) Blotch-shaped mine of the sawfly *Phyllotoma ochropoda* on willow. (C) Cross section through the mine of *Lithocolletis* on aspen; EP, leaf epidermis; PP, palisade parenchyma; SP, spongy parenchyma; SK, insert silk (A-C after Herring, 1951).

mining caterpillars in the genus *Phyllonorcytes* are restricted to the spongy mesophyll layer during the early larval instars, and at these stages their body is dorsoventrally compressed. Older instars expand their feeding preferences to include the palisade layer, which makes for more room, and the body becomes rounder (Miller, 1973). In external feeders, mouthparts typically point ventrally, but in leaf miners they more often point forward to meet the feeding surface.

Mining can hasten leaf senescence (Faeth et al., 1981), but some species extend the nutritional life of the leaf by secreting cytokinins, plant growth hormones. These physiologically active compounds maintain a small "green island" of food around the mine even though the rest of the leaf may turn brown (Englebrecht et al., 1969; Kahn and Cornell, 1983).

Plant parts other than leaves are also susceptible to boring insects. Technically, many of these borers are not herbivorous in the strictest sense because they do not eat living plant tissue but instead subsist on dead pith and heartwood. Those that are true herbivores can devastate their host plant by feeding on cambium and active vascular tissue. Bark beetles in the family Scolytidae excavate galleries in the inner bark of many conifer trees. Infestation of pine and other lumber species by these beetles causes substantial economic loss every year.

Sucking Insects

An alternative to ingesting plant tissue per se is to suck the liquid contents from the plant, leaving the indigestible cellulose and lignin in place. This method is used by all herbivorous Hemiptera and Thysanoptera. Feeding on plant liquids is made possible in these orders through radical departures from typical mouthpart structure. In the thrips, for example, the maxillae and the right mandible are modified into thin, sharp stylets for piercing plant cells (the left mandible is vestigial). In the Hemiptera, phylogenetic relatives of the thrips, the modification is taken even further to form a beak. The maxillary and mandibular stylets form an integrated, tubular bundle that is used to inject salivary secretions into the plant as well as to imbibe plant liquids. A tongue-and-groove mechanism interlocks the two maxillae to form a double-barreled tube with salivary and feeding channels (Fig. 4-5). After the stylets are inserted into the plant, a muscular pump in the digestive tract draws fluid into the gut (Wigglesworth, 1972).

Sucking insects survive on several types of plant fluids. Thrips feed on the liquids they extract from individual plant cells: the cell is penetrated and macerated with the stylets, and then the liquid is sucked up through the proboscis (Lewis, 1973). The leafhopper *Empoasca bifurcata* tears through the leaf's palisade layer and thrusts its beak into cells of the spongy mesophyll. The empty cells it leaves behind appear as white spots on the leaf's upper surface (DeLong, 1971). Other hemipterans, such as the meadow spittle bug *(Philaenus spumarius)*, tap the xylem vessels for food (Wiegert, 1964). Most commonly, however,

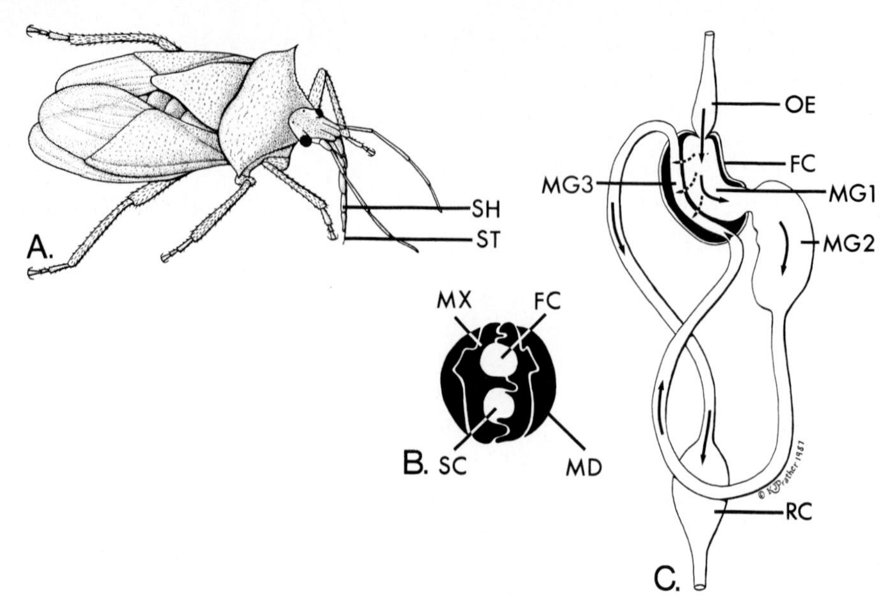

Figure 4–5. Mouthparts and digestive tract of typical sucking insects. (A) Stink bug, showing location of the beak; ST, mandibular and maxillary stylets in bundle; SH, sheath, derived from labiums surrounding bundle (from Ross, 1965). (B) Stylet bundle in cross section; MD, mandible; MX, maxilla; FC, food channel; SC, salivary channel. (C) Hemipteran digestive tract. Incoming plant fluid enters through the oesophagous (OE) into the anterior midgut (MGI). Nutrients flow to the absorptive area of the midgut (MG2) but water is passed through the filter chamber (FC) to the posterior midgut (MG3) for excretion through the rectum (RC). The filter chamber thus concentrates nutrients for efficient absorption and reduces retention time of ingested water (after Snodgrass, 1935).

hemipterans feed on the richer phloem sap (DeLong, 1971) by inserting their beak directly into a sieve tube. Penetration is aided by salivary secretions, which contain not only a lubricant but also enzymes which break down the cementlike substance which binds cell layers together. When the phloem is reached, additional salivary secretions solidify into a lipoprotein sheath around the beak to line the puncture and form a sealed conduit for sap flow directly to the digestive tract. For some phloem feeders, hydrostatic pressure in the plant is sufficient to force the sap to rise up the beak.

Plant sap is a nutritionally dilute food; although it is rich in sugars, it is deficient in nitrogen and other mineral elements. Large quantities must be ingested to sustain growth. In the Hemiptera, the digestive tract is modified so that it can concentrate nutrients while limiting the retention time for the water in which they are dissolved. The foregut and hindgut are closely associated to form a filter chamber (Fig. 4-5). Water and some sugars from incoming sap bypass most of the midgut and are transported rapidly to the rectum (Wigglesworth, 1972). Amino acids and other nutrients are absorbed in the central midgut. The sugary, liquid excrement that results is called honeydew, which may be collected by ants (see Chap. 6).

Gall-makers

Gall-making insects are unique in that they can alter the development of plant tissue to form a tumorlike growth in which the insect gains nutrition and protection from the environment (Fig. 4-6). Gall formation typically begins after an adult female insect injects an egg into the plant; plant tissues undergo rapid cell division and enlargement and eventually surround the larva. Development of the gall phenotype entails the interplay of two genotypes, that of the insect, which codes for the gall-inducing stimulus, and the plant, which codes for the growth response (Weis and Abrahamson, 1986). In cases where several gall-makers utilize the same host species, each induces a gall of distinctive morphology.

Gall-makers constitute a functional group broadly scattered across taxonomic lines, including mites, thrips, aphids, moths, weevils, flies, and wasps. This feeding habit has had its broadest radiation in the two families Cecidomyiidae (Diptera), the gall midges, and the Cynipidae (Hymenoptera), the gall wasps. In North America these two families account for 70 percent of the nearly 1700 insect gall-makers. Although a variety of plant taxa are host to gall-makers, relatively few plant groups account for most of the galls. Over 90 percent occur on dicots, and within this group most galls are found on members of the families Rosaceae (roses), Compositae (sunflowers), and Fagaceae (oaks) (Abrahamson and Weis, 1987). Oaks in particular support a wide variety of cynipids (Cornell and Washburn, 1979). As a rule, gall-makers restrict their oviposition to a single type of structure, such as a vegetative bud, a leaf midrib, or a petiole.

Insect galls typically consist of several tissue types organized into concentric zones (Mani, 1964). Nutritive tissue lines the central chamber, where the insect resides (Fig. 4-6). The cells of this tissue are characterized by high physiological activity; accordingly they have many mitochondria and ribosomes, and are frequently multinucleate (Rohfritsch and Shorthouse, 1982). They are also rich in proteins,

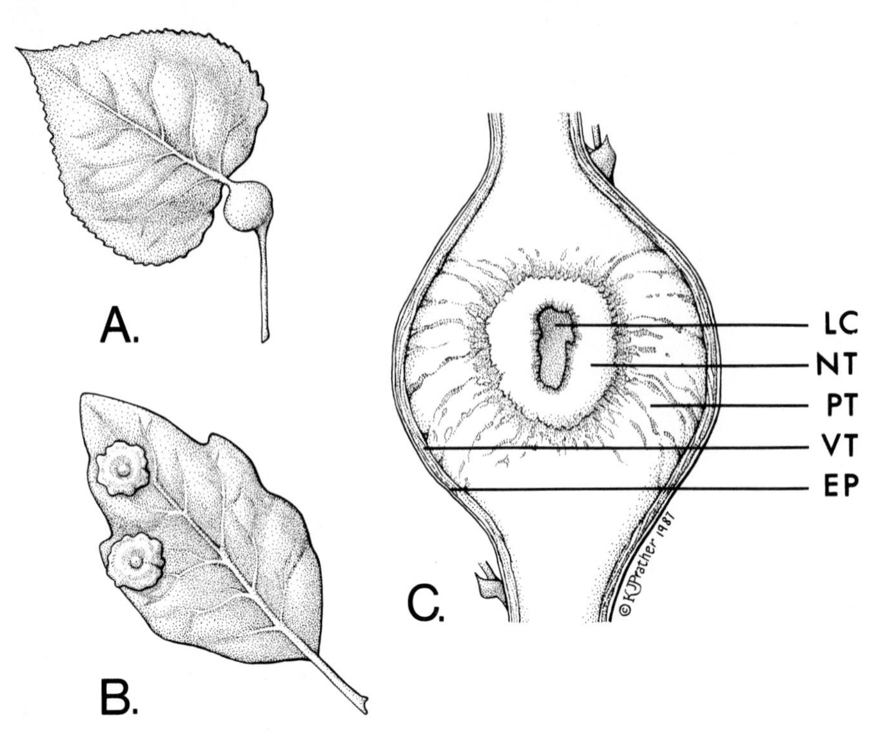

Figure 4–6. (A) Traverse gall of the aphid *Pemphigus populitransversus* on a poplar petiole. (B) Spangle gall of the cynipid wasp *Andricus pattersonae* on blue oak leaf [(A) and (B) from Romoser, 1973]. (C) Longitudinal section of the stem gall induced by the dipteran *Eurosta solidaginis* on the goldenrod, *Solidago altissima;* LC, larval chamber; NT, nutritive tissue; PT, parenchyma tissue; VT, stem vascular tissue; EP, stem epidermis (after Beck, 1947).

sugars, and lipids. Next, there is often a zone of relatively hard, thick-walled lignified cells, with vascular tissue. Although this zone has been dubbed the protective layer, its protective function is still open to question. Exterior to this zone, there is typically a layer of cortical parenchyma cells, which in turn is surrounded by plant epidermal cells, the last of the tissue layers (Abrahamson and Weis, 1987).

The precise nature of the insect's stimulus to the plant remains unknown; however, secretions of the insect are apparently required. Initiation of galls on willows by sawflies results from substances in the ovipositional fluid injected with the egg. In other species, gall formation begins only after the egg hatches and the insect larva begins to feed. The chemicals involved may vary among insects as well. Auxins and auxin precursors have been implicated in some cases, while cytokinins are suggested in others (Abrahamson and Weis, 1987). It has even been suggested that insects inject mutualistic plasmids or viroids into the plant which incorporate into the plant genome and take over gene regulation to produce the gall (Cornell, 1982). Whatever the stimulus, new and undifferentiated plant tissues respond best. Oviposition by some gall-makers is limited to a few weeks during the year when the particular plant part they gall first appears.

In many gall-makers, the anus is fused shut, which probably keeps the gall's central cavity hygenic. However, with no anus, excreta must be stored until the gall-maker emerges as an adult. Since the total volume of stored waste is not especially great, gall-makers must either be extremely efficient at extracting nutrients from the ingested gall tissue, or gall tissue must be extremely nutritious, or both.

Seed Predators

Seeds provide a resource for chewing, sucking, and gall-making insects. Thus, seed eaters are not considered a special guild because of the way they use the plant, but because of the special consequence they have on plant ecology. Members of the other phytophagous guilds can be considered as grazers or parasites (Price, 1980), but seed eaters are predators (see Chap. 3; Janzen, 1969a). Seeds are individual plants, albeit in an embryonic stage. Thus, an insect that consumes a seed kills an individual, just as when a wolf eats a hare.

Bruchid beetles are seed predators par excellence. Many bruchids are specialists on the plant family Leguminoseae (Southgate, 1979). Typically, the female lays an egg on a seed, which she then marks with an oviposition-deterring pheromone, a chemical signal that inhibits other females from laying eggs on the same seed. Larvae then develop in the seed, eventually killing it as they grow.

Plants as Food

In terms of biomass, plant material is the most abundant food source on earth (Table 4-1), yet much of it goes uneaten. Hairston et al. (1960), in a classic paper, maintained the "earth is green" because, while every other form of life is food-limited, herbivores are predator-limited. They reasoned that, since insects rarely demolish their food, the earth remains green because predators and parasites act to reduce insect population levels. In those rare cases where insect feeding kills mature plants, the herbivores have usually been introduced into a novel environment devoid of their normal complement of parasites and predators.

Hairston et al. (1960) made one major logical error in their argument—that is, that everything green is edible. Such has repeatedly been shown not to be the case. Herbivores can starve in the midst of apparent plenty. It is clear that there are a great many problems associated with feeding on plants.

Structural Materials

There are many morphological obstacles that physically hinder ingestion. Trichomes (plant hairs), glands, spines, wax, silica, and lignin act as physical barriers (Isely, 1944). Morphological barriers can be quite subtle and tied into other aspects of plant physiology. Environmental selection pressures, for example, have given rise to two basic patterns of photosynthesis among land plants. The so-called C3 plants, the majority of temperate species, fix carbon dioxide by initially incorporating it into a three-carbon component. However, many plants in tropical and arid habitats instead incorporate carbon dioxide into a four-carbon compound that carries the carbon dioxide into neighboring cells. The difference between C3 and C4 plants extends beyond the nature of the carbon dioxide acceptor in photosynthesis. Anatomical features necessary for C4 metabolism present problems to insect herbivores. For one thing, starch is localized in bundle sheaths, not distributed interveinally. These bundle sheaths are thick-walled, with cells reinforced with indigestible hemicellulose. As a consequence, C4 plants appear to be less preferred by generalized feeders.

Protein

For insect herbivores, nutritional deficiencies are as fundamental as morphological deterrents in adjusting to a plant diet. Plant leaf tissue is generally much lower in protein than animal tissue (Southwood, 1972). As a consequence, foliage-feeding insects that generally lack the ability to move from one plant to another must scramble for protein even more than other herbivores (Bernays, 1982). The efficiency of conversion of ingested food, a rough index of the usability of food (Waldbauer, 1968), generally increases with food nitrogen content, a correlate of plant protein. Insects can compensate for low conversion efficiency by increasing the amount of food eaten. Slansky and Feeny (1977) experimentally demonstrated this compensation with cabbage worms *(Pieris rapae)* fed collards *(Brassica oleracea)* supplemented with fertilizer. Consumption rate was negatively correlated with leaf nitrogen content, a reflection of protein content, with the result that nitrogen intake and, hence, growth rates were the same on fertilized and unfertilized plants. This pattern

Table 4–1 Productivity of plants and herbivores in different habitats

	Productivity (kcal/m²/yr)				
	Tanganyika Grassland	*North Carolina Field*	*Michigan Old Field*	*Georgia Salt Marsh*	*Pawnee Shortgrass Prairie*
Plants	747	1075	1360	6585	3767
All herbivores	3.1	4.7	0.9	81	53
Insects		4.5	0.7	81	1.8*

*Aboveground arthropods.

Sources: Wiegert and Evans, 1967; Southwood, 1972; Scott et al., 1979).

Table 4–2 Mineral content of three trophic levels in a Tennessee old field

	Mineral Content (mg/g ash-free dry weight)		
	Na	Ca	K
Plants	0.43	5.82	14.28
Herbivores	1.18	0.85	1.87
Predators	1.94	0.88	1.96

Source: Van Hook, 1971.

holds among other species as well. Herbivores that feed on low-nitrogen tissues generally consume more food than do herbivores on high-nitrogen plant tissues. Spittle bugs, which feed on xylem sap (containing less than 0.01 percent nitrogen) must consume 100 to 1000 times their body weight per day (Horsfield, 1978).

Much of the feeding behavior of herbivorous insects can be explained in terms of nitrogen acquisition. The hemipteran *Leptoterna dolobrata* switches feeding preferences from leaves of its grass host *Holcus mollis* to flower heads at the end of the third instar (McNeil and Southwood, 1978). This switch occurs at the point in the growing season when the nitrogen content of leaves declines and floral nitrogen increases. Seasonal movements of aphids from tree hosts in the spring to herbaceous hosts in summer is associated with decrease in soluble nitrogen in trees. Aphids return to tree hosts in fall, when soluble nitrogen again increases in senescent leaves.

However, total nitrogen or protein content is not the only consideration in insect nutrition, since it is the balance of amino acids that is critical to growth and development. Up to 20 percent of the weight of insects can be tied up in the aromatic compounds that bind the cuticular proteins together. Thus, insects in general may have a particularly pronounced dietary requirement for aromatic amino acids, such as phenylalanine (Bernays, 1982). The desert locust *(Schistocerca gregaria)* shows a significant increase in growth when phenylalanine is added to a protein-deficient diet.

Micronutrients

Micronutrients are also important in insect nutrition, and, since they occur in plants in limited quantities, they are another factor that make plants less than perfect food. Plant tissue is extremely low in sodium, except in very rare cases (Table 4-2). Mammalian herbivores, particularly grass feeders, are often hard pressed to obtain sufficient dietary sodium (Chap. 5). Chronic sodium deficiency is why conscientious farmers provide a salt lick in horse or cow pastures. To some extent, plant-feeding insects can compensate for low sodium. Butterflies, particularly males, have long been known to form "puddle clubs" at mud puddles, bird droppings, and animal urine. The high level of sodium in these materials may stimulate feeding and also provide the adult requirements for this ion (Arms et al., 1974).

Allelochemicals

Allelochemicals, constituents of all known vascular plants (Chap. 5) make life difficult for herbivorous insects. They can reduce food intake by their inhibitory effect on feeding behavior. They can also reduce the digestibility of the available nutrients in plant tissue. Finally, they can be acutely toxic and kill insects outright. There are abundant examples of secondary substances acting in all three ways. Tannins (polyphenols present in foliage of many trees) are known to act as feeding repellents to both insect and mammalian herbivores. Independent of effects on consumption rate, they can complex with gut digestive enzymes, thus decreasing digestibility of plant tissues (Klocke and Chan, 1982). Finally, in sensitive herbivorous species, tannic acid causes lesions to form in the gut lining, leading to the release of material into a hemolymph and possible septicemia (Steinly and Berenbaum, 1985).

Water

Desiccation is a problem for any insect associated with land plants (Southwood, 1972). The problem grows more acute as the difference between the water content of food and of the insect increases. Scriber (1977) showed that cecropia moth *(Hyalophora cecropia)* larvae grew more slowly and were less efficient in utilizing biomass, energy, and nitrogen when fed excised leaves of their host plant, wild cherry *(Prunus serotina)*. Scriber attributed the decrease in efficiencies to increases in relative maintenance costs brought about by water stress. Scriber (1979) subsequently showed that the effects of leaf water deprivation were more pronounced for tree feeders than for herb feeders. Water deprivation suppressed the relative growth rate of tree feeders 40 percent, but of forb feeders only 25 percent. The reason ostensibly is that tree feeders already work at low efficiency due to the naturally low water and nitrogen in tree foliage. Decreasing water content of the food to an even lower level has a correspondingly greater effect on larval growth rate.

A factor that confounds nutritional problems for insects is that plant resources change in quality over time (Scriber and Slansky, 1981). Protein and water content vary up to an order of magnitude over the season, so an individual insect's digestive system must accommodate for changes occurring over the duration of its lifetime.

Herbivore Efficiency and Ecosystem Dynamics

Given that plants are less than perfect food, how efficient insect herbivores are can depend on a number of things. One thing, however, that efficiency does not necessarily depend on is diet breadth (how specialized an insect is on a particular food plant). A logical assumption in studying insect–plant relations was that "a Jack-of-all-trades is a master of none"—that is, generalists are less efficient at feeding on a particular plant species than a specialist on that same species would be. Scriber and Feeny (1979) tested this idea examining growth rates and efficiencies of food and nitrogen utilization by larvae of 20 Lepidoptera. The magnitude of differences in parameters for specialists vs generalists was *much* less than the magnitude of differences for tree vs forb feeders. On some food plants, generalist species were capable of growth rates comparable to specialists. Nonetheless, larval growth on trees was invariably slower and less efficient than growth on forbs, irrespective of specialization. This slower growth on tree foliage is a reflection of many factors, including low water, low nitrogen, high fiber, and high tannins in tree foliage.

General trends can be described over all available types of plant food. For example, food high in nitrogen is associated with high growth rates and high digestive efficiencies. In addition, the higher the proportion of refractory material in the diet (e.g., tannins or lignins), the lower the digestive efficiency. Thus, sucking insects feeding selectively on phloem sap, entirely lacking in indigestible tannins, lignins, or cellulose, generally have high digestive efficiencies.

Surprisingly, on the ecosystem level, insect herbivores can compare favorably with much larger mammalian herbivores. The measure traditionally used to compare trophic level efficiency is the ratio of production to assimilation, or net productivity. When net productivity is compared (i.e., kcal produced/kcal assimilated), grasshoppers and spittlebugs range from 36 to 41 percent, whereas mammalian herbivores (such as meadow mice and voles) living in the same old-field habitat all fall below 10 percent (Table 4-3). The total energy flow through the insect herbivores (in terms of kcal/m^2/yr) was three to four times higher than energy flow through the mammalian herbivores (Petrusewicz, 1967). Invertebrates may be more efficient due to lower maintenance costs; as poikilotherms, they do not expend energy to maintain a constant body temperature.

Table 4–3 Secondary productivity of insect and mammalian herbivores

Population	kcal Produced/ kcal Assimilated	Total Energy Flow (kcal/m^2/yr)
Saltmarsh grasshopper	36.73	
Old-field grasshopper	36.22	25.6–64
Alfalfa field spittlebug	41.45	
Old-field mouse (*Peromyscus*)	1.79	
Old-field mouse (*Microtus*)	1.80	
Meadow vole	2.95	6.7–16.7
Uganda kob (antelope)	1.46	62.4
African elephant	1.46	

Source: Petrusewicz, 1967; Price, 1984.

Estimates for the proportion of the net primary production removed by foliage feeding by insects also compare favorably with those for mammals. Insect herbivores eat from about 2 to 15 percent of primary production in forests, and between 4 and 24 percent in old fields and grasslands (Slansky, 1974). During outbreaks, the proportions rise. For example, for 3 years during a gypsy moth outbreak in an oak grove in Russia, gypsy moths consumed 40, 50, and 74 percent of the foliage (Slansky, 1974). While the total amount of vegetation consumed may be low, certain plant species may be disproportionately hard hit. Grasshoppers in New Zealand consumed only 1 percent of the total plant biomass but up to 59 percent of certain species. This level of damage, however, is unusual and is associated with population outbreaks.

Estimates for sap-feeding insects are harder to come by. Van Hook et al. (1980) estimated that a *Macrosiphum liriodendri* aphid population on tulip poplar *(Liriodendron tulipifera)* consumed 1 percent of annual photosynthate production, an amount approximately equal to 17 percent of the annual production of foliar nitrogen. Other estimates for sap feeders range from 1 to 6 percent primary production.

One other aspect of ecosystem dynamics in which insect herbivores figure prominently is in nutrient transfer. In any ecosystem, nutrients are exchanged between living organisms and the physical environment. As in the case with energy, plant production is the most important component in making such essential elements as carbon, nitrogen, phosphorus, sulfur, and potassium available to animals. Herbivorous insects selectively store many such elements in their bodies during their lifetime. In one North Carolina forest, for example, total arthropod biomass was only 0.07 percent as great as foliage biomass, yet arthropods contained 12.5 percent as much sodium, a disproportionately large amount (Schowalter et al., 1981).

Insects are important in nutrient dynamics in an ecosystem not only when they die and release those nutrients back to the ecosystem via decomposition, but also when they eat and process plant tissue. For example, in 1 year in an oak woodland, insect frass (excrement) contributed 1.4 percent of the calcium, 1.8 percent of the carbon, 1.8 percent of the magnesium, 2.4 percent of the sodium, 4.5 percent of the potassium, 5.3 percent of the nitrogen, and 8.2 percent of the phosphorus to the total litter fall in the forest (Carlisle et al., 1966, as cited in Slansky, 1974). Moreover, in northern temperature forests, frass fall may contribute significantly to nutrient cycling in that it occurs primarily during the summer months, when other litter fall is minimal (Reichle et al., 1973). In an outbreak year, when caterpillar population densities are extraordinarily high, their contribution to nutrient cycling in a forest ecosystem is correspondingly greater.

Ecological Effects of Insect Herbivores on Plant Populations and Communities

Although insects consume only a fraction of plant primary production, they can nonetheless have a profound effect on plant population dynamics. A classic example in support of this statement is the case of the biological control of St. John's wort *(Hypericum perforatum)* by the Klamath beetle *(Chrysolina quadrigemina)* (Huffaker and Kennett, 1959). Introduction of this herbivore decreased the range of this introduced weed and continues to exclude it from open, sunny, well-drained slopes where it grows best. The plant now is restricted to shady areas which the beetle tends to avoid.

Plants are susceptible to insect attack at all life stages, and the effect on plant population structure and dynamics varies with the stage attacked. For instance, the effect of herbivory on mature trees is sometimes evidenced by decreased growth (Kuhlman, 1971). Karban (1980) detected smaller growth rings in oak trees during the year following emergence of 17-year cicadas when compared with pre-emergence years. The reduced growth was caused by newly hatched cicada nymphs, which burrowed into the ground to feed on (and kill) the fine rootlets. When nymphs were intercepted before they could enter the soil (by spreading bed sheets under the egg-laden trees), trunk growth proceeded normally (Karban, 1982).

Occasionally, herbivory results in the death of mature plants, as shown in a study by Rausher and Feeny (1980) on the pipe-vine swallowtail butterfly *(Battus philenor)* and its host plant, the Texas Dutchman's-pipe *(Aristolochia reticulata)*. They kept caterpillars off experimental vines with screen cages and manual removal, and allowed natural infestation on control plants. More than 90 percent of all vines survived the 2-year study period, but exposed vines suffered a morality rate that was four times that for protected vines. Exposure also led to a 50 percent decrease in the number of leaves produced.

Phytophagous insects can have a large effect on mature plants by reducing fecundity. In a 4-year study of the English oak *(Quercus robur)*, Crawley (1985) demonstrated that even moderate insect attack depresses seed production (Table 4-4). Typically 5 to 15 percent of the leaf area of this tree is lost to herbivory every year. Applications of insecticide at 2- to 4-week intervals on experimental trees kept defoliation below 5 percent, while water-sprayed control trees suffered twice that amount. The amount of growth, determined from tree rings, did not differ between the two treatments, but the number of acorns produced per shoot was consistently higher on insecticide-treated trees. The effect of herbivory in one season may carry over for several years when it reduces allocation to storage tissues. In the case of the clonal goldenrod and the stem-galler *Eurosta solidaginis* (Hartnett and Abrahamson, 1979), galled shoots produced about half as many seeds per unit biomass than ungalled shoots. But, in addition, only 60 percent as much biomass was allocated to rhizomes in galled versus ungalled shoots. Since the number of the next year's shoots in each clone depends in part on the size of the current rhizome (McCrea and Abrahamson, 1985), infestation by the gall-maker in one year reduces the number of seeds produced in the following year (Abrahamson and McCrea, unpublished data).

Table 4–4 Effects of insect exclusion by insecticide on growth and reproduction in *Quercus robur**

Parameter	Sprayed	Unsprayed
Percent defoliation†	3.00–5.00	8.00–11.00
Tree ring growth (cm)§	0.34–0.56	0.32– 0.54
Acorns produced per shoot†	0.36–0.86	0.08– 0.26

*Range in means over 4 years.
†$P < 0.05$.
§Not significantly different.
Source: Crawley, 1985.

Negative effects of insect herbivores on mature plants, though palpable, are seldom fatal. Herbivory is more often life threatening for plants attacked at earlier developmental stages. A grasshopper's bite, for example, is much less significant to a mature shrub than to a seedling (Parker, 1985). Even before germination the plant embryo is vulnerable to attack, first as it develops on the maternal parent and then as it awaits germination as a seed in the soil. Louda (1982) made a thorough study of the effects of flower- and fruit-feeding insects on the demography of the coastal California shrub, *Haplopappus squarrosus* (Compositae). Experimental plots of this shrub were regularly sprayed with insecticide, beginning when flowers bloomed and ending when seeds matured. Control plots were sprayed with water or unsprayed. All treatments initiated the same number of flowers (Fig. 4-7), but many flowers were destroyed by insects in the control groups, thus reducing seed set. Of those seeds that matured, seed feeders on the control plants reduced the number to reach the soil by an order of magnitude compared to the insecticide treatment. In the subsequent years, fewer seeds in the soil led to fewer seedlings and yearling plants in the unsprayed plots. In terms of biomass, insects did not eat much of this shrub, but, because attack occurred at a vulnerable point in the shrub's life cycle, insects were shown to have a regulatory effect on plant population density.

Insect herbivory can alter the competitive regime among plants and thus influence community composition. Bentley and Whittaker (1979) examined competition between two species of dock, *Rumex obtusifolia* and *R. crispus,* with and without the leaf beetle *Gastrophysa viridula. R. obtusifolia* was a better competitor but was more susceptible to the effects of the beetle. Its competitive advantage was eliminated because the beetle reduced the number of *R. obtusifolia* seeds but not those of *R. crispus.*

Plant community composition can be altered when one plant species acts as a reservoir of phytophagous insects for attack on another plant species. A plant species that can tolerate insect attack can support a resident insect population which can then wreak havoc on neighboring species that have little or no resistance. Such is the case described by Parker and Root (1981). While studying the effect of the grasshopper *Hesperotettix viridis* on the demography of the shrub *Gutierrezia sarothrae,* they noticed that the herb *Machaeranthera canescens* was usually killed when it grew near the shrub. It seemed that the larger *Gutierrezia* better withstood defoliation than did *Machaeranthera.* They performed an experiment in which 25 *Machaeranthera* were transplanted next to *Gutierrezia* shrubs; 13 transplants were fully enclosed by screen cages, while dummy cages were placed over the other 12. It took only 12 hours for the first 4 exposed plants to be attacked and, by 3 days, all 12 exposed plants had been attacked. The average exposed plant was totally defoliated within a week, some with as many as 10 grasshoppers per plant. All exposed plants died, compared to a 25 percent death rate for protected plants. Parker and Root concluded that herbivory by *Hesperotettix* limits the local distribution of *Machaeranthera* to areas far removed from *Gutierrezia* and its associated insect fauna. The insult to this injury is that *Machaeranthera* is not even an important component in the grasshopper's diet.

Evolutionary Response of Plants to Insect Attack

Over evolutionary time, plants have developed adaptations to the selection pressures exerted by insect herbivores. Escape in time and space is one route—that is, plants evolve mechanisms that function before feeding begins. This stands in contrast to resistance mechanisms, which function to reduce growth rate, survivorship, or fecundity of insects that *do* feed on the plants. Escape mechanisms result in the elimination or exclusion of certain plants from an insect's host range. A cautionary note to consider, however, is that not all plant traits associated with reduced herbivory are necessarily specific defensive adaptations. Many traits evolved in response to other factors but may fortuitously free a plant from its insect problems. In other words, a trait may be "adaptive" in the face of some environmental pressure without having evolved in response to that pressure (Dobzhansky, 1968).

Escape in Time

One obvious way to reduce the probability of encounter between plant and herbivore is by temporal displacement. The English oak is the principal host of the winter moth *(Operophtera brumata),* a foliage-feeding inchworm caterpillar (Feeny, 1970). Females lay their eggs on twigs in the crown; eggs hatch in spring and first instar larvae can only feed on oak buds that have just opened (they can't penetrate buds or eat leaves a few days old). Defoliation is rare because it occurs only when egg-hatch and bud-break coincide. Usually larval mortality is very high due to lack of coincident timing.

Escape in Space

Habitat location is another means by which plants escape from insect herbivores. An example was described by Hicks and Tahvanainen (1974), in which two species of mustard-feeding flea beetles *(Phyllotreta* and *Phylliodes)* have strong habitat preferences regardless of host plant. Transplantation of toothwort *(Dentaria diphylla),* a normally shady-habitat plant in the mustard family, into a sunny open field resulted in heavy damage by three of the flea beetle species, all open-field species that do not normally occur on toothwort in nature. Growing in shade, toothwort escapes from a number of otherwise competent herbivores.

Morphological Escape

Morphological escape is also a mechanism for avoiding herbivory. It is particularly effective against visually orienting herbivores. In the passion flowers (Passifloraceae), some species have modified leaf stipules that resemble egg masses of zebra butterflies, the principal herbivores associated with the Passifloraceae. Female butterflies will not oviposit on plants already containing eggs and are "fooled" into passing up the plants with the mimic eggs (Gilbert, 1975).

Crypsis or camouflage is another mechanism of morphological escape. Australian mistletoes, plants that are parasitic on other plants, develop leaf shapes that resemble those of their host plants and are thus difficult for herbivores to distinguish. Barlow and Wiens (1977) estimate

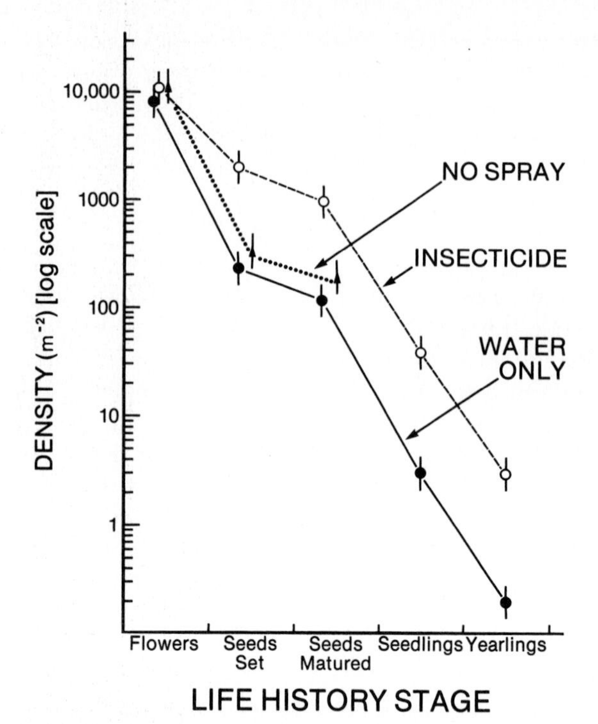

Figure 4–7. Survivorship of *Haplopappus squarrosus* offspring in plots where insects were excluded by pesticides, sprayed with water, or unsprayed. Excluding insects from the plants during the flowering and seed set stage of the parental plants led to higher recruitment of seedling and yearling plants. After Louda (1982).

that over 75 percent of the Australian mistletoes are mimics. One species, *Dendrophatae shirleyi*, mimics hosts with three different leaf shapes that resemble different hosts: flattened, linear-lanceolate leaves that resemble *Eucalyptus* and acacias; thick, rounded leaves that resemble mangroves; and linear, compressed leaves that resemble several other trees of the Australian forest. Mistletoe-feeding butterflies and possums are thought to be involved in maintaining this diversity of leaf shapes within the mistletoe. By resembling the foliage of its inedible host-plants, mistletoe avoids detection by its own highly specific, visually orienting herbivores.

Chemical Escape

Plants can escape from herbivores that orient primarily by olfactory cues in two ways. One way is by production of repellents. Considerable evidence shows that many volatile components of plant oils can deter herbivores without necessitating physical contact with the plant. Some plant oil components are even toxic via fumigant action. As it evaporates, apiole, an odor component of anise, dill, and parsley leaf oil, is toxic to fruit flies (Marcus and Lichtenstein, 1979).

The other principal means of escaping from herbivores that orient by olfaction is by absence of attractants. Many insect herbivores orient to a particular chemical "token stimulus" or kairomone (host recognition cue) in their host plant. Plants that lack the specific cue are in effect "invisible" to such herbivores. Feeny (1977) suggests that this phenomenon accounts for the absence of typical mustard herbivores on field pennycress *(Thlaspi arvense)*. When leaf tissue is torn or damaged on most mustard plants, the allylglucosinolate they contain hydrolyzes to allylisothiocyanate, a flea beetle attractant. However, in pennycress, it hydrolyzes to the unusual isomer allylthiocyanate, which flea beetles cannot detect. When vials of the normal isomer are placed in the field adjacent to pennycress plants, they are discovered and attacked by the flea beetles.

Resistance

Resistance involves mechanisms that reduce herbivory or the effects of herbivory after contact or ingestion. The modus operandi can be physical or chemical. Physical defenses include hooks, spines, hairs, sticky glands, and the like (Levin, 1973). They act by immobilizing the insect or by puncturing the body wall so that death occurs by slow starvation or loss of hemolymph.

Chemical defenses act either in acute or chronic fashion. Acute toxins act immediately and can cause death at low doses; they are also known as qualitative toxins. Chronic toxins are those whose effects are not immediately exerted but rather are manifested over the growth period as lower fecundity, lower digestive efficiency, or slower growth; these have been called quantitative toxins (Feeny, 1976). The qualitative-quantitative distinction is unsatisfactory because the terms were originally coined to describe the relatively low toxicity of quantitative toxins at low concentrations. However, all drugs have dosage-dependent effects and even qualitative toxins become more toxic at increasing dosages (Blau et al., 1978).

Another distinction can be made in defensive strategy regarding the distribution and abundance of a chemical defense. Allelochemicals can be present at effective concentrations at all times (constitutive) or synthesized de novo or translocated when the need arises (induced). The phenomenon of induction has been long known in plant pathology; infection by pathogenic fungi promotes plant production of toxic substances called phytoalexins. Similar results have been obtained with insect damage (Haukioja and Niemela, 1979). Some tree species respond to insect attack with increased phenolic and tannin production, and decreased protein and mineral nutritional suitability. Schultz and Baldwin (1982) demonstrated that trees defoliated by gypsy moths produce foliage in a subsequent year less suitable for supporting gypsy moth growth. In herbaceous plants, proteinase inhibitors in tomato, cucurbitacins in squash, and terpenes in sweet potato are all locally induced by insect-feeding damage, often within an interval of several hours (e.g., Tallamy, 1985).

Recently, several ecologists have suggested that actual physical contact may not be necessary to induce chemical defenses in plants. They suggest that herbivore encroachment within a plant population can be detected via pheromones (intraspecific chemical signals) released by attacked trees and responded to by unattacked trees, allowing the unattacked trees to gear up their defenses and ward off the attackers without suffering any losses. Baldwin and Schultz (1983) used potted poplar and sugar maple in a growth chamber and demonstrated that mechanical damage of leaf tissue promotes synthesis of phenolics in both damaged trees and undamaged neighbors within 52 hours. The jury, however, is still out on this idea; experimental design problems in this and related experiments (Rhoades, 1983a) make interpretation of the data difficult (Fowler and Lawton, 1985).

One more way for a plant to deal with insect herbivores is to "stand there and take it." Many plants respond to insect damage simply by outgrowing the damage. With such compensatory growth and reproduction, the overall effect on fitness can be minimal. The wild parsnip *(Pastinaca sativa)* is attacked by the parsnip webworm *(Depressaria pastinacella)*, which webs together and eats flowers and developing seeds early in the season. Wild parsnip can offset webworm damage by increasing seed set in the flowers produced later in the season, after the webworm attack has ceased (Hendrix and Trapp, 1981).

Evolutionary Responses of Insects to Plant Defense

Adaptation to Escape in Time

Insects have adapted to both plant escape and plant resistance. Adaptation to escape in time is illustrated in the fall cankerworm *(Alsophila pometaria)* (Schneider, 1980), an inchworm that feeds on a variety of deciduous trees in eastern North America. Eggs overwinter and hatch in spring; larvae feed for up to a month, drop to the ground to pupate and emerge, then mate and lay eggs in November and December. In this set of characters, the fall cankerworm resembles the winter moth. Like the winter moth, the fall cankerworm suffers greatest mortality when egg-hatch and bud-burst are out of synchrony. Hatching before bud-burst results in starvation, dispersal, and mortality; hatching long after burst results in low fecundity due to a decrease in the nutritional quality of leaves. There is, then, considerable selective pressure to synchronize hatch time with bud-burst.

Populations of fall cankerworm are unusual because they consist of several morphologically identical but generally distinct parthenogenetic (asexual) clones that occur along with intermittent sexual forms. Differences in the average date of hatch of egg masses in different localities are due to differences in the distribution of clones (Mitter et al., 1979). A set of cankerworm life-history traits has led to host plant-specific distribution of clones. Females tend to oviposit on the same tree they grew up on—prepupae and wingless females can't move very far and so remain physically associated with a suitable host tree. Their offspring, genetically identical to the mother, share whatever genetic mechanisms exist that determine oviposition preference and continue to seek out and use the same host species. The association between a clone and a plant can continue for many generations because trees that leaf out early in any given year tend to leaf out early in other years. These three factors allow for close, precise tracking of foliation and reduction in the efficacy of plant escape in time.

Adaptation to Plant Morphological Escape

Insects have also adapted to plant morphological escape; one startling example of this adaptive process is search-image formation (Courtney, 1983), the ability to selectively seek out a particular shape or item with experience. Vertebrate examples of search-image formation are legion, but there is suggestive evidence that herbivorous insects display comparable behaviors. In east Texas, the pipevine swallowtail has two host plants, *Aristolochia reticulata,* with heart-shaped leaves, and *A. serpentaria,* with linear, grasslike leaves. While searching for host plants with broad leaves like *A. reticulata,* many females fly right past the perfectly suitable grass-shaped *A. serpentaria.* However, later in the season when the heart-shaped plants lignify and become unsuitable for caterpillar growth, the majority of female butterflies switch their behavior and search exclusively for the grass-shaped *Aristolochia* (Rausher, 1978).

Adaptation to Plant Physical Resistance

Neither physical nor chemical defenses are absolute. One salient case was put forth by Gilbert (1971) in a paper entitled "Butterfly–Plant

Coevolution: Has *Passiflora adenopa* Won the Selectional Race with Heliconiine Butterflies?" Heliconiine (zebra) butterflies are the major herbivores associated with passionflower *(Passiflora)* species throughout Central and South America. *P. adenopa* is, for the most part, immune to damage. Scanning electron micrographs revealed that the surface of the entire plant, including tendrils, is covered with hooked trichomes. They are spaced out such that three to five could hook into each proleg of the caterpillar. Hemolymph drains out through punctures and the immobilized caterpillar slowly starves to death. Mortality is understandably high; thus, the title of the paper. However, the selection race may not yet be won. Rathcke and Poole (1975) found a related caterpillar that avoids trichomes in another plant family by gregariously spinning a silken web that acts as scaffolding on which larvae can crawl and feed in safety. Trichomes, then, cannot be considered the ultimate plant defense against insect herbivores.

Adaptation to Plant Chemical Defense

As effective as chemical defenses may seem, they may nonetheless be breached. There are three primary means of dealing with plant secondary chemicals. Behavioral adaptation to chemical defense are those in which an insect, due to a particular behavioral response, fails to encounter a lethal dose. Biochemical adaptations involve metabolic alteration by the insect of the chemicals so as to render them harmless. Physiological adaptations are those that prevent the ingested chemical, which is not structurally altered, from reaching sensitive tissues or organs.

Among behavioral adaptations, selective feeding is the most ubiquitous mode. Herbivores simply avoid plant parts or plant individuals rich in toxins. The tarnished plant bug *(Lygus lineolaris)*, for example, is responsible for reducing germination success and seed yield of various members of the carrot family (Umbelliferae). Detailed observation (Flemion et al., 1954) revealed that this bug has highly flexible mouthparts that travel intercellularly to reach seed embryos. The degree of control exercised over the stylets allows the tarnished plant bug to avoid encountering the external oil ducts on the seed coat. These ducts contain furanocoumarins, secondary substances with insecticidal properties, as well as toxic essential oils.

The abilities to detect deterrents, to perceive toxic substances at low concentrations, and to react with an avoidance response are highly developed in many generalized feeders (Bernays and Chapman, 1977). The African migratory locust *(Locusta migratoria)* can detect various secondary chemicals in concentrations well below those occurring in plants and can respond to them as deterrents. Inorganic and organic acids, alkaloids, sulfur-containing components, glycosides, phenolics, monoterpenoids, sesquiterpenoids and triterpenoids, steroids, and aristolochic acids can all be detected and avoided in concentrations as low as 0.000001 percent. Many insects also display flexible behaviors when faced with inducible plant defenses. The squash beetle *(Epilachna tredecimnotata)* feeds on foliage of plants in the cucumber family (Cucurbitaceae); many species in this family respond to insect-feeding damage by synthesizing cucurbitacins, triterpenoid compounds which are among the most bitter and toxic known. The squash beetle is substantially unaffected by cucurbitacins due to its unique method of feeding. Before it ingests foliar tissue, the squash beetle cuts a circular trench through mesophyll and vascular tissue to isolate one corner of a leaf. This trench prevents the plant from moving cucurbitacins into that section of the leaf, thereby allowing the squash beetle to feed on cucurbitacin-free foliage (Tallamy, 1985).

Biochemical Resistance

Biochemical resistance is largely associated with a suite of oxidative enzymes known as the mixed-function oxidases (MFOs). These membrane-bound enzymes, associated with the midgut of caterpillars and other herbivorous insects, basically effect a series of coupled oxidation-reduction reactions on a wide variety of substrates. The end result of these transformations is that fat-soluble substances are oxidized to water-soluble substances. Several of the substrates for MFOs are endogenous; they are involved, for example, in the metabolism of ecdysone, or molting hormone. However, many of the substrates are derived from the environment. These include such foreign substances as organic insecticides and allelochemicals from host plants. The utility of the MFO enzyme system in metabolizing these toxicants is that the conversion from fat solubility to water solubility not only makes a foreign compound less reactive toward lipid-rich tissues and organs, but also makes it more excretable. Thus, the MFO system is of great importance to many herbivorous insects in processing host-plant allelochemicals. One additional feature of many MFO enzymes is that they are inducible —that is, the appropriate enzymes are synthesized de novo only when a particular foreign substance enters the body (Brattsten et al., 1977). This inducibility may present a metabolic advantage to a herbivore, which produces enzymes for metabolism only when the need arises. Inducibility of MFO enzymes, however, is a two-edged sword.

Some plant chemicals (as well as some synthetic insecticides) are "bioactivated" by MFOs in that the hydrophilic conversion actually enhances toxicity. This is true for organophosphate insecticides such as parathion, which is converted to the highly toxic paraoxon. It is also the case for naturally occurring pyrrolizidine alkaloids. Insects that feed on alkaloid-containing legumes such as rattlebox *(Crotalaria)* do not respond biochemically by inducing MFO activity. One such insect, the bella moth *(Utetheosia ornatrix)*, which feeds exclusively on rattlebox, has vanishingly low levels of these enzymes and thus keeps the alkaloid in the unmetabolized nontoxic state. Another means of coping is illustrated by the southern armyworm *(Spodoptera eridania)*, a generalist feeder with extraordinarily flexible midgut enzymes. Its enzymes are induced by plant chemicals in carrot foliage but suppressed by the bioactivated chemicals in rattlebox seed (Brattsten, 1979a).

Mixed function oxidases have been implicated in the metabolism of pyrethrins, rotenoids, nicotine, and a number of other secondary substances, but they are by no means the only biochemical ways in which insects handle secondary substances. There are epoxide hydrases, hydrolases, cytosolic reductases, as well as group transfer enzymes which conjugate toxins to produce an inactive product. There are even a few highly specific enzymes, such as the group transfer enzyme rhodanase that converts the toxic cyanide ion to the relatively nontoxic thiocyanate (Brattsten, 1979b).

Along with changes in amounts of enzyme produced, biochemical resistance can be achieved simply by a change in the activity of an enzyme or in the rate of conversion from toxic target compound to nontoxic metabolite. The black swallowtail *(Papilio polyxenes)* and fall armyworm *(Spodoptera frugiperda)* can both break the furan ring double bond in furanocoumarins, toxins found in many plants of the carrot family (Umbelliferae) to metabolize them. Swallowtails, however, are capable of breaking it almost 50 times faster than are the armyworms (Ivie et al., 1983) and thus can and do feed extensively on furanocoumarin-rich umbellifers, which armyworms of necessity avoid.

Highly specific detoxification processes can be found in some highly specific herbivores. Over 95 percent of the free amino acid nitrogen in the leguminous shrub *Dioclea megacarpa* is contained in L-canavanine, a structural analogue of arginine (Rosenthal et al., 1982). L-canavanine is highly toxic when incorporated into protein by most insects. *Caryedes brasiliensis*, a Central American bruchid beetle, however, lives entirely within the seed of the legume. Not only does the arginyl-tRNA synthetase of *C. brasiliensis* fail to "recognize" canavanine, the beetle has highly active arginase and urease enzymes, which catalyze the hydrolysis of canavanine to urea. Urea is further catabolized to obtain nitrogen for synthesizing protein. Not only does the former toxin fail to protect the plant, it provides an additional source of nitrogen to hungry bruchids.

Yet another example of "turnabout is fair play" involves phenolic compounds, which are generally growth inhibitors or feeding inhibitors for most insects. The tree locust *(Anacridium melanorhodon)* has the unusual capacity to increase growth and survival in the presence of the phenolic compound tannic acid. After digestion, radiolabeled phenols show up in the insect's integument with a specific activity over four times that in any other tissue. Using dietary phenols in cuticle sclerotization instead of tyrosine, the normal precursor, conserves amino nitrogen. This is advantageous since the plants *(Acacia)* on which the locusts feed are very low in nitrogen (Bernays and Woodhead, 1982).

Physiological Adaptation to Plant Chemicals

Physiological adaptation to plant chemicals generally involves modification of systems for transporting secondary chemicals within the insect body. For example, nicotine is a highly insecticidal alkaloid found in the foliage of tobacco plants. However, at the physiological pH of the guts of the specialist tobacco hornworm *(Manduca sexta)*, it exists primarily in its ionized form. In this polar form, it cannot cross into the nervous system where it normally exerts its toxic effects (Self et al., 1964). Gut pH can also influence the ability of dietary tannin to complex with protein. The greater the amount of complexing, the lower the nutritional quality of the food. Insects that feed habitually on tanniniferous plants appear to have on average a high gut pH (range 8.5 to 9.0). At these levels, tannin-protein complexes break down (Berenbaum, 1980). Moreover, insect gut fluid has high surface tension, due to the presence of surfactants (Martin and Martin, 1984). Lysolecithin and linoleoylglycine are two such surfactants in insect gut fluids that interfere with the formation of tannin-protein complexes.

Sequestration is perhaps the most common means of physiological adaptation to secondary chemicals. Sequestration, loosely defined in an ecological context, is the deposition of secondary substances into specialized tissues or glands of an insect (Duffey, 1980). It is a widespread phenomenon; Rothschild (1972) recorded sequestration in 43 species from six orders of insects. Sequestration also involves at least six classes of secondary compounds, including cardenolides, alkaloids, aristolochic acids, mustard oils, cucurbitacins, and cyanogens. Sequestration is a function of both intrinsic physiochemical factors (e.g., size, structure, charge, and stability of the chemicals) and extrinsic biological factors (toxicity, solubility, and metabolizability in the body of the insect). Casual sequestration can result from toxin-solubility characteristics, through the nonspecific accumulation of lipophilic compounds in fatty tissues. If the sequestered allelochemicals are unchanged in structure from the plant source, they are likely to be unchanged in function as well. These sequestered chemicals, still as physiologically active as they are in intact plant tissue, can function in protecting a herbivorous insect by conferring toxicity to that insect.

Plant Effects on Insect Natural Enemies

The elaborate defense mechanisms mounted by plants against herbivores and the equally elaborate counterdefenses seen in insects are seemingly complex. Yet when plants and herbivores are considered in isolation, the story is incomplete. Virtually all terrestrial communities based on green plants include three trophic levels: plants, herbivores, and the natural enemies of herbivores. There are many remarkable ways that a host plant can affect the attractiveness and vulnerability of a phytophagous insect to an insectivorous bird, arthropod, or microbial pathogen (Price et al., 1980).

Plant Traits Inhibiting Natural Enemies

Sequestering toxins can be an effective way to survive on a poisonous plant, but it can also have the added advantage of deterring insect predators. For example, the monarch butterfly *(Danaus plexippus)* sequesters cardiac glycosides from its hosts, milkweeds. These chemicals are heart poisons, but at sublethal doses they act as powerful emetics. Chickadees exhibit violent retching and vomiting within 15 minutes of ingesting milkweed-reared monarchs (Brower et al., 1967), but there is no reaction by birds when the insects are reared on cabbage, a nonhost which lacks cardiac glycosides. The milkweed bug *(Oncopeltus fasciatus)* sequesters the same toxins from milkweeds. These toxins have an emetic effect on praying mantids, which discard milkweed bugs after a few bites (Berenbaum and Miliczky, 1984).

Monarchs and milkweed bugs, as well as other milkweed associates, are conspicuously colored (Table 4-5) and feed in exposed locations. By contrast, most phytophagous insects are cryptic. Thus, milkweed associates are aposematic, that is, they use bright colors to advertise their distastefulness to potential enemies. Predators learn after several encounters to associate the bright color patterns with the unpleasant experience and thereafter avoid these insects.

Plant-derived compounds may also have deleterious effects on other types of herbivore enemies, such as parasitoids. These are carnivorous insects that spend their larval stage as parasites inside an insect host but are free-living as adults. Unlike true parasites, parasitoids kill the host as a matter of course. Living internally, parasitoids are vulnerable to plant toxins circulating in the host's blood. This is the case with the parasitoid

Table 4–5 Herbivorous insect community on milkweeds (Aslepiadaceae) in temperate North America that are mostly aposematic*†

Order	Family	Species	Coloration
Coleoptera	Cerambycidae	*Tetraopes teraophthalmus*	Red with black spots
		T. femoratus	Red with black spots
		T. quinquemaculatus	Red with black spots
	Chrysomelidae	*Labidomera clivicollis*	Red with black pattern
	Curculionidae	*Rhyssomatus lineaticollis*	Black
Lepidoptera	Danaidae	*Danaus plexippus*	Adult: orange and black Larva: black, yellow, and white stripes
	Arctiidae	*Euchaetias egle*	Adult and larva: white, yellow, and black
		Cycnia iopinatus	Adult: white, yellow, and black Larva: orange with gray hairs
Hemiptera	Lygaeidae	*Oncopeltus fasciatus*	Red and black
		Lygaeus kalmii	Red and black
Homoptera	Aphididae	*Aphis nerii*	Yellow and black

*Other insects feed sporadically on milkweeds but are not found predominantly on these plants.

†*Source:* Price, 1984.

wasp *(Hyposoter exiguae)* which parasitizes the tomato fruitworm *(Heliothis zea)*. Caterpillars fed on tomato foliage absorb the alkaloid tomatine. In a laboratory experiment, parasitoids developing in hosts fed artificial diets rich in tomatine were more likely to die before they reached pupation. Those that survived took longer to develop, were stunted as adults, and had shorter adult life-spans (Campbell and Duffy, 1979). In a similar fashion, feeding on tomato decreases the susceptibility of the Colorado potato beetle *(Leptinotarsa decemlineata)* to infection by the fungus *Beauveria bassiana*. In the field, beetle larvae feeding on tomato had infection rates substantially lower than beetles feeding on potato and horse-nettle, which lack this toxin (Hare and Andreadis, 1983). Fungal spores are found on the surface of all these leaves and are ingested during normal feeding. It is thus possible (though not proved) that alkaloids from tomato leaves mix with and kill the spores in the beetle's stomach. These examples indicate that plant breeding programs to increase plant resistance to herbivores through allelochemicals occasionally may be incompatible with biological control efforts. Ironically, if the natural enemies are more susceptible to the toxin than the herbivore, increased toxin levels may make it possible for the herbivore to "escape" population control.

Such a scenario, involving a physical plant defense, has been played out in the Netherlands, where fresh vegetables are grown commercially in greenhouses. The greenhouse whitefly is a common pest which was controlled on tomato and eggplant with the parasitoid wasp *Encarsia formosa*. However, whitefly outbreaks on cucumber were common and costly. The hairy leaved cucumber varieties being used were less suitable for whitefly development than smooth-leaved varieties, yet whitefly population growth still outpaced the regulatory ability of the wasp (Hulspas-Jordan and van Lentren, 1978). Surprisingly, biological control was successful when smooth-leaved cucumbers were grown. The dense mat of leaf trichomes on the hairy varieties greatly hindered this tiny wasp as it walked the leaves in search of hosts. Further, the honeydew produced by the hemipterous host accumulated on the trichomes and then stuck to the wasp, forcing it to spend an appreciable amount of time grooming. The net effect was that the walking speed of the wasp was three times faster on glabrous cucumber leaves than on hairy leaves, which increased parasitoid efficiency and made biological control feasible on smooth-leaved varieties.

Plant Traits Fostering Natural Enemies

Plant characteristics in some cases increase herbivore vulnerability to their enemies in direct ways. The volatile chemicals that attract phytophagous insects to the host plant can also attract their natural enemies (Read et al., 1970). *Diaeretiella rapae*, a major parasitoid of the cabbage aphid, *Brevicoryne brassicae*, responds to the odor of cabbage leaves. Allylisothiocyanate, a chemical widespread in the mustard family, was identified as the attractant. Another aphid, *Myzus persicae*, occasionally feeds on cabbage and other mustards, since it freely feeds on plants from many families. As might be predicted, it is attacked by *D. rapae* when it feeds on odoriferous mustards.

Plant damage caused by herbivores may also attract visually orienting enemies. Leaf miners occupying leaves damaged by other folivores are more likely to be discovered by parasitoids (Faeth, 1985). Parasitoids of gall-makers are attracted by the gall itself (Askew, 1961), although in some cases galls physically protect the gall-maker (Weis and Abrahamson, 1986).

Insectivorous birds may use the jagged outline of partially eaten leaves as a clue when searching for caterpillars. In one experiment (Heinrich and Collins, 1983), black-capped chickadees were released into an aviary supplied with branches bearing entire leaves (controls) and branches with damaged leaves (experimentals) to see if the birds would learn to search damaged branches preferentially. Food was supplied only on the damaged branches; either mealworms were attached to the branch or cryptic green caterpillars were placed on the leaves. On consecutive foraging trials, birds spent increasing amounts of time searching on the branches with damaged leaves. After this behavior had been learned, all insects were removed from the cages. Birds still preferentially searched the damaged branches, indicating that they key in on the damage and not the insects themselves. Many forest caterpillars chew off the remaining portion of a leaf after they have finished eating, and thus "cover their tracks."

Plants may indirectly increase the success of natural enemies by altering the physiological state of phytophagous insects. This may be particularly true when insects feed on host plants with digestibility reducing compounds. Although these compounds are not fatal, they may increase herbivore susceptibility to disease (Price et al., 1980). One effect of feeding on poor-quality diets is increased development time. The longer a larva takes to develop, the more time it is exposed to the attack of predators, parasites, and pathogens (Moran and Hamilton, 1980).

Plants And Insect Herbivores: The Coevolutionary Question

As the two most abundant life forms on earth, it is not surprising that plants and insects interact. Documenting the players in these interactions has been a straightforward process; discovering the rules of the game, however, has proved to be an elusive goal. By examining the patterns of host-plant use among insects, some inferences on the evolution of specificity should be possible. In what is arguably the most influential paper on plant–insect interactions yet written, Ehrlich and Raven (1964) examined the phylogenetic relationships among butterflies, a group that contains many specialist herbivores, and their host plants. In brief, they found that closely related groups of butterflies (i.e., subfamilies or tribes) tend to feed on closely related plants. For instance, the Danaidae (monarchs and their relatives) feed on the Asclepiadaceae (milkweeds). Members of the subfamily Pierinae (Pieridae) (which includes the cabbage white butterfly) feed on members of the Capparidaceae in the tropics and the closely related Cruciferae in the temperate zones. Exceptions to this pattern led to the discovery of a rule of sorts. Some of the Pierinae feed on the Resedaceae, Salvadoraceae, and Tropaeolaceae. These exceptional host plants are similar to the Capparidaceae and Cruciferae in one important respect: they all contain mustard oils. Thus, biochemical or behavioral specificity in overcoming chemical defenses may be an important force in the evolution of feeding specificity. At the same time, selective pressures exerted by herbivores may be an important force in directing the evolution of plant secondary chemicals. Ehrlich and Raven (1964) envisioned a five-step process for the evolution of butterfly feeding habits:

1. By a random genetic event, an angiosperm produces a novel secondary compound or class of compounds.
2. By chance, these compounds render the plant less suitable as food for insects.
3. Freed from herbivore pressure, these plants undergo evolutionary radiation in a new adaptive zone.
4. By a random genetic event, an insect evolves resistance to the secondary compound.
5. Able to exploit a plant resource previously unavailable to herbivores, the adapted insect enters a new adaptive zone and undergoes its own evolutionary radiation.

Ehrlich and Raven brought the term coevolution into general usage to describe this scenario.

A strong case for coevolution can be made from an examination of the interaction between plants in the family Umbelliferae on the one hand, and swallowtail butterflies and oecophorid moths on the other. Biochemical evidence indicates that in the Umbelliferae, novel secondary chemicals have arisen sequentially; these are hydroxycoumarins, linear furanocoumarins, and angular furanocoumarins. These compounds are sequentially less suitable to polyphagous insects, yet some oligophagous species have counteradapted to furanocoumarins through behavioral and biochemical means. Genera within the Umbelliferae containing these chemicals have more species than those lacking coumarins, and the insect taxonomic groups that can handle these chemicals are more diverse than related groups that cannot (Berenbaum, 1983).

The intuitive appeal of Ehrlich and Raven's scenario led to its wide acceptance. Frequently, cases in which an insect can detoxify or sequester a plant toxin have been uncritically accepted as an example of coevolution. Actually, such a relationship between plant and herbivore need not arise by coevolution, but simply by adaptation by the insect to a part of its environment, i.e., the chemicals in its host plant (Futuyma, 1983). Because of this, Janzen (1980a) has argued that the term coevolution should be used in a very restricted sense. He defined coevolution as a change in one species (e.g., a plant) in response to selection imposed by a second species (e.g., an insect) followed by a change in the second species in response to the change by the first. Thus, in this sense, coevolution is a *reciprocal* process between pairs of interacting species. In this narrow sense, even Ehrlich and Raven's scenario does not describe a coevolutionary process. Production of novel compounds, such as mustard oils or furanocoumarins, may have been favored because they reduce damage by a variety of plant enemies, including herbivores (insect and vertebrate), bacteria, and fungi. The contemporary insect specialists that overcome these defenses may have contributed little or nothing to the selection regime which originally led the evolution to those traits. Janzen (1980a) and others have suggested the term "diffuse coevolution" be used to indicate when reciprocal evolution has occurred between suites of plants and suites of herbivores.

Studies of plant–insect interactions have for the most part tried to infer the course of past evolution by examining contemporary species. Recently, some investigators have switched from the retrospective approach to a prospective one. Using methods from quantitative population genetics (Falconer, 1981), they are attempting to evaluate how rapidly "interaction traits" may evolve, by measuring the amount of genetic variation in these traits and the strength of the selective pressures acting on that variation. For instance, Berenbaum et al. (1986) have examined the genetics of furanocoumarin production in the wild parsnip with respect to its major herbivore, the parsnip webworm. Plant genotypes that produce high levels of two different furanocoumarins have comparatively lower fitness when grown in insect-free environments. However, when insects are present, these plant genotypes are protected from herbivory and show higher fitness. Thus, within plant populations, the levels of secondary compounds may change as herbivore populations fluctuate and different plant genotypes are favored.

In another study, Weis and Abrahamson (1986) used quantitative genetic methods to investigate the interaction between goldenrod and the goldenrod ball gall fly. They found that the size of the gall induced by this insect varies with both the insect's and the plant's genotype. From the insect's perspective, selection should favor the gall-maker to stimulate the production of larger galls because big galls protect the gall-maker from parasitoid wasps. From the plant's perspective, selection should favor lowered reactivity of goldenrod to the gall-maker's stimulus, since galls drain energy away from plant reproduction (Hartnett and Abrahamson, 1979; McCrea et al., 1985; Abrahamson and McCrea, 1986). Changes in gall sizes over time will reflect the balance in the intensity of the selective forces operating on the two interacting species. The approach taken by these investigators is new, and only further work will determine its utility in understanding interactions and coevolution of plants and herbivorous insects.

Interactions between insects and plants have implications beyond the population-genetic level at which coevolutionary changes are believed to occur; these various and sundry trophic interactions have a great impact in structuring communities and ecosystems. There have been several efforts, varying in their generality and predictive ability, to account for patterns of plant–insect interactions at the community level. Feeny (1976), perhaps the first to attempt a general theory, elaborated on the importance of herbivores in general and insect herbivores in particular in determining the chemical composition of a plant community by describing two ends of a continuum. At one end are plants that can, to some extent, escape from herbivory by a set of life-history traits. Early successional plants generally are unpredictable in time and space and thus are "unapparent" to insects lacking specialized adaptations for finding hosts. These unapparent plants, at least partly dependent upon escape in time and space, depend less heavily on chemical resistance. Secondary chemicals from these plants tend to be low molecular-weight compounds occurring in low concentrations; these are "qualitative toxins." In contrast, long-lived species such as trees are "bound to be found" over the course of their lifetime and thus depend, to a great extent, on nonspecific chemical defenses in high concentrations present at all times (constitutive); these chemicals, increasingly more effective as concentrations increase, are "quantitative toxins." The case studies from which Feeny drew his insights were his own work with glucosinolates in Cruciferae or cabbage plants, typical qualitative toxins, and tannins in oak trees, typical quantitative toxins. Rhoades and Cates (1976) advanced a theory similar to Feeny's the same year, although they distinguished chemicals principally by mode of action; they described toxins as being analogous to qualitative toxins and digestibility-reducing substances as being analogous to quantitative toxins.

Herbivore selection pressure was an essential factor in both the

Feeny and Rhoades and Cates arguments. Nutrient limitations were soon perceived as an influential factor as well, particularly with respect to latitudinal and successional gradients. Janzen (1974a) suggested that, in nutrient-poor soils, plants can less-afford to lose biomass to herbivores, since the cost in nutrients of replacing tissues is so great; as a result, such plants should "invest" more heavily in chemical defenses. Coley et al. (1985) incorporated nutrient limitation into another theory of plant–insect relations. Growth rate, as controlled by nutrient availability, determines the concentration of plant secondary compounds. Fast-growing plants in resource-rich habitats have lower quantities and different kinds of chemicals than do slow-growing species of resource-poor environments. Characteristically, fast-growing species experience greater herbivory. Secondary chemicals of slow-growing plants have low turnover and occur in high concentrations. Conservation of resources is the factor dictating patterns of secondary chemical production; moreover, it dictates the impact of resource removal by herbivores.

This approach differs only in degree from the ideas advanced by Feeny and others. Coley et al. argue that apparency theory attributes differences in chemical defense to differences in selection pressure by generalist vs specialist species, yet point out that the proportions of generalists and specialists in the contemporary fauna on apparent vs unapparent plants are not those predicted by apparency theory. For Coley et al., herbivory is a complementary factor rather than a driving force for the system. Changes in the pattern of chemical allocation in plants can derive solely from physiological factors; patterns of insect herbivory thus reflect rather than determine patterns of chemistry. Coley et al. (1985) recognize two types of defensive chemicals. Mobile defenses are nitrogen-based with high turnover, and immobile defenses are carbon-based with low turnover. The resource pool available to other plants determines the chemistry; when nitrogen is limiting, nitrogen-based compounds are rare or absent. They argue that phosphorus is so limiting to plants that there are no known phosphorus-based defensive chemicals. Since most leaf nitrogen is present in the form of the major photosynthetic enzyme ribulosebisphosphate decarboxylase, available nitrogen is directly related to photosynthetic ability as well. Photosynthetic efficiency can determine how much carbon is available for synthesis of carbon-based compounds. These statements are supported by a recent finding by Lincoln et al. (1986) that carbon dioxide enrichment increases the ratio of carbon to nitrogen in plant tissues.

Theories of plant–insect relations are particularly valuable, not so much because they can explain natural phenomena without exception, but rather because they provide a useful paradigm for directing research. Each theory has associated with it certain predictions; if the theory is an accurate description of reality, then research centers around describing these patterns and testing these predictions—which is certainly as it should be. No contemporary theory is wholly satisfactory with respect to predicting and describing the almost innumerable ways that insects and plants interact. These theories do provide a framework for conducting research and gathering additional data to be used at some future juncture, not only to modify the paradigm, but also possibly to generate entirely new paradigms for future testing.

Selected References

Berenbaum, M. R., A. R. Zangerl, and J. K. Nitao. 1986. Constraints on chemical coevolution: wild parsnips and the parsnip webworm. *Evolution* 40:1215–1228.
Brattsten, L. B. 1979b. Biochemical defense mechanisms in herbivores against plant allelochemicals. pp. 199–271. In: G. A. Rosenthal and D. H. Janzen, eds. Herbivores: their interaction with secondary plant metabolities. Academic Press, NY.
Ehrlich, P. R., and P. H. Raven. 1964. Butterflies and plants: a study in coevolution. *Evolution* 18:586–608.
Feeny, P. 1970. Seasonal changes in oak leaf tannins and nutrients as a cause of spring feeding by winter moth caterpillars. *Ecology* 51:565–581.
Price, P. W., C. E. Bouton, P. Gross, B. A. McPheron, J. N. Thompson, and A. E. Weis. 1980. Interactions among three trophic levels: influence of plants on interactions between insect herbivores and natural enemies. *Annu. Rev. Ecol. Syst.* 11:41–65.
Strong, D. R., J. H. Lawton, and R. Southwood. 1984a. Insects on plants. Harvard University Press, Cambridge, MA.

5

Mammalian Herbivore–Plant Interactions

RICHARD L. LINDROTH

Department of Entomology
University of Wisconsin
237 Russell Laboratories
1630 Linden Drive
Madison, WI 53706

Introduction

Mammals are one of the most important animal classes in terms of the number and types of herbivorous individuals and their impact on the evolutionary history and ecology of plants. The most available and widespread foods for terrestrial mammals are plants and insects. Roughly half of the nearly 4000 species of living mammals are primarily herbivorous. Of the 16 orders of terrestrial mammals, one is associated with plant nectar and pollen, two feed mainly on plant seeds, and seven consume mostly vegetative parts (Southwood, 1985). This chapter provides an overview of the interactions between mammalian herbivores and plants, including the evolutionary adaptations and ecological responses to one another, and the implications of these for population dynamics, community organization, and ecosystem function.

The manuscript benefited from reviews by George Batzli, May Berenbaum, and Ann Johnson. A National Science Foundation Postdoctoral Fellowship (BSR-8503464) provided support during preparation of this chapter.

Nonmammalian Vertebrate Herbivores

Although the chapter focuses on plant–mammal interactions, this does not mean that other vertebrates are unimportant in the ecology of multitrophic level interactions. For example, many species of birds are herbivorous. Of these, only the Anatidae (ducks and geese) are true grazers, whereas others such as gallinaceous birds (e.g., grouse, quail, turkey), pigeons, parrots, and many perching birds (songbirds) are mixed feeders that eat fruits, buds, and young shoots. Birds are especially important to plants with respect to the processes of pollination (see Chap. 2) and seed dispersal (see Chap. 3). Reptiles probably strongly influenced plant communities during the Mesozoic Era (230 to 63 million years BP), but today only a few herbivorous species remain. These include the marine iguana, which feeds on seaweeds, and some turtles and tortoises. In some cases these vertebrates may greatly influence their food sources as evidenced by the strong selection pressure of Galapagos tortoises on the growth forms of *Opuntia* cacti (Racine and Downhower, 1974).

Adaptive Radiation of Mammalian Herbivores

The earliest mammals appeared late in the Triassic Period (approximately 190 million years BP) and were descendants of the therapsid reptiles known as cynodonts. At that time, the continental masses were still a connected series of plates, and their flora was dominated by conifers and cycads (Eisenberg, 1981). Herbivory as a dominant form of foraging evolved early in the evolutionary history of mammals. Eisenberg (1981) suggests that exploitation of trees for their invertebrate fauna and fruits by early mammalian insectivores and carnivores could have led to increased feeding on plant parts and ultimately to specialization on leaves themselves as a nutrient source.

Although now extinct, members of one of the most successful of all mammalian orders, the Multituberculata, were primarily herbivorous. The multituberculates were a specialized side branch apart from the main direction of mammalian evolution. They are often described as the "rodents of the Mesozoic," ranging in size from that of small mice to small rabbits. Clemens and Kielan-Jaworowska (1979) proposed that multituberculates evolved after entering a new adaptive zone as vertebrate herbivores early in their evolution. Competition with reptilian herbivores may have been reduced by their smaller size and use of alternative food sources. Their major evolutionary radiation during the Cretaceous (135 to 63 million years BP) was related to the rapid spread and diversification of angiosperms. In the latter part of the Cretaceous, these mammals often comprised more than half of the mammalian species occurring in Holarctic faunas. Extinction of the multituberculates may have been caused by competition with eutherian herbivores.

Another phylogenetic line, the Theria, experienced many early radiations of now-extinct mammals, and eventually led to the two major groups of extant mammals today, the Metatheria (marsupials) and Eutheria (placentals). Marsupials originated during the late Cretaceous, probably in the Americas. They radiated into a variety of forms in South America and later in Australia. In South America, an insectivore/omnivore ancestral group led to the formation of carnivorous species and herbivorous or granivorous species in the arid southern regions, while in Australia the marsupial radiation led to insectivores, carnivores, omnivores, and specialized herbivores such as arboreal folivores, browsers, and grazers (Eisenberg, 1981).

Placental mammals (eutherians) arose during the early Cretaceous and were probably insectivores. By the late Cretaceous these mammals, especially in North America, exhibited considerable adaptive variation. Insectivore/carnivore forms specialized in different lines, while herbivorous forms, such as browsers and grazers developed in many lineages and persist today, especially in the orders Lagomorpha (e.g., pikas, rabbits), Rodentia (e.g., squirrels, mice), Artiodactyla (e.g., pigs, deer, antelope), and Perissodactyla (e.g., rhinoceroses, horses, zebras) (Eisenberg, 1981). Large and diverse assemblages of grazing mammals developed in the Miocene (25 million years BP) during the rapid radiation and expansion of the Gramineae (grasses) into extensive grassland ecosystems such as the Great Plains of North America and Serengeti Plains of eastern Africa.

Differences between Insect and Mammalian Herbivores

Although the primary metabolic processes in all animals are very similar, mammals are not simply "large furry insects." They differ from insects physiologically, behaviorally, and ecologically, and these differences have implications for how each group influences, and is influenced by, plants (Lindroth, 1988). First, insects and mammals differ with respect to some dietary requirements. Insects require sterols, but mammals synthesize their own, whereas the requirement of mammals for iron and calcium (to produce hemoglobin and bone) is greater than that of insects. Many aposematic insects (i.e., those with warning coloration) require particular plant-derived toxins that are sequestered for defense against predators, but this adaptation is not known among mammals. Second, detoxication systems differ both qualitatively and quantitatively between insects and mammals. The glucuronidation pathway, important for the detoxication of many plant compounds by mammals, does not occur in insects, and the activity of enzymatic pathways that the two groups share in common is usually higher (per unit protein) in mammals. Third, the low conversion efficiencies (resulting from maintenance of homeothermy) and large body sizes of mammals require them to consume large amounts of food, which in turn forces them to adopt generalist feeding strategies. Specialist feeders are common among insects but very rare among mammals, occurring only when a food plant is available in large concentrations (e.g., giant pandas feeding on bamboo or koalas feeding on *Eucalyptus*). Moreover, mammals must maintain considerable plasticity in their food habits to adjust to substantial seasonal and annual variation in the quality and quantity of available food (Lindroth et al., 1986). These factors have less impact on insects, which have shorter life-spans and often pass through adverse periods in dormancy. Finally, the capacity of mammals for sophisticated learning and long-term memory are greater than that of insects and this difference is probably reflected by differential foraging strategies in the two groups.

Influence of Mammalian Herbivores on Ecosystem Processes

Mammalian herbivores substantially influence the function of many ecosystems due to their impact on primary production, decomposition of organic matter, and redistribution of nutrients (Batzli, 1978). Many ecology textbooks discuss the two major ecosystem functions—energy flow and nutrient cycling—as if they were separate, discrete processes, when in fact they are closely coupled. Energy moves into and out of ecosystems as light and heat, but within ecosystems it moves from one unit to another linked with nutrients as living organisms or dead organic matter. Both energy and nutrients follow the flow of organic matter from producing organisms through herbivore and/or detritivore food webs. As Batzli (1978) explained, energy and nutrients are required for the production of new organic matter at any trophic level, and the availability of each influences the utilization of the other. Thus, interactive processes for feedback between energy flow and nutrient cycling are common in ecosystems. Herbivores, for example, may directly affect the rate of production of their own food.

Impact on Primary Production

The proportion of net primary production consumed by herbivores varies greatly both in space and time. The little evidence available indicates no clear relationships between primary productivity and the amount of production consumed by mammalian herbivores. In the moderately productive Serengeti grasslands, removal of annual primary production by grazing mammals averaged 66 percent, and ranged from 15 to 94 percent (McNaughton, 1985a). In an ecosystem with much lower productivity, the arctic tundra, grazing by lemmings in years of high population densities reduced the total yield of grasses and sedges by 50 to 90 percent (Schultz, 1964).

Estimates of the effect of mammalian herbivores on plant production simply in terms of the amount of vegetation consumed are confounded by the fact that grazing itself affects rates of primary production. In his review of the responses of plants to defoliation, Jameson (1963) concluded that above ground, seed and root production decrease when a major portion of a plant's biomass is removed, but what constitutes a major portion varies according to species and environmental factors, such as precipitation. Vickery (1972) found that net primary production of pastures was greater when sheep grazed at moderate densities (20/ha) than at lower (10/ha) or higher (30/ha) densities. Moderate levels of grazing by large mammals on the Serengeti grasslands stimulated productivity up to twice that of ungrazed (control) plots (McNaughton, 1985a), and productivity was maintained at control levels even under intense short-term grazing (McNaughton, 1979).

The long-term effects of grazing on primary production may differ from short-term effects. Exclosure studies in the arctic tundra have shown that in spite of intensive grazing during years of high lemming population densities, the production of grasses and sedges outside 20-year-old exclosures was double that inside the exclosures. Batzli (1975, 1978) attributed the reduced production to accumulation of dead organic material within the exclosures, which slowed the rate of nutrient cycling.

Impact on Decomposition

Mammalian herbivores alter the production and nutrient cycling of ecosystems by their influence on decomposition of organic matter. The availability of soil nutrients to plants is limited by both the accumulation of nutrients in plant and animal biomass and the rate of decomposition of organic matter, so excretion of nutrients by herbivores may accelerate the rate of nutrient replenishment in soils (Ruess and McNaughton, 1984). Soil nutrient levels in turn influence the nutrient status and growth of plants.

Depending on the animal species, plant species, and other factors involved, mammals fail to digest from 20 to 80 percent of the food they ingest (Batzli and Cole, 1977; Robbins, 1983), and generate large quantities of feces and urine. A 350-kg cow, for example, produces 10 to 25 L of urine and approximately 34 kg of feces per day (Spedding, 1971). Batzli's (1978) review indicates that nutrients are often released more rapidly from the feces of herbivores than from the plants from which they are composed, and decomposition of feces may also accelerate the decomposition of adjacent plant litter. However, in some cases the release of fecal nutrients (including, for example, calcium, phosphorus, magnesium, and iron) may actually be slower than from plant litter, especially if detritivores do not mix the feces with soil. Urinary constituents (e.g., nitrogen, phosphorus, and sulfur), however, are readily available to plants, so herbivory increases the rate of cycling for these nutrients.

Consider, for example, the effects of brown lemmings on nutrient cycling. Population densities of these arctic rodents fluctuate dramatically with a cycle of 3 to 6 years (Schultz, 1964, 1969). Schultz found that concentrations of calcium, phosphorus, potassium, and nitrogen in grazed vegetation cycled in phase with lemming population densities. Concentrations of the same constituents from exclosed (ungrazed) vegetation showed no such response. Apparently the availability of nutrients decreases during years of low population densities because nutrients become inaccessible in the accumulating dead plant material, and this litter in turn reduces the depth of thaw in the soil, decreasing the soil volume exploitable by roots. Nutrient release and subsequent uptake by plants is facilitated in years of high population densities by the clipping of standing dead vegetation, increased thawing of the soil, and deposition of large amounts of feces and urine by lemmings.

The impact of mammals on decomposition rates may influence plant productivity as well as nutrient concentrations. Studies by McNaughton and his colleagues (Ruess and McNaughton, 1984; McNaughton, 1985a) on the Serengeti grasslands have shown that not only is plant nutrient (especially nitrogen) status improved, but productivity is stimulated by the grazing of large mammals. They attribute these responses largely to accelerated nutrient recycling through the deposition of dung and urine by animals.

Redistribution of Nutrients

Mammalian herbivores also influence the movement and spatial distribution of nutrients in ecosystems. Animals that forage over large areas but defecate over small areas may substantially concentrate nutrients. Hilder

(1966) found that about one-third of the fecal output by sheep was deposited on less than 5 percent of their paddocks. In the arctic tundra, soil nutrients (especially phosphorus) and plant production are highest in microtopographic units known as troughs, where lemmings deposit large concentrations of feces near their winter nests (Batzli, 1978). Distribution of the dung of large mammals in the Serengeti is clumped in space, and fresh dung is more likely to be excreted near other recent deposits (with an active dung beetle fauna) than near old dung (McNaughton 1983a, 1985a).

Mammalian herbivores may also expedite the flow of nutrients between habitats. Large mammals in the Serengeti transfer great quantities of nutrients from the understories of tall grass savanna to adjacent open grasslands. This nutrient drain may explain the unusually low production of tall grass sites in relation to the rainfall they receive (McNaughton, 1985a). Similarly, the behavior of hippopotamuses, which consume large quantities of vegetation on land at night but defecate in the water during the day, may cause a substantial nutrient loss to the terrestrial community (Lock, 1972, and references therein).

Adaptations of Plants to Mammalian Herbivores

It's not easy bein' green.
[Kermit the frog, singing Joe Raposo's Bein' Green]

On the average, 10 percent or more of the annual production of plant communities is consumed by herbivores (Crawley, 1983; Coley et al. 1985, and references therein), an amount greater than the average allocation to plant reproduction (Mooney, 1972). Herbivores strongly influence the fitness of plants by reducing growth or reproduction and increasing mortality. Thus, it is not surprising that plants have evolved a variety of adaptations to reduce or compensate for the detrimental effects of herbivores. This section describes the variety of plant adaptations presumably elicited as evolutionary responses to browsing and grazing by mammalian herbivores, and provides a foundation for understanding the ecological responses of plants to herbivory to be discussed later.

McNaughton (1983b) and McNaughton et al. (1985) have argued that the evolution of plant–herbivore interactions has proceeded along two main, divergent pathways. These pathways originated with shifting climatic patterns toward greater terrestrial aridity that began in the Cretaceous and became established in the Eocene and Pleistocene. One pathway involved the adaptive radiation of gymnosperms, dicotyledonous and some monocotyledonous plants, and most of the herbivorous insect taxa; the other, more recent pathway involved radiation of most monocotyledonous plants (especially grasses) and associated large grazing mammals and orthopteran insects. Of course, numerous exceptions exist, such as the many herbivorous mammals that feed on the foliage of dicotyledons. But in general, the patterns seem to hold true. As we shall see, these two categories of plants developed substantially different adaptations to herbivory.

Physical Defenses

Structural Feeding Deterrents

One common adaptation of plants against herbivores has been called the "barbed-wire syndrome": the possession of hooks, bristles, prickles, spines, and thorns (Grant, 1984). The larger of these structures—spines and thorns—ostensibly evolved as a defense against vertebrate herbivores because they are avoided by insect herbivores with relative ease.

Silicification (accumulation of silica) of exposed tissues is a major, physical form of defense adapted by graminoids (grasses and grasslike plants). McNaughton et al. (1985) summarized the biological effects of silica: it accelerates tooth wear (its crystals occur in the geometric forms likely to promote abrasion of mouthparts), it reduces the digestibility of plant tissue, it contributes to the development of esophageal cancer, and it may cause fatal silica urolithiasis due to formation of calculi (concretions of minerals) in the urinary tract. Within plants, silica is deposited in locations to protect especially vulnerable tissues such as leaf blades and inflorescences. Moreover, grazing appears to be a factor in natural selection for accumulation of silica. Among Serengeti field sites, silica concentrations in plants were greater in the more intensely grazed vegetation, and this pattern was maintained in laboratory-grown plants collected from the various sites (McNaughton et al., 1985).

Growth Form

Herbivory by mammals has played a role in the evolution of grazing-resistant growth forms in plants. McNaughton (1983b) suggests that different types of herbivory select for different growth forms. Chronic herbivory by large mammals should select for prostrate genotypes that emphasize vegetative propagation and lateral spread. Acute bouts of intense herbivory with substantial growth periods between should select for more erect genotypes that compete well for light but with less capacity for vegetative propagation.

The ultimate grazing-resistant growth form has been achieved in the graminoids (grasses and sedges). The meristems of these plants grow close to ground level, enclosed by basal leaf sheaths and often covered by layers of plant litter or moss. This growth form provides the basal meristematic tissues a spatial refuge from most grazing mammals. Another characteristic contributing to the grazing-resistant growth form of graminoids is the maintenance of a large amount of belowground biomass (high root/shoot ratio) which enables the plants to accumulate substantial carbohydrate and nutrient reserves (Bryant et al., 1983). These reserves facilitate the rapid regrowth of leaf tissue following grazing.

Evolutionary adaptations of growth form in response to herbivory by mammals are not as apparent in dicotyledons. Individual plants, however, may adopt resistant growth forms in response to grazing or browsing, as is discussed later.

Chemical Defenses

Plant secondary compounds are biochemicals that do not function directly in the primary metabolic processes that support the growth, development, and reproduction of plants (e.g., the synthesis of protein, oxidative metabolism of carbohydrates, or replication and transcription of DNA). Secondary compounds have been used by humans for their pharmacological and toxic properties since the dawn of time. But the function of these plant constituents eluded the understanding of most scientists until well into the 20th century, in part because of the unfortunate connotation of "secondary" as "unimportant."

Several scientists in the late 1800s and early 1900s suggested that secondary chemicals in plants may protect them from many herbivores. But this idea aroused little interest until Fraenkel's (1959) classic paper, which asserted that the compounds strongly influence the host-finding and gustatory behavior of insects, and that indeed, the very reason for their existence in plants is defense against herbivores. In recent years a large body of evidence has accumulated to support a primarily ecological role for secondary plant compounds. They function in the interactions of plants with their pathogens, competitors, and herbivores, as well as protecting plants from physical stresses such as ultraviolet radiation and desiccation. Some compounds may also play important roles in the internal metabolism of plants, functioning as regulators of biochemical processes and storage compounds for certain elements (e.g., nitrogen and carbon) that can later be recycled into primary metabolic pathways (Seigler, 1977).

The major pathways of secondary metabolism are outlined in Fig. 5-1. Three principal components of primary metabolism are used as building blocks for secondary compounds: shikimic acid, the amino acids, and acetyl coenzyme A. There are also three major routes for the biosynthesis of secondary compounds: the shikimic acid, acetate (polyketide), and mevalonate pathways. These pathways are present in all plants and lead to the formation of some primary metabolites (e.g., fatty acids, amino acids) as well. Several recent textbooks consider the biochemistry of these pathways in detail (e.g., Mann, 1978; Vickery and Vickery, 1981; Torssell, 1983; Luckner, 1984). An excellent review of the biology and chemistry of the major groups of secondary compounds can be found in Rosenthal and Janzen (1979), and a concise compilation is presented in Table 5-1.

Erythrose-4-phosphate from the pentose phosphate cycle and phosphoenol pyruvate from the glycolytic pathway combine to form shikimic acid, from which the shikimic acid pathway derives its name. This pathway is the major metabolic route for the synthesis of aromatic compounds (those containing a benzene ring) in plants, and leads to the production of a diverse array of ecologically important secondary

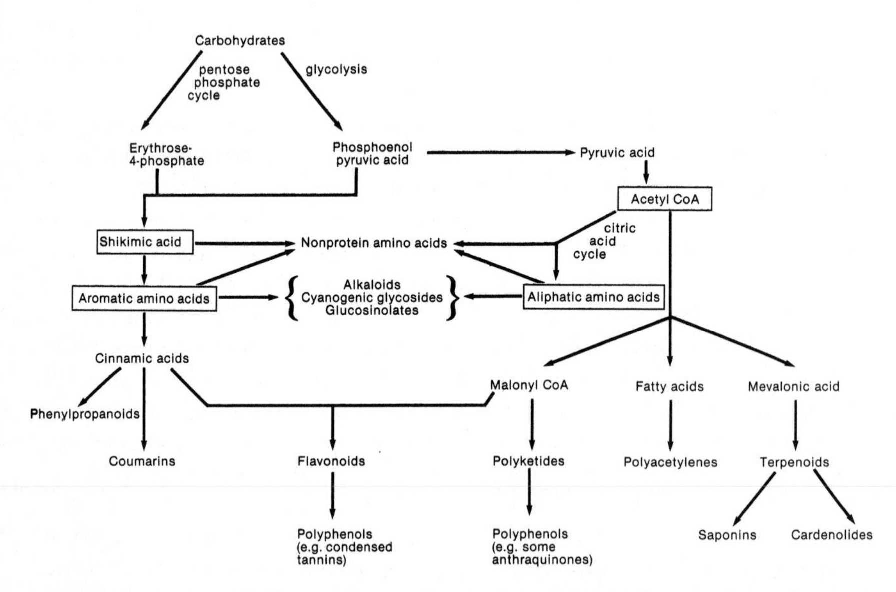

Figure 5–1. Metabolic pathways for major groups of secondary plant compounds. Principal building blocks from primary metabolic pathways are shown in boxes.

compounds. These include simple phenolic compounds, flavonoids, coumarins, tannins, and lignins.

Acetyl coenzyme A, derived primarily from the oxidation of pyruvate from the glycolytic pathway, is the major precursor for the acetate pathway. This pathway proceeds by the stepwise addition of two-carbon units and subsequent modification to form fatty acids, polyacetylenes, polyketides, and some polyphenols. A few products of this pathway enter the shikimic acid pathway as precursors for the synthesis of flavonoids and some tannins.

Combination of three acetyl coenzyme A molecules produces mevalonic acid, a six-carbon molecule. Mevalonic acid is subsequently decarboxylated to yield a five-carbon isoprene unit, the basic building block for the mevalonate pathway. Condensation of these C_5 units gives rise to the terpenoids, the largest and most diverse group of secondary plant compounds. The terpenoids are categorized into classes based on the number of isoprene units they contain (monoterpenes, C_{10}; sesquiterpenes, C_{15}; diterpenes, C_{20}; triterpenes, C_{30}; carotenes, C_{40}). Modification of triterpenes produces several other important groups of compounds, including cardenolides, saponins, and steroids.

Amino acids are metabolized via a variety of pathways to produce several other important classes of secondary compounds. The largest and most diverse group is the alkaloids. All alkaloids contain nitrogen, almost always in a heterocyclic ring structure. In addition, they typically exhibit pharmacological activity (e.g., the effects of caffeine, cocaine, and morphine). Amino acids are also converted to cyanogenic glycosides, glucosinolates, and nonprotein amino acids.

Finally, as indicated earlier, some secondary compounds have a mixed biosynthetic origin. In these cases a metabolite from one pathway serves as a substrate for production of another metabolite through a second pathway. For example, the flavonoids are products of both the shikimic acid and acetate pathways, the cannabinoids (active constituents in marijuana) come from the acetate and mevalonate pathways, and indole alkaloids derive from the shikimic acid and mevalonate pathways.

To utilize toxins as a defense against herbivores, plants were required to overcome a major obstacle, namely, how to poison biochemical and physiological systems that were fundamentally similar to their own. Plants use a variety of means to protect themselves from their own toxins. Some secondary compounds are localized in specialized vacuoles, plastids, or glands while others accumulate in the cell wall or on the surface of leaves and stems. Some compounds are stored as inactive precursors (often by conjugation with sugar molecules to form glycosides) which become activated only when the tissue is damaged. For example, in plants containing cyanogenic glycosides, the glycosides and activating enzymes (glycosidases) are spatially separated in plant tissues.

Table 5–1 Major classes of secondary plant compounds and their effects on mammalian herbivores

Class	Approx. number of Compounds	Distribution	Physiological Activity*
Nitrogen compounds:			
Alkaloids	5500	15–20% of all vascular plants, one-third of angiosperm plant families, especially dicotyledons	Diverse toxic effects, including organ lesions and inhibition of enzyme function, neurotransmission, membrane transport, and DNA, RNA, and protein synthesis; a few carcinogenic or teratogenic
Amines	100	Universal in higher plants	Exhibit diverse toxic effects including organ lesions and inhibition of enzyme function; some hallucinogenic
Cyanogenic glycosides	30	100 plant families, especially the Araceae, Compositae, Euphorbiaceae, Graminae, Leguminosae, Passifloraceae, and Rosaceae	Disrupt cellular respiration
Glucosinolates	80	Common in the order Capparales, especially the families Capparidaceae, Cruciferae, and Resedaceae. Also in the families Caricaceae and Euphorbiaceae	Thyroid enlargement, liver and kidney lesions
Nonprotein amino acids	200	Sporadic across many plant families, especially common in the Leguminosae	Diverse toxic effects, including organ lesions, hair loss, hypoglycemia, infertility, inhibition of enzyme function; a few teratogenic

Phenolic compounds:			
Simple phenols and flavonoids	2200	All vascular plants	Diverse toxic effects, including disruption of cellular respiration, inhibition of enzyme function and membrane transport; estrogenic activity interferes with reproduction
Coumarins	800	Common in higher plants, especially in Araliaceae, Rutaceae	Diverse toxic effects including organ lesions, phototoxicity, hemorrhagic and anticoagulant effects, estrogenic activity
Tannins	Many (unknown number)	All vascular plants	Disrupt digestive processes, toxic effects include organ lesions but are mostly unknown
Polyacetylenes	750	Mostly in Compositae and Umbelliferae	Phototoxic, neurotoxic
Terpenes:			
Monoterpenes	1000	Common in higher plants	Disrupt digestive processes, inhibit cellular respiration
Sesquiterpene lactones	1000	Mostly in Compositae	Disrupt digestive processes, cause organ lesions and allergic dermatitis, inhibit enzyme function
Diterpenes	1000	Common in higher plants	Disrupt digestive processes, cause organ lesions and dermatitis, some are cocarcinogenic
Saponins	500	Most higher plants	Cause bloat, organ lesions, and inhibit enzyme activity
Cardenolides	150	Sporadic, especially common in the Apocynaceae, Asclepiadaceae, and Scrophulariaceae	Inhibit Na^+-K^+-ATPases, leading to cardiac arrhythmia and arrest

*Demonstrated by at least some members of the chemical class against some mammals.

Source: Scheline, 1978; Rosenthal and Janzen, 1979; Conn, 1981; Harborne, 1982; Luckner, 1984.

Upon crushing a leaf, the two components mix, releasing free hydrogen cyanide (Conn, 1979; Fig. 5-2). Other plant compounds, such as the pyrrolizidine alkaloids, become toxic only upon ingestion and subsequent metabolic activation by herbivores.

In addition to varying among species, secondary chemicals vary qualitatively and quantitatively both within populations and individual plants. Cyanogenic and acyanogenic morphs coexist in single populations of several species of plants (Crawley, 1983, and references therein). Within individuals, temporal changes in secondary chemistry probably result from a variety of interacting factors, including changes in the relative value (in terms of fitness) of certain tissues, rates of herbivory, and availability of nutrients. Secondary compounds that always occur in appreciable quantities in plants are known as "constitutive," those that are synthesized and accumulated in response to damage by pathogens or herbivores are labeled "induced" (Levin, 1972).

Physiological Defenses

An alternative to the evolution of strong physical or chemical armaments by plants has been the evolution of physiological mechanisms that appreciably enhance a plant's ability to compensate for herbivory. Grazing tolerance has evolved in a number of plant taxa, especially graminoids. Why have plants not adopted a combination of these strategies? Most likely because limited energy and nutrient budgets preclude the development of mechanisms to support both an extensive array of chemical defenses and the capacity to rapidly regrow following browsing or grazing.

Indeed, the type of evolutionary response of plants to herbivory must be understood in the context of constraints imposed by the availability of energy and nutrients in the environment. Bryant et al. (1983) and Coley et al. (1985) have argued that resource availability is the major evolutionary determinant of the type of defense employed by woody plants. In general, low resource availability (poor nutrient sites) selects for inherently slow photosynthetic and growth rates associated with low levels of leaf protein. Because they occur in environments where lost nutrients are not readily replaced, these plants tend to have long-lived leaves and twigs, and avoid herbivory by investing heavily in constitutive chemical defenses. Boreal forest evergreens, especially spruce, are prime examples.

In contrast, high resource availability (nutrient-rich sites) selects for species with the capacity for fast photosynthetic and growth rates, which enhance competitive abilities. Because photosynthetic rates decline with age, turnover of leaves is rapid. These plants allocate fewer resources to chemical defense and show a greater capacity for induced

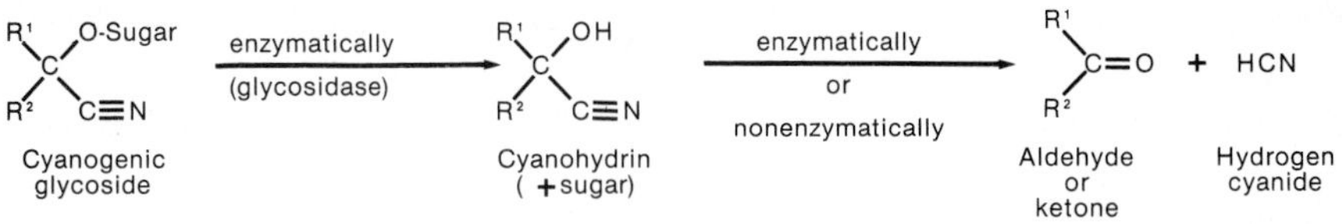

Cyanogenic glycoside ($R^1R^2C(\text{O-Sugar})C\equiv N$) —enzymatically (glycosidase)→ Cyanohydrin (+ sugar) ($R^1R^2C(OH)C\equiv N$) —enzymatically or nonenzymatically→ Aldehyde or ketone ($R^1R^2C=O$) + HCN (Hydrogen cyanide)

Figure 5–2. Release of hydrogen cyanide from cyanogenic glycosides. R^1 and R^2 may be any of a variety of alkyl or aryl groups; in lotaustralin, for example, $R^1 = CH_3CH_2$, $R^2 = CH_3$.

chemical responses. Nutrient losses associated with rapid turnover of plant parts and herbivory are not as detrimental to plant fitness in nutrient-rich environments that allow for rapid growth. Examples of these types of plants include early successional boreal forest trees such as willow and birch. Unlike woody plants, graminoids retain large nutrient reserves below ground in both nutrient-poor and nutrient-rich environments, enabling them to respond to herbivory by compensatory growth in all habitats.

Of course herbivory elicits physiological responses in all plants (no physical or chemical defenses are impenetrable to all herbivores), but they are most pronounced in those species that are evolutionarily adapted to respond to grazing primarily by compensatory growth. McNaughton (1979, 1983b) and Hilbert et al. (1981) summarize a variety of physiological processes that promote growth after defoliation. These include: (1) increased photosynthetic rates in residual or new plant tissues; (2) reallocation of substrates from other plant parts and increased allocation of new photosynthates to production of new leaves; (3) increased rates of cell division and elongation; (4) stimulation of vegetative propagation (e.g., tillering in grasses following activation of meristems rapidly expands photosynthetic area); and (5) delay of plant senescence.

Impact Of Mammalian Herbivores On Plants

"Captain," said Bluebell, "do you know what the first blade of grass said to the second blade of grass?"
. . . Holly replied, "Well?"
"It said, 'Look, there's a rabbit! We're in danger!'"
[Conversation among rabbits, from Richard Adams, Watership Down]

Plant populations have developed a variety of adaptations to herbivory over evolutionary time and, as has already been noted, these are exhibited in the responses of individual plants to herbivory over ecological time. These responses in turn influence the dynamics of plant populations and communities.

Effects on Plant Structure

Physical Defenses

Individual plants may respond to browsing and grazing by mammals by increasing their physical defenses. When browsed by goats, *Acacia raddiana* produced stunted branches with stiff thorns that appear to provide protection from further grazing (Seif el Din and Obeid, 1971). Abrahamson (1975) found that the prickles on dewberries *(Rubus trivialis)* were larger and sharper in areas where they were eaten by cattle than in areas where they were not eaten. McNaughton et al. (1985) showed that artificial defoliation more than doubled the silica content of a grass species that was grazed by large mammals in the Serengeti.

Growth Form

Mammalian herbivores may also alter plant growth forms. Browsing of shrubs often produces densely compacted, "hedged" plants, that may be more resistant to further herbivory. In years of high population densities, snowshoe hares in Alaska produce strikingly sharp and extensive browse lines on deciduous trees. Indian elephants and giraffes distort and sculpture the crowns of trees (Crawley 1983, and references therein). Hickey's (1961) investigation of the effects of grazing by cattle on the growth form of crested wheatgrass *(Agropyron desertorum)* showed that in ungrazed areas the growth form was mostly erect, but became increasingly horizontal with increased grazing pressure. In areas with intense grazing, the plants were prostate.

Effects on Plant Chemistry

Nutrient Content

As plants respond to herbivory with increased photosynthetic rates and redistribution of stored reserves and new photosynthates, changes in tissue nutrient concentrations occur. These changes are due largely to shifts in tissue phenology. Young tissues, undergoing rapid protein synthesis, typically have relatively high levels of protein and minerals. As they mature, concentrations of protein and minerals decline as a consequence of dilution by accumulation of structural materials (cellulose and lignin) in cell walls. By promoting the growth of new tissues and allocation of stored reserves to residual tissues, herbivory may improve the overall nutrient status of aboveground tissues. Of course, depletion of stored reserves decreases a plant's capacity to respond similarly to future instances of herbivory.

Numerous studies have documented increases in plant protein content resulting from grazing or mowing of forage crops (Jameson, 1963). In natural systems, grazing by large mammals increased protein content of Serengeti grasses (McNaughton, 1985a); browsing by moose increased protein content of birch leaves (Danell and Huss-Danell, 1985); and browsing by snowshoe hares increased protein and phosphorus concentrations in adventitious shoots of aspen, balsam poplar, birch, and alder (Bryant, 1981). However, grazing or browsing by mammals does not always increase plant nutrient content. Moss et al. (1981) found that the effects of feeding by mountain hares *(Lepus timidus)* and red deer *(Cervus elaphus)* on protein and mineral concentrations in heather varied in relation to soil type and possibly the type and amount of herbivory.

Secondary Compounds

Herbivory by mammals may also influence production of secondary compounds in plants. This occurs either as a consequence of reversion to an earlier phenological state, or because of a true induction of secondary biochemical pathways. In the latter case, the particular biochemical and physiological processes that translate tissue damage into defensive responses are largely unknown. But Cramer et al. (1985) showed that when exposed to a fungus, bean cell cultures induced production of messenger RNAs that code for several enzymes utilized in the shikimic acid pathway. These researchers suggested that similar biochemical changes may be involved in the responses of plants to other stresses, such as mechanical damage.

Reversion of plants to the secondary chemical profiles of earlier phenological stages is best documented by the work of Bryant (1981). He found that when severely browsed by snowshoe hares, alder, aspen, birch, and balsam poplar produced adventitious shoots highly resistant to browsing. Juvenile shoots contain significantly higher concentrations of phenolic resins and terpenes than twigs of mature trees. Although not as well documented, herbaceous plants may have similar responses. When substantially defoliated in midsummer, foxglove *(Penstemon digitalis)* produced leaves similar in appearance and chemical makeup (alkaloids, phenolics, and protein) to young spring leaves (Lindroth, unpublished data).

Several studies have shown that induction of biochemical pathways involved in the synthesis of secondary compounds occurs in response to artificial herbivory. Repeated cuttings of alfalfa may increase saponin concentration (Birk, 1969). Phenolic concentrations increased in a tundra sedge, *Carex aquatilis,* in response to clipping (Rhoades, 1979), and mechanical defoliation of pine seedlings increased both the number and concentration of phenolic constituents (Wagner and Evans, 1985). Very few researchers have investigated the chemical responses of plants induced by grazing by mammals. Oksanen and Oksanen (1981) discovered that phenol/nitrogen ratios nearly doubled in bilberry *(Vaccinium myrtillus)* in the 2 years encompassing the peak and decline of lemming population densities. Lindroth and Batzli (1986) found that when vole populations were high, levels of phenols were significantly higher in grazed than in ungrazed alfalfa in an alfalfa old field.

Finally, an especially intriguing response of plants to herbivory may be the ability to transfer information from attacked to unattacked plants via airborne chemical substances. Rhoades (1983b) found that undamaged willows underwent induced chemical responses when nearby trees were defoliated by insects. Using individually potted maple and poplar trees, Baldwin and Schultz (1983) defoliated some plants and subsequently measured increased phenolic concentrations in damaged plants and undamaged plants sharing the same growth chamber. However, the validity of these and related studies, and indeed the concept itself, are currently issues of lively debate (Fowler and Lawton, 1985; Rhoades, 1985).

Effects on Plant Production

Growth

As already noted, an evolutionary adaptation of some plants (especially graminoids) to herbivory has been development of the capacity for compensatory growth. But the impact of herbivory on the growth of plants in general is highly variable, and is determined by characteristics of the plant (e.g., species, nutrient status), the environment (resource availability), the consumption process (type, timing, severity, and longevity of herbivory), and interactions among these factors (e.g., rates of nutrient cycling).

If generalities from a large and conflicting body of literature can be drawn, they are best done so in relation to plant growth form. Defoliation of woody plants usually decreases growth rates, and typically more so for evergreens than for deciduous plants (Archer and Tieszen, 1980; Bryant et al., 1983). For example, moose in a Russian forest damaged the leading shoots of 61 to 72 percent of the young pine trees, which reduced their growth rates for several years (Dinesman, 1967). Browsing by hares and red deer depressed growth in heather, an evergreen shrub (Moss et al., 1981). Deciduous trees of the boreal forest (e.g., willow, birch, aspen) show some capacity for compensatory growth following damage, but browsing still reduces growth rates, especially of young trees (Bryant et al., 1983). In contrast, rangeland shrubs that are frequently browsed by large mammals often show increased production in response to clipping (Ellison, 1960). Perennial herbaceous dicotyledons also tend to respond to grazing with compensatory growth, but usually this does not entirely offset the damage incurred. For example, leaf size of the clonal herb *Aralia nudicaulis* was significantly smaller in damaged plants than in undamaged plants a year after clipping (Edwards, 1985). Considering grasses, Jameson (1963) concluded that total yield is reduced by clipping during the growing season. But McNaughton (1979, 1983c) has argued that in many cases, clipping and grazing at low to moderate levels actually increases the total growth of grasses. Most experiments showing greater than 100 percent compensation in growth, however, have been relatively short-term. The responses of these plants may differ after many successive years of defoliation, when underground storage reserves become depleted.

In recent years, the discovery that mammalian saliva may promote the growth of grasses has captured the interest of ecologists. Saliva is a rich source of growth-promoting compounds, such as thiamine (McNaughton, 1985b, and references therein). Cattle saliva and thiamine increased the growth rates of artificially clipped sideoats grama *(Bouteloua curtipendula)* (Reardon et al., 1972). The same plant species grew faster when grazed by cattle, sheep, or goats than when manually clipped (Reardon et al., 1974), and annual ryegrass *(Lolium perenne)* grew faster in response to grazing by the hispid cotton rat *(Sigmodon hispidus)* than it did to clipping (Howe et al., 1982). However, extensive experiments with bison *(Bison bison)* saliva revealed no growth-promoting ability (Detling et al., 1980).

Reproduction

Mammalian herbivores can exert a strong, direct influence on the reproduction of plants by consuming seeds. In the desert southwest of the United States, heteromyid rodents may remove greater than 75 percent of all seeds produced (Hay and Fuller, 1981, and references therein). More than 41 and 22 percent of the seed fall from Douglas fir *(Pseudotsuga menziesii)* and western hemlock *(Tsuga heterophylla)*, respectively, were destroyed by small mammals in an Oregon coniferous forest, and small mammals consumed 25 to 75 percent of the seeds produced by ash *(Fraxinus excelsior)* in Britain (Crawley, 1983, and references therein).

Browsing and grazing by mammals indirectly affect the reproduction of plants by influencing vegetative propagation, the probability and timing of flowering, and the number and size of seeds produced. Grazing mammals often increase vegetative propagation by plants (especially graminoids) by promoting tillering. In general, however, damage to plants adversely affects sexual reproduction (Jameson, 1963). The intensity of this effect is determined by the same plant, animal, and environmental characteristics that influence the growth response of plants to herbivory. For example, in most studies with range plants, clipping reduced seed yield (Ellison, 1960). Dewberry plants subjected to cattle grazing had lower seed production and more vegetative growth than did ungrazed plants (Abrahamson, 1975). Relatively few studies have been done on wild mammals and their food plants, but Edwards (1985) showed that simulated moose grazing on *Aralia nudicaulis*, a preferred food, reduced fruit production in the year of clipping and the probability of flowering the year after clipping. Batzli and Pitelka (1970) reported that seed production of annual plants preferred by California voles *(Microtus californicus)* was 70 percent less in grazed areas than in ungrazed areas.

Effects on Plant Populations

Virtually no study has documented the long-term dynamics of plant populations in response to herbivory by mammals (Crawley, 1983, and references therein). Of course the large literature on changes in plant community composition resulting from herbivory (or the lack thereof) attest to the fact that mammals must influence plant population dynamics. But these studies have typically compared population densities over only a few points in time rather than monitoring the dynamic responses of populations to known and variable rates of herbivory over long time spans. Moreover, the spectacular impact on vegetation of mammalian herbivores at high population densities also suggests that these animals must affect plant population dynamics (Fig. 5-3).

Mathematical models of plant–herbivore population dynamics may provide insight into the effects of mammals on plant populations. Of course these models do not come close to mimicking all the complexities of interactions in the natural world, but are of great value in crystallizing our understanding of both the significance of individual factors and how these factors interact in natural processes.

A host of mathematical models, varying from simple to quite complex, have been developed to describe the dynamics of plant–herbivore interactions. These models typically consist of paired equations: the first describes the rate of change in the plant population and the second describes the rate of change in the herbivore population. As an introduction to this mathematical approach, a simple model adapted from Crawley (1983) will be examined. More complex and realistic

Figure 5–3. Results of browsing by moose in a Scots pine *(Pinus sylvestris)* forest in northern Sweden (photograph provided by Dr. Kjell Danell, Umeå, Sweden).

models are presented in Caughley and Lawton (1981) and Crawley (1983).

Suppose that in the absence of herbivores a population of plants undergoes logistic growth, with the carrying capacity *(K)* set by resource limitation. Growth of the plant *(P)* and herbivore *(H)* populations can be modeled as the following differential equations:

$$\frac{dP}{dt} = aP\left(1 - \frac{K}{P}\right) - bHP \tag{5.1}$$

$$\frac{dH}{dt} = cHP - dH \tag{5.2}$$

The plant population's intrinsic rate of increase is expressed by a; b is a constant describing the efficiency of foraging by the herbivores, and $-bHP$ is the effect of consumption on plant population growth (Eq. 5.1). The efficiency with which herbivores turn their food into new herbivores (progeny) is described by the constant c, and when multiplied by both the plant and herbivore population densities (cHP) gives the maximum growth rate of the herbivore population (Eq. 5.2). This maximum growth rate is never achieved, however, because of the deaths (d) of herbivores due to a variety of other factors (e.g., predation). (Note that the parameters a, b, c, and d are per capita rates, so they must be multiplied by the population densities P or H to obtain population rates.)

Effects of the basic components of the equations on plant and herbivore population densities become more clear when the factors affecting equilibrium densities are highlighted. By definition, at equilibrium dP/dt and dH/dt equal zero. Thus, the equilibrium plant population density ($\overset{*}{P}$) can be calculated by setting Eq. 5.2 equal to zero and solving to find

$$\overset{*}{P} = \frac{d}{c} \tag{5.3}$$

Similarly, the equilibrium herbivore population density ($\overset{*}{H}$) can be calculated by setting Eq. 5.1 equal to zero, substituting d/c for P, and solving to give

$$\overset{*}{H} = \frac{a}{b}\left(1 - \frac{d}{cK}\right) \tag{5.4}$$

From Eqs. 5.3 and 5.4, it is clear that the equilibrium population density of plants is unaffected by the intrinsic rate of increase of the plant population a, whereas the equilibrium population density of herbivores is a linear function of a. Foraging efficiency (b) does not affect plant equilibrium density. But as b increases, the equilibrium density of herbivores decreases, because a particular plant population can support fewer efficient, than inefficient, herbivores. Factor c reflects herbivore foraging efficiency; thus as c increases, the equilibrium population density of plants decreases, while that of herbivores increases. In contrast, as the death rate (d) of herbivores increases, the equilibrium density of plants increases, but that of herbivores decreases. Stability of the model results from density-dependent control of plant population densities as determined by the carrying capacity, K. Control of plant and herbivore populations increases as $\overset{*}{P}$ approaches K. Thus stability is enhanced with decreasing values of c and increasing values of d. Depending on the values selected for the different parameters, a diverse family of curves can be obtained (Crawley, 1983, Fig. 5-4).

Noy-Meir (1975) developed a simple but elegant set of graphic models of grazing systems. In these models, the growth rate of vegetation is measured as the change in total biomass, which is the sum of individual and population growth. Growth is assumed to be maximum at an intermediate level of biomass; above this level growth declines as a result of competition and the shifting balance between photosynthesis and respiration. Total consumption rates increase with increased population densities of herbivores, and for any one population density, consumption rates may follow different saturation curves (Fig. 5-5). An important outcome of these models is that plant and herbivore populations may coexist at more than one stable equilibrium.

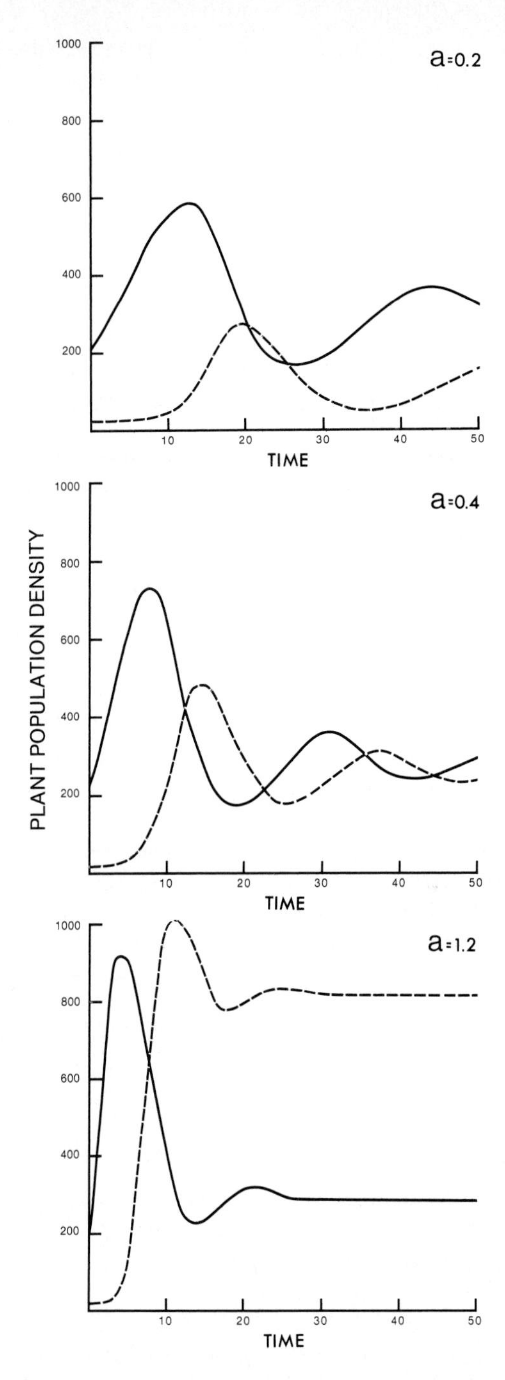

Figure 5–4. Dynamics of interacting plant (—) and herbivore (---) populations. For this model, a increases from 0.2 to 1.2, enhancing stability and increasing equilibrium densities of herbivores. Other parameters are $b = 0.001$, $c = 0.001$, $d = 0.3$, $K = 1000$ (herbivore population densities in arbitrary units, from Crawley, 1983, used by permission).

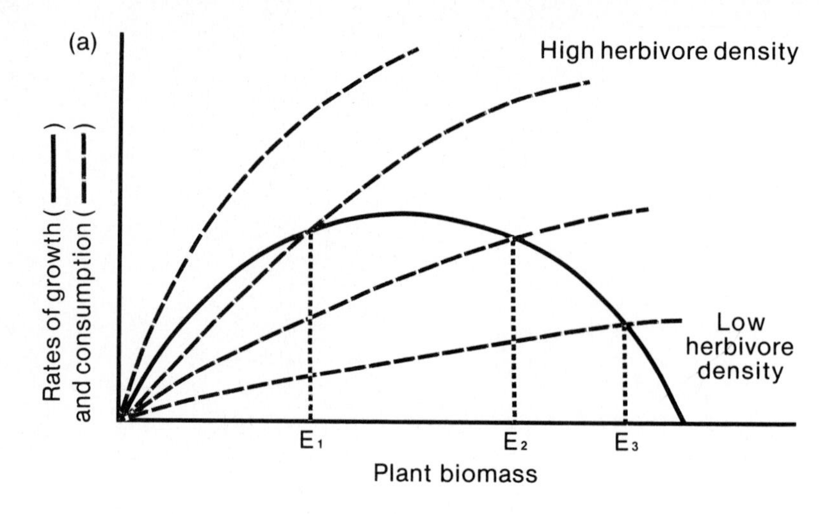

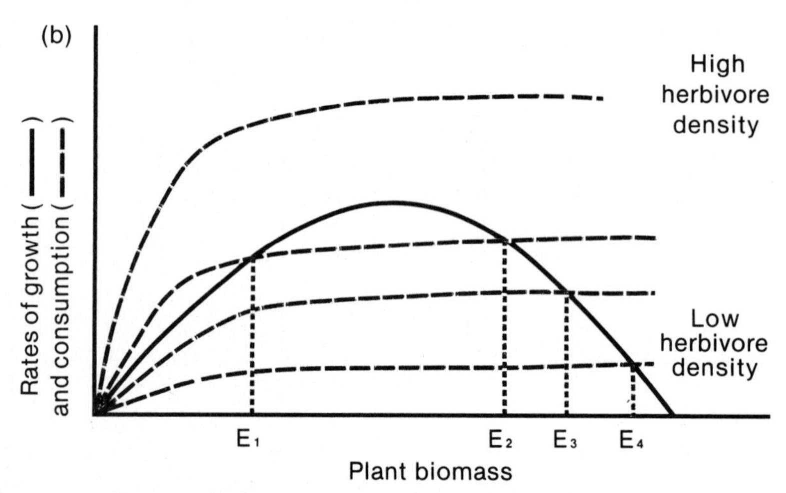

Figure 5–5. Rates of plant growth and herbivore consumption (at several population densities) as a function of plant biomass. (a) Stable equilibria occur at biomasses of E_1, E_2, and E_3. (b) Altered consumption rate curves produce an unstable equilibrium at a biomass of E_1, and stable equilibria at E_2, E_3, and E_4 (after Noy-Meir, 1975).

In the simplest model (Fig. 5-5a), for each consumption curve, plant biomass increases when initial biomass falls below the equilibrium point (growth is greater than consumption), but decreases above the equilibrium point (consumption is greater than growth). Consequently, each equilibrium is stable and the level of stable biomass decreases with increasing herbivore population density. When consumption exceeds growth for all levels of biomass, the plant population is eaten to extinction.

Figure 5-5b illustrates a more complex model, in which the herbivore consumption curves rapidly approach saturation. As in the previous model, low herbivore population densities produce stable equilibria (E_3, E_4), whereas high densities cause extinction. But at intermediate population densities, the consumption curve crosses the growth curve at two points, producing two different equilibria. Plant biomass now decreases both below point E_1 and above point E_2, and increases between E_1 and E_2. Thus, E_1 represents an unstable equilibrium; initial plant biomass below E_1 will be eaten to extinction whereas biomasses above E_1 will increase until they reach the stable equilibrium at E_2. Clearly, small changes in particular herbivore population densities or plant biomasses may lead to dramatically different outcomes in this model.

The models presented here and related ones have been substantially elaborated by ecologists to make them increasingly realistic. Depending on the type and intensity of factors included, modeled interactions may be unstable (leading to the extinction of one or both populations), stable, in which the return to equilibrium usually follows a series of damped oscillations, or have stable limit-cycle behavior, in which populations follow cyclic trajectories and return to them after perturbation. It is clear though, that what is needed most at this time are thorough, empirical studies on the demographics of interacting plant and mammal populations.

Effects on Plant Communities

Mammalian herbivores substantially influence the dynamics of plant communities, altering species diversity (both richness and relative importance of species), three-dimensional structure, and successional processes. These effects in turn affect the diversity and activity of other herbivores, predators, and parasites in the community (Fig. 5-6).

Direct and Indirect Effects

Mammals directly influence community dynamics by feeding selectively on various plant species. Selective feeding by generalist herbivores in particular may cause the local extinction of a plant species because herbivore population densities may not decline appreciably even as a preferred plant becomes scarce.

Even more pervasive and important, however, are the indirect effects of herbivores on community dynamics. By feeding selectively and altering the cycling and dispersion of nutrients, mammals modify the

Figure 5–6. A grazing exclosure on the upper Gallatin River (Gallatin National Forest, Montana) constructed in 1948, enlarged in 1958, and photographed in 1960. Dense browse in older portion of enclosure consists of willow. Browsing by elk during winter has strongly hedged willows outside the exclosure (foreground) and eliminated willow from much of the upper Gallatin Valley (photograph provided by the National Agriculture Library, Forest Service Photo Collection).

competitive relationships among plant species. Soil disturbance via trampling and burrowing creates new microsites for the germination of seeds.

Plant Species Diversity

Harper (1969) concluded that no simple generalizations about the impact of herbivores on plant species diversity can be made. Depending on the plant species involved and the type and intensity of herbivory, herbivores may decrease or increase species diversity.

Herbivores tend to decrease species diversity when they avoid feeding on competitively dominant plants. Harper (1969) summarized results from several studies of sheep grazing. In one experiment in which the dominant pasture species were unpalatable, species diversity decreased because of removal of less common plant species. Low to moderate levels of grazing by sheep have promoted the spread of the unpalatable grass *Nardus stricta* in British pastures. Unless the available forage becomes limiting, sheep feed extensively on grasses such as *Agrostis* and *Festuca* but avoid *Nardus*, which comes to increasingly dominate the plant community (Crawley, 1983, and references therein). Watt (1981a, 1981b) found in a long-term study of British plant communities that protection from grazing by rabbits eventually nearly tripled the number of species in the community. Most of the species added were known to be sensitive to grazing. Finally, the most striking examples of the fact that mammals can dramatically lower species diversity can be found in the vast areas of overgrazed rangelands world wide. Such overgrazing is a major contributor to the desertification of formerly productive ecosystems, an environmental problem of tremendous importance.

Perhaps the more common scenario in natural communities is that dominant species are eaten by herbivores. Subsequent decline in their competitive abilities allows for the growth of less competitive species and an increase in species diversity. In contrast to Watt's (1981a, 1981b) results, other studies of rabbit grazing in Britain, summarized by Harper (1969), indicated that grazing promoted vegetation diversity. In the absence of rabbits a few species of grass became dominant and the total number of species, especially dicotyledons, rapidly declined. Ericson (1977) conducted an extensive study of the effect of voles and lemmings on the vegetation of the floor of a Swedish coniferous forest. He found that during years of high mammal densities, dominant species (dwarf shrubs, grasses, and mosses) declined but some less competitive species (e.g., perennial herbs) increased in importance. Another study showed that feeding by black-tailed prairie dogs *(Cynomys ludovicianus)* increased the species diversity of vegetation in mixed-grass prairies (Coppock et al., 1983a). Species diversity was highest on prairie dog towns of intermediate age (3 to 8 years). In an area colonized for 26 years, forbs and shrubs dominated, whereas on an uncolonized site, grasses dominated. Both sites had lower species diversities than the intermediate-aged communities. The researchers suggested that diversity is maximized in grazing systems where herbivores reduce, but do not eliminate, the dominant plant species.

Plant species diversity may also be altered by the effects of mammals on nutrients and soils. High concentrations of nutrients (from urine and feces) near small mammal burrows or the carcasses of caribou on the arctic tundra led to increased importance of graminoids and decreased importance of lichens and sometimes shrubs (McKendrick et al., 1980). Overall species diversity of an annual grassland community was enhanced by the creation of soil mounds by western pocket gophers, *Thomomys bottae* (Hobbs and Mooney, 1985). Differential timing of seedfall of plants in the community led to differential colonization of gopher mounds, and increased spatial patterning and species diversity.

Community Structure

Feeding by mammalian herbivores may alter the structure of plant communities, sometimes drastically. These changes may in turn influence the dynamics of interactions between herbivores and other members of the biological community. For example, during years of high population densities lemmings near Point Barrow, Alaska, severely uproot and clip vegetation, resulting in a 50 to 90 percent decrease in standing crop (Schultz, 1964). Such destruction of the vegetative cover increases the susceptibility of lemmings to predation; indeed, during periods of high population densities, predatory birds such as pomarine jaegers and snowy owls move in, establish territories, and attack the exposed lemmings (Batzli et al., 1980). Research on migratory ungulates in the Serengeti has shown how grazing by one species may change vegetation structure to the benefit of another species. Grazing by wildebeest *(Connochaetes taurinus albojubatus)* migrating through the Serengeti Plains reduced green biomass and height by 85 and 56 percent, respectively, and stimulated productivity of remaining grasses. Thomson's gazelle *(Gazella thomsonii)* followed a month later, and preferentially grazed in areas heavily defoliated by wildebeest. In these areas, gazelle have improved access to the especially nutritious grass shoots and dicotyledons that grow close to ground level (Bell, 1971). Similarly, bison preferentially graze around prairie dog towns because of the structural, compositional, and nutritional changes in vegetation caused by prairie dogs (Coppock et al., 1983b).

Succession

The activities of mammalian herbivores can influence the rate at which plant species are added to or removed from communities, and so affect the process of succession. Ellison (1960) reviewed the effects of grazing by domestic herbivores on succession of rangelands. As expected, he found that the direction of succession (toward woody or herbaceous vegetation) depends on the type and severity of grazing. With regard to natural systems, grazing by bison inhibited the invasion of woody plants in some areas of the Great Plains. Anderson and Loucks (1979) showed that browsing by whitetail deer *(Odocoileus virginianus)* in Wisconsin reduced recruitment of shade-tolerant hemlock *(Tsuga canadensis)*. Hemlock is rapidly replaced by sugar maple *(Acer saccharum)*, which readily resprouts after browsing. Authorities differ as to whether hemlock or sugar maple is the climax forest species, so browsing by deer may accelerate or delay succession in this instance (Anderson and Loucks, 1979, and references therein). Moderate to severe browsing by mammals, especially whitetail deer, delayed successional development in a pine *(Pinus resinosa)* forest in northern Minnesota. In this region succession follows a sequence from pines, aspen, and oak to birch and maple (Ross et al., 1970).

Adaptations of Mammalian Herbivores to Plants

First he ate some lettuces and some French beans; and then he ate some radishes; and then, feeling rather sick, he went to look for some parsley.
[Beatrix Potter, The Tale of Peter Rabbit]

The problem of transforming plant tissues that are poor in protein and nutrients but rich in fiber, physical defenses, and chemical defenses into animal tissues rich in protein and nutrients but free of toxins is a formidable one. Mammalian herbivores have evolved a variety of adaptations to accomplish the task of utilizing an abundant (usually) but poor-quality food source.

Behavioral Adaptations

The fact that herbivores do not consume plant species and tissues in the same proportion as they are encountered is evident to even the most casual observer of natural (or semi-natural) systems. Branches and trunks of trees remain untouched even after severe defoliation by insects; bull thistles *(Cirsium vulgare)* stand with apparent impunity in heavily grazed pastures. But what factors determine a herbivore's choice of food, and how does it "know" what is good or not good to eat?

Diet Selection

Plant species that are consumed in proportions greater, similar to, or less than their proportions in the environment are classified as "preferred," "nonpreferred," or "avoided," respectively. Thus, plant availability is an integral component of any estimate of preference. Plant species that are avoided by herbivores may still constitute substantial portions of the diet if they are very abundant in the habitat.

Plant nutrient (especially protein and mineral) concentrations are one of several major factors influencing the selection of forage by mammals. Deer preferentially feed or browse on herbaceous plants with high levels of protein and minerals; moose on Isle Royale, Michigan, select plants with high sodium levels (Lindroth, 1979, and references therein); and voles prefer plants with high protein levels (Lindroth and Batzli, 1984a).

Physical defenses of plants also affect their use by herbivores. Selectivity is influenced by the presence of silica, fiber, spines, and thorns, and the ease with which plants can be harvested and swallowed (Lindroth, 1979, and references therein). Thus, as food resources become limiting, range managers burn the spines off cacti to promote their consumption by cattle. Unfortunately, few ecological studies have investigated the response of mammals to physical defenses in potential food plants.

Chemical defenses may be the most important plant characteristic influencing food choice by many mammals. Bryant and Kuropat (1980) concluded that for snowshoe hares, moose, and several species of gallinaceous birds, food selection was not correlated with protein or energy content, but strongly and negatively correlated with resin (containing terpenes and phenolics) content. Browsing by three species of ruminants in a South African savanna was strongly deterred by condensed tannins in woody plants, and only weakly correlated with protein and mineral content (Cooper and Owen-Smith, 1985). Food habits of microtine rodents (voles and lemmings) in the arctic also appear to be influenced more by plant chemical defenses than by nutrient concentrations (Batzli and Jung, 1980; Rodgers and Lewis, 1985).

However, it is naive to think that food preferences of mammals can be explained by one or a few forage characteristics; more likely they are determined by a complex of interacting factors (Lindroth, 1988). Herbivores are probably adapted to achieve the best mix of nutrients given their nutritional requirements and the quality of available forage (Lindroth, 1979). In some instances they may select plants with low nutritional quality in order to avoid chemical and physical defenses; in others they may select plants with high levels of defenses because the costs of doing so are more than offset by high levels of nutrients. Moreover, preferences change in relation to changing nutritional needs of herbivores and changes in relative availability of forage. Herbivores may continually sample available food items in their habitats in order to monitor changes in food quality (Westoby, 1974).

Learning Mechanisms

How do herbivores "know" what is or is not good to eat, given that their own nutritional needs and the quality of available food change over time? It is unclear at this time whether the preferences of mammals are primarily innate or learned. Mammals appear to have innate preferences for certain scarce but essential nutrients (e.g., sodium) and innate aversions to some plant chemicals that exhibit broad spectrum toxicity to mammals (e.g., many alkaloids). The heritability of food preferences of wild mammals has been poorly studied, but some research indicates that broad preference patterns are inherited, e.g., the preference of brown lemmings *(Lemmus sibiricus)* for grasses and mosses and of collared lemmings (*Dicrostonyx groenlandicus*) for shrubs and herbaceous dicotyledons (Rodgers and Lewis, 1985). However, mammals probably do not have physiological mechanisms for the detection of every conceivable nutrient or toxin. Thus, a learning-based model probably provides a better explanation for most foraging behavior.

Many mammals can form (learn) associations between ingested foods and sickness or recovery from sickness. These associations can be made on the basis of only a single trial even when long time intervals elapse between consumption and sickness or recovery, can be remembered for weeks or months, and influence subsequent foraging behavior. Such "long-delay learning" explains the feeding behavior of rats quite well. They rapidly learn to avoid diets that contain toxins or are nutritionally deficient, and to preferentially consume diets that enable them to recover from previous dietary deficiencies (Lindroth, 1979, and references therein). Little research has addressed the suitability of this model for wild mammals, but preliminary evidence is promising. The opossum *(Didelphis virginiana)* associates an innocuous chemical in mushrooms with delayed illness caused by a mushroom toxin, and bank voles *(Clethrionomys glareolus)* associate the nutritional quality of diets with cues such as taste or color, and alter subsequent feeding preferences accordingly (Lindroth, 1988, and references therein). Mammalian herbivores may indeed learn to forage in ways to avoid or compensate for toxic or nutritionally deficient diets.

Mammalian Morphological Adaptations

The teeth of most mammalian herbivores are adapted to diets that are high in fiber or silica but low in nutrients. High crowned molars, and in some mammals, evergrowing molars evolved in response to rapid tooth wear. Specialized grinding surfaces reduce plant material to fine particles, allowing maximum surface area for digestion (Vaughan, 1972; Robbins, 1983).

The gastrointestinal tracts of herbivores exhibit a variety of adaptations for the utilization of plants as food. The small intestine is the major site for enzymatic digestion of food and absorption of nutrients, and the large intestine is important for synthesis of vitamins, absorption of nutrients, and in many herbivores, fermentative digestion of fiber. To promote these functions, herbivores have developed especially long small and large intestines (Robbins, 1983, and references therein). Most herbivores also have evolved specialized chambers for fermentation of plant material by symbiotic microorganisms. Many herbivores, most notably the ruminants, have compartmentalized stomachs, while other herbivores, most notably the lagomorphs (rabbits and hares) and equids (horses), have greatly enlarged cecae. Microorganisms in these chambers digest cellulose (mammals lack the enzymes required to do so) and produce essential amino acids, fatty acids, and vitamins.

Gastrointestinal morphology of individual animals may acclimate to changes in forage quality or a mammal's nutritional requirements. Gross et al. (1985) found that when food fiber increased, the sizes of large intestines and cecae increased in prairie voles *(Microtus ochrogaster)*. When their energy requirements increased (under low temperatures) the sizes of small intestines and cecae increased.

Physiological Adaptations

Ingested food is subjected to the physiological processes of digestion, absorption, and transport, which in turn are related to the overall metabolic rates of mammals. These processes have also been molded by evolution to enhance the ability of herbivores to utilize plants as food.

Digestive Physiology

The activity of digestive, absorption, and transport processes may be induced or suppressed as the quantity and quality of ingested food changes. The most important response involves the pH regulation system (Kaufmann et al., 1980). Changes in dietary composition cause shifts in gastrointestinal pH, which in turn influence activity of gut microorganisms and rates of absorption of specific nutrients. Although many nutrients are passively absorbed from the gastrointestinal tract into the bloodstream, some require active transport systems. Specialized carrier systems exist for the transport of some sugars, amino acids, and minerals, and all of these are under physiological control.

A unique physiological mechanism evolved to minimize the loss of nitrogen (an especially important nutrient because animals cannot store protein) in herbivorous mammals is urea cycling. Urea enters the bloodstream via the kidneys and is recycled back into the gastrointestinal tract by diffusion or via saliva or digestive enzymes. The urea can then be used by gut microorganisms to synthesize amino acids and proteins, which are made available to the herbivore (Robbins, 1983). Finally, although little work has addressed this possibility, it is likely that the digestive physiology of mammalian herbivores also functions to inhibit the activity of digestion-reducing compounds and slow the absorption of toxins (Lindroth, 1988).

Metabolic Rates

When the effect of body size on basal metabolic rates (BMR) of mammals is factored out, much of the residual variation is correlated with food habits (McNab, 1983). High BMRs are associated with short gestation times, rapid postnatal growth, and high fecundity. So why do not all mammals have high BMRs? McNab (1983, 1986) argues that BMRs have adjusted evolutionarily to fit the food habits of mammals. Food characteristics that may have caused decreases in BMR include seasonal periodicity (rates lowered to the maximum level possible during the poorest season), low levels of digestible nutrients (rates lowered to the maximum level that can be maintained given the energetic and physical constraints on consumption rates), and high concentrations of defensive chemicals (rates lowered to the maximum level that can be maintained given the costs of detoxifying the compounds). For example, mammals (such as some monkeys) that feed on leaves of trees typically encounter high levels of secondary compounds in their foods. They have lower metabolic rates than do grazing mammals of similar size, which eat plants (especially grasses) with lower levels of secondary compounds.

Biochemical Adaptations

All adaptations of mammals are ultimately linked to chemical processes, but some adaptations of mammals to plant diets involve primarily chemical mechanisms. Alterations of basic chemical processes (i.e., the enzymatically moderated biochemical pathways) serve the same purpose as do the forms of adaptations previously discussed. They tend to optimize the efficiency with which food nutrients are utilized, and reduce or circumvent the harmful effects of food constituents.

Metabolism of Nutrients

The metabolic pathways of different mammals have adapted to the particular end products of their digestive systems. Thus, the function of certain enzyme systems has been enhanced or whole systems deleted in the course of evolutionary adjustments to different diets. In addition, the enzyme systems utilized by mammals have the capacity to be induced or suppressed.

For example, herbivores exhibit biochemical adaptations for energy metabolism. Although their diets are high in carbohydrates (especially polymers such as cellulose), microbial fermentation of these compounds to volatile fatty acids results in little glucose being absorbed from the gastrointestinal tract. This is especially true for ruminants, and may hold as well for nonruminants that rely heavily on fermentative processes. But glucose is required for cellular metabolism. Thus, ruminants have adapted to low inputs of glucose by two means. First, glucose is utilized conservatively; acetate rather than glucose is used for synthesis of fatty acids. Second, they maintain high rates of gluconeogenesis (Morris and Rogers, 1982).

Mammalian herbivores (like all mammals) can also modify protein metabolism in response to changes in their diets. When consuming foods with low levels of protein, the activity of enzymes involved in amino acid catabolism decreases, thereby conserving protein. But when foods have high levels of protein, the activity of these enzymes is induced to allow for maximum utilization of excess amino acids (Morris and Rogers, 1982). Enzymes involved in the urea cycle may also be induced or suppressed to moderate loss of nitrogen.

Metabolism of Toxins

Metabolic detoxication of secondary plant compounds by mammals involves a variety of enzymatic pathways in a two-phase system. In phase I, compounds are oxidized, reduced, or hydrolyzed; in phase II they are conjugated with endogenous compounds such as amino acids, glucuronic acid, or sulphate. These reactions are not detoxication reactions per se; occasionally they may activate compounds, such as the pyrrolizidine alkaloids. But the metabolized compounds are more hydrophilic and thus more readily excreted by animals (Scheline, 1978; Brattsten, 1979b). Compounds may be excreted after modification by phase I, phase II, or both phase I and II reactions.

Gut microorganisms may be considered an herbivore's first line of active defense against plant toxins, and this function may have been important in the evolution of mutualistic associations between herbivores and their microbes (Lindroth, 1988, and references therein). Gut symbionts transform plant compounds by hydrolytic, reductive, and degradative reactions. For example, ruminal microbes in sheep hydrolyze the pyrrolizidine alkaloids heliotrine and lasiocarpine to nontoxic products. In some cases, however, microbes may activate plant toxins, such as in the hydrolysis of cyanogenic glycosides to release hydrogen cyanide. Gut microorganisms can accommodate not only a diversity of plant compounds, but also temporal variation in the amounts and types of compounds. Adaptations occur by changes in the activity or composition of the microbial community (Lindroth, 1988, and references therein).

Plant toxins absorbed from the gut are metabolized by specific tissue enzymes located in most organs, but which are especially concentrated in the liver. In contrast to the reactions of gut microorganisms, the most important tissue reactions are oxidative and conjugative. The mixed-function oxidase (MFO) enzymes are especially important; they catalyze many oxidative reactions (Fig. 5-7), act on a great variety of compounds, and their activity can be rapidly induced by exposure to toxic constituents (Brattsten, 1979b).

Impact of Plants on Mammalian Herbivores

There is a nutritional factor here, as if the juices of green growing plants were essential to the diet before fertile matings take place.
[Ronald Lockley, The Private Life of the Rabbit]

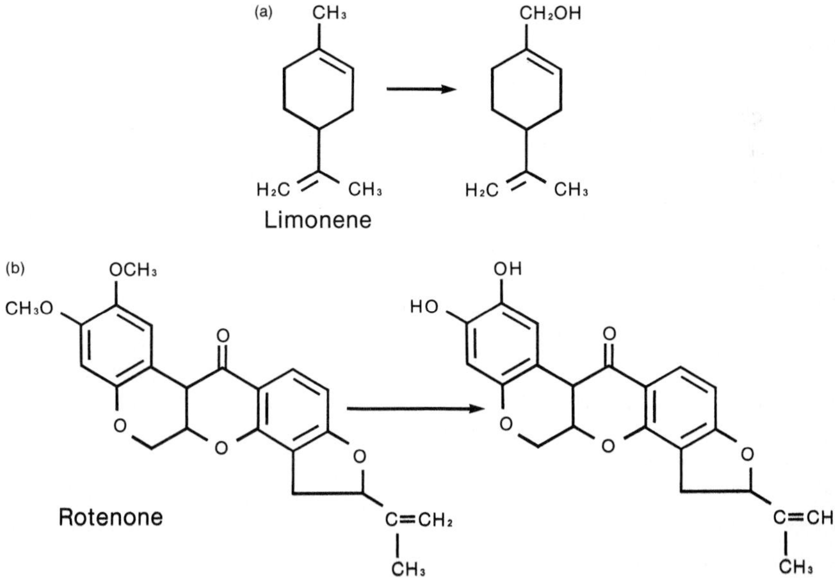

Figure 5–7. MFO-catalyzed detoxication of (a) limonene, a monoterpene, and (b) rotenone, an isoflavonoid.

Just as plants respond to the impact of herbivores over ecological time, so herbivores respond to the quality and quantity of food over time. The influence of plants on the performance of individual herbivores has repercussions for the dynamics of herbivore populations and communities.

Effects on Individual Performance

Performance Parameters

The quality and quantity of available food affect each of the common measures of herbivore performance, viz., survival, growth, and reproduction. Evidence of herbivore mortality resulting from inadequate food resources is rare. Most malnourished animals probably succumb to disease or predation before starvation, and if they do die, the telltale signs are quickly removed by scavengers. Still, enough evidence exists to implicate insufficient food resources as a major cause of mortality in some situations. Death rates of juvenile snowshoe hares, wild sheep, and many species of deer increase with decreasing food availability (Keith, 1983; Crawley, 1983, and references therein; Fig. 5-8). Survival of prairie voles (Cole and Batzli, 1979) and young prairie dogs (Rayor, 1985) was better in habitats with high-quality forage rather than low-quality forage.

The quality of available forage may also inhibit the growth of herbivores, either because of low nutrient concentrations or because of physical or chemical defenses. Cole and Batzli (1979) found that prairie voles grew faster in laboratory feeding trials when fed natural diets with high nutrient concentrations, and in the field growth rates were greatest in the habitat with highest quality forage. Similarly, White (1983) reported that growth of reindeer *(Rangifer tarandus)* calves was more rapid on high nutrient ranges, and Rayor (1985) showed that growth of prairie dogs was faster in a habitat with greater diversity and quantity of vegetation. Growth of prairie voles is inhibited by increases in dietary levels of plant secondary compounds under both laboratory and field conditions (Lindroth and Batzli, 1984b, 1986). In habitats of marginal quality, the most important food items may have levels of nutrients so low or levels of secondary compounds so high that growth of mammals is appreciably below the maximum level possible (Lindroth, 1988, and references therein).

The quality of available food can also directly and indirectly influence reproduction. Reproductive success of voles, deer, sheep, and cattle is correlated with the nitrogen (protein) content of food (Crawley, 1983, and references therein). Sodium may limit reproduction by moose on Isle Royale (Belovsky and Jordan, 1981); calcium and sodium may

Figure 5–8. Severely browsed juniper (note the high browsing line) and dead mule deer in the western United States (photograph provided by the National Agriculture Library, Forest Service Photo Collection).

limit reproduction by California voles; and calcium and phosphorus may limit reproduction by brown lemmings (Batzli, 1985, and references therein). Cinnamic acids (secondary compounds) inhibit reproduction by montane voles *(Microtus montanus)* in laboratory studies, and reproduction by voles in the field ceases as levels of cinnamic acids increase in plants in autumn (Berger et al., 1977). Another secondary compound (6-methoxybenzoxazolinone) triggers the onset of reproduction in voles in the laboratory and may cue voles in the field to commence breeding in spring (Berger et al., 1981; Sanders et al., 1981).

Poor food quality may indirectly reduce reproduction by slowing growth rates and delaying the onset of maturity. In many, if not most, mammals reproductive maturity is determined more by the attainment of a certain minimum body size than a minimum age (White, 1983; Rayor, 1985).

Interactions among Factors

It is clear then, that particular plant characteristics can affect any or all of the performance parameters for individual herbivores. But the impact of plants on the nutritional status of herbivores is rarely simple and direct; rather, it is the product of an extremely complex set of interacting plant and animal characteristics (Lindroth, 1988; Fig. 5-9).

A mammal's response to a food plant is influenced by its genetic constitution, age, sex, and previous experience with the same and other food plants. High levels of nutrients affect the nutritional status of an

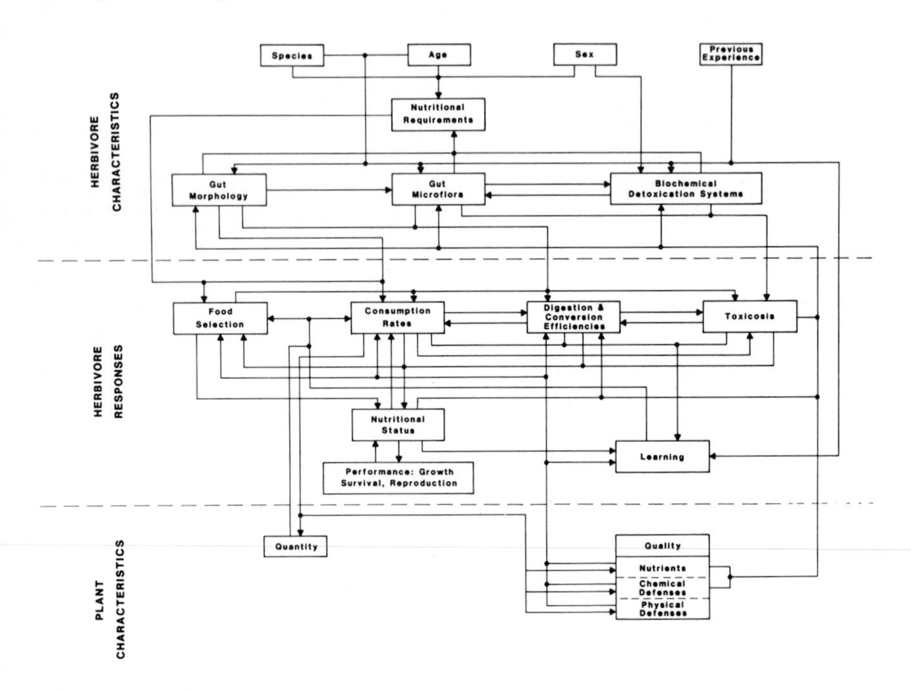

Figure 5–9. Direct and interactive relationships among biotic factors affecting the performance of mammalian herbivores.

animal by altering its food selection and consumption behaviors, and increasing digestion and conversion efficiencies. Chemical defenses also alter an animal's food selection and consumption patterns and digestion and conversion efficiencies, and may cause various levels of toxicosis. Alternatively, plant nutrients and chemical defenses affect herbivores indirectly by changing their gut morphologies and microbial communities, and the efficacy of their biochemical detoxication systems. These effects will feed back to influence herbivore responses such as food selection and consumption rates. Thus the responses of voles (e.g., survival, growth, consumption rates, digestion and conversion efficiencies, and toxicosis) to different foods vary according to vole species and according to the concentrations of both nutrients and secondary plant compounds in their foods (Lindroth, 1988, and references therein).

Interactions among herbivore responses (especially foraging patterns) may, via "multiplier effects," greatly increase their performance (White, 1983). Based on an experimental study of reindeer grazing on the arctic tundra during summer, White (1983) calculated that selective grazing increased food digestibility by 14 percent and food intake by 27 percent, which combined to produce a 268 percent increase in growth compared to reindeer that did not graze selectively. Similarly, relatively small changes in body size of many ungulates prior to mating may greatly increase their probability of conception.

Effects on Herbivore Populations

The influence of plants on the dynamics of herbivore populations depends on the demographic characteristics of both populations and the timing, type, and degree of their density dependence (Crawley, 1983). Fluctuations in herbivore populations may result from many factors other than food, e.g., weather, disease, and predation. But numerous studies have indicated that food resources do play an important role in the regulation of many herbivore populations. In fact, it is almost universally accepted that food resources determine the carrying capacity of a habitat for mammalian herbivores (Crawley, 1983).

For example, population densities of several large ungulates in the Serengeti appear to be regulated by food supplies. Sinclair (1974) concluded that African buffalo *(Syncerus caffer)* populations were regulated by adult mortality, caused by inadequate nutrition during periods of poor food quality. The shortage of good quality food itself resulted from climatic changes and competition among buffalo, and between buffalo and wildebeest. Later Sinclair et al. (1985) reported that populations of Serengeti wildebeest were also limited by their food supply. The food supply and mortality rate for adult wildebeest interacted in a density-dependent manner to regulate population densities. In the deserts of southern California, the cactus *Opuntia bigelovii* provides woodrats *(Neotoma lepida)* with four resources: food, water, materials for construction of dens, and protection from predators. Brown et al. (1972) found that 80 percent of the variability in woodrat population densities was attributable to variation in the population densities of cacti.

Population densities of some mammalian herbivores exhibit dramatic and regular cycles in abundance, with periods of 3 to 4 years (voles and lemmings) or 10 years (hares). Hypotheses regarding the underlying cause(s) of these cycles have been hotly debated for most of this century, and have yet to be resolved. Tests of the many variations of the food hypothesis have usually shown that food certainly influences the dynamics of cycling populations, but is not the single causative factor. In summarizing the relevant literature Batzli (1985) concluded that voles and lemmings can reach high population densities typical of cyclic peaks only when high quality forage is available, but that neither absolute shortages of food nor relative shortages of specific nutrients (e.g., protein and minerals) are necessary for population declines. In other words, the nutrient quality of available vegetation may strongly influence some demographic parameters (e.g., the intrinsic rate of increase, mortality and fecundity rates, peak population densities) but is not the single driving force behind the cycles. More recently, several researchers (Haukioja, 1980; Rhoades, 1983b) suggested that population cycles might be caused by induction and relaxation of plant chemical defenses in response to grazing. Indeed Batzli's (1983) theoretical model of plant–lemming interactions indicated that induced secondary plant compounds could lead to oscillatory population densities. Lindroth and Batzli (1986) tested this hypothesis for prairie voles and found that although grazing at high population densities did induce a defensive response in forage, the response was short-lived and could not account for the multiyear cycling phenomenon.

Changes in the quality or quantity of forage can better explain population cycles of hares, especially snowshoe hares. During periods of high population densities, snowshoe hares in Alaska severely browse deciduous trees and shrubs. Afterward, these plants produce adventitious shoots that are heavily defended by terpene and phenolic resins for 2 to 3 years. The shoots are highly unpalatable and will not support the survival of hares. These findings are consistent with the suggestion that defensive responses of deciduous trees drive the 10-year cycle. The periodicity of the cycle is probably a result of time-delayed negative feedback related to the recovery of vegetation following extensive browsing (Bryant, 1981; Bryant et al., 1983). Alternatively, Pease et al. (1979) and Keith (1983) argue that absolute shortages of food in winter, rather than changes in food quality, drive hare population cycles. Based on their estimates of food requirements of hares, there simply was not sufficient food available during cycle peaks and for several years following to sustain large populations of hares. Whether it is due to changes in food quality, food quantity, or both, changes in available vegetation best explain the dynamics of snowshoe hare populations.

Effects on Herbivore Communities

The dynamics of interspecific relationships among mammalian herbivores are strongly influenced by food resources. This discussion addresses the two most common of such relationships: competition and commensalism. Mutualistic associations among herbivorous mammals have not been observed, and amensalism has received little attention by ecologists. [Crawley (1983), however, considers asymmetric competitive interactions as amensalisms.]

Competition

Either the evolutionary role of competition or its present action can explain many of the current patterns of resource partitioning and composition in herbivore communities (Linzey, 1984, and references therein). Plant chemical defenses have had a profound influence on the evolution of the dynamics of rodent communities on the coastal plain of northern Alaska. Food habits of the three major species differ markedly. Brown lemmings *(Lemmus sibiricus)* eat mainly monocotyledons (> 80 percent of diet during summer), which have relatively low nutrient concentrations. Collared lemmings *(Dicrostonyx torquatus)* eat mostly dicotyledons (> 76 percent of diet during summer), which have higher nutrient concentrations but are also more strongly defended by chemicals. Tundra voles *(Microtus oeconomus)* eat both monocotyledons and dicotyledons, but occur in restricted areas (e.g., disturbed sites). Brown lemmings and collared lemmings perform poorly on each other's foods and on diets coated with extracts containing some of the secondary compounds from each other's foods. Apparently the two species have evolved detoxication systems that allow them to specialize on certain types of plants and reduce competition. Restricted distribution of the less specialized voles may be a result of competition with lemmings (Batzli and Jung, 1980).

Numerous studies have shown that competition influences the dynamics of communities in contemporary time as well. For example, the structure of heteromyid rodent communities in North American deserts is strongly influenced by competition for seeds. Seed size selection is positively correlated with body size. In habitats where seed production is high, species that utilize similar seeds coexist, and herbivore species diversity is high. In less productive habitats, some ecologically similar species are excluded, and herbivore species diversity is lower (Brown and Lieberman, 1973). A study by Linzey (1984) provides strong evidence for competition between meadow voles *(Microtus pennsylvanicus)* and southern bog lemmings *(Synaptomys cooperi)* in the southern Appalachians. Lemmings are excluded from preferred habitats by voles, and occupy areas where cover is sparse and food has poor nutritional value. When vole populations declined either naturally or after removal, lemmings moved into the preferred habitats and exhibited food habits similar to those of voles. Finally, Belovsky (1984) used field data and competition models to demonstrate competition between moose and snowshoe hares on Isle Royale. The interaction is strongly asymmetric; hares are highly susceptible to competition from moose, whereas moose are only slightly affected by competition from hares.

Commensalism

Some interactions between herbivores are commensal in nature; one species benefits but the other is unaffected. A classic example involves the interactions of zebra, wildebeest, and Thomson's gazelle in the Serengeti. These ungulates are migratory, and their movements follow a characteristic succession in which grazing by one species alters the vegetation structure such that the remaining vegetation is more effectively utilized by the migratory species that follows. Zebra move first in the succession, eating the tall, protein-poor stems and sheaths of grasses. Large herds of wildebeest follow, consuming grass sheaths and the more nutritious grass leaves. Last to migrate are Thomson's gazelles, which feed on the nutritious regrowth of grasses and protein-rich herbs. Thus temporal partitioning of the food resource enables the activities of one species to facilitate the flow of energy and nutrients into the following species (Bell, 1971; McNaughton, 1976).

Evolution or Coevolution?

Clearly, both plants and mammalian herbivores have evolved adaptations in response to the influence of each on the other. But to say they have coevolved is another matter. In its most narrow usage, coevolution refers to closely integrated, reciprocal evolutionary adjustments between two interacting populations. But this narrow definition excludes mammals and generalist insects because these herbivores feed on a variety of plant species. Thus, ecologists introduced the notion of "diffuse coevolution," involving gradual evolutionary adjustments among complex assemblages of plants and animals (Lindroth, 1988, and references therein). But again, coevolution requires reciprocal evolutionary responses in plant–herbivore systems, and the degree of reciprocity certainly varies among populations in different communities. In some (probably most), the adaptations of mammalian herbivores evolved in response to plant adaptations that themselves evolved in response to insect herbivores. Thus, the mammals have evolved in response to, but not coevolved with, the plants. At the other extreme, in some communities the evolution of plants and mammals is more tightly coupled, and insects play a less significant role.

Exploitative Interactions

Because of the difficulty of unraveling complex evolutionary histories of assemblages of organisms, conclusive evidence supporting diffuse coevolution between plants and mammals is sparse. But evidence from some systems is provocative at least.

Chemically mediated coevolution may explain the present-day characteristics of several plant–mammal systems. As already mentioned, winter browsing by snowshoe hares exerts a strong selective pressure on deciduous trees of the boreal forest. Protective chemical defenses (deterrent resins) probably evolved in response to such selection, and these in turn may strongly influence hare populations. Herbivory by lemmings can drastically affect tundra vegetation during years of high population densities, and in turn, the food selection and performance of these animals is strongly influenced by secondary plant compounds. Finally, Mead et al. (1985) suggested that the coevolution of some plants and several species of herbivorous marsupials in Australia may have been chemically mediated. They suggest that numerous species in the Leguminosae evolved fluoroacetate (a potent inhibitor of the citric acid cycle) in response to herbivory by mammals. But the mammals counteradapted by evolving resistance to fluoroacetate. Presently, populations of kangaroos and bush rats whose ranges include the fluoroacetate-producing plants are highly resistant to the compound. But the same species of bush rat and related species of kangaroos whose ranges do not encompass the plants are susceptible. For each of these systems, however, additional information is needed to more fully substantiate the possibility of coevolved interactions.

Diffuse coevolution between plants and mammals is best documented with respect to grasslands and large grazing mammals (Mack and Thompson, 1982; McNaughton, 1983b, 1985a). The two groups have closely associated evolutionary histories, exert strong selection pressures on one another, and have developed a variety of adaptations for avoiding, tolerating, or abetting herbivory.

Mutualistic Interactions

No discussion of plant–mammal interactions would be complete without considering the notion that herbivory benefits plants. This view has been vigorously promoted by Owen (1980) and Owen and Wiegert (1981), but its acceptance by the larger scientific community has been cool at best (e.g., Silvertown, 1982; Crawley, 1983; Belsky, 1986). Proponents of this view suggest that the responses of grasses to grazing (e.g., delayed senescence, compensatory growth, increased vegetative propagation) increase their genetic fitness. Thus, Owen and Wiegert (1981) assert "grasses do not 'defend' themselves from grazers; rather they encourage them." Opponents argue that herbivory is most unlikely to increase fitness because it causes an appreciable nutrient and energy drain. The fact that in some short-term experiments, grazing increased reproduction does not mean that fitness was enhanced, because fitness is a compilation of many factors in addition to reproduction. Moreover, the presence of physical defenses by grasses (e.g., high and inducible concentrations of silica, sharp awns) and counter-adaptations by mammals (e.g., high crowned teeth) would suggest that grazing is an antagonistic, not a mutualistic, interaction. McNaughton's (1985a) study of the effects of grazing in the Serengeti concluded that herbivory was not advantageous to plants, and reminds us that "highly interactive organisms can be interdependent . . . without being mutualistic."

Selected References

Bell, R. H. V. 1971. A grazing system in the Serengeti. *Sci. Am.* 225:86–93.
Crawley, M. J. 1983. Herbivory: the dynamics of animal-plant interactions. University of California Press, Berkeley, CA.
Eisenberg, J. F. 1981. The mammalian radiations: an analysis of trends in evolution, adaptation, and behavior. The University of Chicago Press, Chicago, IL.
Lindroth, R. L. 1988. Adaptations of mammalian herbivores to plant chemical defenses. In: K. C. Spencer, ed. Chemical mediation of coevolution. Academic Press, NY (in press).
McNaughton, S. J. 1983b. Physiological and ecological implications of herbivory. pp. 657–677. In: O. L. Lange, P. S. Nobel, C. B. Osmond, and H. Ziegler, eds. Physiological plant ecology. III. Responses to the chemical and biological environment. Encyclopedia of Plant Physiology; New Series, vol. 12C. Springer-Verlag, NY.
Rosenthal, G. A., and D. H. Janzen. 1979. Herbivores: their interaction with secondary plant metabolites. Academic Press, NY.
Robbins, C. T. 1983. Wildlife feeding and nutrition. Academic Press, NY.

6

Ant–Plant Interactions

KATHLEEN H. KEELER
School of Biological Sciences
University of Nebraska–Lincoln
Lincoln, NE 68588-0118

Introduction

This chapter concerns the interaction of ants and higher plants. With a few exceptions, there are no direct feeding links between ants and plants; the vast majority of ants (social insects of the order Hymenoptera, family Formicidae) are carnivorous and consume animals smaller or slower than themselves. Vascular plants, however, are autotrophic, manufacturing their own food from simple raw materials such as carbon dioxide and water. Despite their dissimilar life styles, ants and plants often interact in direct mutualistic relationships.

The mutualisms between ants and plants emerge as solutions to problems in the lives of the participants. For example, without being able to move, plants must escape herbivores and find suitable sites for their seeds to grow. Interaction with animals can provide these services, repelling herbivores or dispersing seeds. Plants can provide nest sites and/or food for ants and so, when ants defend these resources from other

I thank S. Collinge, L. Powell, R. Bertin, F. Rickson, R. Buckley, A. Johnson, and W. Abrahamson for critical reading of the chapter.

animals, they are also defending their host plants. Seeds or pollen carried by animals, intentionally or unintentionally, will be transported to other places.

At the same time, ants use plants to solve ant problems. Ant colonies need a protected nest for their vulnerable eggs, larvae, and pupae. Nest sites are usually in soil, but in many ecosystems good nest sites are in short supply, so that plants that provide nest sites are almost always occupied by ants. Since ant foragers always search for food, wherever food is available from plants, ants will gather it. Thus, many different ant–plant mutualisms have evolved as ants and plants use each other to increase their own fitness.

Ants form an important group of plant mutualists because of their abundance and distribution. Plants have mutualistic interactions with many kinds of animals, but ants are probably more numerous and widespread than any other single type of animal. Ants inhabit virtually every ecosystem in the world and within ecosystems, the total number of ants can be very great, with literally thousands of workers per square meter (Pisarski, 1978). The social structure of ant colonies enhances their effectiveness as mutualists: A single ant colony may have foragers searching hundreds of places at once, or send many different individuals to the same place. As a result, ant–plant mutualisms include relationships in which plants are defended from herbivores, are fed nutrients, or have their seeds and pollen transported. In return, plants provide ants with shelter and/or food.

A few ant groups have evolved methods for eating plants. Most ant species consume some seeds, but harvester ants eat predominantly seeds and are major herbivores in some areas (e.g., arid regions). Ants in the New World tropical tribe Attini consume plants indirectly by feeding them to a fungus and eating the resulting fungal mycelia. Both seed-harvesting and fungus-gardening ants maintain colonies with hundreds of thousands of workers, and greatly affect the composition of the plant community wherever they occur.

Indirect interactions also occur between plants and ants. For example, ants may indirectly injure plants by protecting plant-feeding insects such as aphids and membracids. Ants tending membracids have in some cases been shown to provide indirect benefit to plants by chasing off herbivores which cause more damage to the plant than do the membracids (e.g., Messina, 1982). Ant nest-building activity turns the soil, aerating it, and differentially benefiting particular plant species. These indirect interactions of ants and plants form an important, rapidly expanding area of ecology that is already too great to cover in this chapter [but see Buckley (1982), Beattie (1985), and Jolivet (1986) for recent reviews, and the section "Floral Larceny" in Chap. 2 for another example of such interactions].

This chapter describes both mutualistic and herbivorous interactions between ants and plants. First, ant characteristics are briefly summarized (a description of plants can be found in Chaps. 2 and 3). Then the mutualistic interactions are considered, including ant-inhabited plants (myrmecophytes), defense at nectaries, ant-dispersed seeds (myrmecochory), and pollination, and finally the herbivorous ants, harvester ants, and fungus-gardening ants are discussed.

Ant Biology

Ant colonies contain a reproductive individual, the queen, and a large number of nonreproductive individuals, the workers. At times there are a few male reproductives and immature queens within the nest as well. Male ants are haploid, produced by the queen from unfertilized eggs while female ants, queens and workers, are diploid. Eggs that develop into workers are treated with different hormones and foods than eggs that develop into queens, leading to inhibition of reproduction of workers which is usually irreversible.

Ants go through a typical insect life cycle of complete metamorphosis, passing through egg, larva, pupa, and adult stages. In order to develop normally, larvae require a diet rich in protein. The other stages have lesser food requirements, since eggs and pupae do not feed, relying on stored tissue for food, and workers can frequently subsist on a diet almost entirely composed of carbohydrate (sugars) and water.

Ant foragers generally gather mainly small arthropods, many of which are captured alive. Any invertebrate corpse will be collected, or a vertebrate corpse dismembered and then collected. Other diet items include seeds, nectar (see Chap. 2) and honeydew (sugar water produced by insects such as aphids) (Wheeler, 1910; Wilson, 1971).

Ant foragers depart daily from the nest, searching the environment in three dimensions for food. Species vary greatly in their behavior upon finding food. For example, large ants or those of small colonies tend to carry off whatever each forager can manage. Smaller ants or those from large colonies return to the nest, often leaving a pheromone (chemical) trail behind them, and recruit other workers who assemble in large numbers to break up the food and remove it. Food-gathering tactics also vary with the nature of the food; for example, liquids must be sucked up by individuals while a group of ants may jointly drag a dead grasshopper to the nest entrance.

Estimates of the number of workers in colonies range from 4 to 22 million, depending on the species (Wilson, 1971). Many ant species form small colonies which never have more than a dozen members, but on the other extreme, colonies of both the highly predacious army ants (e.g.,

Eciton) and the herbivorous leaf-cutter ants (*Atta* spp.) contain more than 1 million workers. It is likely that the majority of ant colonies contain relatively few workers (twenty to several hundred).

Many ant colonies do not stay long in one place, either because the ants move frequently or because colonies die out and are replaced by new ones. Culver and Beattie (1978) found woodland ant colonies (e.g., *Formica subsericea, Myrmica puntiventris*) averaged only 20 days at a site, while on the other extreme *Formica* nests in Europe lasted 25 to 60 years (Wilson, 1971). These *Formica* nests were potentially immortal since, in this species, new queens can be accepted into an established colony. The interaction of ant colony and plant may thus be very brief or may continue for many years.

In a colony, only the queen lays eggs. All the other ants routinely found in the nest are sterile female workers. Winged males are produced at intervals, as are winged virgin queens. Any winged form, regardless of sex, is referred to as an alate. Ant colonies outcross when members of distant colonies mate in the air during a mating flight. Although the males die soon afterward, the fertilized queens begin building nests, lose their wings, and lay the hundreds or millions of eggs that become the workers.

The evolutionary unit is the colony, not the individual ant, since of all the ants found there, only the queen reproduces. Workers are conveniently thought of as cells or organs in a discontinuous body since they are nonreproductive participants in colony survival. The peculiar genetics of ants (females are diploid, males haploid) results in nonreproductive workers' fitness increasing with colony success, in a form of kin selection. Workers share more genes with their sisters in the nest, including new alate queens, than they would with their own offspring (Hamilton, 1964; Wilson, 1971). Thus, at least part of the cooperative behavior that is the key to colony success is explained because cooperation increases each individual worker's fitness.

Ant Distribution

Because of the large number of workers and the large number of colonies present, ants are a major component in most communities (see Brown and Taylor, 1970; Peţal, 1978). In forest ecosystems, 60,000 to 100,000 ant colonies may occur per hectare (Culver and Beattie, 1978). Often more energy passes through ant populations than through the homoiothermic vertebrates living in the same habitat [compare 58 to 75 kJ/m^2 for *Pogonomyrmex badius* alone to 17 kJ for sparrows and 31 kJ for mice in the same South Carolina old field (Golley and Gentry, 1964)]. The energy flows through ants in such quantities because even though individually small, they collectively gather and consume a huge volume of prey. Thus, ants are important animals in the structure of most communities.

One perennial problem in the study of ants is the difficulty in measuring abundance. Censuses of ants can count workers, colonies, or species. The latter two are relatively straightforward, but worker censuses are not since worker numbers fluctuate throughout the season. Furthermore, some workers are foraging outside the nest at any given time so that whether nests or foragers are collected, some workers are usually missed. Evaluation of colony size requires a sustained effort over a full year (or more) and is not well understood (Pisarski, 1978; Briese, 1982a, 1982b). Good species lists and colony censuses exist, but worker estimates are generally not available. Since workers, not colonies or species, interact with plants, this problem means that few estimates of ant abundance can be applied to understanding ant–plant interactions.

Ants are presently found on every continent and large island on earth. Even before human movement redistributed them, ants were virtually ubiquitous. Only a few areas previously lacked ants: (1) Hawaii and islands in the Central Pacific (apparently because these islands were too far from a source of ants; Wilson and Taylor, 1967); (2) Antarctica, because it is presently too cold to support ants; (3) mountain tops, especially in the tropics, presumably because these are also too cold for ants [but, because ants are seasonally present at high elevations in temperate zones (e.g., 3000 m) the distribution of ants at high elevations deserves further study]; and (4) aquatic and marine environments, because ants do not voluntarily swim. Since these are the chief exceptions to a worldwide distribution of ants, a primary reason for the abundance of ant–plant interactions is likely the frequency of encounter between ant foragers and plants.

Plant Distribution

Primary aspects of plant biology affecting plant–ant interactions are the plant's need for transportation of pollen and seeds (see Chaps. 2 and 3), losses to herbivores (see Chap. 4), and nutrient limitations in some habitats. Vascular plants are even more widespread than ants; they are absent only from incredibly dry or extremely cold locations (e.g., desert sand dunes and Antarctica).

Mutualism between Ants and Plants

Ants are particularly good potential mutualists for plants because they rarely eat plants or destroy living plant tissue. By comparison, the ant–aphid mutualism is precarious; if the aphid fails to produce honeydew, the edible aphid is carted off as prey. Plants that fail to provide adequately for their ant mutualists suffer only neglect.

Table 6-1 Summary of ant–plant mutualistic exchanges

Interaction	Plant Provides	Ant Provides	Example
Myrmecophyte*	Shelter	Defense	Ant–acacias
Myrmecophyte*	Shelter	Nutrients	*Myrmecodia*
Myrmecophyte*	Shelter	Dispersal	*Myrmecodia*
No established name	Food	Defense	*Ipomoea*
Myrmecochory	Food	Seed transport	*Viola*
Pollination	Food	Gamete transport	*Diamorpha*

*Actually describes the plant rather than the interaction.

Note: As indicated, one interaction may include more than one exchange between the species.

Ant–plant mutualisms involve the exchange of a variety of services (Table 6-1). Plants that provide shelter and/or food usually receive defense or transport of pollen or seeds in return. Those plants that receive nutrients from interacting with ants provide only shelter, not food, for the ants. Mutualists are probably never altruists, since the mutualist enhances its own fitness by aiding another. The crucial questions are when and why this occurs.

Ant–plant mutualisms may be very specific and symbiotic, or may be quite casual with many ant species interacting facultatively with many plant species (Boucher et al., 1982). All known symbiotic ant–plant interactions are confined to the tropics. Other ant–plant mutualisms, such as defense by nectary visitors and myrmecochory, are found in virtually every ecosystem (Keeler, 1981b; Buckley, 1982; Beattie, 1983, 1985; Jolivet, 1986).

Many ant–plant interactions are little-studied and poorly understood. In addition, there is a broad spectrum of relationships, with each category intergrading into several others. Some plants participate in several different types of mutualism with ants, and vice versa. The result is that there are many different classification schemes and terminologies, each reflecting the purposes of the author. For example, Beattie (1985), focusing on function (what is exchanged), arranged things differently from Jolivet (1986), who used a structural classification emphasizing morphology. This chapter uses a conservative arrangement which makes no attempt to be definitive. In this chapter, ant–plant mutualisms are arranged first by the plants' contribution to the mutualism, since there are fewer alternatives, and within each of those, by the ants' contribution.

Mutualisms in Which Plants Provide Shelter

Myrmecophytes

The term myrmecophyte, meaning simply "ant–plant," is usually restricted to interactions in which living plant tissue is regularly occupied by ants. Myrmecophytes, found in both the Old and New World tropics, include tree, shrub, and epiphytic life forms. (Epiphytes, for example orchids and bromeliads, are plants that live on other plants, generally taking nutrients from the air and rain and not parasitizing the plant upon which they grow). Myrmecophytes include species in at least 103 genera of 36 vascular plant families (Jolivet, 1986). Some examples are given in Table 6-2.

Myrmecophytes range from plants with complex adaptations to house and feed specialized ants to the relatively casual occupation of partially hollow twigs and stems by various ant species. The obligate pairs are generally believed to be mutualistic since Janzen's thorough investigation of ant acacias (Janzen, 1966, 1967a, 1967b, see the following). However, in most cases no definitive studies have tested the nature of the interaction. The less well-known relationships are interpreted variously by different researchers (is it a mutualism or a form of parasitism?) and many interactions have scarcely been described (see Buckley, 1982; Beattie, 1985; Jolivet, 1986, for recent reviews).

Defense by Ants

Ant Acacias. *Acacia,* a genus of trees and shrubs (Leguminosae, subfamily Mimosoideae) contains at least 750 species world wide. Ant-inhabited acacias appear to have evolved separately in the tropics of America and Africa. In Central America, ants hollow out the enlarged thorns (modified stipules) to use as nest sites (Janzen, 1966, 1974b; Fig. 6-1A). African acacia ants inhabit stem swellings called galls just below or surrounding the stipular thorns (Hocking, 1970). The similarities in ant acacias on the two continents are interpreted as evolutionary convergence (Hocking, 1970; Janzen, 1974b). Evolutionary convergence is the production of similar adaptations separately in two different evolutionary lineages, usually ascribed to similar evolutionary pressures working on the two groups. This chapter considers only the New World species, but Hocking (1970) gives a good description of African ant acacias.

In New World ant acacias, the thorns are greatly enlarged to an internal volume of as much as 7.0 cm^3. The *Pseudomyrmex* ants (subfamily Myrmicinae) hollow out these thorns and distribute the

Table 6-2 Examples of plants reported to be Myrmecophytes*

Plant Family by Location	Genus	No. of Spp.	Habit	Exchange	Structure Involved†	Ant Genus	Refs.‡
Central and South America:							
Boraginaceae	*Cordia*	11	S	s × p?	*Stem*	*Azteca, Pseudomyrmex, Allomerus*	2,3
Bromeliaceae	*Aechmea*	1	E	s × n?	*Leaf bases*	?	6
	Tillandsia		E	s × n?	*Leaf bases*	*Crematogaster*	6
Euphorbiaceae	*Macaranga*	>2	T	fs × p?	*Bc*, stems	*Crematogaster*	7
Leguminosae	*Acacia*	14	T	fs × p	*Bl, N, thorns*	*Pseudomyrmex*	3
	Tachialia	13	T	s × p	*Swollen petioles*	*Pseudomyrmex*	1,2,3
Moraceae	*Cecropia*	27	T	fs × p	M, P, hollow internodes	*Azteca*	2,5
Africa:							
Apocynaceae	*Epidtaberna*	1	S	s × p	Swollen internodes	*Pachysima*	1
Combretaceae	*Terminalia*	2	T	fs × pn	Hollow stems		1,2
Euphorbiaceae	*Macaranga*	2	T,S	ns × p or n	*Swollen stipules*	*Crematogaster*	1
Leguminosae	*Acacia*	16	T	fs × p	*N, "gall"*	*Crematogaster*	4
	Leonardoxa	1	T	fs × p	Hollow	*Petalomyrmex Cataulacus*	9

Rubiaceae	*Duroia*	3	T,S	s × ?	Leaf blade base	Small ants	2
	Plectronia	6	T,S	s × ?	Hollow stem	*Crematogaster*	1,8
Sterculiaceae	*Cola*	3	S	s × n?	Leaf pouches	*Engramma, Plagiolepsis*	1
Asia and Australia:							
Asclepiadaceae	*Dischidia*	29	E	s × n?	*Leaf*	*Iridomyrmex*	6
Euphorbiaceae	*Endospermum*	2	T	s × ?	Hollow stem, trunk	*Camponotus*	1
Palmae	*Korthalsia*	1	T	s × p	Stems	*Camponotus*	1
Polypodiaceae	*Lecanopteris*	9	EF	s × n?	*Rhizome*	*Iridomyrmex*	6
Rubiaceae	*Hydnophytum*	46	ES	s × n	*"Tuber"§*	*Iridomyrmex*	6
	Myrmecodia	25	ES	s × n	*"Tuber"§*	*Iridomyrmex*	6

*Key: EF = epiphytic fern, ES = epiphytic shrub, S = shrub, T = tree; f = food, n = nutrients, p = protection, s = shelter; Bc = Beccarian bodies (lipid), Bl = Beltian bodies (protein), M = Müllerian bodies (glycoprotein), N = extrafloral nectary (carbohydrate), P = pearl bodies (lipid).

†Italics indicates that the structure appears to have evolved in response to the interaction.

‡References: (1) Bequaert, 1922; (2) Wheeler, 1942; (3) Janzen, 1966; (4) Hocking, 1970; (5) Rickson, 1971; (6) Huxley, 1980; (7) Rickson, 1980; (8) Buckley, 1982; (9) McKey, 1984.

§Hollow or inflated roots.

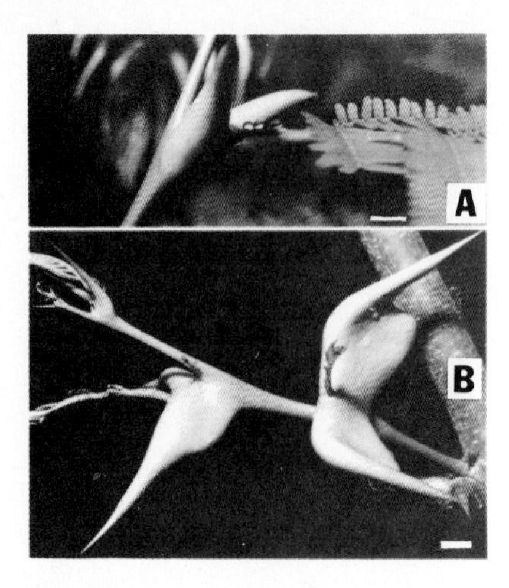

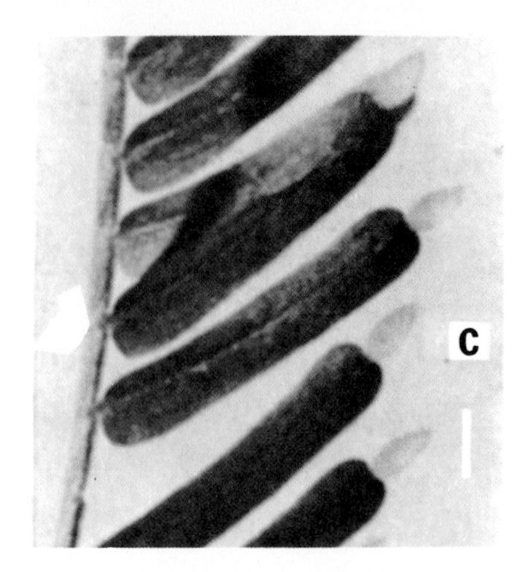

Figure 6–1. Ant acacia. (A) Petiolar nectaries of *Acacia collinsii* (Leguminosae). Note droplets of nectar. Scale = 0.5 cm. (B) Swollen thorns and new leaves of *Acacia collinsii.* Note *Pseudomyrmex* ant on stem. Scale = 1 cm [from Janzen, 1974c, by permission of Smithsonian Institution Press from Smithsonian contributions to Botany Number 13, *Swollen-thorn acacias of Central America* by Daniel H. Janzen. Figure 97, parts (B) and (C). Smithsonian Institution, Washington DC, 1974]. (C) Beltian bodies on tips of leaflets of ant acacia. Scale = 0.1 cm (from Janzen, 1966).

colony among the numerous thorns of a tree. Ant workers patrol the tree 24 h/day, both guarding the colony against predators and searching for food. Since the tree contains their nest, obligate acacia ants react strongly to any disturbance of the tree, for example, by browsing deer. If alarmed, foragers will recruit others from within the thorns to attack the intruder. Workers pour out of the thorns to attack, defending the acacia by both biting and inflicting a painful sting on any animal in contact with the tree. Keen eyes and a good sense of smell let them respond to even potential herbivory. A human standing less than 1 m from a well-defended ant acacia, for example, results in the ants massing on the near side.

Ant acacias are adapted for feeding as well as housing acacia ants. Like the vast majority of *Acacia* species, swollen-thorn acacias have extrafloral nectaries at the petiole (Fig. 6-1B; see also section on defense nectaries). Ants feed on the solution of water, sugars, and amino acids secreted by these nectaries. Ant acacias also provide protein for the ants by producing Beltian bodies at the leaf tips (Fig. 6-1C; Rickson, 1969). Named for Thomas Belt, who first described the ant–acacia interaction (Belt, 1874), Beltian bodies contain proteins and lipids. Ants collect

Beltian bodies as soon as they are available, just as the leaflets open. Consequently, ant attraction to Beltian bodies promotes defense of the *Acacia* by the ants since it encourages ant foragers to visit the newest leaves, where herbivore damage is the most injurious. Modified leaf tips containing protein are unusual in *Acacia* (or any plant) and must be the result of substantial interaction and coevolution with the ants (see Table 6-3).

Acacia ants keep the plant free not only of insect and vertebrate herbivores, but also of other plants and fungi. Ants rarely attack plant material, but Janzen (1966) showed that in ant acacias the removal of encroaching branches is crucial to preventing small trees from being overwhelmed and killed by vines. Plant-removing behavior is an important adaptation of *Pseudomyrmex* to interaction with the acacia.

Pseudomyrmex colonies can contain more than 30,000 ants and spread across several trees more than 10 m apart. The ants cross the ground to reach other trees that are encompassed by a single colony. In those *Pseudomyrmex* species that accept multiple queens, the colony may extend over several hundred acacia trees and their shoots (Janzen, 1966).

All of the ant species found living on ant acacias are not equally aggressive defenders of the acacia. Ant inhabitants range from obligate highly adapted *Pseudomyrmex* species to obligate *Pseudomyrmex* species that provide adequate but less vigorous defense to species (e.g., *Crematogaster*) that inhabit the thorns but do not defend the acacia. Many arboreal ant species will inhabit swollen-thorn acacias, if acacia ants do not exclude them.

This interaction is an obligate mutualism for a number of species of *Acacia* and *Pseudomyrmex* because acacia ants are found only on ant acacias, and ant acacias do not survive if deprived of defending ants (Janzen, 1966). The origins of the mutualism are unknown, but it is clearly very old, since both ants and acacias appear to have evolved in response to the interaction (Janzen, 1966; Table 6-3). For example, while most species of *Pseudomyrmex* are arboreal ants that sting, only obligate acacia-inhabiting species show aggressive defense of the tree and insert the sheath as well as the stinger when stinging. Similarly, most acacias have thorns and petiolar nectaries, but only ant acacias have swollen thorns and protein-containing Beltian bodies.

It is believed that after the establishment of the mutualism, selective advantage caused evolution of both species to increase the effectiveness of the interaction. The scenario may have been that ants first inhabited the trees casually, providing incidental but advantageous defense. Ants foraged more on plants with more active petiolar nectaries, resulting in greater fitness for those plants, selecting for continual nectary activity.

Table 6-3 Coevolution of ant acacias and acacia ants

Acacia Traits Related to the Ant–Acacia Coevolution

General features of acacias important to the interaction	Specialized features of swollen-thorn acacias (coevolved traits)
1. Woody shrub or tree forms	1. Woody but with very high growth rate
2. Reproduce from suckers	2. Rapid and year-round sucker production
3. Plants of dry areas	3. Plants of moister areas
4. Ecologically widely distributed	4. Ecologically very widely distributed
5. Leaves shed during dry season	*5. Year-round leaf production
6. Shade-intolerant, sometimes covered by vines	6. Shade-intolerant and free of vines
7. Thorns	7. Longer-persisting thorns, woody with soft pith
8. Bitter-tasting foliage	8. Bland-tasting foliage
9. Each species with a group of relatively host-specific phytophagous insects able to feed in the presence of the physical and chemical properties of the acacia	9. Each species with a few host-specific phytophagous insects, able to feed in the presence of the ants
10. Foliar nectaries	*10. Enlarged foliar nectaries (Fig. 6-1B)
11. Compound unmodified leaves	*11. Leaflet tips modified into Beltian bodies (Fig. 6-1C)
12. Not dependent upon another for survival	12. Dependent upon another species for survival

Pseudomyrmex Traits Related to the Ant–Acacia Coevolution (Worker Traits unless Otherwise Indicated)

General features of *Pseudomyrmex* important to the interaction	Specialized features of obligate acacia ants (coevolved traits)
1. Fast and agile runners, not aggressive	1. Very fast and agile runners, aggressive
2. Good vision	2. Good vision
3. Independent foragers	3. Independent foragers
4. Smooth sting, barbed sting sheath not inserted	4. Smooth sting, barbed sting sheath often inserted
5. Ignore living vegetation	5. Maul living vegetation contacting ant acacia
6. Arboreal, highly mobile, colony	6. Arboreal, highly mobile, colony
7. Larvae resistant to mortality by starvation	7. Larvae resistant to mortality by starvation
8. Colonies small	8. Colonies large
9. Diurnal activity outside nest	9. 24-hour activity outside nest
10. Few workers per unit plant surface	10. Many workers active on small plant surface area
11. Discontinuous food sources and unpredictable new nest site	11. Continuous food source and predictable new nest sites
12. Founding queens forage far for food	12. Founding queens forage short distances for food
13. Not dependent on another species	13. Dependent on another species group

*Essential to the interaction.
Source: After Janzen, 1966.

Where ants inhabited the plant, defense was better, so that selection favored enlarged thorns to house the ants, and finally Beltian bodies so that all ant foraging was contained within the acacia. With each plant change, the efficacy of ant defense improved.

Conversely, ants that better defended the *Acacia* received better food and shelter because greater vigilance and more aggressive attacks on insects and encroaching vegetation resulted in healthier acacias which produced more Beltian bodies, extrafloral nectar, and inhabitable thorns to feed or house them. One interesting aspect of the ant–plant mutualism is the apparent loss of defensive chemicals by ant acacias. Rehr et al. (1973) showed that, compared to acacias that are not associated with obligate ants, ant acacia foliage is very poor in secondary chemicals (whose function probably includes antiherbivore defense; see Chaps. 4 and 5). Thus, ant defense appears to have substituted for chemical defense. Both partners have certainly changed in response to mutualism with the other (Janzen, 1966).

Other Ant-Defended Myrmecophytes. Other tropical plants have obligate or regular ant inhabitants that provide protection. *Cecropia* (Moraceae), a New World genus whose hollow stems are regularly inhabited by ants of the genus *Azteca* (subfamily: Dolichoderinae), provides Müllerian bodies and pearl bodies as food for ants. Müllerian bodies, continuously produced, are filled with glycogen (Rickson, 1971), while pearl bodies contain lipid and carbohydrate (Rickson, 1976). *Azteca* maintains large numbers of honeydew-producing insects (Homoptera, sap-feeding insects, the most familiar of which are the aphids) within the stems, so that with Müllerian bodies, pearl bodies, and Homoptera, all the ants' food is contained within the tree. The details of the trade-offs for *Cecropia* between feeding the homopterans and being defended by the ants have not been studied explicitly, but defense by *Azteca* is believed to be important to *Cecropia*'s success (Janzen, 1969b). In the Caribbean Islands, where *Azteca* ants are absent, presumably because they have never survived the crossing from the mainland, *Cecropia* populations vary more for traits relating to the ants: Many have lost plant structures that facilitate ant occupation, such as the trichilia in which Müllerian bodies are produced (Janzen, 1973a; Rickson, 1977). This implies that ongoing selection for the interaction is responsible for the maintenance of these traits.

Of the scores of myrmecophytes for which protection has been suggested (e.g., Jolivet, 1986), only *Acacia, Barteria* (Janzen, 1972), *Cecropia, Leonardoxa* (McKey, 1984; Fig. 6-2), and *Ochroma* (O'Dowd, 1979, 1980) have been studied in any detail in the last half century, and the nature of the interaction was tested experimentally only in *Acacia*.

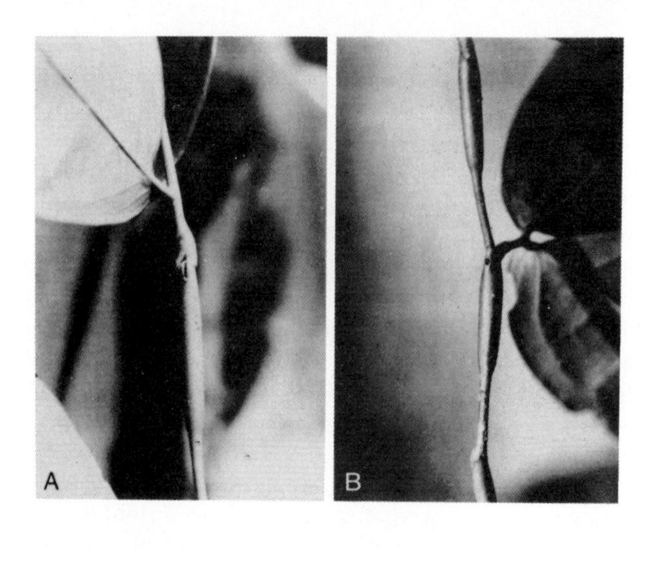

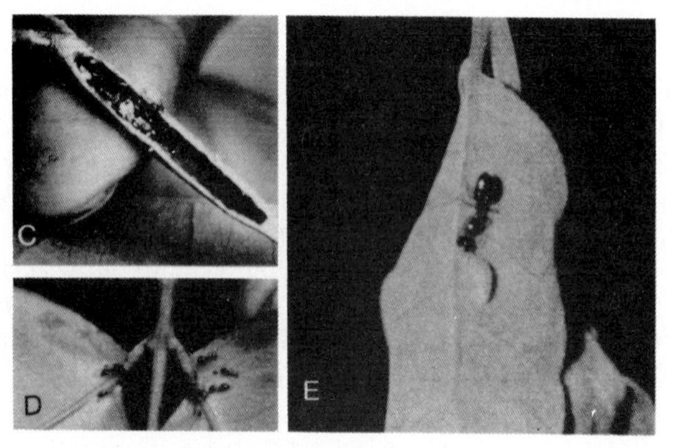

Figure 6-2. African myrmecophyte, *Leonardoxa africana* (Leguminosae: Caesalpineae), a small tree of the rainforest understory in equatorial Africa, studied by McKey (1984) had an obligate relationship with ants that inhabit the swollen internodes. (A) Slit in *Leonardoxa* internode used as entrance to domatium by *Petalomyrmex phylax* ants. The swollen internode is approximately 5 cm long. (B) A second internode entrance, this one made by *Cataulacus mckeyi*. The different shapes of the entrance holes are characteristic of the ant species that made them. (C) Dissected internode showing *Petalomyrmex* workers and brood inside. (D) *Petalomyrmex* foragers gathered around extrafloral nectaries on the underside of *Leonardoxa* leaf. (E) *Petalomyrmex* worker demonstrating the aggressiveness of these ants: it is attacking the microlepidopteran caterpillar that was placed on this young leaf (from McKey, 1984).

Nutrients from Ants

A second group of myrmecophytes appears to receive nutrients rather than protection from their ant inhabitants. These plants, woody epiphytes of wet, nutrient-poor habitats, known as ant epiphytes (Huxley, 1982), provide nest sites in a habitat where dry shelters are rare (Janzen, 1974a; Huxley, 1980, 1982). The plants produce hollow or inflated roots (often called tubers), hollow rhizomes (horizontal stems), or folded leaves for ant nests. The botanical terminology is complex because applying the proper term requires knowing the tissues from which a modified structure is derived (root or stem, for example) and for many myrmecophytes, that has not been determined. A general term for a plant structure regularly inhabited by ants is *domatium* (plural domatia). (See also Chap. 4 on mite domatia). In some species, the ants must excavate the tissue; in others, the plant produces hollow structures and ants need only gain access. The ants living in these shelters deposit their fecal matter, refuse, and possibly gathered food items within the plant. Left in a leaf, tuber, or rhizome, the refuse decays: the nutrients released by decomposition are taken up by the plant.

The specifics of uptake have received more speculation than experimentation. Huxley (1978) demonstrated that sulfur and phosphorus in honey fed to ants appeared in tissues of *Myrmecodia,* and Rickson (1979) showed that carbon from ant prey items was transferred to *Hydnophytum* tissues. The presence of numerous stomates lining the chambers suggests that CO_2 can be absorbed from the chambers, and other forms of uptake have been suggested (see reviews by Huxley, 1980, 1982; Buckley, 1982). Adventitious roots grow into the chambers containing ant refuse and absorb nutrients directly (Janzen, 1974a; Huxley, 1980, 1982). The full extent of nutrients that myrmecophytes receive from the ants they shelter and how important these are to the nutritional needs of the plant have not been worked out for any species, although it is crucial to the understanding of the interaction.

Myrmecodia. The best-studied nutritional myrmecophyte is *Myrmecodia tuberosa* (Rubiaceae) (Fig. 6-3), an epiphytic shrub found from southeast Asia to northern Australia. This plant has an enlarged tuber containing large empty cavities, some with light-colored, smooth-textured interior walls and others with dark, rough-textured interior surfaces. The ants keep their brood in the smooth-walled chambers, and deposit refuse in the rough-walled chambers. Huxley (1978) showed that radioactive sulfur and phosphorus fed to the ants could be found in plant tissues, particularly the rough-walled chambers. Uptake of nutrients from the chamber contents is further enhanced by the growth of adventitious roots into the rough-walled chambers. The chambered tubers are produced whether or not ants are present, but root growth into the rough-walled cavities is initiated or at least enhanced by ant presence (Huxley, 1978).

Myrmecodia's tuber is interpreted as having evolved at least partly in response to ant inhabitation. The most frequent alternative explanation for the tuber was that it caught rainwater, but, as Janzen (1974a) noted, the opening is positioned horizontally, which is not suitable for catching rain. The growth of roots inside the tuber (i.e., roots on the inside of a plant structure) is very unusual and certainly part of the plant's adaptation to ant inhabitants (Bequaert, 1922; Janzen, 1974a; Huxley, 1980, 1982).

Within the tuber, ants find a dry, protected nest site in an environ-

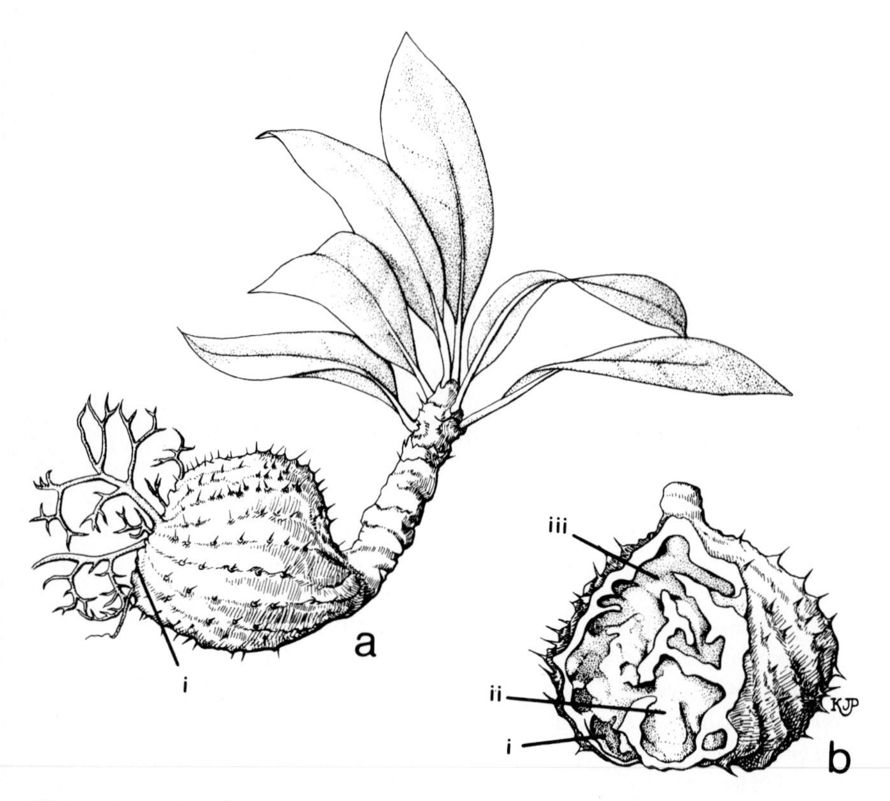

Figure 6–3. Epiphytic myrmecophyte, *Myrmecodia tuberosa* (Rubiaceae). (a) Whole plant, pictured as attached to tree branch. Scale = 5 cm. (b) Tuber of young plant showing internal chambers: i = entrance hole of ants, ii = inner smooth-walled chamber used by ants as nest site, iii = dark, rough-walled chamber used by ants as a refuse dump. Scale = 5 cm (redrawn from Huxley, 1982).

ment with very heavy rainfall. In Sarawak, *Myrmecodia* tubers had occupancy rates as high as 90 percent, and there were few alternate nest sites for ants (Janzen, 1974a). This is consistent with the suggestion that a protected nest is important to the ants.

Although a variety of ant species live in *Myrmecodia* tubers, only a series of allopatric species in the genus *Iridomyrmex* are effective mutualists (Huxley, 1982). Whether the plant-inhabiting populations of *Iridomyrmex* represent species distinct from the generalist *Iridomyrmex* of the region is doubtful; there is little morphological distinction. However, the plant ant shows some behavioral specialization to the myrmecophyte. For example, *Iridomyrmex* living in *Myrmecodia* deposit their refuse inside the plant, rather than dropping it off the plant, as is more common for arboreal ants. *Iridomyrmex* colonies of myrmecophytes gather *Myrmecodia* seeds and place them on their covered runways. For an epiphyte like *Myrmecodia,* this amounts to the ants planting the seeds. How specialized these ant behaviors are, and whether they represent adaptation to inhabiting epiphytic ant plants is unclear (Janzen, 1974b; Huxley, 1980, 1982). An understanding of the degree to which these are ant adaptations to the interaction would help clarify the relationship between *Iridomyrmex* and *Myrmecodia.*

Myrmecodia seeds possess an elaiosome (oil-rich appendage) which is highly attractive to *Iridomyrmex*: the ants collect the seeds, remove and consume the elaiosome, and build the seeds into the structures they construct over their runways, thereby planting the seed of this epiphyte. The ant-attracting seed is believed to have evolved as a dispersal mechanism rather than in relation to the myrmecophytic interaction (Janzen, 1974a; Huxley, 1982; Buckley, 1982). In summary, *Myrmecodia* is a myrmecophyte that provides shelters for ants but not food, and in return receives ant refuse from which it extracts nutrients.

Other Myrmecophytes that are Fertilized. Other ant-fertilized myrmecophytes include the epiphytic fern *Lecanopteris* (Polypodiaceae) and the shrub *Dischidia* (Asclepiadaceae) (Table 6-2). In the former, the rhizomes are partially hollow and the ants further enlarge them. Ants maintain their colonies in older rhizomes and abandon old rhizomes for new ones as soon as they fill with debris. In general, tubers or rhizomes of myrmecophytes are darkly pigmented, which is interpreted as an adaptation which makes a normally pale or green structure more attractive to ants as a nesting site (Janzen, 1974a; Huxley, 1980).

Dischidia species show increasing specialization for ant inhabitation (Huxley, 1980). In the most complex, some "ant leaves" grow in the form of closed hollows and are generally found inhabited by ants; other leaves are normal. The ant leaves are far better supplied with stomates on the *interior* surface than on the exterior, presumably to gather CO_2 released by ants and by decay of refuse (Huxley, 1980). Adventitious roots grow into the ant leaves, remaining poorly developed unless debris is present (Janzen, 1974a). On the other hand, Weir and Kiew (1986) argue that despite ant leaves, *Dischidia astephana*, at least, is not part of an ant–plant mutualism but rather parasitizes an ant–tree mutualism between *Crematogaster* ants and *Leptospermum* trees (Myrtaceae), growing best where ants are present but providing no service to the ants. Thus, most species of presumed myrmecophytes deserve serious investigation.

Generally, myrmecophytic epiphytes (1) are inhabited by small and nonaggressive ants, (2) provide no food for ants, (3) possess leaves that are chemically protected from herbivores (Janzen, 1974a; Huxley, 1978, 1982), and (4) are found in habitats that are nutritionally poor for epiphytes, either because no overstory exists to drip nutrients or because of low overall productivity in the area (Janzen, 1974a; Huxley, 1980, 1982; Thompson, 1981). These interactions are therefore believed to involve exchange of a nest site for the ants with nutrients for the plants. The interactions described are from Asia, but bromeliads (Bromeliaceae) and other epiphytic myrmecophytes elsewhere probably have similar relationships.

Distribution of Myrmecophytes

Living plants that are inhabited by ants are believed to be confined to the tropics. Bequaert (1922) suggested that the thin walls of plant structures, whether tubers, thorns, or stems, restricted the interaction to regions in which severe cold never killed the ant colony. Epiphytes, thorns, and hollow stems do occur outside the tropics, as do arboreal ant species which nest in trees during the growing season. Compelling evidence for Bequaert's argument is lacking, but few alternative explanations have been proposed.

Ant acacias occur only where the trees can put out the new leaves that bear food for ants all year round (Janzen, 1966). Unless the plant is continuously active, the ants starve. However, this limit does not explain the distribution of ant-fertilized myrmecophytes. Plant-fed, plant-dwelling ants may be confined to aseasonal environments because there is little food available in the season in which plants are inactive, but this should not affect the distribution of ant-fertilized, ant-inhabited pairs.

Myrmecophytes are found in both the Old and New World tropics but the taxa usually differ. For example, the genera *Randia* and *Clerodendron* are found in both the Old and New World but are only myrmecophytic in Africa (Bequaert, 1922; Jolivet, 1986). Although *Acacia* species are very abundant in all tropical regions, ant acacias occur in Africa and America but not in Australia (Brown, 1960). These myrmecophytes must have had multiple independent origins. In *Acacia*, Brown (1960) argued that grazing mammals drove the evolution of ant acacias since the absence of grazing mammals is the most dramatic difference between Australia and other tropical regions. It is interesting that the myrmecophytic plants of Australia all appear to receive nutrients rather than defense from ants (Huxley, 1982).

Jolivet (1986) points out that myrmecophytes are much more diverse in the Americas and southeast Asia than in tropical Africa, and attributes the difference chiefly to the greater floral diversity and age of the American and Asian tropics. Generalizations to explain the distribution of myrmecophytes are greatly restricted by taxonomic differences between the continents. For example, the bromeliads are found only in the Americas where they form one of the major groups of epiphytic myrmecophytes (see also Chap. 7). Thus, it is difficult to determine how much of the difference between abundance and distribution of myrmecophytes in different regions is attributable to the specific evolutionary history of particular plant lineages and how much to different climatic or biotic interactions.

To date, only epiphytes (including epiphytic shrubs) are known to be ant-fertilized myrmecophytes (Huxley, 1982). They are found in places where epiphytes are severely nutrient-limited, and their interaction with ants seems to extend their distribution. Epiphytes, characteristic of humid aseasonal forests, should interact with ants where humidity and seasonality are suitable and nutritional limits make ant refuse a valuable plant nutrient source. Conversely, tree-dwelling ants should most strongly interact with epiphytes where few nest sites are available. The distribution of such habitats outside of southeastern Asia deserves further study.

In contrast, ant-defended myrmecophytes are generally plants of successional sites in aseasonal environments. These sites are very productive places where the plants can provide abundant food for the ants. In these interactions, ants provide antiherbivore defense and, at least in New World ant acacias, remove encroaching vines. The plants provide both shelters and food and receive defense in return. Defense mutualisms exist for a large number of plant species that do not house ants (see the following), many of them outside the tropics. Since ant-based defense is so widespread, it would seem reasonable to expect a broader variety of trees and shrubs, not just those of successional sites, to have evolved to house ants, but it has not been reported. Perhaps only the rapidly growing trees of disturbed sites possess both continuous leaf production to provide food and woody structures substantial enough to house ants. The limits of myrmecophytes remain poorly understood.

Mutualism in which Plants Provide Food

Defense by Ants

Protein and Oil Rewards

Some plants exchange food for defense by ants. The most specialized providers of food are myrmecophytes, such as the ant acacias, which are a source of carbohydrate, protein, and lipid to the ants. Only a very few plants supply protein for ant mutualists, and all appear to be obligate myrmecophytes (Rickson, 1969; Buckley, 1982; Beattie, 1985; Jolivet, 1986). Free amino acids, however, are commonly found in extrafloral nectars in small amounts (Baker et al., 1978).

Plants provide oils to ants in foliar structures called Beccarian bodies, Beltian bodies, Müllerian bodies, pearl bodies, and in elaiosomes, which are discussed separately in the section on myrmecochory. Oil is a highly attractive substance to ants and important to their diet (see discussion in Beattie, 1985). Leaf structures containing lipids include Beccarian bodies of *Macaranga* (Euphorbiaceae), Beltian bodies of *Acacia*, and Müllerian bodies of *Cecropia* (Rickson, 1969, 1971, 1976, 1980). Pearl bodies or, simply, food bodies are more general terms for a variety of single or multicellular processes containing lipid and occasionally protein or carbohydrate. Not known to occur in monocotyledons, pearl bodies are found in 50 genera of dicotyledons in 19 families (O'Dowd, 1982). Ants harvest pearl bodies (and Beccarian, Beltian, and Müllerian bodies) so thoroughly that they are rarely seen on plants in nature. Pearl bodies are inferred to form the basis of ant defense of the plant, leading to reduction of herbivory (O'Dowd, 1980, 1982).

Sugar Rewards: Defense Nectaries. Plant glands that secrete sugar water are called nectaries. The best-known are floral nectaries, which furnish one type of incentive for animals to pollinate flowers (see Chap. 2). Many other nectaries, not involved in pollination, occur elsewhere on the plant, for example on petioles, on leaf undersides, on the outsides of buds, along pedicels, and on the surfaces of developing fruit (Zimmermann, 1932; Bentley, 1977a; Elias, 1983; Fig. 6-4). The vast majority of these nectaries function in a mutualism, usually with ants, which reduces herbivory. These glands are called extrafloral or defense nectaries, the former being a structural and the latter a functional definition. The two terms are not precisely synonymous because some extrafloral nectaries reward pollinators (e.g., the leaf nectaries of *Acacia terminalis* in Australia; Knox et al., 1985) and some defense nectaries occur on floral parts [e.g., nectaries of the sepals in morning glories (*Ipomoea*); Keeler, 1980].

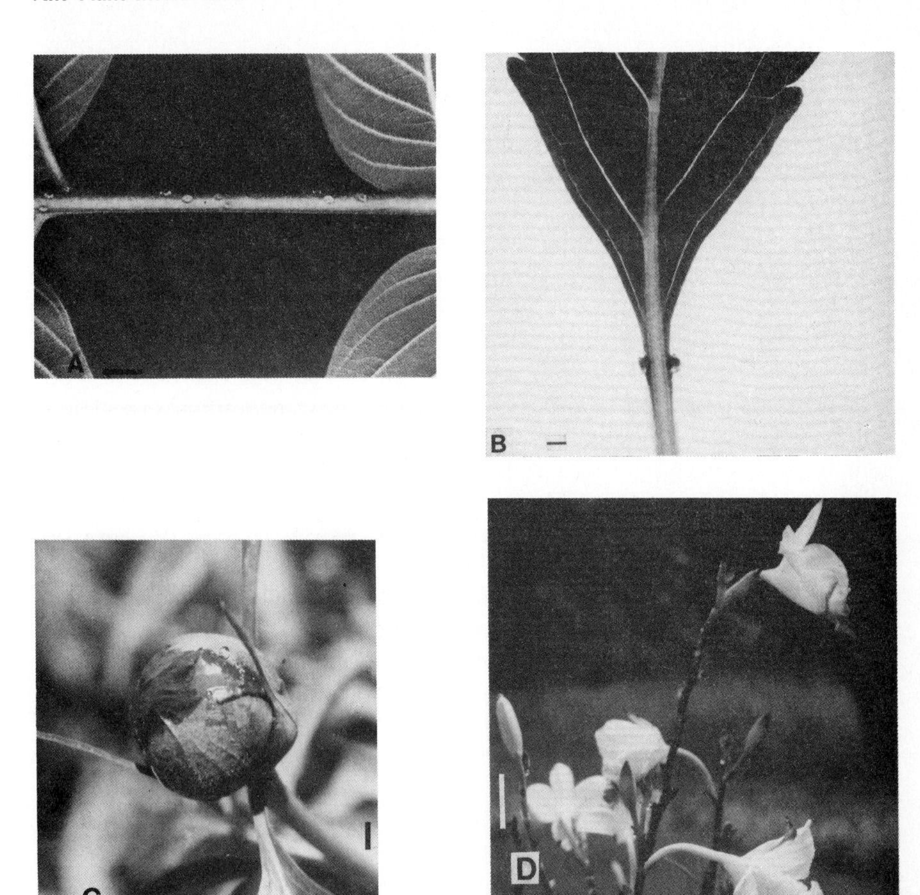

Figure 6–4. Extrafloral nectaries. (A) Nectaries of leaf rachis between the paired leaflets of *Cipadessa baccifera* (Meliaceae), a tropical shrub. These nectaries have an irregular distribution and are relatively simple structurally. Scale = 0.5 cm (from Lersten and Pohl, 1985). (B) Paired nectaries of petiole of *Turnera ulmifolia* (Turneraceae), a tropical herb. Note size of nectaries. Scale = 0.1 cm (from Elias et al., 1975; with permission). (C) Extrafloral nectaries of bud of *Paeonia* sp. (Paeoniaceae), a temperate herb. Nectaries have little structure but secrete copious nectar, attracting ants to the developing buds. Scale = 0.5 cm (photo by R. B. Kaul). (D) Nectaries on pedicel of *Arundina bambusifolia* (Orchidaceae), a tropical herb. Buds, flowers, and developing fruit are visited by ants gathering extrafloral nectar. Scale = 5 cm (photo by K. H. Keeler).

Plants have utilized the attraction of ants to sugars to establish the antiherbivore mutualism at defense nectaries. The nectaries secrete sugar, water, and usually trace amounts of free amino acids (Baker et al., 1978). This food source increases the frequency of ant visits to the plants, because while ant foragers probably visit every plant and every leaf eventually, if food is present the ants come more often and stay longer. Ant presence results in plant defense because ant workers will take as prey insects they encounter, and, on a plant, these are likely to be herbivores of the plant. Plant defense at nectaries can also occur when gregarious ants show "ownership behavior" and drive other insects from the plant (Carroll and Janzen, 1973), but the nest-defending behavior that makes ants of at least some myrmecophytes so vicious does not occur in defense nectary mutualisms. In these cases, the plant provides sugar water but not shelter and receives antiherbivore defense.

While not all extrafloral nectaries act as defense nectaries, many do. Defense of plants by ants feeding at extrafloral nectaries has been demonstrated in a number of cases (Table 6-4). These include both temperate and tropical annual and perennial plants, and defense against generalist and specialist herbivores. Although many extrafloral nectaries are associated with leaves, most demonstrations of antiherbivore defense at nectaries involve nectaries associated with flowers and fruit because of the more direct relationship to fitness as measured by seed set (see Chap. 2).

An example of a defense nectary system effective against both generalist and specialist herbivores is *Ipomoea leptophylla* (Convolvulaceae), the bush morning glory of North American prairies. The large pink flowers which open only for a single morning are rapidly eaten by grasshoppers and beetles. Ants visiting the nectaries on the sepals significantly reduced grasshopper damage to flowers and increased subsequent seed set (Keeler, 1980): Ants at the nectaries repeatedly attacked the feet of grasshoppers, causing the grasshoppers to leave the flower.

Those morning-glory flowers that escaped grasshopper damage were attacked by bruchid beetles (*Megacerus discoidus,* Coleoptera: Bruchidae). The larvae of these small specialist beetles bore into the seeds, consuming one or more. Plants with more ants had significantly fewer seeds destroyed by bruchids. Ants probably were not fast enough to catch and kill the ovipositing bruchid females, but nonetheless interfered with bruchid oviposition, and took eggs or larvae as prey. In *Ipomoea leptophylla,* ants feeding on extrafloral nectar were effective in reducing plant damage from both generalist (grasshoppers) and specialist (bruchid beetle) herbivores.

Other studies have shown effective defense against other types of

Table 6-4 Antiherbivore defense by ants attracted to angiosperm extrafloral nectaries*

Ecosystem	Plant Form	Species (Family)	Reference
Tropical	Trees	*Inga densiflora* (Leguminosae)	Koptur, 1984
		I. punctata (Leguminosae)	Koptur, 1984
		Ochroma pyramidale (Bombacaceae)	O'Dowd, 1979
	Shrubs	*Bixa orellana* (Bixaceae)	Bentley, 1977a
	Vines	*Ipomoea carnea* (Convolvulaceae)	Keeler, 1977
		Passiflora quadrangularis (Passifloraceae)	Smiley, 1986
		P. vitifolia (Passifloraceae)	Smiley, 1986
	Herbs	*Aphelandra deppeana* (Acanthaceae)	Deuth, 1977
		Calathea orandensis (Marantaceae)	Horovitz and Schemske, 1984
		Cassia fruticosa (Leguminosae)	Deuth, 1980
		Costus woodsonii (Zingiberaceae)	Schemske, 1980b
Temperate	Trees	*Catalpa speciosa* (Bignoniaceae)	Stephenson, 1982a
		Prunus virginiana (Rosaceae)	Tilman, 1978
	Vines	*Campsis radicans* (Bignoniaceae)	Elias and Gelband, 1975
		Ipomoea pandurata (Convolvulaceae)	Beckmann and Stucky, 1981
		Passiflora incarnata (Passifloraceae)	McLain, 1983
	Herbs	*Cassia fasciculata* (Leguminosae)	Boecklen, 1984; Barton, 1986
		Helianthella quinquenervis (Compositae)	Inouye and Taylor, 1979
		Ipomoea leptophylla (Convolvulaceae)	Keeler, 1981b
		Mentzelia nuda (Loasaceae)	Keeler, 1981a
		Vicia sativa (Leguminosae)	Koptur, 1979
		Yucca glauca (Liliaceae)	Boettcher, 1979
Desert	Cacti	*Ferocactus acanthodes* (Cactaceae)	Ruffner and Clark, 1986
		F. gracilis (Cactaceae)	Blom and Clark, 1980
		Opuntia acanthocarpa (Cactaceae)	Pickett and Clark, 1979

*Studies of plants offering additional rewards or nest sites (myrmecophytes) are not included.

insects consuming dissimilar plants. In the wild plum, *Prunus americana* (Rosaceae), Tilman (1978) excluded ants from young plants by surrounding the stem with a sticky resin. This provided a test of the ants' effectiveness since the ants (*Formica obscuripes*) crawled up from the ground and the herbivorous tent caterpillars (*Malacosoma americanum*, Lepidoptera; Lasiocampidae) were already on the plant. On plants near a large ant colony, the ants captured all small tent-caterpillar instars, eliminating the colony. Since tent-caterpillar infestation can kill small plum trees, this defense greatly increased plum fitness (Tilman, 1978).

The nectar of defense nectaries generally differs in its composition from that of floral nectaries, even on the same plant (Baker et al., 1978). Since floral nectaries have been shown to be adjusted very specifically to the nutritional needs of pollinators (see Chap. 2), it seems reasonable that the visitors to defense nectaries, ants in particular, have had appreciable selective effects on the composition of defense nectars. However, this has not been shown.

Distribution and Ecology of Defense Nectaries. The taxonomic pattern of defense nectaries is peculiar. At least 15 species of ferns are known to have foliar nectaries (Koptur et al., 1982; Fig. 6-5). Antiherbivore defense has not been demonstrated, and their function in bracken (*Pteridium aquilinum*) is the subject of vigorous investigation (Tempel, 1983; Heads and Lawton, 1985; Lawton and Heads, 1985). Extrafloral nectaries are unknown in gymnosperms and their absence is unexplained. Among angiosperms, more than 2200 species (almost 1 percent of the 315,000 species) are known to have extrafloral nectaries, (Cronquist, 1981; Keeler, unpublished); the actual number is probably much higher. Extrafloral nectaries occur in both monocotyledons and dicotyledons. In dicotyledons, extrafloral nectaries appear to be more common in the advanced families (e.g., Compositae) than in more primitive families (e.g., Magnoliaceae), although there are scattered occurrences in many families. However, primitive monocotyledons (e.g., the Liliaceae) do have extrafloral nectaries, as do advanced monocots (e.g., the Orchidaceae).

Aquatic plants appear to completely lack extrafloral nectaries. Presumably this is because they are inaccessible to ants, but it may also reflect differences between herbivores on terrestrial and aquatic plants, since much of the serious damage to aquatic plants is inflicted by aquatic organisms, rather than terrestrial insects.

Vines, on the other hand, are rich in defense nectaries. Bentley (1981) demonstrated that vines were more likely to have defense nectaries than trees, shrubs, or herbs in the dry tropical forests of Costa Rica. Ant defenses are probably especially available to vines since ants travel through trees using the vines as highways to get around.

Figure 6–5. Nectaries of ferns. (a) *Polypodium myriolepis.* (b) *P. thyssanolepis.* (c) *P. rosei.* (d) *P. rosei* closer view. (e) *Drynaria rigidula.* (f) *Polybotrya osmundacea.* Bars represent 0.5 cm. From Koptur et al. (1982).

Suggestions that woody, rather than herbaceous, plants and that species in successional, rather than climax, communities possess defense nectaries have some support (Bentley, 1977a; Tilman, 1978; Boucher et al., 1982; Keeler and Kaul, 1984), but more data are needed. Proposed explanations of these patterns generally suggest that defense nectary mutualisms are more likely where ants are more numerous, where herbivore damage by insects is more serious, and where water, sugars, and/or amino acids are less expensive to plants and more valuable to ants (Bentley, 1977a; Tilman, 1978; Keeler, 1981b; Thompson, 1982; Beattie, 1985; Heads and Lawton, 1985). The relationship to other defenses is not clear, because it is possible that trade-offs exist so that defense nectaries replace mechanical or chemical defenses. However, it is also possible that each defense is effective against a somewhat different suite of herbivores, so that under some environmental circumstances multiple defenses are favored (Keeler, 1981b; Keeler and Kaul, 1984; Lawton and Heads, 1985). A thorough understanding of the conditions for a particular defense and its relation to other antiherbivore defenses will require much further study.

A great deal of variation exists in the effectiveness of defense nectaries (O'Dowd and Catchpole, 1983; Tempel, 1983; Heads and Lawton, 1985; Barton, 1986; Smiley, 1986), which Beattie (1985) suggests is inherent in a behavioral rather than structural defense. This sort of defense, he argues, may be selectively favored only in some microhabitats (places), under certain climatic conditions (times), or at particular prey densities (conditions). This variation is currently the subject of intense investigation.

Myrmecochory: Dispersal by Ants

Another problem for plants is the need to get seeds to new locations, that is, dispersal (see Chap. 3). Dispersing the seeds away from the parent plant provides a number of advantages that are not mutually exclusive: (1) dispersal allows colonization of distant habitats, (2) herbivory may be more severe under the parent plant than at any other location in the habitat, so that seeds that are dispersed away from the parent will have improved survival, and (3) the dispersed seedling escapes competition with the parent and siblings. In theory, the distribution of safe sites for germination and establishment will determine the type of dispersal that is most beneficial to a particular species. If, for example, herbivores can be satiated, then dispersal advantage (2) may be unimportant, and low dispersibility might be successful. In reality it is often much more complex (see Chap. 3).

Ant dispersal of seeds is called myrmecochory (myrmeco- from *myrmex*, "ants"; -chory from *chorein*, "to wander"). The number of myrmecochorous species has been estimated to be more than 1500 in Australia, more than 1000 in southern Africa, but fewer than 300 in the rest of the world (Buckley, 1982). However, little is known of ant dispersal in tropical Africa and South America. Myrmecochory is wide spread in angiosperms and is found in families as diverse as the Marantaceae, Violaceae, and Compositae (Buckley, 1982).

Dispersal by ants has different characteristics from, for example, dispersal by birds (ornithochory), because ants do not travel very far. In one study, the average ant-dispersed violet seed was carried only 75 cm (Culver and Beattie, 1978). Thus, although in some systems ants carry seeds further (Westoby et al., 1982), myrmecochory is not very effective for long-distance dispersal. However, ants provide one advantage over most other dispersers: they generally plant the seed in a nutrient-enriched area. Attracted to the oil-rich elaiosome, seed-dispersing ants carry the seeds of myrmecochores to their nest where they either store them in chambers within their shallow nests for future use or remove the elaiosome at the nest entrance and discard the seed there. Since ants enrich the soil about their nests with feces and discarded prey remnants, and they aerate the soil with their tunnels, whether it is stored or discarded, the seed is placed in a very favorable site for germination and establishment. Deposition in a protected or nutrient-rich microsite is called inhumation (from the Latin *humare,* to cover with earth). Thus, a fourth possible advantage of myrmecochory is inhumation.

If ants disperse seeds, there is always the risk that the highly nutritious seed will be eaten by the ants. Most myrmecochorous plants protect their seeds by the joint strategy of providing a tasty and easily removed elaiosome for the ants to eat, and a seed coat that resists consumption by being extremely tough and frequently so smooth and rounded that it is hard for an ant to crack with her mandibles. The elaiosome feeds the ants, and ant removal of the elaiosome often scars the hard seed coat enough to enhance germination. The reward for the ant is variable, ranging from very lipid- and fatty acid-rich elaiosomes to elaiosomes relatively poor in nutrient content (Buckley, 1982; Beattie, 1985). This diversity may represent adaptation to the habits of particular ant groups or may reflect taxonomic patterns within plants.

The ants that carry myrmecochorous seeds are generalists, collecting any attractive seeds they encounter. None of the known ant-dispersal interactions are specialized for a single ant species or a single plant species: in all cases a community of plants and ants interact. Consequently, one constraint on the seeds of myrmecochorous plants is that they must be of a size and shape that are attractive to ants and easily carried by diverse ants. Other factors, such as provision of ample reserves for the seedling, are also involved in the evolution of seed size, but the observation that larger ants carry more of the large *Sanguinaria canadensis* (Papaveraceae) seeds and smaller ants more small violet (*Viola* spp., Violaceae) seeds (Culver and Beattie, 1978) suggests ants do impose size constraints.

Many myrmecochores have evolved a complex of characters to enhance ant dispersal. These include the elaiosomes just discussed, stalks that bend to present the seeds at ground level, release of seeds at times of day that favor discovery by ants rather than seed predators (Turnbull and Culver, 1983), and the shaping of the seed and elaiosome to facilitate carrying by ants.

The benefits to plants vary with the species. *Carex pedunculata* (Cyperaceae), an ant-dispersed woodland sedge, was shown by Handel (1976, 1978) to be ephemeral relative to other *Carex* species in the same habitat. Ants transported *C. pedunculata* seeds to rotting logs where there was little competition from other sedge species, which favored germination because of the enriched substrate in the ant nests. For *Viola* species, inhumation seems to be the chief benefit of myrmecochory. Although rodent predation on seeds is very strong under the parent, *Viola* has a ballistic mechanism that throws ripe seeds farther from the parent (at least 120 cm) than ants normally carry them (35 to 75 cm), so the elaiosome's function does not seem to be seed transport. However, germination and establishment of *Viola* has been shown to be much greater for seeds carried to ant nests than for seeds planted or dropped elsewhere in the environment (Culver and Beattie, 1978; Heithaus, 1981; Turnbull and Culver, 1983). *Datura discolor* (Solanaceae) benefits from myrmecochory chiefly because seeds carried by ants escape rodents who search for food under the parent plant (O'Dowd and Hay, 1980). Part of the advantage to South African myrmecochores is in escape from the heat of wildfires (Bond and Slingsby, 1984). Thus, the benefits of myrmecochory are complex and diverse even though most myrmecochores studied have been shown to receive increased fitness from inhumation in addition to other advantages (see also Chap. 3; Buckley, 1982; Westoby et al., 1982; Beattie, 1985). However, this generalization has recently been challenged for Australian myrmecochores (Rice and Westoby, 1986).

Some plants depend heavily on ant dispersal agents to maintain their populations. For example, in *Viola* and *Datura,* virtually 100 percent of the seeds that remained under the parent canopy were eaten by rodents (Culver and Beattie, 1978; O'Dowd and Hay, 1980). Even more dramatically, South African plants in the family Proteaceae are suffering widespread reproductive failure because the native seed-dispersing ants have been replaced by the Argentine ant, *Iridomyrmex humilis.* Individuals of *I. humilis* are not strongly attracted to myrmecochorous seeds, so they are left undispersed. The result has been massive losses to predators and fire which were previously avoided when native ants removed and buried the seeds (Bond and Slingsby, 1984).

The best-studied myrmecochores interact with ants only for dispersal. However, many myrmecophytes are also myrmecochorous. For example, *Myrmecodia* (discussed above) has elaiosome-bearing seeds. *Codonanthe cordifolia* (Gesneriaceae), an epiphytic tropical vine which houses ants in its roots, also has ant-dispersed seeds with elaiosomes (Kleinfeldt, 1978). Certainly, interacting with ants at one life stage should favor evolution of ant–plant interactions at another life stage, but it would be interesting to know which interaction came first.

Myrmecochory is widespread in xeric habitats in Australia and Africa, common in European and American temperate forest herbs, but is not well studied elsewhere (Buckley, 1982; Beattie, 1985). In Australian and South African communities paired by community structure, myrmecochory predominated on the drier sites with poorer soils, while fleshy fruits dispersed by birds were more common on the richer, more mesic sites. Milewski and Bond (1982) interpreted this to mean that myrmecochory is common on poorer soils because myrmecochory is cheaper than bird dispersal since the elaiosome requires fewer nutrients and less water to produce than a fleshy fruit (see Chap. 3).

This explanation of myrmecochory does not address the benefits of inhumation. Myrmecochory may be especially important where the greater nutrients, moisture, and aeration of the ant nest are crucial for successful establishment, including poor, dry soils in Australia and South Africa (Milewski and Bond, 1982; Beattie, 1985). The efficacy of myrmecochory is probably also limited by ant behavior. Not all ant species are attracted to seeds, even those with huge elaiosomes. In communities dominated by such ants, myrmecochory will not be favored. Furthermore, ants that are normally seed-gathering may shift their priorities. For example, Culver and Beattie (1978) showed that *Viola* seeds produced late in the season were much less often transported by ants than earlier seeds, despite larger elaiosomes. This shift was apparently due to changes in the status of the ant colonies (e.g., more pupae than larvae) and in the foods available to ants (more insects available).

Little has been said about the importance of myrmecochory to ants. The seed-carrying ants show no specialization to particular seeds or vice versa. While presumably the food from elaiosomes makes up an important part of the diet of ant colonies collecting many myrmecochores, it seems unlikely to be so important as to determine ant distributions (Berg, 1972; Pijl, 1982; Westoby et al., 1982; Beattie, 1985). However, Beattie (1985) points out that ants receive some lipids from elaiosomes that appear to be essential for ant nutrition, so the relationship may be tighter than generally appreciated. On the other hand, many ant species in the same habitat survive very well without ever eating an elaiosome. Much remains to be done to understand the role of myrmecochorous seeds in ant ecology.

In summary, ant dispersal may lead to many different benefits for plants, the chief of which appears to be planting in an enriched microsite. In successful myrmecochores, the dispersed diaspore possesses both structures to attract ants and to protect it from being eaten by the ants. Often a whole suite of characters assists in attracting ants. In some ecosystems such plants comprise a critical part of the flora (Beattie, 1985) and many ant species participate.

Pollination by Ants

Ants gather floral nectar but have been considered poor pollinators because of their low mobility (i.e., they crawl and cannot fly). In addition, they exhibit poor fidelity, engaging in many activities other than visiting flowers during a foraging trip. Another problem with ants as pollinators is that at least some ants' epidermis is toxic to pollen (Beattie et al., 1984), and in any case pollen adheres poorly to ants (Hocking, 1975). Despite these deficiencies, ants can be effective pollinators in certain environments. Plants that are primarily ant-pollinated are reported from arctic and alpine tundra, hot deserts, and scattered situations elsewhere (Hickman, 1974; Hocking, 1975; Buckley, 1982).

Hickman (1974) suggested that for efficient ant pollination, the desirable situation is one where small plants occur in dense stands of uniform height, so that ants can easily walk among flowers. Ant pollination is further facilitated by the presence of nectar that is readily accessible to ants and where a limited amount of aggregated pollen is attached to each flower-visiting ant. Hickman (1974) and Wyatt (1981) (Fig. 6-6) found ant pollination in dense populations of low-stature herbs growing in shallow soils. Here, in populations of restricted geographic distribution, pollination by crawling insects worked effectively. Furthermore, the impoverished growing conditions were seen as favoring small plants with minimal production of pollen, nectar, or floral display. Poor rewards are less likely to deter ant visitors than flying insects because of the lower energy costs for crawling insects. Low, dense plants blooming in large numbers are certainly found in tundra, hot deserts, and in very shallow soil (see Chap. 2).

However, exclusive relationships between open flowers blooming en masse and ants seem unlikely. These would be expected to attract a diversity of generalist insects any of which might pollinate. Indeed, Beattie (1985) has questioned whether any of the ant-pollinated systems that have been described truly represent ant-adapted plants or simply plants utilizing small generalist insects, including ants. Beattie feels definitive experiments are lacking. If ant-pollinated plants are actually specialized for ant-visitation, then ant-adapted traits are expected. One trait that might distinguish ant-pollinated plants from generalists is the nectar reward. Other things being equal, "ant nectar" should be more like that of defense nectaries than of flowers attracting bees or flies. At present not enough is known to evaluate this.

Figure 6–6. Ant-pollinated plant. *Diamorpha smallii* (Crassulaceae). (1) Population of *Diamorpha smallii* growing in shallow soil in a depression in granite. Diameter of population about 0.5 m. (2) Close up of population showing high density of plants within the population. Each plant is about 4 cm high. (3) Ant (*Formica subsericea,* center) walking on flowers of *D. smallii.* Note dense flowers form a relatively level and solid canopy. Flowers are about 3 mm in diameter. (4) *Formica schaufussi* taking nectar from *D. smallii* flower. Nectar is located at the base of the stamens. Ant is about 5 mm long (from Wyatt, 1981).

Nonmutualists: Herbivores

Ants consume plants in two ways: by eating seeds and by feeding leaves to fungi which the ants then eat. Both methods circumvent the problem of digesting cellulose, the inert carbohydrate that makes up the majority of plant biomass.

Harvester Ants

While most ants eat some seeds, seeds make up approximately 80 percent of the diet of "harvester ants." Seed harvesters are a very diverse group which has clearly converged upon a seed-feeding lifestyle. These ants include at least eight genera, some of which are exclusively harvesters [e.g., *Pogonomyrmex* (Fig. 6-7), *Veromessor*], while others are not (e.g., *Monomorium, Pheidole*) (Wilson, 1971; Buckley, 1982). Generally, species are designated harvester ants if they build extensive granaries within their nests in which large numbers of seeds are stored.

Harvesters by no means eat only seeds; they also gather insect material when it is available.

Harvester ants are found in warm, arid areas of Australia, North America, and India, where the temperatures allow ant activity most of the year, but where insect foods are not predictable. In these ecosystems, seeds provide a reliable food source when other prey is scarce. Plants, especially the annuals, shed vast numbers of seeds which may lie exposed for months, until the next growing season. Harvester ant foragers carry single seeds back to the nest where they are eaten or placed in seed storage chambers until needed.

There has been considerable discussion as to whether seeds ever germinate out of seed storage chambers, making the relationship seed dispersal rather than predation. In most species, the storage chamber is too deep underground for seeds to reach the surface even if they do germinate. Thus, deposition of a seed in a deep harvester granary should be regarded as fatal. Of course, many seeds are dropped by the ants, so harvesters do provide some inadvertent seed dispersal (see Chap. 3).

Figure 6–7. Colony of western harvester ant, *Pogonomyrmex occidentalis.* The roughly conical mound of pebbles lies over the nest itself and the bare area cleared around the nest is characteristic. The nest entrance is seen as a dark spot near the base of the mound. This picture was taken in shortgrass prairie in western Nebraska. The diameter of the lens cap was about 7 cm (photo by K. H. Keeler).

Harvester ants change the plant community by reducing the absolute abundance of plants through reduction of the number of seeds available to germinate (Brown et al., 1979). They also modify relative plant abundances by foraging selectively (Tevis, 1958; Whitford, 1978; Briese, 1982b). However, the importance of seed harvesting to the plant community varied greatly in studies of different communities in different years. Whether the gathering of seeds by harvester ants normally regulates the abundance of plants in arid ecosystems is unclear, but it is clear that they sometimes have a critical impact (e.g., Tevis, 1958; Davidson, 1977; Whitford, 1978; Brown et al., 1979; Briese, 1982b).

Defenses against harvesters are not commonly described. Plants may be constrained by other limits requiring that abundant seeds be available until suitable growing conditions occur. This makes unprotected seeds highly vulnerable to harvester ants. Alternatively, harvesters may provide enough inadvertent dispersal by dropping seeds that there is no selective pressure for major suites of antiharvester defenses.

Leaf-cutting Ants

Ants in the tribe Attini are unique among ants in that they cultivate fungi for food. Entirely confined to the New World tropics, these approximately 200 species in 12 genera probably have a single origin (Wilson, 1971; Weber, 1972). Attine leaf-cutting ants remove leaves and flowers which they carry back to their large, fungi-containing, underground nest chambers. The ants prepare leaves for the fungi by cutting them into tiny pieces, chewing them to soften them and remove waxy layers, and then placing the mascerated leaf tissue on the mound of fungus. The fungus feeds on this food, and the ants eat the resulting fungal mycelia, the vegetative tissues of the fungus, composed of a mass of hyphae.

Many fungi of attine ant gardens cannot be assigned names because they have never been observed to produce the fruiting bodies that are the basis of fungal classification (Weber, 1972). There is no doubt that fungus and ant have had a long and intimate relationship, since the fungi have virtually ceased sexual reproduction. In fact, the mutualistic fungus varies: different ant genera cultivate different fungus species (see Table 2 of Jolivet, 1986).

Ants tend the fungus carefully, removing plant refuse and dead tissue, and clearing away spores or mycelia of other fungi. "Weeding" other fungi is important because in the absence of ants, the fungi of ant gardens are quickly overgrown and outcompeted by invading fungi. Apparently mechanical cleaning of the fungus garden is sufficient to keep the nest free of competing fungi, since searches for a fungicide secreted by the ants have been unsuccessful (Wilson, 1971; Weber, 1972).

The fungi in ant gardens produce specialized mycelial structures, called staphylae, which the ants collect and eat. These are not known in other fungi, and are clearly an adaptation to ant gardening. The relationship is obligate since neither fungi nor ants are found without the other in nature. In fact, most of the fungi have never been grown successfully in culture. Alate queens of attine ants carry an inoculum of the fungus with them when they leave the nest and establish a new fungal garden immediately upon digging their first nest.

Not all plants are equally suitable for use by fungus-gardening ants. Leaf-cutters are very selective herbivores, choosing plants based on the needs of both the ants and the fungus. For example, plants with resins that cause pieces of leaf to stick to each other and to the ants are not often cut, however nutritious they might be as fungus food. Likewise, plants poisonous to ants are avoided (Weber, 1972; Buckley, 1982; Hubbell et al., 1984). Of the array of plants that are cut, some are toxic or otherwise unsuitable for supporting the fungus. The degree to which the obvious variation in ant-cutting of plants reflects fungal preference is an area of great interest. Colonies that provide better food for the fungus will prosper relative to colonies that cut plants at random. Even though the feedback of fungal growth to ant behavior is indirect, selection should lead to the evolution of ant foraging that reflects the food preferences of the fungus, or at least avoids gathering species strongly deleterious to the fungus.

Leaf-cutter colonies are such large generalist herbivores that they must cause a shift in the abundance of plants in the community in which they live. For example, single *Atta* colonies may contain 1 to 2 million ants, the nest extending over an area 8 m in diameter, and the workers foraging in a territory with a 250-m radius (Weber, 1972; Lewis et al., 1974). While broad in their feeding habits, they have very strong preferences (Rockwood, 1973, 1976; Buckley, 1982; Hubbell et al., 1984), which change, by differential herbivory, the success of the plant species present.

Conclusion

Direct ant–plant interactions are probably mutualistic because of the dissimilarity of the two organisms. Since ants cannot digest cellulose, only very specialized ants (harvesters and leaf-cutters) can feed on plants. The active foraging habits of scavenging and carnivorous ant species bring them into regular contact with the plant community. Because of this, some species interact mutualistically with the plant providing food or shelter for the ants and, in return, the plant receiving protection, seed dispersal, pollination, or nutrients. Despite the many excellent studies of ant–plant interactions, there are many cases that have never been studied, and others in which contradictions abound.

It is apparent that the mutualists exchange diverse services: a nest site for crucial nutrients, dispersal for oil, nectar for defense. Probably the interaction occurs most readily where each species can easily provide something the other needs, or where the need is very great so that the investment that secures the mutualism is more than repaid (see Boucher et al., 1982; Thompson, 1982). In addition, if alternative mechanisms can efficiently provide the service (e.g., antiherbivore defense), interaction with ants may have no selective advantage. There is an interplay of all these factors in the evolution of mutualistic interactions. Recent studies have increasingly shown that the relationships are variable in time and space. Thus, a complete understanding of ant–plant interactions will require examining particular interactions in diverse spatial and temporal situations to learn their dynamics and their impact on the demography of the interacting species and the surrounding communities (see Buckley, 1982; Beattie, 1985; Jolivet, 1986).

The mutualisms must increase fitness, but many, such as extrafloral nectary–ant mutualisms, are not critical for plant or ant survival, nor do they seem likely to determine the distribution of the pair of species. Myrmecophytes, on the other hand, generally can be seen as extending the distribution of both ants and plants into previously uninhabitable areas. For example, domatia extend ant ranges in habitats where nest sites are limiting, and fertilization by ants allows epiphytes to grow in nutrient-poor areas. Pollination by ants occurs chiefly in marginal habitats and may extend and increase plant distributions in such environments. Transportation of seeds (myrmecochory) certainly redistributes the plants within the community, and in some cases has been shown to be crucial to plant abundances (e.g., Handel, 1978; Bond and Slingsby, 1984), but its importance to the ants involved is less certain.

Since seed harvesters and fungus-gardening ants feed on plants, their distribution is determined by plant distributions, just as all predators can be seen as limited by their prey. Conversely, both ant types influence the abundance and distribution of the plants they prey upon. The outcome of these trophic interactions are very complex in time and space and are receiving a good deal of attention.

Plants have evolved more than ants in response to ant–plant mutualisms. In defense nectaries, myrmecochory, and ant pollination, no changes in ant structure or behavior have been demonstrated, while the nectary, elaiosome, perhaps floral structure, and numerous associated traits have evolved to fit the plant to interaction with ants. Only the obligate ants of myrmecophytes (e.g., *Pseudomyrmex*) show adaptations specialized to mutualism with plants. Perhaps this represents a meaningful pattern, in which an asymmetrical mutualism is established between ants and plants. It is also possible, given the currently available information, that behavioral or nutritional ant adaptations exist that simply have not yet been detected.

In summary, ant–plant relationships run the gamut from obligate pairs never found apart to very casual interactions in which benefit can be demonstrated but varies greatly in time, space, and participants. As a result of this great diversity, much remains to be learned about these widely occurring interactions.

Selected References

Beattie, A. J. 1985. The evolutionary ecology of ant–plant mutualisms. Cambridge studies in ecology. Cambridge University Press, Cambridge.

Buckley, R. C., ed. 1982. Ant–plant interactions in Australia. Dr. W. Junk, Publishers, The Hague.

Janzen, D. H. 1967. Interaction of the bull's horn acacia (*Acacia cornigera* L.) with an ant inhabitant (*Pseudomyrmex ferruginea* F. Smith) in Eastern Mexico. *Univ. Kansas Sci. Bull.* 47:315–558.

Jolivet, P. 1986. Les formis et les plantes: un exemple de coevolution. Societé Nouvelle des Editions Boubee, Paris.

Pijl, L. van der. 1972. Principles of dispersal in higher plants. Springer, NY.

Wilson, E. O. 1971. The insect societies. Belknap Press, Cambridge, MA.

7

Ecology and Evolution of Carnivorous Plants

THOMAS J. GIVNISH

Department of Botany
University of Wisconsin
Madison, WI 53706

Introduction

Carnivorous plants represent one of the great anomalies of the natural world. Whereas most plants are eaten, in whole or in part, by animals acting as pollinators, seed-dispersers, herbivores, or defensive agents, carnivorous plants have turned the evolutionary tables on animals and consume them as prey. As a result, carnivorous plants also possess the potential for unique interactions with animals as competitors for prey, as digestive symbionts, and perhaps as prey mutualists. Furthermore, by having evolved the capacity to augment their nutrition using specialized structures that lure, capture, and digest prey, carnivorous plants appear

I would like to thank Thomas Gibson, Hugh Iltis, Susan Knight, and Steven Solheim for suggesting references and providing helpful comments on the manuscript. Susan Knight, together with Thomas Frost, provided information on their unpublished study of allocation to bladders in *Utricularia vulgaris*. Thomas Gibson and Theodore Weiss kindly granted permission to cite important findings from their unpublished Ph.D. dissertations. The University of Chicago Press provided permission to reprint portions of an earlier article (Givnish et al., 1984), which appeared in the *American Naturalist*.

to have gained the ability to compete successfully in nutrient-poor environments, though perhaps at the expense of reduced competitive ability elsewhere.

This chapter examines some of the facts and important unresolved questions regarding the ecology and evolution of this extraordinary group. First, a definition of carnivory is presented and the taxonomic and ecological distribution of carnivorous plants reviewed. Second, the natural history of prey attraction, capture, and digestion is briefly described for representative species. Evidence is offered for specialization on different prey types, and for digestive mutualisms and antagonisms between carnivorous plants and animals living in traps. Then, the experimental support is reviewed for the nature and context-specificity of the nutritional benefits associated with carnivory.

This information is incorporated in a cost–benefit model and used to predict trends in the ecological distribution and seasonal activity of carnivorous plants. The paradoxical association of carnivorous plants with fire, and with the enrichment of certain soil nutrients, is discussed in terms of this model. Selective pressures on trap structure in sundews and pitcher plants are analyzed, and the intriguing possibility of mutualism between certain kinds of carnivorous plants and prey is raised. Finally, the benefits and costs of carnivory are compared with those associated with nitrogen fixation and ant-fed myrmecophily, and potential evolutionary pathways to carnivory are discussed. The chapter concludes with a shopping list of research needs for a deeper understanding of the ecology and evolution of carnivorous plants. Although the form and physiology of carnivorous plants have been studied carefully since they first attracted the attention of botanists, many of the most important ecological issues, particularly those bearing on carbon economy and the basis of their competitive ability in different habitats, remain open and potentially fertile areas for future research.

Definition of Carnivory

A plant should fulfill two requirements to be classified as carnivorous (Givnish et al., 1984). First, it must be able to absorb nutrients from dead animals next to its surfaces, and thereby obtain some increment to fitness in terms of increased growth, chance of survival, pollen production, or seed set. Second, it must have some morphological, physiological, or behavioral feature whose primary effect is the active attraction, capture, and/or digestion of prey.

The first requirement is needed to differentiate carnivory from purely defensive adaptations that immobilize or kill potential animal enemies without leading to substantial nutrient absorption or consequent increases in plant survival and reproduction (Lloyd, 1942; Benzing, 1980). Examples of such defensive traits might include the glandular, viscid stems of several species of *Silene* (catchfly) that trap potentially destructive insects crawling toward their flowers; the fused, cup-shaped leaf bases of *Dipsacus*, which impound water and form a moat about the stem in which many such insects drown; and the specialized, hooked trichomes on the leaves of *Glycine*, which immobilize and ultimately kill herbivorous leafhoppers (Darwin, 1874; Pillemer and Tingey, 1976). The first proviso also separates carnivores from plants (e.g., species of *Aristolochia, Arum, Nymphaea*) with flowers that entrap animals—usually, though not invariably, without killing them—as part of a mechanism promoting cross pollination, not energy or nutrient capture (Schmucker and Linnemann, 1959).

The second requirement is needed because many unspecialized plants can passively profit by absorbing some nutrients from dead animals (e.g., nematodes, earthworms) decomposing in the soil or on leaf surfaces (Benzing, 1980). A survey of the 18 genera of recognized carnivorous plants shows that adaptations for all three processes of active

Table 7-1 Presence (+) or absence (−) of adaptations for active prey attraction, capture, and digestion in genera of carnivorous plants

Genera (no. of spp.)	Attraction*	Capture	Digestion	Type of Trap
Catopsis (1)	−	Passive†	−	Pitfall
Brocchinia (2)	+	Passive	−	Pitfall
Heliamphora (6)	+	Passive	−	Pitfall
Darlingtonia (1)	+	Passive	−	Pitfall
Sarracenia (9)	+	Passive	±	Pitfall
Cephalotus (1)	+	Passive	+	Pitfall
Nepenthes (70)	+	Passive	+	Pitfall
Genlisea (35)	−	Passive	+	Lobster-pot
Byblis (2)	−	+	+	Passive flypaper
Drosophyllum (1)	+	+	+	Passive flypaper
Triphyophyllum (1)	−	+	+	Passive flypaper
Drosera (90)	−	+	+	Active flypaper
Pinguicula (35)	−	+	+	Active flypaper
Aldrovanda (1)	−	+	+	Steel trap
Dionaea (1)	+	+	+	Steel trap
Biovularia (1)	−	+	+	Bladder trap
Polypompholyx (2)	−	+	+	Bladder trap
Utricularia (280)	±	+	+	Bladder trap

*A conservative approach has been taken in evaluating attractants that may be merely incidental aspects of other traits (e.g., glistening digestive glands in *Drosera*).

†Possesses a highly flocculent cuticle that has been shown to enhance prey capture (see text).

After: Givnish et al., 1984.

prey attraction, capture, and digestion are not required to qualify a plant as carnivorous on logical or historical grounds (Table 7-1). For example, many bladderworts (*Biovularia, Polypompholyx, Utricularia*) and butterworts (*Pinguicula*) lack obvious attractants for prey (Lloyd, 1942; Meyer and Strickler, 1979; Slack, 1979). Pitcher plants in the families Cephalotaceae, Nepenthaceae, and Sarraceniaceae entrap prey passively in water pitfalls, although these possess clear morphological specializations (e.g., one-way passages of downward-pointing hairs, zones of readily detached wax) for that function. Finally, some pitcher plants (e.g., *Darlingtonia, Heliamphora*) lack digestive glands and apparently rely on bacteria and other organisms to break down prey (Lloyd, 1942; Adams and Smith, 1977; Slack, 1979; Bradshaw and Creelman, 1984; Joel and Gepstein, 1985). Thus, specialized digestive glands are not required for carnivory, although the research of Heslop-Harrison and her colleagues (Heslop-Harrison, 1975, 1976, 1978; Green et al., 1979; Heslop-Harrison and Heslop-Harrison, 1980) has demonstrated the presence of characteristic secretory cells in most carnivorous genera.

Some plants are capable of absorbing nutrients from dead animals, but lack active means of prey attraction and prey digestion and possess neither motile traps nor passive structures (like one-way surfaces whose primary result is immobilization of animals near plant surfaces). Such plants should be considered saprophytes and not carnivorous plants. Many tank bromeliads (Bromeliaceae) that impound water in pools among their overlapping leaf bases, into which insects or other animals can blunder and drown, probably fall into this latter category and must be studied with care before a given species is claimed to be carnivorous (Benzing, 1980; Givnish et al., 1984).

Taxonomic Occurrence

Approximately 538 species of carnivorous plants occur in 18 genera and eight families (Table 7-1), including the Bromeliaceae (*Brocchinia, Catopsis*), Byblidaceae (*Byblis*), Cephalotaceae (*Cephalotus*), Dioncophyllaceae (*Triphyophyllum*), Droseraceae (*Aldrovanda, Dionaea, Drosera, Drosophyllum*), Lentibulariaceae (*Biovularia, Genlisea, Pinguicula, Polypompholyx, Utricularia*), Nepenthaceae (*Nepenthes*), and Sarraceniaceae (*Darlingtonia, Heliamphora, Sarracenia*) (Lloyd, 1942; Fish, 1976a; Green et al., 1979; Givnish et al., 1984). Taxonomic isolation is a striking feature of carnivorous plants: with but three exceptions, each is found in a genus and family composed entirely of carnivores. Monotypic *Triphyophyllum* is a carnivorous vine in the tropical West African family Dioncophyllaceae, which contains two other monotypic noncarnivorous genera, *Dioncophyllum* and *Habropetalum* (Green et al., 1979). The bromeliad genera *Brocchinia* and *Catopsis* each contain at least one carnivorous species and belong to a family composed almost entirely of noncarnivores; *Brocchinia* is known with certainty to contain noncarnivorous as well as carnivorous species, and the same is likely to be true of *Catopsis* (Fish, 1976a; Ward and Fish, 1979; Frank and O'Meara, 1984; Givnish et al., 1984; Benzing et al., 1985). These three genera, and particularly the highly variable genus *Brocchinia*, offer unique opportunities for the study of evolutionary pathways to carnivory.

Based on the phylogenetic affinities of the families in which carnivory is known, carnivory seems to have arisen independently at least six times: twice in the superorder Magnoliidae (Aristolochiales: Nepenthaceae; Sarraceniales: Sarraceniaceae), once in the Dilleniidae (Violales: Dioncophyllaceae), once in the Rosidae (Rosales: Bylidaceae, Cephalotaceae, Droseraceae), once in the Asteridae (Scrophulariales: Lentibulariaceae), and once in the Commelinidae (Bromeliales: Bromeliaceae) (see Metcalfe and Chalk, 1950; Jay and Lebreton, 1972; DeBuhr, 1975a; Carlquist, 1976, 1982; Heywood, 1978; Nicholls et al., 1985). Pitcher plants with remarkably convergent traps appear to have evolved independently at least three times, based on familial relationships (cf. Cronquist, 1981) and the very different means by which trap leaves develop in the Cephalotaceae, Nepenthaceae, and Sarraceniaceae (Frank, 1976).

Geographic and Ecological Distribution

Carnivorous plants occur in almost every region of the world, with the exception of the high Arctic, Antarctica, and extreme deserts (Fig. 7-1). Several genera (*Brocchinia, Byblis, Cephalotus, Darlingtonia, Dionaea, Drosophyllum, Heliamphora, Triphyophyllum*) are endemic to narrowly restricted areas, whereas *Drosera* and *Utricularia* are nearly cosmopolitan. Centers of generic diversity include (1) the Guayana Highlands of northern South America, with six sympatric genera (*Brocchinia, Catopsis, Drosera, Genlisea, Heliamphora, Utricularia*), at least four of which are syntopic (i.e., can be found in the same habitat) (Givnish, personal observation); (2) the southeastern United States, with five genera (*Dionaea, Drosera, Pinguicula, Sarracenia, Utricularia*), at least four of which are syntopic (Roberts and Oosting, 1958); and (3) southwestern Australia, with five genera (*Byblis, Cephalotus, Drosera, Polypompholyx, Utricularia*), at least three of which are syntopic (Erickson, 1978). These regions correspond to three of the six extensive

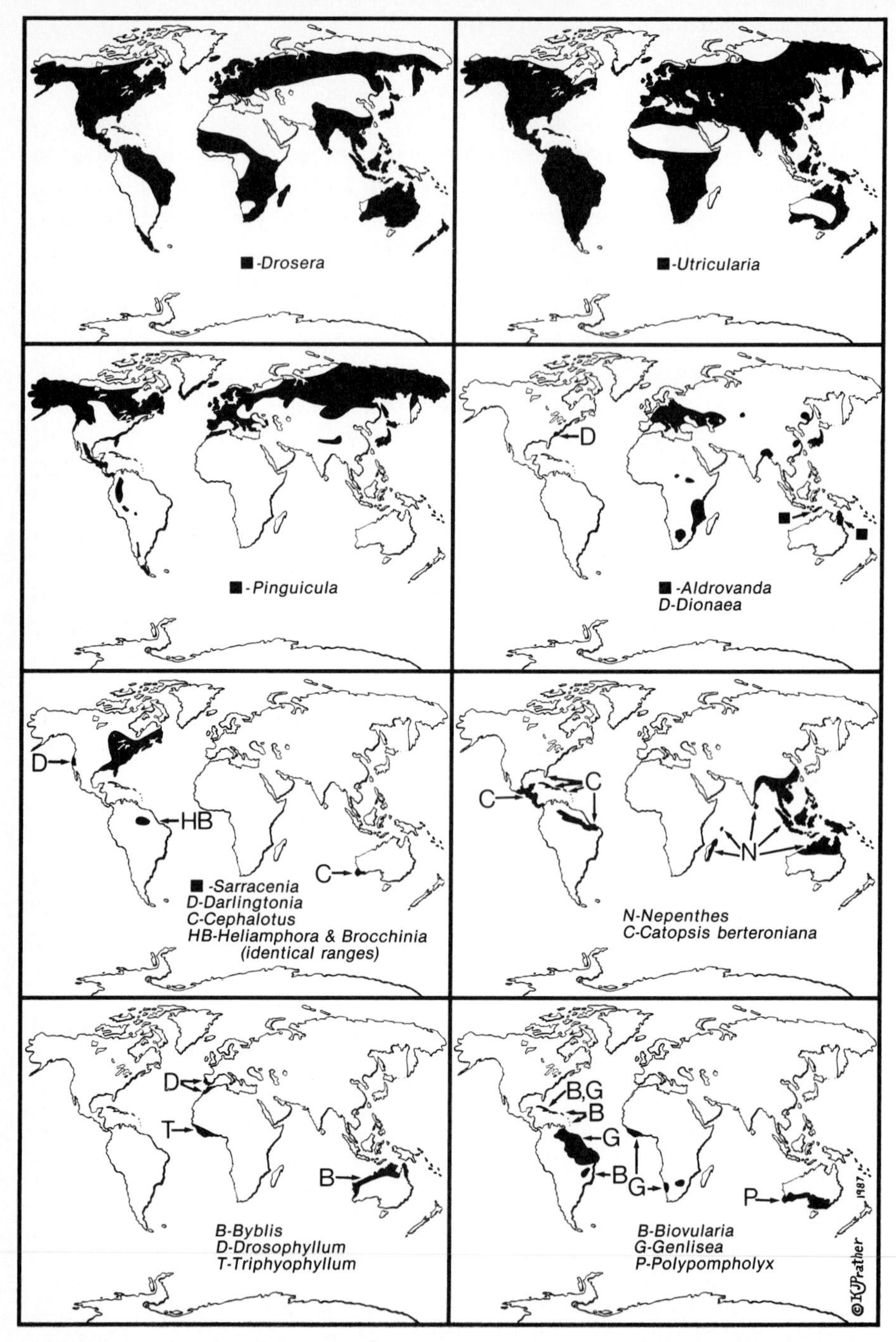

Figure 7–1. Geographic ranges of genera of carnivorous angiosperms (data from Diels, 1906; Macfarlane, 1908; Wherry, 1935; Lloyd, 1942; Roberts and Oosting, 1958; Taylor, 1964; Caspar, 1966; McDaniel, 1971; Erickson, 1978; Heywood, 1978; Green et al., 1979; Smith and Downs, 1974, 1977).

global belts of sandstone or sandy soils in areas of moderate to high rainfall; such belts also occur in South Africa, tropical West Africa, and Borneo. Southwestern South Africa, with only two carnivorous genera (*Drosera, Utricularia*) and vast areas of infertile sandstone in a climate similar to that seen in southwestern Australia, appears anomalously poor in such plants. Bitypic *Roridula* (Roridulaceae), however, is endemic to the Cape region, has leaves with numerous glandular tentacles, and was considered carnivorous until Lloyd (1934) noted that its glandular secretions are resinous rather than mucilaginous, and hence unlikely to bear proteases; its status should be confirmed using modern techniques.

Extensive areas of sandstone also occur on Borneo, which supports three genera (*Drosera, Nepenthes, Utricularia*) and is a center for a massive adaptive radiation in *Nepenthes* (Macfarlane, 1908). Tropical West Africa supports four carnivorous genera (*Drosera, Genlisea, Triphyophyllum, Utricularia*). The circumpolar boreal zone, with high rainfall relative to evaporation and highly leached, infertile soils, supports three genera globally (*Drosera, Pinguicula, Utricularia*), with the addition of *Sarracenia purpurea* in eastern North America. Five genera (*Brocchinia, Genlisea, Heliamphora, Nepenthes, Triphyophyllum*) are wholly tropical; of these, the three pitfall genera *Brocchinia, Heliamphora,* and *Nepenthes* are most abundant and/or speciose in cool montane habitats with heavy rainfall, high humidity, and strongly leached podzols (Smythies, 1964; Givnish et al., 1984; Steyermark, 1984). Almost all the remaining pitfall taxa (*Sarracenia* of the southeastern U.S. Coastal Plain, *Darlingtonia* of the northern California–southern Oregon coast, *Cephalotus* of the southwest Australian coast) occur in temperate climates marked by unusually heavy rainfall (1500 to 3000 mm).

Darwin (1874) appears to have been the first to recognize that carnivorous plants are generally restricted to nutrient-poor habitats. Until recently, however, there has been little explicit recognition of the fact that most carnivorous plants grow in sites that are also sunny and moist at least during the growing season, mostly in pocosins, bogs, swamps, and aquatic habitats (Thompson, 1981; Lüttge, 1983; Givnish et al., 1984), where they are often dominant (Fig. 7-2). Such sites are occupied by *Aldrovanda, Biovularia, Brocchinia, Byblis, Cephalotus, Darlingtonia, Dionaea,* some *Drosera, Genlisea, Heliamphora, Pinguicula, Polypompholyx, Sarracenia,* and most *Utricularia.*

There are only a few exceptions to this general rule (Lloyd, 1942; Chandler and Anderson, 1976a; Erickson, 1978; Slack, 1979; Givnish et al., 1984; Wilson, 1985). *Drosophyllum lusitanicum* remains active in arid upland sites during the dry Mediterranean summer. Two shade-loving sundews (*Drosera adelae* and *D. schizandra*) live in the understory of Queensland rain forests, but show signs of losing the

(a)

(b)

(c)

carnivorous habit in having few tentacles per leaf. *Catopsis berteroniana* is epiphytic on shrub and treetop perches subject to periodic desiccation in low- to mid-elevation forests of the Caribbean, Central America, and South America. *Nepenthes* vines usually inhabit forest openings on nutrient-poor sites (Richards, 1936a, 1936b; Smythies, 1964), but 6 of 71 are epiphytic and a few grow under closed canopies. Most *Utricularia* are aquatic or grow on moist open ground, but 12 of 280 species are epiphytic in moist cloud forests (Taylor, 1964; Madison, 1977). Several *Drosera* of southwestern Australia are well-known to occupy open, mineral-poor, upland sites but are not an exception because they are active principally during the moist winter and spring (Dixon and Pate, 1978; Erickson, 1978; Slack, 1979). Finally, *Drosera whittakeri* grows on substrates relatively rich in cations (Chandler and Anderson, 1976a), and a few gypsum-inhabiting species of *Pinguicula* thrive on carbonate-rich soils (McVaugh and Mickel, 1963).

On sterile soils, carnivores are frequently among the most abundant plants of early succession. In southern New Jersey and the southeastern U.S. Coastal Plain, *Drosera* and *Utricularia* are often among the first and most numerous plants to invade recently cut paths in damp pine forests, or recently exposed sandy lake bottoms (Shetler, 1974; Givnish, personal observation); a similar trend is seen among Australian sundews and bladderworts (Erickson, 1978). On the humid summits of the tepuis, or sandstone table mountains, of the Guayana Highlands, *Brocchinia reducta* is an important element of primary succession, almost the only vascular plant that can invade open rock surfaces (Brewer Carias, 1976). Finally, many carnivorous plants increase dramatically in coverage and sexual reproduction following disturbance by fire. For example, several *Sarracenia* species (e.g., *S. flava, S. leucophylla, S. psittacina, S. rubra*) that inhabit savannas and open pine woods in the southeastern United States increase sharply after fire, and can disappear from areas that remain unburnt for several years (McDaniel, 1971; Eleutarius and Jones, 1969; Folkerts, 1982). The Venus flytrap (*Dionaea*) occurs only in transitional zones between pocosins and wet savannas that are kept open by frequent fires; its sheathing leaf bases protect the apical meristem from damage (Roberts and Oosting, 1958). Species of *Heliamphora* are among the few plants atop tepui summits that vigorously resprout after fire, its rhizome and meristems protected by sheathing, water-filled pitchers (Givnish et al., 1986). *Byblis gigantea* of southwestern Australia

Figure 7–2. Habitats dominated by carnivorous plants: (a) *Darlingtonia californica* in a coastal sphagnum bog, southwestern Oregon. (b) *Drosera filiformis* growing on sphagnum, sedgey peat, and sand, Buck Run, New Jersey, and (c) *Brocchinia reducta* in a moist savanna on decomposing sandstone, 1400 m, Gran Sabana in southeastern Venezuela (photographs by T. J. Givnish).

is highly fire-adapted, burgeoning in abundance and blossom after fire, and having seeds with thick coats that germinate readily only after exposure to flame temperatures (DeBuhr, 1975b; Slack, 1979). Several Australian species of *Drosera* also increase in abundance and flowering after fire (Erickson, 1978; Dixon and Pate, 1978).

The association of carnivorous plants with fire is somewhat paradoxical in that many soil nutrients increase in abundance following fire, at least in the short term, due to their mobilization from plant material to ash to leachate. It should be pointed out, however, that nitrogen is largely volatilized by burning and usually does not increase in soils after fire (Eleutarius and Jones, 1969; Pate and Dixon, 1978; Folkerts, 1982). Any analysis of the basis for the competitive ability of carnivorous plants must explain this paradoxical association with fire and higher nutrient levels, as well as the overall association with nutrient-poor sites that are at least seasonally moist and sunny.

Mechanisms of Interactions between Carnivorous Plants and Animals

Carnivorous plants can interact with animals in an unusually wide variety of ways. This section reviews some of their specific mechanisms of interaction with animals acting as (1) prey, (2) digestive symbionts, (3) competitors, (4) herbivores, and (5) pollinators. Emphasis is given to the definitive interactions of carnivorous plants with their prey, including their means of prey attraction, capture, and digestion, as well as evidence for their specialization on different prey types. Discussion of a potential mutualism between certain kinds of carnivorous plants and their prey is deferred to a later section dealing with selective pressures on trap structure.

Interactions with Prey

Prey Attraction, Capture, and Digestion

For ease of discussion, carnivorous plants may be divided into six functional groups based on their mechanisms of prey capture (Lloyd, 1942; Slack, 1979). Three of these classes involve plants with passive traps which, although they often employ ingenious means of seducing their victims, do not involve any movement of plant parts. The remaining three classes involve *active* traps that do use movement to help secure prey capture or digestion (Fig. 7-3). The salient features of plants in these classes, arranged in order of increasing complexity of their traps, can be summarized briefly as follows:

1. The *pitfall trap,* seen in all five genera of pitcher plants and in the two bromeliad genera containing carnivorous species, is the simplest passive trap. In the variant of this trap type seen in the pitcher plants, arthropods are attracted to the trap by nectar, often advertised by vivid coloration of the pitcher (Slack, 1979). Prey find their way to the richest nectar source, which lies at or just within the mouth of the pitcher and is protected from dilution by rain (or sun-driven evaporation) by an overhanging hood (except in *Sarracenia purpurea*). Below this is a zone with unsure footing, onto which the prey wander while feeding, lose their footing, and then plummet into the watery abyss, from which escape is prevented by vertical pitcher walls, downward pointing hairs, slippery wax surfaces, and/or a paralyzing agent (coniine) in the pitcher fluid (Lloyd 1942; Juniper and Buras, 1962; Miles et al., 1974; Mody et al., 1976; Lüttge, 1983). Digestive glands are present only in *Cephalotus, Nepenthes,* and certain species of *Sarracenia* (Adams and Smith, 1977). These glands participate in both enzyme secretion and absorption of

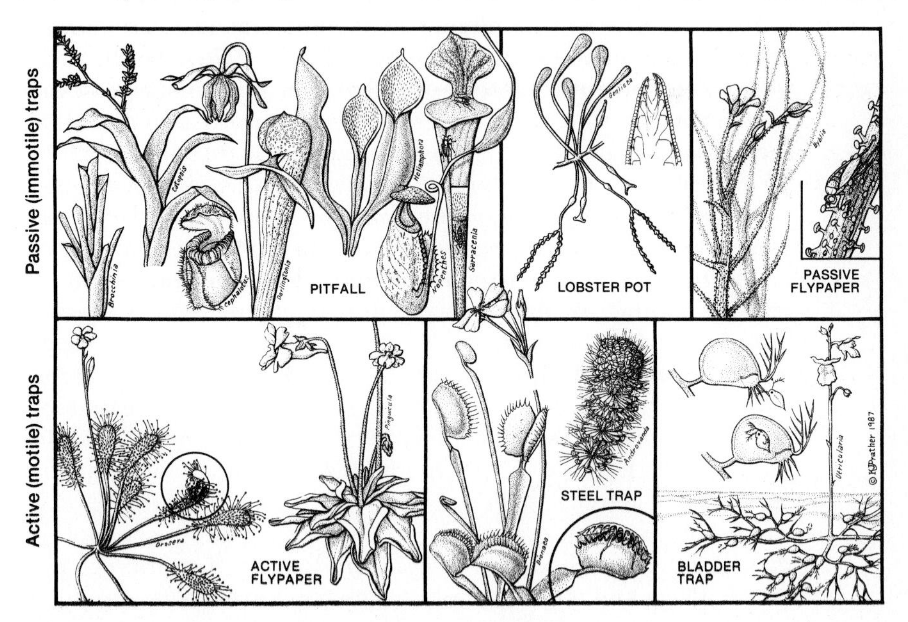

Figure 7–3. Classification of trapping mechanisms occurring in carnivorous plants, with illustrations of representative genera.

ions and amino acids from the resulting digest (Lüttge, 1965, 1983; Parkes and Hallam, 1984). In other pitcher plants, bacteria and other organisms apparently are important in breaking down prey, and nutrient absorption generally occurs through the leaf epidermis (Hepburn et al., 1920; Schmucker and Linnemann, 1959; Lüttge, 1983).

In the variant of the pitfall trap seen in the carnivorous bromeliads *Brocchinia reducta, B. hechtioides,* and *Catopsis berteroniana,* insects are probably attracted to the trap by the bright yellow coloration of the leaf rosette. *Brocchinia* also emits a sweet, nectarlike scent from the tank fluid impounded in its cylindrical rosette (Givnish et al., 1984). Carnivorous species of both genera bear a layer of chalky-white, waxy dust on their highly inclined, inner leaf surfaces, providing a slippery, constantly disintegrating foothold which ultimately precipitates hapless victims into the tank field. Fish (1976a) has suggested that the high ultraviolet reflectivity of the leaf powder on *Catopsis* helps render it invisible against the horizon to many insects, but this seems unlikely given the location of the powder. Furthermore, the powder's reflectivity at a wide range of wavelengths is probably a by-product of the finely divided nature of its threadlike scales, adapted primarily to enhance prey entrapment. (Indeed, the slippery zone of specialized, detachable wax scales inside *Nepenthes* pitchers is notably glaucous, or whitish, as is a similar zone inside *Cephalotus;* Slack, 1979). No digestive glands are known in these bromeliads, but *Brocchinia* possesses specialized trichomes (leaf hairs) that can absorb amino acids at high rates (Givnish et al., 1984; Benzing et al., 1985).

2. The *lobster-pot trap* occurs in *Genlisea,* which employs a one-way passage through which prey can enter, but from which they cannot find a way out. *Genlisea* is a curious tropical aquatic found on both sides of the Atlantic, apparently little-studied in its natural habitat. In the Guayana Shield, it grows in wet sand, often with standing water, with its trap leaves in the substrate (Givnish, personal observation). It is a member of the bladderwort family, and like *Utricularia,* shows no clear distinction between leaves and shoots; it has no true roots. It possesses aerial photosynthetic leaves, which are narrow, erect, and spoon-shaped; and achlorophyllous trap leaves, which are borne underwater and/or underground. Each trap leaf resembles a divaricately branched root 2 to 15 cm long, with spiral grooves on each arm and a hollow bulb in the main axis above the point where the arms unite (Fig. 7-4). The grooves are lined with hairs that point toward the branchpoint and the entrance is found

there. This entrance is lined with overlapping, inward-pointing scale cells that form a series of overlapping cones, similar to those seen in certain lobster or fish traps. This one-way passage ends in a cavity found within the swollen bulb of the main axis, which is lined with digestive glands. The spiral grooves may act as collectors for prey moving in many directions over a large spatial volume (Lloyd, 1942); as long as the prey remain alive, they should tend to move in the direction of the entrance and digestive cavity. Little is known of the prey, although aquatic microcrustaceans are likely to be among them.

3. *Passive flypaper traps* appear in the rainbow plants (*Byblis*) and dewy pine (*Drosophyllum*), strongly convergent members of the Byblidaceae and Droseraceae, as well as in the trap leaves of *Triphyophyllum* of the Dioncophyllaceae. Winged insects are allured by brilliant points of light refracted by glistening droplets on stalked glands; in *Drosophyllum* the glands are also colored bright red, and emit a rich, honeylike scent (Lloyd, 1942). The droplets secreted by the stalked glands immobilize the prey, though in different ways: in *Byblis*, the secretion is mucilaginous and viscid, holding the prey in a gluey trap, whereas in *Drosophyllum*, the droplets are less viscid and readily detached from the glands, but cluster in globules on the prey as it moves about the plant, until it suffocates (Slack, 1979). The stalked

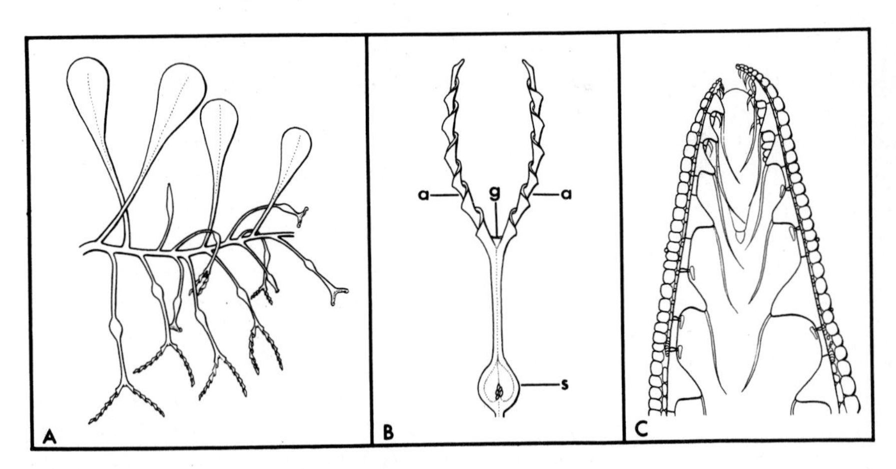

Figure 7–4. *Genlisea.* (A) Schematic drawing of part of a typical plant, showing foliage and trap leaves. (B) Close-up of an enlarged trap leaf, showing hollow, bulblike stomach *s*, neck leading to slitlike gullet *g*, and two arms with spiral grooves *a*, lined with hairs pointing toward the gullet. (C) Cross section through entrance to gullet, showing overlapping, clawlike cells forming lobster-pot trap (see text).

glands of both genera are mushroom-shaped and immobile. Digestive enzymes are secreted copiously by numerous, minute stalkless glands scattered over the linear leaves; in *Drosophyllum* the digestive process is particularly powerful, involving the secretion of protease, acid phosphotase, esterase, and peroxidase —a mosquito can be completely absorbed in 26 hours (Heslop-Harrison, 1978; Lüttge, 1983). Nutrient absorption is also effected by the stalkless glands. The glandular trap leaves of the West African vine *Triphyophyllum* are convergent to those of *Drosophyllum* in most respects, and are borne principally on seedling and juvenile shoots (Green et al., 1979). Mature vines generally lose their trap leaves, and bear two kinds of eglandular leaves with or without a terminal, strongly developed hook adapted for climbing.

4. *Active flypaper traps* are seen in the sundews (*Drosera*) and butterworts (*Pinguicula*). Both initially ensnare prey with a sticky glandular secretion, as do the passive flypapers, but they also employ a further refinement: the localized power of movement. In *Pinguicula,* this ability is limited to the enrolling of the leaf margins and, in certain species, to a "dishing" of the blade under larger prey (Darwin, 1874; Lloyd, 1942; Slack, 1979). The rate of movement is extremely slow, requiring some 2 to 4 hours, and is unlikely to aid in capturing or overcoming prey. Minute prey are instead retained by secretions of stalked glands; their tiny drops of clear mucilage impart a glistening sheen and greasy touch to the yellowish, tongue-shaped leaves, giving rise to the name "butterwort." Movement of the leaf blade serves to bring more digestive surface into contact with the prey and to prevent the escape of digestive fluid from the leaf surface while digestion proceeds. Although the stalked glands secrete some digestive enzymes, notably amylases, the main secretory outflow is provided—as in *Byblis* and *Drosophyllum*—by stalkless, sessile glands flush with the leaf surface (Heslop-Harrison and Knox, 1971; Heslop-Harrison, 1978). Resorption of the pool of digestive fluid occurs via mass flow through the same sessile glands that released their stores of proteases, acid phosphotases, esterases, and ribonucleases; each gland can fire but once, and the resulting disruptions to the cell wall and plasmalemma are irreversible (Heslop-Harrison, 1978).

In *Pinguicula,* no means of prey attraction has definitely been established, although some report the leaves emit a weak fungoid odor (Lloyd, 1942). In *Drosera,* the glistening droplets borne by the conspicuous, reddish tentacles probably serve to entice prey. These mucilaginous secretions, which remain abundant through the heat of the day and give rise to the name "sundew," are the mechanisms of initial prey capture. Within seconds, nearby peripheral tentacles begin to bend toward the one holding the prey, bringing to bear their additional adhesive and digestive powers. The movement is elicited by both mechanical and chemical stimuli, and can be completed in as little as 18 minutes (Darwin, 1874; Lloyd, 1942; Lüttge, 1983). The motion is the result of cell growth, with cells on the side of the tentacle away from the prey expanding during the initial phase. After the victim has been digested, growth of cells on the other side restores the tentacle to its original posture. The process can be repeated up to three times (Lloyd, 1942). A slower growth response, involving a folding of the leaf blade about the prey, occurs in some species over a period of hours or days, and is probably important only in prey digestion (Darwin, 1874; Slack, 1979; Bopp and Weiler, 1985). The secretory cells atop *Drosera* tentacles release several digestive enzymes and participate, together with numerous minute hairs or sessile glands on the tentacles and leaf blade, in resorption of the digest (Lloyd, 1942; Heslop-Harrison, 1975, 1976, 1978; Chandler and Anderson, 1976b; Lüttge, 1983).

5. Functionally similar *steel traps* are seen in the closely related Venus flytrap (*Dionaea muscipula*) and waterwheel plant (*Aldrovanda vesiculosa*), the former of which is terrestrial and occurs in frequently burnt, sandy shrub bogs (pocosins) of the Carolina coastal plain, whereas the latter is aquatic and displays a most unusual multiple-disjunct range encompassing parts of Europe, east Asia, Timor, Queensland, India, and central Africa (Fig. 7-1). Their leaves possess twin lobes held ajar like the jaws of an open steel or spring trap. Prey creeping about on these lobes can trip trigger hairs that elicit an extremely rapid closure of the trap. In *Dionaea*, the leaf lobes are roughly 0.5 to 2 cm in diameter, joined to each other along a continuation of the midrib of the winged petiole, and are fringed with incurving teeth that interlock when the trap is sprung. Their inner surfaces are bright red in sunlit plants, and secrete abundant nectar. Three trigger hairs lie in a triangle on the inner surface of each lobe. In most cases, two touches of a single hair, or of two separate hairs, within 20 seconds is required to spring the trap. The trap shuts sufficiently within 1 to 3 seconds of being triggered that the interlocking teeth form a cage and prevent the escape of all but large prey. The mechanism of closure is an acid growth response: The outer walls of the lobes expand by as much as 28 percent in length following stimulation of the trigger hairs, concurrent with

the rapid, ATP-driven release of H+ ions, which acidify the cytoplasm to below pH 4.5 and trigger rapid expansion of cell wall components while turgor declines (Williams and Bennett, 1982). Provided the prey does not escape, the trap further narrows over a period of hours to achieve a tight seal in response to chemical stimuli highly specific to insects (such as uric acid, the principal component of insect excretions) (Robins, 1976). Sessile glands then secrete digestive enzymes, and later resorb and digest (Lloyd, 1942; Heslop-Harrison, 1978). Trap closure can be repeated three or more times, and the breakdown of glandular cell walls and plasmalemma during secretion is at least partly reversible (Lloyd, 1942; Scala et al., 1968, 1969).

Aldrovanda possesses traps that are much smaller (2 to 3 mm in diameter) and are borne underwater in whorls of eight. Remarkably, its trap requires less than 0.2 seconds to close, even though the expulsion of incompressible water is involved (Lloyd, 1942). The trigger hairs are large relative to the trap, and number some 20 per lobe; mucilage-secreting glands are present near the lobe rims, and may play some role in preventing the dilution of the digestive fluid. No evidence of alluring adaptations exists to date.

6. Finally, highly sophisticated *bladder traps* are seen in the closely related bladderwort genera *Utricularia*, *Biovularia*, and *Polypompholyx* of the Lentibulariaceae. The traps are bean-shaped to spherical, hollow, and minute, from 0.25 to 5 mm in length; their size and extreme rapidity of action helped to delay recognition of a trapping mechanism even more remarkable than that of the Venus flytrap. The traps are borne underwater, in wet sand, mud, or epiphytic moss, or in bromeliad tanks (Lloyd, 1942). The entrance to a trap is complex and guarded by two flexible valves: a so-called door and a smaller velum (Fig. 7-5). When the trap is being set, four-armed sessile hairs inside the bladder (quadrifids) absorb water and create a partial vacuum. Once the trap is set, the lower edge of the dog-legged door lodges against a stop in the velum and forms a tight, albeit exquisitely sensitive, seal. Triggering of the trap is purely mechanical. Four bristles project from the lower half of the door; pressure on any of these by prey swimming nearby serves to swing the dog-legged door past the stop in the velum, causing it to swing back under the vacuum, sucking water and nearby prey into the bladder. Once the vacuum is filled, mechanical tension in the door snaps it shut; digestive enzymes are then secreted (Heslop-Harrison, 1978), and the quadrifids absorb the digest

and later reset the trap. Bifid hairs just inside the door apparently help form a mechanical barrier to prevent prey escape (Darwin, 1874; Lloyd, 1942). The entire time required for a trap of *U. vulgaris* to engulf a victim is less than 35 milliseconds (Lloyd, 1942).

Aquatic prey (insect larvae, microcrustaceans, and the very occasional fish larva or tadpole) may be attracted to the trapdoor by glands secreting mucilage and sugars, though an alluring function has not been proved. Experiments by Meyer and Strickler (1979), however, have shown that the antennae found near the trapdoor in *U. vulgaris* (as in most species) serve to enhance prey capture. Excision of the antennae reduces the capture rate; apparently, the antennae funnel prey toward the door where they are more likely to trigger the trap—certain microcrustaceans habitually graze along strands of filamentous algae which the antennae resemble.

Two additional trapping mechanisms have been proposed, but the plants bearing them have not generally been accepted as being carnivorous. First, the root pellicle of shepherd's purse (*Capsella bursa-pastoris*) becomes sticky and mucilaginous upon germination, and numerous microbes and nematodes adhere to it (Barber, 1978). Although some nutrient absorption does take place, the crucial question is whether the mucilage exists primarily to ensure effective hydraulic contact with the surrounding soil, which seems likely, or to ensnare soil organisms. Second, Rees and Roe (1980) suggest that the giant Andean

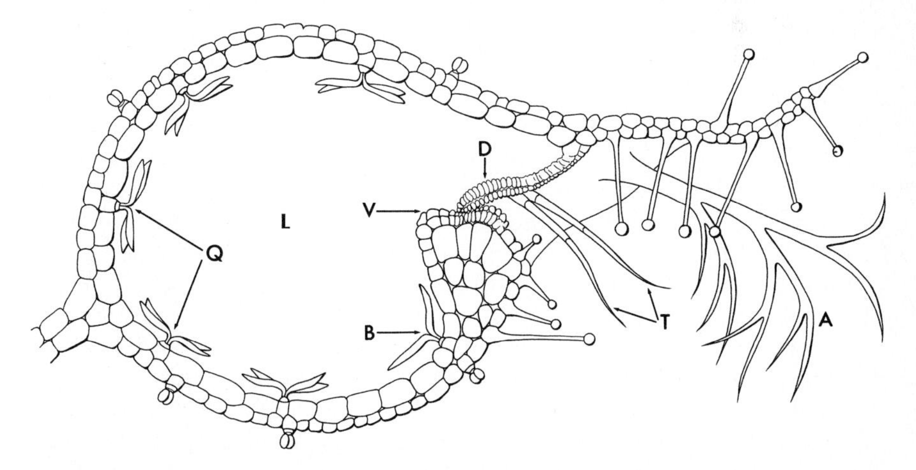

Figure 7–5. *Utricularia.* Schematic drawing of a longitudinal section of a typical trap, showing: trap door D; velum V; trigger hairs T; bladder lumen L; quadrafid hairs Q; bifid hairs B; and antennae A.

bromeliad, *Puya raimondii,* may be carnivorous because small birds that nest among its densely packed leaves are often found impaled on its recurved leaf spines. However, Givnish et al. (1984) conclude that *Puya*'s spines evolved mainly to deter consumption of its single terminal inflorescence by Andean spectacled bears, which destroy the inflorescences of 90 percent of some populations of related *Puya* species, or possibly to deter consumption by elements of the extinct South American megafauna. The recurvature of the spines, which Rees and Roe (1980) emphasize as a trait unlikely to be adapted to antiherbivore defense, might easily be understood as an antibear defense: When a massive ursine shinnies up a *Puya*, its leaves would flex under the weight, bringing the recurved spines to bear (so to speak), and driving them directly into the paws. Benzing (1980) and Givnish et al. (1984) rebut suggestions by several botanists (e.g., Picado, 1913) that many epiphytic tank bromeliads are carnivorous.

Prey Specialization

Differences among species in trap structure often appear related to differences in the kinds of prey they attract and capture. For example, butterworts possess minute stalked glands, a relatively weak mucilage, and no obvious visual or olfactory lures; thus, their diet is largely restricted to small insects like gnats and springtails (Slack, 1979). On the other hand, the trapping mechanism of the Venus flytrap—including bright coloration, nectaries, and trigger hairs and cilia spaced far apart so that small prey are unlikely to elicit closure or subsequent narrowing and digestion—ensures that it mainly traps larger prey. Given that each trap leaf may function only a few times, a triggering mechanism that permits specialization on large prey may be highly adaptive (Darwin, 1874).

The prey ensnared by certain carnivorous species seems conditioned as much by the spectrum of prey available in different habitats as by differences in trapping mechanisms. Rain forest-dwelling *Triphyophyllum,* for example, captures mainly small beetles on its glandular tentacles, whereas bladderworts entrap aquatic creatures (Lloyd, 1942; Green et al., 1979).

A weakness of many studies of prey taken by carnivorous plants is that most fail to characterize the range of insects present in the habitat. Without such data, it is difficult to determine whether plants with different diets are actually specialized or merely sampling different faunas at random. When several different carnivorous species coexist in the same habitat and capture different kinds of prey—as do butterworts, bladderworts, the Venus flytrap, and *Sarracenia* pitcher plants in the southeastern United States (Schnell, 1976; Gibson, 1983)—this problem is largely obviated.

Among pitcher plants, many species capture large numbers of ants and possess traits seemingly adapted to enhance their capture. In *Sarracenia,* the myrmecophages *S. minor* and *S. rubra* have an "ant-trail" of nectaries leading from the ground up the frontal wing on the exterior of the upright pitcher (Fish, 1976b; Schnell, 1976; Slack, 1979). This wing marks the fusion of the two edges of the leaf that forms the pitcher. In turn it leads to the richest nectar source, the nectar roll, which surrounds the brink of the abyss. Ants also form a large part of the diet of *Nepenthes* species (Macfarlane, 1908; Shetler, 1974; Slack, 1979). Their pitchers possess nectaries on two ciliated frontal wings, which are not homologous with the single wing of *Sarracenia* (Frank, 1976); the wings lead to the corrugated pitcher rim with its abundant nectaries and the nearby slippery zone inside the pitcher of specialized, readily detached wax scales with narrow bases (Juniper and Buras, 1962). Several species have numerous extrafloral nectaries on their woody, climbing stems, which doubtless would also tend to attract ants. Interestingly, one species, *Nepenthes bicalcarata,* appears to be both a carnivore and a myrmecophyte (see Chap. 6), providing a domicile in a hollow swelling below its aerial pitchers (Macfarlane, 1908). The biological role of the ants that reside there is as yet unknown.

Although there is a strong trend toward myrmecophagy in pitcher plants, several species of the genus *Sarracenia* show divergent adaptations to other prey. Species closely related to the myrmecophages (*S. flava, S. leucophylla, S. oreophylla*) have tall, brightly colored pitchers, less well-developed frontal wings and nectaries, and tend to attract a higher proportion of flying insects like bees and wasps (Slack, 1979; Gibson, 1983). In *S. purpurea,* the pitchers recline on the ground and present a watery pitfall visible at the trap entrance; flies, crickets, and grasshoppers, perhaps drawn by scintillations of light off the trap fluid, figure prominently in the diet (Judd, 1959; Slack, 1979; Folkerts, 1982). Finally, reclining pitchers are also seen in the parrot pitcher plant, *S. psittacina,* in which the lobes of the lid or hood fuse to form a dome. Translucent areas in the dome confuse insects once they are inside, making it more likely that they will fail to find the exit and instead collide with the walls and downward-pointing hairs, and fall prey to the pitfall; a similar system occurs in the erect pitchers of *Darlingtonia.* In *S. psittacina,* however, the downward-pointing hairs are dense and extend from near the dome to the base of the pitcher, and may act as a secondary trapping mechanism when its pitchers are submerged. *S. psittacina* grows in low-lying sandy savannas and pocosins that flood frequently. Jones (1935) reported that sometimes its pitchers contain large numbers of small water beetles, perhaps ensnared by the lobster-

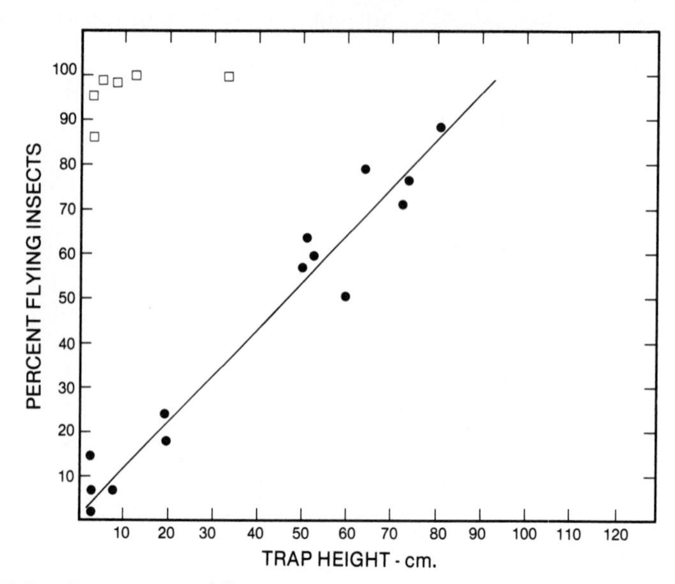

Figure 7–6. Percentage of flying insects among prey as a function of trap height in *Sarracenia* species and the Venus flytrap (Gibson, 1983). Solid circles are population means based on 7490 insects contained in 1007 traps; the line represents the linear regression of percentage of flying prey against trap height across plant species ($r = 0.97$, $p < 0.01$ for 12 d.f.). This trend, however, is not general: the six *Drosera* and *Pinguicula* species studied (hollow squares) are short in stature but mainly entrap flying prey.

pot mechanism the hairs generate. Extensive studies by Gibson (unpublished data), however, indicate that small crawling, terrestrial arthropods (spiders, mites, and ants) are much more common as prey.

In *Sarracenia*-dominated habitats in the southeastern United States, something akin to resource partitioning appears to be operating, with one species of the taller *S. flava/leucophylla/oreophylla* group coexisting with *S. psittacina*, *S. purpurea*, and *S. minor/rubra* (McDaniel, 1971; Schnell, 1976; Folkerts, 1982). Gibson (1983) has indeed shown that the prey taken by *Sarracenia* species varies systematically with trap height (Fig. 7-6), and that the rate of prey capture per trap decreases with trap density among plants of similar height (Fig. 7-7), even though insect availability (as measured by sticky traps) and total prey capture per unit ground area increases with plant density. These phenomena suggest a possible evolutionary basis for character displacement in trap height, and for species of dissimilar height to coexist more frequently than expected (Gibson, 1983). It remains to be seen, however, whether this trend for *Sarracenia* with dissimilar prey to coexist is actually the result of competition, or instead a sequence of allopatric speciation among related and functionally similar taxa.

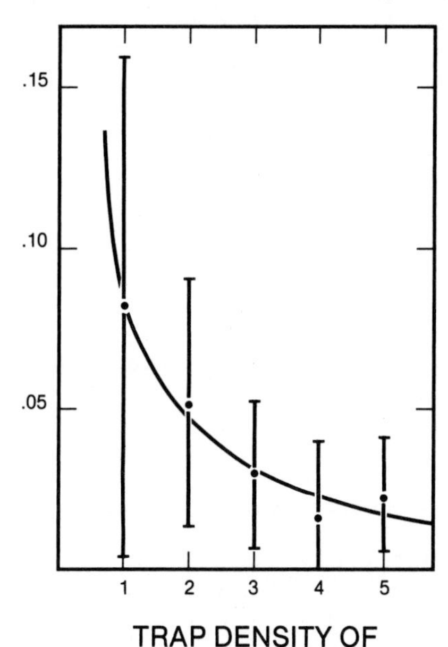

Figure 7–7. Prey capture rates of *Sarracenia alata* traps as a function of the trap density of a potential competitor of comparable height, *S. leucophylla* (Gibson, 1983). Potted plants were transplanted to a bog habitat lacking *S. alata* at Lake Carr, Florida; individual *S. alata* traps were surrounded by 1 to 5 traps of *S. leucophylla*. Capture rates decrease significantly with increasing density of *S. leucophylla* ($y = 0.085\ x^{-0.936}$, $r = 0.94$, $p < 0.02$ for 3 d.f.). Bars indicate 95 percent confidence limits for prey capture rates.

Mutualism with Digestive Symbionts

Pitcher plants are known to harbor a rich fauna of nonprey animals, or inquilines, within their trap leaves. For example, the trap fluid of *Sarracenia purpurea* supports a variety of bacteria, protozoa, rotifers, small crustaceans (Hepburn et al., 1920; Judd, 1959; Addicott, 1974), aquatic mites (Hunter and Hunter, 1964), and the larvae of three dipterans found only in *Sarracenia* pitchers. These latter include the sacrophagid *Blaesoxipha fletcheri,* which feeds mainly at the surface on freshly killed prey; the chironimid midge *Metriocnemus knabi,* which feeds on dead victims at the base of the leaf; and the culicid *Wyeomyia smithii,* which grazes on suspended particulate matter including bacteria and protozoa (Judd, 1959; Fish and Hall, 1978; Bradshaw and Creelman, 1984). The fluid of *Nepenthes* pitchers harbors an even richer fauna, with 150 macroscopic species known for the genus, of which roughly two-thirds are mosquito larvae, with the rest comprised of other dipterans, lepidopterans, parasitoid hymenopterans, harpaciticoid copepods, and the occasional errant land crab (Beaver, 1983, 1985).

It has long been suggested that certain of these inquilines may play a mutualistic role in promoting digestion, particularly in species like *Sarracenia purpurea* that appear to lack digestive glands (Schimper, 1882; Hepburn et al., 1920; Plummer and Jackson, 1963; Joel and Gepstein, 1985). This must certainly be true of bacteria and protozoans in such plants, but the case for the larger inquilines is less clear. Bradshaw (1983) and Bradshaw and Creelman (1984) have shown experimentally that *Metriocnemus* and *Wyeomyia* larvae do accelerate the breakdown of prey and the rate of nitrogen release, based on comparisons of nitrogen levels in actual *Sarracenia* leaves and glass jars with neither, one, or both of these dipterans. *Metriocnemus*, by mechanically breaking down prey carcasses, strongly stimulates nitrogen release during the first 4 to 6 days following prey capture; *Wyeomyia*, presumably by filter-feeding on the resulting particulates and bacterial populations and preventing microbes from tying up the nitrogen pool, stimulates nitrogen release less dramatically from 4 to 12 days after prey capture (Bradshaw, 1983; Bradshaw and Creelman, 1984). These two species often peak in a successional sequence in pitchers after they open (Fish and Hall, 1978). It is not obvious, however, that these dipteran larvae are a net benefit to their host, insofar as Bradshaw and Creelman (1984) find that they inhibit net nitrogen release to the pitcher fluid after about 14 days. As Benzing (1980) has pointed out in connection with an analogous argument for a nutritional mutualism between insect larvae and tank bromeliads that trap vegetable debris, such "mutualists" eventually remove a considerable amount of nutrients from the system when they mature and disperse. The crucial question is whether the economic benefits of more rapid nutrient release—perhaps in terms of a greater amount of short-term plant growth and trap production —outweigh the reduction of the total nutrient load released from a given amount of prey. This important issue has yet to be resolved.

Metriocnemus, *Wyeomyia*, and other inquilines of *Sarracenia purpurea* have relatively high respiratory rates, yet the pitcher fluid remains highly aerobic (Cameron et al., 1977; Bradshaw and Creelman, 1984). Joel and Gepstein (1985) have shown that this is a result of the highly unusual presence of photosynthetically active chloroplasts throughout the epidermis of the submerged inner pitcher surface; this epidermis is, perhaps more remarkably, isolated by nonchlorophyllous tissue from the photosynthetic tissue on the external surface of the pitcher. These unique morphological features, which result in conditions favorable to inquilines, suggest that a certain amount of coevolution may have taken place between the host and the alleged mutualists. This argument would be strengthened considerably if it were shown that *Sarracenia* species with digestive glands lack or have fewer chloroplasts in their inner epidermis. In any case, the relationship of *Sarracenia purpurea* to its inquilines is nonobligate, limiting the potential for close coevolution: In heavily loaded pitchers, anaerobic photosynthetic bacteria (*Rhodopseudomonas palustris*) appear to replace dipterans and aerobic bacteria in breaking down prey (Bradshaw and Creelman, 1984).

Tentative support for the role of bacterial symbionts in promoting digestion has also been obtained for carnivorous plants with digestive glands. In *Nepenthes,* leucine aminopeptidase was found only in open pitchers and not in the sterile fluid of closed pitchers, whether or not they were fed axenically (free of other living organisms) with casein or albumin, strongly suggesting a microbial origin for this enzyme (Lüttge, 1964, 1983). Similarly, Chandler and Anderson (1976b) compared the activity of various digestive enzymes in *Drosera binata* grown under field and axenic conditions. They found no chitinolytic activity and a single pH optimum for proteolytic activity in the axenic plants, whereas the field-grown plants showed chitinase and a wide range of proteases, suggesting microbial action in this instance as well. Chandler and Anderson (1976b) also found that prey-enhanced growth disappears when *Drosera binata* is grown axenically and when the tentacles of *D. whittakeri* are treated with an antibiotic, implying that, in these species, the nutritional benefits of carnivory may rely on microbial action. None of these experiments, however, are wholly conclusive: Immature *Nepenthes* may not be capable of secreting leucine aminopeptidase, and the side effects of axenic culture or antibiotics may impair plant as well as microbial secretory ability.

Competition with Animals for Prey

In addition to the symbionts just discussed, there are several animals that unilaterally exploit carnivorous plants and compete directly with them for prey, untempered by any possibility of mutualism. Several species of Capsidae (Heteroptera) move about the sticky leaves of Australian *Drosera* and *Byblis* without difficulty or harm, sucking the juices of recently ensnared victims. Members of the same group scavenge dead insects from the glandular tentacles of *Roridula* in South Africa (Lloyd, 1942; Erickson, 1978; Watson et al., 1982). Larvae of the plume moth *Trichoptilus parvulus* eat insects trapped by *Drosera capillaris* in Florida. They also feed on the sundew's stalked glands and leaf blades (Eisner and Shepard, 1965). Both capsids and *Trichoptilus* appear to avoid ensnarement through agility rather than special immunity to mucilage.

Several spider species live within or on plants of *Brocchinia, Byblis, Darlingtonia, Drosera, Nepenthes,* and *Sarracenia,* and compete for prey drawn to their host (Lloyd, 1942; Judd, 1959; Givnish et al., 1984). In *Brocchinia, Nepenthes,* and *Sarracenia,* these "terrestrial" predators sit in the tank or pitcher above the fluid, seizing insects as they enter the plant's trap; aquatic predators, such as the sarcophagid *Blaesoxipha,* intercept recently drowned prey.

The greatest trophic complexity is probably seen within the pitchers of *Nepenthes*. Geographic variation in their food web structure generally follows theoretical expectations. In small, outlying regions with few *Nepenthes* host species (e.g., the Seychelles and Sri Lanka), there is a general impoverishment of the pitcher fauna, involving most notably a complete loss of terrestrial and aquatic predators at the top of the food chain, and a sharp reduction in the numbers of detritivore and filter-feeding taxa (Beaver, 1983, 1985). The richest food webs studied were those in West Malaysia, near the center of radiation and diversity for *Nepenthes* on Borneo. In Sabah on Borneo, pitcher food webs are further enriched by the activities of the small, insectivorous tarsier (*Tarsius spectrum*), which visits the pitchers of various species, perches on the lip, and then scoops out and devours recently trapped prey. Burbridge (1880) reported that *Tarsius* raids the pitchers of *N. rafflesiana* with impunity, but not those of *N. bicalcarata,* which are similar in every respect except that they each have two sharp, fanglike processes that project below the lid. Burbridge's suggestion that "the very sharp spurs are so arranged that the *Tarsius* is certainly held and pierced when he inserts his head to see what there is in the pitcher" has yet to be tested.

Finally, some opportunistic predators and herbivores have behaviors that result in prey being excluded from carnivorous plants, and thus, are in some form of indirect competition for prey. For example, the wasp *Chlorion harrissii* builds its multitiered nest within pitchers of *Sarracenia* and thus indirectly excludes potential prey (Shetler, 1974). Larvae of the pitcher plant moths (*Exyra*), which are solitary herbivores on various species of *Sarracenia,* display a remarkable behavior that also results in indirect prey exclusion (Jones, 1921). Older, midsummer larvae spin a horizontal diaphragm of silk across the upper throat of each pitcher they occupy, which keeps out rain, insects, and predators and parasitoids, while they feed in concealment on the inner surface of the pitcher. Larvae in early spring instead cut a narrow ringing groove near the top of the pitcher; above this, the pitcher wall dies, collapses, and hardens, forming a sheltering barrier. Certain insectivorous birds have developed the ability to use the collapsed pitchers as a cue to the presence of succulent larvae or pupae within. Perhaps as an evolutionary countermeasure, one insect species (*Exyra semicrocea*) moves to an unblemished pitcher before spinning its pupal cocoon (Jones, 1921).

Herbivory

Perhaps not surprisingly, relatively few herbivores are known to attack carnivorous plants. Larvae of the moth genus *Exyra* feed on *Sarracenia* leaves (see above), while caterpillars of the moth *Papaipema appassionata* feed in burrows in their rhizomes (Lloyd, 1942). Aphids are known to attack the inflorescences and unexpanded leaves of

Drosera. Mature leaves and glandular tentacles are eaten by the plume moth *Trichoptilus parvulus* (see above), weevils, and the snail *Cochlicella ventrosa* (Watson et al., 1982). The dangers that carnivorous plants present to insects, and the low nutrient levels that may prevail in the leaves of plants from infertile habitats, may help explain what appears to be a relatively low incidence of herbivory and herbivores specialized on carnivorous plants.

Pollination

Although carnivorous plants possess a wide range of floral syndromes and insect pollinators (e.g., see Schnell, 1976; Slack, 1979), their most striking floral feature is that all bear flowers on leafless (or essentially leafless) scapes. The evolutionary basis for this convergence should be self-evident: the placing of (trap) leaves on a flowering stem might deprive carnivorous plants of potential pollinators, whereas separating the flowers spatially from the traps permits, in principle, the separate recruitment of prey and pollinators to traps and flowers, respectively. Scapes in species with traps close to the ground (e.g., *Cephalotus, Dionaea*) may also serve to attract pollinators to an otherwise inconspicuous plant growing among taller neighbors. The specific cues used to attract pollinators to flowers and not to traps apparently have not been studied; many carnivorous species use many of the kinds of traits normally employed to attract pollinators (e.g., nectar and scent production, bright coloration, ultraviolet reflectance patterning) to attract prey (Joel et al., 1985).

Nutritional Benefits of Carnivory

Following Charles Darwin's (1874) conclusive demonstration that various carnivorous plants do indeed capture and digest animals, his son Francis Darwin (1878) was the first to demonstrate that prey capture enhances growth and seed production in such plants, based on experimental feeding of *Drosera rotundifolia*. Similar results for the same species were soon published by Kellerman and von Raumer (1878) and Büsgen (1883). Prey capture has now been shown to increase growth and/or reproduction through experimental lab feeding of *Drosera binata* and *D. whittakeri* (Chandler and Anderson, 1976a), *Pinguicula lusitanica* and *P. vulgaris* (Harder and Zemlin, 1967; Aldenius et al., 1983), and *Utricularia exoleta* and *U. gibba* (Harder, 1963; Sorenson and Jackson, 1968); and through exclusion or addition of prey, in field populations of *Drosera intermedia* (Wilson, 1985), *D. filiformis* (Gibson, 1983), *Pinguicula planifolia* (Gibson, 1983), *Sarracenia flava* (Christiansen, 1976; Weiss, 1980), and *S. leucophylla* (Gibson, 1983).

Prey-enhanced growth and reproduction might result from (1) the absorption of mineral inorganic nutrients limiting photosynthesis and/or the conversion of photosynthate to new tissue; (2) the input of mineral nutrients that directly limit flower or seed production; or (3) uptake of carbon skeletons, and the partial replacement of autotrophy with heterotrophy (Givnish et al., 1984). Evidence for each of these possibilities can be summarized as follows.

Absorption of Mineral Nutrients

Carnivorous plants are known to absorb amino acids, peptides, and various mineral cations from their prey (Hepburn et al., 1920; Plummer and Kethley, 1964; Chandler and Anderson, 1976a, 1976b; Christiansen, 1976; Dixon et al., 1980; Aldenius et al., 1983). Given the low concentrations of nitrogen and phosphorus in the substrates inhabited by many carnivorous plants, and the fundamental importance of these nutrients for plant metabolism, several investigators have suggested that their absorption might provide the key to prey-enhanced growth and the apparent competitive advantage of carnivores on sterile sites.

The key findings in support of this hypothesis are that carnivorous plants (1) can complete their life cycle without capturing prey, if provided with access to a complete solution of inorganic nutrients (Harder and Zemlin, 1967; Small et al., 1977; Lüttge, 1983; cf. Pringsheim and Pringsheim, 1962); (2) absorb mineral nutrients from prey, with effects on plant nutrient contents and/or tissue concentrations; (3) show differential growth responses to nitrogen, phosphorus, and other nutrients added by prey; (4) increase growth in response to prey capture more on nutrient-poor than on nutrient-rich substrates; and (5) display increased, or prolonged, peak photosynthetic rates in response to prey capture. Empirical support for the latter four points follows.

Nutrient Uptake

Christiansen (1976) showed that *Sarracenia flava* in unfertilized vermiculite accumulated significantly higher leaf concentrations of both nitrogen and phosphorus when exposed to prey than if prey were excluded by glass wool plugs inside the pitchers; concentrations of potassium, calcium, and magnesium were unaffected by prey availability. Application of a nitrogen–phosphorus–potassium–calcium–magnesium fertilizer eliminated any significant effect of prey availability on tissue concentrations of these elements. Chandler and Anderson (1976a) found that feeding *Drosophila* to the Australian sundew *Drosera whittakeri* significantly increased the tissue concentrations and total plant content of phosphorus, regardless of whether plants in sand culture were irrigated with distilled water, a complete inorganic nutrient solution, or solutions

Table 7-2 Effect of insects, nitrogen supply, and phosphorus supply on the dry weight increase, total nitrogen and phosphorus contents, and tissue nitrogen and phosphorus concentration of *Drosera whittakeri*

Inorganic salts added to sand	Insects applied	Dry wgt* increase (mg)	Phosphorus		Nitrogen	
			μg/mg	μg/plant	μg/mg	μg/plant
Nil	–	13.5A	0.68	15.7D	12.6	278G
Nil	+	17.2B	0.99	26.6EF	10.3	276G
Complete	–	20.2C	0.81	24.1E	12.2	365H
Complete	+	20.6C	1.02	30.1F	12.0	355H
Complete minus N	–	12.5A	0.84	18.4D	14.3	313GI
Complete minus N	+	27.6B	1.16	31.7F	11.3	307GI
Complete minus P	–	18.9BC	0.63	17.9D	11.6	329HI
Complete minus P	+	19.7BC	0.92	27.2EF	12.3	367H

*Values followed by the same letter are not significantly different at the 95 percent confidence level, based on analysis of variance.

After: Chandler and Anderson, 1976a.

lacking nitrogen or phosphorus. Surprisingly, total plant nitrogen was unaffected by prey availability, and leaf nitrogen concentration tended to decrease in fed plants (Table 7-2).

Pate and Dixon (1978) found that feeding the Australian tuberous sundew *Drosera erythrorhiza* four *Drosophila* per week significantly increased tuber concentrations of nitrogen and phosphorus. This species is one of several southwestern Australian sundews which is active mainly on upland sites during winter and flowers prolifically following fire (Dixon and Pate, 1978). Pate and Dixon (1978) showed that ash added to soil from its habitat significantly increased tuber levels of nitrogen, phosphorus, magnesium, potassium, and zinc. During seasonal senescence, a total of 88 percent of leaf and stem phosphorus and 79 percent of nitrogen were withdrawn and stored in the underground tubers for recycling in following years, suggesting that these nutrients may possess particular significance to the plant.

Dixon et al. (1980) fed ^{15}N-labeled *Drosophila* to *Drosera erythrorhiza,* and found ^{15}N enrichment of the leaves, stems, and tubers. *D. erythrorhiza* absorbed 76 percent of the labeled N from the insects; by the end of the growing season, a new set of tubers had stored 70 percent of the total ^{15}N applied. These results were combined with data on prey capture in field populations to estimate nitrogen input via carnivory. The observed season's catch of 0.25 to 0.39 mg nitrogen per plant provided the equivalent of 14 to 21 percent of the nitrogen transferred to the season's tubers and 11 to 17 percent of the net nitrogen uptake during a growing season; roughly 27 percent of the nitrogen catch is diverted to leaves. Plant nitrogen content increases by about threefold with the annual expansion of the leaf rosette, apparently as the result of both prey capture and root absorption. Watson et al.

(1982) observed much higher prey capture rates for *D. erythrorhiza*, which could account for well over 100 percent of the total plant nitrogen and phosphorus pools; however, even at these higher rates, carnivory makes only a slight contribution to total plant potassium content. Interestingly, Dixon et al. (1980) note that prey capture by *D. erythrorhiza* is less effective during rainy seasons, presumably because the glandular secretions of the leaves wash off.

Finally, Aldenius et al. (1983) examined the consequences of feeding field-grown *Pinguicula vulgaris* with insects in northern Sweden. Plants supplied with insects showed greater increases in dry weight, leaf number and length, and nitrogen and phosphorus content than did those deprived of prey grown on comparable natural substrates, whether undisturbed or fertilized with a complete nutrient solution. However, the amount of nitrogen absorbed when insects were added to fertilized plants was more than the insects contained; insect phosphorus content, however, was sufficient to account for the phosphorus absorbed. Aldenius et al. (1983) inferred that *P. vulgaris* absorbs both nitrogen and phosphorus from its prey, but that a substance other than nitrogen —most likely phosphorus—is absorbed and stimulates root nitrogen uptake. Karlsson and Carlsson (1984) tested this idea by feeding *Pinguicula* agar blocks laced with nitrate, phosphate, and/or combined micronutrients in a complete factorial design. Only phosphorus had a significant effect on plant biomass. As expected, the ratio of root to leaf biomass increased as a result of the nitrogen treatment, and decreased as a result of the joint application of nitrogen, phosphorus, and combined micronutrients. Karlsson and Carlsson (1984) thus concluded that phosphorus is the most important mineral nutrient obtained by carnivory in *Pinguicula*.

Differential Response to Various Nutrients

The relative importance of different inorganic nutrients obtained through carnivory in stimulating growth has been assessed by providing prey to plants on substrates with or without particular nutrients. Both fertilization and feeding tend to increase growth for plants grown on nitrogen- or phosphorus-deficient media (Chandler and Anderson, 1976a; Christiansen, 1976; Aldenius et al., 1983; Karlsson and Carlsson, 1984). Often, but not always (e.g., Aldenius et al., 1983), plants supplied with prey on nutrient-deficient substrates perform about equally or slightly better than plants supplied with prey on fertile substrates. Excessive nutrient levels in the soil may negatively influence the growth of carnivorous plants accustomed to soils with low cation concentrations, just as extraordinarily large amounts of prey on their leaves can lead to leaf death or partial decay (see Lloyd, 1942).

Aldenius et al. (1983) and Karlsson and Carlsson (1984) concluded that phosphorus was the most important limiting nutrient absorbed via carnivory in *Pinguicula,* but other nutrients might be more important in other situations. Plummer (1963) suggested that carnivory might provide an important supply of metallic cations, which are also in short supply in the soils inhabited by many *Sarracenia* species. However, the results of Christiansen (1976) argue against this proposition: Plants that were fertilized or provided access to prey did not sequester higher tissue concentrations or total plant contents of calcium, potassium, or magnesium.

Folkerts (1982) proposed that carnivory might provide a source of the important micronutrient molybdenum, which tends to complex with iron in acid soils. This suggestion is difficult to evaluate because carnivory itself may create a requirement for molybdenum. Studies by Chandler and Anderson (1976a) and Small et al. (1977) show that *Drosera* secretes large amounts of nitrogen reductase, an enzyme containing molybdenum. Results obtained by Chandler and Anderson (1976a) on *Drosera binata* and *D. whittakeri* seem to exclude the possibility that prey provide an important source of molybdenum, or the other micronutrients boron, copper, manganese, and zinc. However, Sorenson and Jackson (1968) found that feeding *Paramecium* to *Utricularia gibba* partially or totally reversed the adverse effects of magnesium or potassium deficiency on internode production or plant length. Surprisingly, experimental feeding on magnesium- or potassium-deficient media increased not only the total number of traps per plant, but also the ratio of traps to internodes; that is, fed plants increased their proportional allocation to carnivory, rather than decreasing it.

Dependence of Prey-Enhanced Growth on Substrate Fertility

Chandler and Anderson (1976a) provide the most comprehensive study to date on the relative importance of different nutrients obtained from prey in stimulating growth, and on the near-disappearance of prey-enhanced growth under fertile conditions. As noted previously, they starved or fed *Drosophila* to *Drosera whittakeri,* a tuberous sundew from southeastern Australia, grown on sand irrigated with distilled water, a complete nutrient solution, a solution lacking nitrogen, or a solution lacking phosphorus. When *Drosera* did not receive prey, growth in dry weight was equally low in plants watered with distilled water or the nitrogen-free solution, whereas plants receiving either the phosphorus-free or complete solutions showed more vigorous and statistically indistinguishable growth (Table 7-2). Furthermore, feeding with prey caused a significant increase in growth only in plants watered with distilled water or nitrogen-free solution. This strongly suggests that nitrogen, not phosphorus, is the element most limiting growth in *D. whittakeri.* The data on tissue nutrient contents seem, however, to belie this. Provision with prey never results in a significant increase in total plant nitrogen content or tissue concentration, whereas an increase in phosphorus content and concentration always results (Table 7-2). Watering with a complete nutrient solution eliminated any significant impact of prey consumption on dry weight growth (Table 7-2).

Sorenson and Jackson (1968) also found that provision with a complete-nutrient solution eliminated prey enhancement of growth in *Utricularia gibba.* Christiansen's (1976) results on *Sarracenia flava* are consistent with these findings, insofar as fertilization eliminated or negated any increase in tissue nitrogen or phosphorus concentration due to feeding. Finally, the data of Aldenius et al. (1983) show no fertilizer-associated decrease in the dry weight growth of *Pinguicula vulgaris* provided with prey. Prey-fed plants on fertilized substrates showed proportional gains in dry weight, compared with unfed controls, similar to those of fed plants on unfertilized substrates. However, fertilization did reduce or negate the proportional increase in tissue nitrogen and phosphorus concentration.

Effects on Photosynthesis

All carnivorous plants investigated thus far possess the C_3 photosynthetic pathway (Lüttge, 1983). There is, however, surprisingly little information on the effect of feeding on the rate of photosynthesis in carnivorous plants. The most important study is that of Weiss (1980), who found that nutrient input through carnivory increased and prolonged the seasonal duration of peak photosynthetic rates in *Sarracenia flava.* Such an increase is certainly consistent with the increase in leaf nitrogen concentration seen after feeding in many studies on carnivorous plants given that, across many species and growth forms, photosynthetic rates per unit leaf mass increase with leaf nitrogen concentration (Field and Mooney, 1986). Further work is needed, however, to determine if the observed increase in leaf nitrogen concentration in carnivorous plants after feeding is a result of increased photosynthetic, rather than digestive, enzyme activity.

Direct Diversion of Nutrients to Reproductive Structures

Apparently, not a single study has evaluated the role that sexual reproduction or the production of flowers, seeds, and ancillary structures plays in the nutritional budget of carnivorous plants. Studies by Benzing and his colleagues (Benzing and Renfrew, 1971; Benzing, 1973, 1976,

1980; Benzing et al., 1976) on *Tillandsia,* however, provide some potential analogs insofar as these epiphytic bromeliads grow on exceedingly nutrient-poor substrates comparable, at least in this regard, to those inhabited by many carnivorous plants. These studies show that epiphytic *Tillandsia* divert large fractions of their nitrogen and phosphorus budgets preferentially into flowers and seeds, implying that nutrients may directly limit reproduction (vis-à-vis any indirect limitation caused by their impact on photosynthesis). Demonstration that this occurs in carnivorous plants is, however, wanting.

Autotrophy vs Heterotrophy

Finally, prey capture may serve to replace autotrophy partly with heterotrophy in carnivorous plants. Most studies point against this last possibility: plants obtain primarily minerals, not carbon, from carnivory. Dixon et al. (1980) found uptake of carbon from ^{14}C-labeled *Drosophila* fed to *Drosera erythrorhiza,* but the form in which it was absorbed was not studied (see Chandler and Anderson, 1976b). The carbon absorbed seems likely to be in nitrogen-bearing amino acids rather than readily converted carbohydrates. This is indicated by studies by Darwin (1874) and several subsequent investigators (e.g., Lüttge, 1964, 1983; Robins, 1976; Clancy and Coffey, 1977) that have shown that glandular secretion in *Dionaea* and *Drosera* occurs only in response to nitrogenous material. Furthermore, Chandler and Anderson (1976a) have shown that *Drosera* fed prey in a darkened room show little growth, and no difference from plants supplied instead with a complete mixture of inorganic salts under the same light conditions, implying a negligible role for carbon heterotrophy. This is consistent with the absence of pale or achlorophyllous carnivorous species (Schmucker and Linnemann, 1959). Nevertheless, one study (Harder, 1970) has shown that addition of sucrose and/or acetate to an axenic nutrient medium stimulates growth of *Utricularia;* this investigation should be replicated.

Cost–Benefit Analysis of the Carnivorous Habit

To analyze why carnivorous plants are mainly restricted to sunny, moist, nutrient-poor habitats, are common on certain fire-swept sites, and are rare in epiphytic situations, the energetic benefits and costs of carnivory in various habitats must be considered. Carnivory should evolve if these benefits exceed the cost of a small investment in carnivorous adaptations, because plants with mutations for such investments should have an energetic advantage in competing with other plants. The following analysis of these costs and benefits draws heavily on previously published work by Givnish et al. (1984).

Potential Benefits and Costs

There are three potential energetic benefits associated with carnivory vs noncarnivory which parallel those just discussed regarding the demonstrated advantages of prey consumption for plants that are already carnivorous. First, carnivory may increase a plant's total rate of carbon gain as a result of increased mineral absorption, through (1) an increased rate of photosynthesis per unit leaf mass; (2) an increase in the efficiency of converting photosynthate to new leaf tissue; and/or (3) a reduction in the allocation of photosynthate to unproductive root tissue, in connection with increased foliar absorption of nutrients. Medina (1970, 1971) showed that increasing a plant's supply of available soil nitrogen can increase photosynthesis and enable it to develop higher concentrations of ribulose bisphosphate (RuBP) carboxylase-oxygenase, the photosynthetic enzyme responsible for carbon dioxide capture; fertilization studies with other mineral nutrients (Natr, 1975; Longstreth and Nobel, 1980) have produced similar increases in photosynthetic output. Weiss (1980) has confirmed that nutrient input through carnivory in *Sarracenia flava* can elevate the rate and seasonal duration of photosynthesis per unit leaf mass.

Second, carnivory may result in an increased level of nutrients in seeds or increased seed production. Finally, carnivory may serve to replace autotrophy with heterotrophy as a source of chemical energy. As discussed previously, the last possibility seems quite unlikely. The second, while more plausible, has at present no empirical support and could easily be considered part of the first benefit to growth: If the mineral output of traps produced early in development is diverted into photosynthetic machinery that results in increased rates of growth and more traps and leaves, then the output of a few later added traps can be allocated to reproduction (Givnish et al., 1984). Thus, for the sake of the following model, the primary benefit of carnivory is assumed to be enhanced photosynthesis, or enhanced conversion of photosynthate to leaf tissue.

Predicted Trends in Ecological Occurrence

How should the benefits and costs of carnivory vary with environmental conditions? Consider a plant with a given amount of biomass in leaves and roots. As the amount of energy devoted to carnivory (in terms of lures, photosynthetically inefficient traps, or digestive enzymes) increases, there should be a corresponding increase in the amount of nutrients absorbed. As a result, the effective rate of photosynthesis per unit leaf mass—measured in terms of the leaf rate of replication, or grams of leaf per grams of leaf per second (g leaf/g leaf/s)—should increase (Fig. 7-8). This could result from (1) an increase in the absolute rate of photosynthesis (g C/g leaf/s); (2) an increase in the rate of conversion of carbon skeletons to new leaf tissue, by virtue of additional mineral availability; (3) and/or a decline in the fractional allocation of energy to unproductive, nutrient-absorbing roots with a corresponding increase in fractional allocation to leaves. Furthermore, as the amount of energy devoted to carnivory and the resulting mineral input continue to increase, the effective photosynthetic benefit should plateau as factors other than nutrients limit photosynthesis or the conversion of photosynthate into leaves.

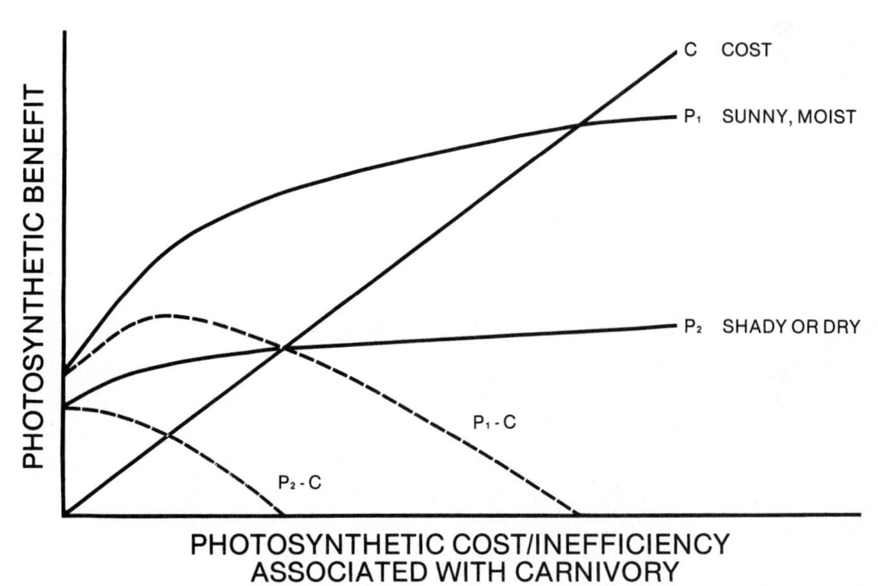

Figure 7–8. Photosynthetic benefits and costs associated with differential levels of investment in carnivorous adaptations in a nutrient-poor site, as a function of environmental conditions (after Givnish et al., 1984). Enhancement of the effective photosynthetic rate (g leaf/g leaf/s) resulting from added nutrients should be more rapid, and show less tendency to plateau, in a well-lit and moist environment (P_1) than in sites where light or water (or other factors) are more likely to limit photosynthesis or the conversion of photosynthate to new leaf tissue (P_2). Dashed lines correspond to the net difference between photosynthetic benefits and cost (C) of obtaining nutrients through carnivory. Carnivory should evolve whenever the benefit of a small investment in it exceeds the cost, i.e., when the net profit curve slopes upward near $C = 0$.

The extent to which photosynthesis, or leaf replication rate, can be enhanced by increased mineral input clearly depends on environmental conditions. By definition, the effective rate of photosynthesis is unlikely to increase unless nutrients are in short supply and limit photosynthesis or the conversion of photosynthate, so the greatest benefit is expected in mineral-poor sites. The studies by Sorenson and Jackson (1968) on *Utricularia* and by Chandler and Anderson (1976a) on *Drosera* confirm that the usual increase in growth of carnivorous plants supplied with prey on nutrient-poor substrates largely disappears as nutrient availability in the substrate increases. Furthermore, if factors such as light or water are in short supply, then they can limit photosynthesis and the extent to which nutrients added by carnivory (or other means) can elevate photosynthesis (Givnish et al., 1984). For example, the data of Gulmon and Chu (1981) on *Diplacus aurantiacus* exposed to different amounts of fertilizer show that photosynthesis increases more slowly with leaf nitrogen content at low light intensities than at high light intensities. The hypothetical benefit curves shown in Fig. 7-8 are similar in form to those actually obtained by Gulmon and Chu (1981) for the response of photosynthesis to increased leaf nitrogen content induced by fertilization.

Similar trends would be expected if increased mineral supply increases not the mineral content per leaf, but the rate of conversion of photosynthate into new leaves at constant mineral content. As nutrient capture increases, the rate at which new leaves can be produced will depend less on limiting minerals like nitrogen or phosphorus, and more on the availability of carbon skeletons. The latter clearly depends on the limitation of photosynthesis by light or water availability, so the rate of conversion should also rise most quickly and plateau most slowly in well-lit, moist, nutrient-poor sites (Givnish et al., 1984). Alternatively, as nutrient capture increases, the decline in the fraction of energy allocated to roots will slow and finally stop as water uptake becomes a limiting factor. (Many carnivorous plants are known to have unusually small root systems; Schmucker and Linnemann, 1959.) In this case, leaf replication should again rise most quickly and plateau most slowly in sunny, moist, nutrient-poor sites which would allow the greatest reduction in required root mass if a nutrient subsidy were provided, and which would also support the highest leaf photosynthetic rates at constant leaf mineral content.

The difference between the benefit and cost curves in Fig. 7-8 can be used to predict both whether carnivory should be favored in a given habitat and, if so, what would be the optimal level of investment in carnivorous traits. If the difference between benefits and costs at low levels of investment is positive, then carnivory should be favored. The optimal level of investment in carnivory is determined by the point at which the difference between benefits and costs are maximized (Fig. 7-8).

Association with Habitat, Successional Status, and Terrestrial Habit

Carnivory should have its greatest impact in sterile, sunny, and moist habitats, and the effective photosynthetic gains resulting from added investments in carnivory should rise quickly and plateau slowly (Fig. 7-8). In sterile habitats that are shady and/or dry during the period of photosynthetic activity, light or water is more likely to limit growth, and the gains from carnivory should be smaller and plateau more rapidly. Thus, the differences between photosynthetic benefits and costs are more likely to be positive at low levels of investment in carnivory, and hence promote its evolution in nutrient-poor habitats that are also sunny and moist during the period of photosynthetic activity. Sunny, moist, nutrient-poor conditions may also increase insect abundance and the benefit accruing from a given investment in carnivory. Such conditions characterize most of the habitats and early successional sites in which carnivorous plants frequently grow.

This model partly explains why carnivory is rare in epiphytes in general (Thompson, 1981; Givnish et al., 1984) and bromeliads in particular (Benzing, 1980), because most epiphytes occur either on shady perches or on better-lit branches subject to frequent desiccation. *Nepenthes* vines probably avoid this problem by maintaining contact with more reliable stores of moisture in the soil. Epiphytic *Utricularia* usually grow in wet cloud forests where water stress is rare even high in the canopy, or in pools impounded at the expense of other plants (often bromeliads) by the hosts' own inefficient, overlapping photosynthetic structures (Givnish et al., 1984).

Significance of Seasonal Heterophylly

The model also explains many aspects of seasonal heterophylly in carnivorous plants (Givnish et al., 1984). Many carnivores fail to produce traps, or produce leaves with a higher proportion of photosynthetically efficient tissue, during "unfavorable" seasons when factors other than nutrients may limit growth (Schnell, 1976; Erickson, 1978; Slack, 1979). *Sarracenia flava, S. oreophila,* and *S. leucophylla* produce trapless phyllodes during late summer droughts (Christiansen, 1976; Schnell, 1976; Weiss, 1980; Iltis, personal communication). *Dionaea* produces leaves with broad photosynthetic petioles and small traps in winter. *Cephalotus* produces phyllodes, not traps, during the cool winter in southwestern Australia; its reed swamp habitat remains damp through the spring and dry summer when the pitchers are active (Lloyd, 1942). *Nepenthes* often fails to produce pitchers in overly dry or shady conditions (Slack, 1979). The seasonal flush of glandular leaves in *Triphyophyllum* just precedes the onset of the rainy season and may correspond to the yearly flush of insect availability. The long-term shift in *Triphyophyllum* from glandular, insectivorous seedlings and juveniles to eglandular, adult vines remains enigmatic for lack of data on the microenvironments experienced by each life stage.

Relation to Fire and the Paradox of Enrichment

The ecological association between many carnivorous plants and disturbance by fire would seem to be, at least in part, paradoxical. It is true that fire removes competing, taller vegetation, reduces transpiration, and volatilizes some soil nutrients (notably nitrogen), thereby creating the sunlit, moist, nutrient-poor conditions favorable to carnivorous plants. However, fire also can increase—at least in the short term—the soil supply of many nutrients released from ash, including calcium, potassium, sodium, phosphorus, and magnesium (Pate and Dixon, 1978). Hence, the question arises as to why enrichment of these soil minerals does not encourage the growth of noncarnivorous species more than it does carnivorous species.

One possible explanation is that two or more nutrients may limit plant growth in a given habitat. If the supply of one nutrient (e.g., phosphorus) is augmented by fire, it may increase the amount by which the effective photosynthetic rate would be increased by inputs of another nutrient (e.g., nitrogen) available mainly or only through carnivory. Thus, complementarity of nutrients obtained through fire and through carnivory might help favor carnivory, and/or an increased allocation to traps in carnivorous plants, in sites recently fertilized by ash.

Such an explanation clearly requires careful testing, because the presence or absence of nutrient complementarity can lead one to predict either an increase or decrease in dominance by carnivorous plants, or in optimal trap allocation, with an increase in substrate "fertility." Nutrient complementarity may lie at the heart of an intriguing pattern uncovered by Susan Knight and Thomas Frost (in preparation). They have found that *Utricularia vulgaris* increases its proportional allocation to traps vs leaves with water fertility (as measured by conductivity and pH) in northern Wisconsin lakes. A related possibility is that carbon per se is a complementary limiting factor to the mineral nutrients (e.g., phosphorus, nitrogen, calcium) that tend to increase with increasing fertility in the lakes surveyed. Carbon dioxide diffuses 10,000 times more slowly underwater than in air, so its local availability may limit underwater photosynthesis in macroscopic plants. Furthermore, in extremely infertile, acid lake waters, there is little calcium bicarbonate available as an alternative source of carbon or—through a buffering chemical equilibrium—carbon dioxide. There is strong evidence that photosynthesis in such lakes is limited by the amount of dissolved inorganic carbon (DIC) (Boston and Adams, 1986). Indeed, some species (*Isoetes macrospora, Littorella uniflora, Lobelia dortmanna*) have evolved special adaptations, such as highly permeable roots or CAM (crassulacean acid metabolism) photosynthesis, to access an alternative DIC source in decomposing sediment.

To the extent that carbon is more limiting to photosynthesis in nutrient-poor lakes, a greater benefit from carnivory would be expected in "fertile" lakes, which would favor a greater allocation to traps vs leaves. Rigorous testing of this idea will require careful measurements of photosynthesis in plants grown in water with varying levels of DIC and provided with varying amounts of prey. Resolution of the "paradox of enrichment" is fundamentally a quantitative issue. Only detailed measurements can determine whether nutrient complementarity explains such behavior in carnivorous plants, or whether it instead contradicts the cost–benefit model for the evolution of carnivory.

Selective Pressures on Trap Structure

Although the literature on carnivorous plants is composed mainly of articles detailing their morphological and physiological "adaptations" for carnivory, not one analysis of optimal trap structure or within-trap resource allocation appears to have been published. Yet selective pressures on trap structure would seem to be an interesting issue, given the tremendous variation in trap form in many carnivorous genera. Two topics that should be explored include the adaptive significance of leaf form in sundews, and the possibility that certain pitcher plants have a mutualistic relationship with their prey.

Leaf Shape in Relation to Light and Nutrient Capture in Drosera

The trap structures of different sundews are especially varied. In southwestern Australia, where more than half of the world's species occur, sundews display a veritable galaxy of growth forms, phenologies, and leaf morphologies, including tuberous and nontuberous species; minute cushions, prostrate rosettes, and meter-tall climbers; winter-active upland species and summer-active lowland species; and species with linear, elliptic, round, petiolate, or sessile leaves (Erickson, 1978).

In the Pine Barrens of New Jersey only three species are found. Although the variation they display is less dazzling, the insights they provide may yield ideas for understanding variation in sundew growth form and distribution elsewhere. The Pine Barrens species include *Drosera rotundifolia,* with round leaves on glandless petioles, generally borne in a nearly horizontal rosette; *D. intermedia,* with more narrowly elliptic leaves borne on long, inclined, glandless petioles; and *D. filiformis,* with erect, linear, sessile leaves that bear glandular tentacles from base to tip. Tentacles are of roughly equal length in each species, and are generally sparse on leaf undersides in *D. rotundifolia* and *D. intermedia.* Transect studies conducted by Givnish, Holt, and several fellow students at Princeton in the early 1970s disclosed that, in bogs and *Chamaecyparis* swamp edges where all three species are found, *D. rotundifolia* tends to occur in partly shaded sites, often under the sparse shade of shrublets growing in well-drained soil atop hummocks (Givnish, 1972). *D. intermedia* occurs mainly in well-lit microsites that suffer considerable seasonal variation in the height of water above the substrate and often have a muck substrate, such as hollows between hummocks or areas near streams. Finally, *D. filiformis* occurs principally on open microsites in seepage zones on sterile sand, in which there is little seasonal fluctuation of the water table above the substrate (Givnish, 1972).

The different leaf forms of these species may be adapted to the different microsites they inhabit. *D. rotundifolia,* with a broad, flat leaf held nearly horizontally, had the greatest ratio of light-trapping surface (i.e., the leaf blade) to insect-trapping surface (i.e., the surface subtended by the tentacles), and the highest potential rate of light capture for a given amount of irradiance from overhead. In plants inhabiting partly shaded sites and potentially limited by light, a high ratio of light- to insect-trapping surface and horizontal leaves should be favored. *D. intermedia,* with a narrower leaf and, as a result, a more nearly cylindrical array of tentacles, had a lower ratio of light- to insect-trapping surface and more steeply inclined leaf surfaces, features that should be better adapted to conditions in which light is less limiting. Furthermore, the erect, long, glandless petioles may serve to prevent glandular secretions from being washed away by many of the frequent rises in water level in the microsites inhabited by *D. intermedia.* Finally, the cylindrical leaf and tentacular array of *D. filiformis* confer on it the lowest ratio of light- to insect-trapping surface as well as the most steeply inclined photosynthetic surfaces. This would appear to be adaptive, in that *D. filiformis* inhabits the microsites in which nutrients appear most limiting relative to light. Furthermore, its possession of digestive glands from ground level to leaf tip is likely to be efficient only in microsites in which the water table rarely rises above the level of the substrate.

It would be extremely interesting to know if the preceding explanations successfully predict the ecological distribution of *Drosera* species elsewhere with linear, spatulate, or round leaves. Southwest Australia, with its diversity of sundews, would be an obvious region for such a study. Although *D. filiformis* is restricted to the U.S. Coastal Plain, both *D. rotundifolia* and *D. intermedia* range throughout almost all of the temperate and boreal zones of the northern hemisphere. Their ranges often overlap with linear-leaved species (e.g., *D. capillaris, D. anglica*) that may be functionally similar to *D. filiformis;* again, it would be useful to study the microdistribution of these species in such areas of overlap to test the proposed hypotheses for the significance of leaf form and arrangement in *Drosera.*

Possible Mutualism between Pitcher Plants and Social Insect Prey

As noted previously, many species of pitcher plants in the Sarraceniaceae, Nepenthaceae, and Cephalotaceae capture ants as their principal prey. It is difficult to understand how this might be so, insofar as ant colonies and pitcher plants tend to be quite common, sedentary, and predictable features of the landscape, and insofar as ants have highly developed communicative abilities that enable colony members to localize other potential enemies. If the relationship between pitcher plants and ants were strictly that of predators and prey, then there should be strong selective pressure on ants to avoid the plants (see Chap. 6). Yet, this is not the case. One possible explanation is that the relationship is actually mutualistic, rather than predator–prey. Pitcher plants may be short on certain inorganic nutrients, and long on fixed carbon, whereas the reverse may be true of ant colonies with access to insect prey. Worker ants avidly forage at pitcher plant nectaries, and only rarely plummet into the watery abyss. Thus, the abundant nectar provided by pitcher plants and gathered by workers might increase colony growth more than it is decreased by the loss of an occasional worker. Similarly, the energetic loss to the plant through nectar production and consumption might be more than offset by the occasional worker, if the nutrients obtained thereby stimulate photosynthesis or the conversion of photosynthate into new leaf tissue. The latter assumption appears tacit in most discussions of pitcher plants; the former possibility seems never to have been recognized. Careful laboratory studies of the marginal rates of return to pitcher plants for nectar, and to ant colonies for workers, are needed to test whether ants and pitcher plants are actually mutualists, unilateral exploiters, or mutual enemies.

Comparison with Other Animal-Based Modes of Nutrient Capture

Many plants can obtain inorganic nutrients through associations other than carnivory with animals or microbes. The most notable of these include symbiotic nitrogen fixation and ant-fed myrmecophily (Janzen, 1974b; Huxley, 1978; Rickson, 1979; Pate, 1986). Nitrogen-fixing prokaryotes (e.g., cyanobacteria, *Rhizobium*) form associations with certain vascular plants, usually in root nodules or leaf bases, with the plant host donating carbohydrates and receiving fixed nitrogen from its microbial symbionts (Pate, 1986). Myrmecophilous plants provide shelter for ant colonies in hollow or swollen nodes, petioles, or leaf bases, and often provide food in the form of nectar or protein bodies (Bequaert, 1922; Wheeler, 1942). In turn, the plants receive protection from herbivores and competitors attacked by their guests, and in certain instances (e.g., the epiphytes *Hydnophytum* and *Myrmecodia*) receive an increased supply of nutrients from food wastes, dead nest mates, and debris packed by the ants into plant recesses (Benzing, 1970; Janzen, 1974b; Huxley, 1978; Rickson, 1979).

Insofar as plants essentially trade carbohydrates for nutrients from animals or microbes in nitrogen fixation and ant-fed myrmecophily, the question naturally arises as to whether the same cost–benefit considerations and expected pattern of distribution apply to species with these associations as to carnivorous plants. With regard to nitrogen fixation, the answer is probably a qualified yes. Sunny, moist, nitrogen-poor conditions are most likely to favor nitrogen-fixing symbioses, as they do carnivores. However, the conditions favoring these two groups should differ in three important respects. First, because highly anaerobic conditions in the soil are inimical to nitrogen fixation in root nodules (Pate, 1986), nitrogen-fixing symbioses are more likely to occur in well-drained or seasonally arid sites than carnivores. Second, legumes and other nitrogen-fixing plants seem to be most competitive on sites that are relatively rich in other limiting nutrients, especially phosphorus (e.g., see Tilman, 1982); carnivores might be expected to have an advantage in soils that lack any nutrient that is abundant in prey carcasses. Third, nitrogen fixation always entails the use of nitrate reductase, and thus, of molybdenum; nitrogen-fixing symbioses should thus be excluded from molybdenum-poor soils, as they indeed are (Pate, 1986). This same exclusion should apply only to those carnivores that obligately produce nitrate reductase.

Carnivorous plants and ant-fed myrmecophytes differ in one important respect: the former are almost always terrestrial herbs, whereas the latter are almost always epiphytes (Thompson, 1981; Givnish et al., 1984). Insofar as the nutrients obtained through ant-fed myrmecophily are probably similar to those obtained through carnivory, the question arises as to why ant-fed myrmecophily is predominant in epiphytes, particularly insofar as they are likely exposed to frequent desiccation.

Perhaps one reason is that myrmecophily yields benefits other than nutrient input, notably defense against herbivores (Givnish et al., 1984). The benefits of such defense are not likely to show the same trends across habitats as those accruing from carnivory. In particular, the benefits of defending leaves may be more important in unproductive shady or dry environments, where leaves are relatively more costly to replace than in more productive sites (Janzen, 1974b). However, in certain ant-fed plants that have been studied carefully (e.g., *Hydnophytum* and *Myrmecodia;* see Janzen, 1974b; Huxley, 1978), a defensive role for the ants seems unlikely, although Janzen (1974b) reports that *Iridomyrmex* will respond aggressively to severe disturbances of their host plant. Thus, caution must be used in exercising this argument regarding the relative advantages of myrmecophily and carnivory.

Thompson (1981) has proposed that epiphytes tend to be ant-fed rather than carnivorous because their access to water is so limited that they cannot produce the glandular secretions associated with carnivory. *Drosophyllum*'s ability to maintain glandular secretions through the arid Mediterranean summer undercuts this argument (Givnish et al., 1984), but it can be incorporated into the cost–benefit model by noting that dry conditions would increase the energetic cost of secreting attractive or digestive fluids, thereby increasing the cost associated with a given photosynthetic benefit. This would decrease the slope of the benefit curve (Fig. 7-8), and render the evolution of carnivory less likely. Thompson (1981) suggests the rate of water loss and costs associated with myrmecophily should be less, because the permeable, nutrient-absorbing surfaces of ant-fed myrmecophytes are usually contained within a cavity. Thus, myrmecophily should be favored over carnivory in drier habitats that are also sunny and nutrient-poor. In addition, ant-fed plants should have an advantage over "basket" epiphytes (e.g., tank bromeliads, staghorn ferns) that passively trap falling debris, in treetop sites and open habitats that have a paucity of such debris and associated minerals.

Evolutionary Pathways to Carnivory

Because almost all carnivorous plants belong to genera and families composed entirely of carnivores, inferring the evolutionary pathway to carnivory has been exceedingly difficult. In his masterful monograph, Lloyd (1942) refused to discuss the matter for lack of evidence. Even today, phylogenetic relationships within most carnivorous families and genera are poorly understood. The only apparent exceptions—and weak ones at that—are the Droseraceae, Sarraceniaceae, and the genus *Brocchinia.*

Drosera appears to have originated in southwestern Australia, its current center of diversity. This region of sterile soils is the only one on earth to support all nine sections of the genus (Diels, 1906), and the endemic *D. paleacea* that occurs there has the lowest chromosome number (n = 6) known for the genus, which appears to be the base number for an extensive polyploid series (Kondo and Lavarack, 1984). Tuberous species in the subgenus *Eragaleium,* adapted to fire and seasonal drought in upland Australian sites, appear to be phylogenetically advanced (DeBuhr, 1977; Marchant and George, 1982). The Mediterranean *Drosophyllum* seems likely to be derived from *Drosera* on morphological grounds, while *Dionaea* appears to be not much more than a condensed, stranded derivative of the widespread aquatic *Aldrovanda* (see Lloyd, 1942). No author, however, seems to have analyzed the nature of the relationship between the *Drosera–Drosophyllum* complex and *Aldrovanda–Dionaea;* additional cytotaxonomic and molecular data would be extremely useful in this regard.

In the Sarraceniaceae, there is general agreement that *Heliamphora* is the most primitive genus. Its pitchers depart little from a simple peltate leaf formed by fusion of leaf margins (Lloyd, 1942; Steyermark, 1984), and it occupies the most ancient land surface of the three genera in the family. *Sarracenia* has evolved additional morphological specializations, including a nectar roll, more advanced lid, and in some species, digestive glands. Certain species have a well-developed ant trail of external nectaries, translucent areoles or "windows" in the lid to confuse prey, and (in *S. psittacina*) a hood formed by the fusion of the margins of the pitcher and lid (Lloyd, 1942; Slack, 1979). *Darlingtonia* almost surely was derived from an ancestor similar to *S. psittacina:* Although its hooded mature pitchers are erect, unlike those of the recumbent *S. psittacina,* its seedling pitchers are prone (Lloyd, 1942).

Brocchinia is currently the only genus in which we have some clear indication of the evolutionary pathway to carnivory, based on the presence of carnivorous and noncarnivorous species within the genus (Givnish et al., 1984). Gross morphology (Smith and Downs, 1974) and trichome structure suggest that the closest relatives of the carnivore *B. reducta* are facultatively epiphytic tank species like *B. tatei.* Epiphytic populations of *B. tatei* in shady cloud forests have a form typical of most tank bromeliads, with a basket-shaped rosette of nearly horizontal green leaves, well-designed for light capture under dimly lit conditions and for passive capture of nutrients from a rain of plant and animal debris (Givnish et al., 1984). Invasion of sunny, sterile savannas and bogs by *B.*

reducta's ancestor presumably would have favored the evolution of *steeply inclined leaves* with strongly reflective *wax cuticles* to reduce light interception and water loss. Mineral poverty may also have fostered chlorosis; in fact, sun-adapted, terrestrial populations of *B. tatei* at high elevations show both of these leaf traits (Steyermark, 1961, 1966).

Brightly colored, vertical, waxy leaves arranged about a tank would be preadaptations for the evolution of carnivory, in that they could coincidentally attract or entrap insects while performing their primary function. The crucial adaptive shift to carnivory probably involved the leakage into the tank fluid of a sweet-smelling, volatile compound stored in the leaf bases of many noncarnivorous *Brocchinia* species. This key trait would have attracted insects to the tank, bringing into play the preadapted functions of the leaves in prey attraction and entrapment, and promoting the evolution of nutrient-absorbing trichomes (Givnish et al., 1984).

The crucial, defining adaptation for carnivory in *B. reducta* is probably active prey attraction via a sweet-smelling compound; other leaf and trichome traits are clearly valuable in promoting carnivory but have other, not readily separated functions that could render their contribution more or less coincidental if considered in isolation. The dense waxy cuticle possessed by *B. reducta* and *Catopsis berteroniana* may also prove to be primarily an adaptation for carnivory, in that it is found mainly on the less exposed, inner leaf surfaces where it would have little effect on reducing leaf temperature or transpiration. However, the same might not be true for other tank bromeliads with more nearly horizontal leaves and less strongly developed cuticle, which nevertheless possess a tank into which insects can occasionally blunder and drown (Picado, 1913).

Givnish et al. (1984) suggest that it would be more parsimonious to call certain plants *protocarnivorous* rather than carnivorous. Such plants have traits that fit them for prey attraction, capture, or digestion, but each of these traits has substantial and inseparable functions for other purposes, with no clear allocation of energy solely to carnivory. Several tank bromeliads might fall into this ambiguous category, and perhaps are on the road to carnivory.

In addition to derivation from tank epiphytism, three additional pathways to carnivory might be posited, even though it is unlikely that direct evidence for them will ever be obtained. These include derivation from (1) glandular defenses against herbivores; (2) myrmecophily; and (3) peltate leaves in areas of heavy rainfall and high humidity. The first of these possibilities is fairly obvious, and may have provided the impetus for the evolution of carnivory in such genera as *Byblis, Drosera, Drosophyllum, Pinguicula,* and *Triphyophyllum*. An important, though not conclusive, test of this hypothesis would be whether removal of digestive glands greatly increases herbivory. If such removal does not stimulate additional grazing, at least by generalist herbivores, then the herbivory hypothesis becomes difficult to sustain. However, if removal does increase grazing, it does not necessarily support the hypothesis, because a species may have secondarily lost defensive compounds after acquiring carnivory.

Carnivory in *Nepenthes* may have arisen from a form of myrmecophily. *Nepenthes* vines bear abundant, highly developed extrafloral nectaries over their stems, leaves, and tendrils (Macfarlane, 1908); similar extrafloral nectaries occur on many plants (e.g., *Passiflora*) known to be defended by ants attracted to these food sources. Furthermore, *Nepenthes* pitchers develop initially as in pouchings of the leaf tip (Macfarlane, 1908), at which stage they resemble in certain gross aspects the leaf domiciles produced by several known myrmecophytes. Finally, at least one species (*N. bicalcarata*) is known to house ants in hollow swellings just below the pitchers (Macfarlane, 1908). It is relatively easy to see how a loose association with ant bodyguards could evolve into carnivory, particularly if the ant–pitcher plant mutualism theory is correct. Field studies on *N. bicalcarata* are crucial to determining whether an evolutionary derivation of carnivory from myrmecophily is plausible.

Finally, carnivory may have arisen in plants with peltate leaves found in areas of high rainfall and humidity. Under such conditions, a small amount of rainwater might accumulate on such leaves and form a primitive pitfall. Provided that a nutritional benefit resulted from such a primitive impoundment, selection might favor progressively greater cupping of the leaf surface, which would result in greater trapping efficiency and lower rates of water loss (Lloyd, 1942). Such a pathway seems likely for the origin of the Sarraceniaceae in *Heliamphora* growing atop the mist-drenched summits of the Guayana Shield. The pitchers of *Heliamphora* and many *Sarracenia* species do indeed fill initially with rainwater, whereas *Nepenthes* and *Cephalotus* initially fill their pitchers by secretion (Slack, 1979). It must be noted that pitcher plants also have an inherent advantage over active and passive flypaper plants in areas of high rainfall, because they have no glandular tentacles whose secretions may be washed away (e.g., Dixon et al., 1980) and because their pitchers are likely to remain filled and active for a longer period. Indeed, in the Gran Sabana region of the Guayana Shield, *Heliamphora* is far more abundant at higher elevations, whereas *Drosera* is relatively more abundant at lower elevations (Steyermark, 1984; Givnish, personal observation). The pinhole "drain" located roughly two-thirds of the way up the pitchers of *Heliamphora* should serve not only to reduce the mechanical load the plants bear (Slack, 1979), but also to prevent rain from washing away prey.

Carnivorous plants with the steel-trap or bladder-trap mechanisms all occur in families that possess active flypaper species with more rudimentary powers of tentacular or leaf movement (Tables 7-1 and 7-2), and presumably have been derived from them. The evolutionary relationships of the lobster-pot mechanism in *Genlisea* to morphologically simpler or more complex traps are totally obscure.

Future Research Directions

Throughout this chapter allusions have been made to studies that are needed to understand more fully the ecology and carbon economy of carnivorous plants. Ten of the most important of these are summarized below, in what can be considered a preliminary agenda for future research.

Nutrient Budgets and Ecological Impact of Prey-Derived Nutrients

1. Comprehensive field experiments are needed to study the importance of specific nutrients, and interactions between different nutrients, in stimulating the growth and reproduction of carnivorous plants. Studies should involve plants from several different genera grown under field conditions and include a combination of prey exclusion; feeding with agar blocks laced with one or more nutrients, distributed in factorial design (e.g., Karlsson and Carlsson, 1984); measurement of growth, reproduction, and tissue nutrient concentrations; and estimation of nutrient inputs from natural prey. Interactions with noncarnivorous competitors should not be excluded from such studies, insofar as Wilson (1985) has shown that they can alter the impact of prey availability on plant performance in *Drosera intermedia*.
2. More experiments like those of Dixon et al. (1980) on the efficiency of uptake of various nutrients from labeled prey, on the relative contribution of prey to the corresponding plant nutrient budgets, and on nutrient cycling would be extremely useful. Matching prey loads to the ranges seen in nature, and growing plants under field conditions would help make such studies more illuminating.
3. Intensive studies are needed of the allocation of various nutrients, including at least nitrogen and phosphorus, to various organs and compounds during the life cycle of carnivorous plants. What fraction of the nitrogen and phosphorus budgets are

diverted to flowers, seeds, and ancillary structures? What proportion of prey-derived supplies of these nutrients are allocated to photosynthetic enzymes, digestive enzymes, or storage during vegetative growth?

Carbon Economy

4. Perhaps the most pressing need is for a detailed study of the effect of feeding on photosynthesis per unit leaf mass as a function of irradiance, water availability, amount of prey, and substrate supply of different nutrients. Such research should be carried out under axenic conditions to eliminate the confounding effects of microbial respiration during digestion, and would be an essential first step toward quantifying benefit curves in the cost–benefit analysis of carnivory. Studies on net carbon uptake as a function of irradiance could finally resolve the question of whether heterotrophy is a significant contribution of carnivory. *Utricularia*, many terrestrial species of which have relatively little leaf tissue, should be carefully studied in this regard.
5. Similarly, investigation of the effect of feeding on root vs shoot allocation and on the efficiency of conversion of photosynthate to new leaf tissue, as a function of irradiance, water availability, amount of prey, and substrate supply of different nutrients, would be extremely useful in helping to quantify the costs and benefits of carnivory.
6. The preceding studies should be combined with a comprehensive investigation of carbon balance and resource allocation in a carnivorous plant in which there is a relatively clear distinction between trap tissue and photosynthetic tissue (e.g., several species of *Genlisea* or *Utricularia*), in order to determine whether carnivory actually yields a net energetic benefit. The elegant costing studies of Pate (1986) and his colleagues on the energetic costs associated with symbiotic nitrogen fixation provide an excellent model for the quantification of cost curves; the studies suggested in points 4 and 5 above would provide the benefit curves.
7. Such an investigation, if carried out under a range of environmental conditions (e.g., irradiance, prey availability, phosphorus supply), might be modified to determine whether the observed pattern of allocation to trap vs photosynthetic tissue is optimal (i.e., maximizes the allocation pattern with little direct effect on plant carbon content, and the impact of such variation on photosynthesis and growth under different conditions could then

be studied. It might, however, be difficult to exclude the possibility of excision causing development trauma.

Paradox of Enrichment

8. The most telling test of the cost–benefit model for the evolution of carnivory would be a study of a carnivorous plant that increases its allocation to trap tissue after fire or in more "fertile" water. Such behavior, depending on whether it results from nutrient complementarity, could either confirm or flatly contradict the predictions of the cost–benefit model. What is needed is a demonstration that the enhancement of photosynthesis as a function of prey capture is greater in substrates fertilized by fire, or in water enriched in dissolved inorganic carbon, that it is on unburnt sites or in DIC-poor water. Susan Knight (personal communication) is currently conducting such a study on *Utricularia vulgaris;* similar studies on other species would be most useful.

Mutualistic Interactions

9. The studies of Bradshaw (1983), Bradshaw and Creelman (1984), and Joel et al. (1985) need to be extended to determine whether dipteran larvae alleged to be digestive symbionts actually increase or decrease the net rate of energy gain of their pitcher plant hosts. Simple field experiments with caged plants, provided with access to prey with or without inquilines, might suffice to determine whether the inquilines stimulate growth. Such experiments, however, should be carried out over a sufficiently long period that the alleged symbionts can eclose and depart with their store of nutrients. More detailed research might include long-term studies on the effect of inquilines on the kinetics of nutrient uptake by their host, together with a theoretical analysis of whether the short-term increase in uptake rates —with its attendant effects on photosynthesis and growth, and potential increase in number of traps—outweigh the long-term decrease in total nutrients released per prey item per trap. Comparative studies on the presence of the specialized photosynthetic epidermis found by Joel and Gepstein (1985) in *Sarracenia purpurea* are clearly needed for other pitcher plants, to determine whether it is found mainly, as expected, in species that lack digestive glands.

10. Finally, it would be interesting to test the proposal made in this chapter that some pitcher plants have a mutualistic, rather than purely predator–prey, relationship with ants. The cost to the plant of obtaining ants might be quantified, in part, by measuring nectar concentration and rates of nectar secretion (neither of which appears to have been investigated previously), as well as rates of ant attraction and ant capture. The benefit to the plant of obtaining ants might be quantified using studies similar to those outlined under points 4 and 5 above. Similarly, the benefit to an ant colony might be quantified by measuring the rate of sugar retrieval per successful visit, and the rate of survival during visits to nectaries. Costs to the ants might be analyzed using techniques similar to those employed by Wilson (1980) to assess the energetic costs of construction and maintenance per worker. Analysis of these costs and benefits should demonstrate whether the relationship between pitcher plants and ants is mutualistic, unilaterally exploitative, or mutually destructive. In this connection, field studies on the role of ants in the myrmecophyte/carnivore *Nepenthes bicalcarata* would be of utmost interest.

Selected References

Bradshaw, W. E. 1983. Interaction between the mosquito *Wyeomyia smithii*, the midge *Metriocnemus knabi*, and their carnivorous host *Sarracenia purpurea*. pp. 161–189. In: J. H. Frank and L. P. Lounibos, eds. Phytotelmata: terrestrial plants as hosts for aquatic insect communities. Plexus Publishing, Medford, NJ.

Chandler, G. E., and J. W. Anderson. 1976. Studies on the nutrition and growth of *Drosera* species with reference to the carnivorous habit. *New Phytol.* 76:129–141.

Givnish, T. J., E. L. Burkhardt, R. E. Happel, and J. D. Weintraub. 1984. Carnivory in the bromeliad *Brocchinia reducta*, with a cost–benefit model for the general restriction of carnivorous plants to sunny, moist, nutrient-poor habitats. *Am. Natur.* 124: 479–497.

Green, S., T. L. Green, and Y. Heslop-Harrison. 1979. Seasonal heterophylly and leaf gland features in *Triphyophyllum* (Dioncophyllaceae), a new carnivorous plant genus. *Bot. J. Linn. Soc.* 78:99–116.

Karlsson, P. S., and B. Carlsson. 1984. Why does *Pinguicula vulgaris* trap insects? *New Phytol.* 97:25–30.

Lüttge, U. 1983. Ecophysiology of carnivorous plants. pp. 489–517. In: O. L. Lange, P. S. Nobel, C. B. Osmond, and H. Ziegler, eds. Encyclopedia of plant physiology, vol. 12C. Springer Verlag, Heidelberg.

Pate, J. S., and K. W. Dixon. 1978. Mineral nutrition of *Drosera erythrorhiza* Lindl. with special reference to its tuberous habit. *Austr. J. Bot.* 26:455–464.

Slack, A. 1979. Carnivorous plants. Ebury Press, London.

Thompson, J. N. 1981. Reversed animal-plant interactions: the evolution of insectivorous and ant-fed plants. *Biol. J. Linn. Soc.* 16:147–155.

Plant Communities as Animal Habitats: Effects of Primary Resources on the Distribution and Abundance of Animals

KERRY N. RABENOLD AND WILLIAM R. BROMER

Department of Biological Sciences
Purdue University
West Lafayette, IN 47907

Introduction

Habitat selection is the process by which animals occupy a subset of possible living conditions and through which species exhibit variable geographic distributions. Few species are so flexible and cosmopolitan that they can inhabit, for example, both oak forest and grasslands. *Habitat* can mean simply the place an organism occupies—its environment—but here it will mean an identifiable type of plant community in which an animal lives. In some cases, mechanisms of active preference for particular plant communities can be demonstrated, and sometimes animals are driven to occupy particular habitats by competitive or predatory pressures. The distributions of plant populations, and therefore the associations that are called communities, are sensitive to growing conditions like climate and soil characteristics. Distributions of animals may be determined directly by this kind of abiotic factor—like the severity of winters or dry seasons. However, where animals live within a range of physiological tolerance is greatly affected by biological relationships like predation and competition. In addition, their distribution can be determined by the suitability of their morphological and behavioral equipment for the vegetation that they inhabit.

Animals that live in plant communities with sharp boundaries, such as marshes or alpine meadows, often show parallel discontinuities in their distributions; they may not inhabit the forests separating marshes or mountain peaks. Even within plant communities, animals often use consistently different parts of the habitat. From one oak forest to another in a particular geographic region, one can predict not only the species of birds and mammals to be found there, but also which will likely make their living near the ground and which in the canopy. This kind of consistent pattern in distribution has prompted biologists to propose several kinds of hypotheses about ecological forces determining where animals live:

1. Different habitats require different physiological, morphological, and behavioral specializations, and natural selection has produced matches of plant and animal characters that restrict successful movement across habitat boundaries.
2. Because even subtle differences in adaptations for exploitation of a habitat can confer a critical advantage in efficiency, similar species of animals needing the same limiting resources will segregate, because of competitive effects, into different habitats or different parts of a habitat.
3. Either of the above two effects will create selection pressures favoring animals that can find habitats where survival and fecundity are highest, so that animals should show distinct behavioral mechanisms of habitat recognition and preference.
4. The spatial and temporal variability of the resources presented by plants in a habitat can strongly affect animal life histories, breeding systems, population dynamics, and movement patterns.
5. Characteristics of the plant community that accommodate a wider variety of animal specializations or ameliorate interspecific competition should increase the variety of animals found there.

This chapter considers evidence bearing on each of the above broad hypotheses about how the adaptations of animals for the exploitation of plant resources, and the nature of those resources, can affect animal communities. The principal observational evidence considered will be patterning of animal distributions among plant communities, their changes in space and time, and the adaptations that animals show for life with particular kinds of plants. This chapter is more one-sided in its consideration of plant–animal interactions than some other chapters. Because plants provide more of the critical resources needed by animals, from substrate to nutrients, than animals provide for plants, the distribution of animals is more likely controlled by plants than vice versa. Plants form much of the physical and biological environment for animals and thereby can greatly affect population dynamics and patterns of distribution and abundance. This chapter explores the issue of whether these effects can be read in the ways that animals move and are distributed in and among plant communities.

Adaptations

Bumblebees with long tongues that enable them to reach into long flower tubes for nectar, or with behavior that enables them to circumvent the tube and bite the back of a flower to "steal" nectar at the source, exemplify morphological and behavioral characteristics that produce a "fit" between animals and the plant resources that they use. When there is evidence that such characteristics have been shaped by natural selection *because* of the advantages they provide (and that these advantages are their main function), they can be considered adaptations for exploitation of those plant resources. Many examples of similar adaptations are given in the preceding chapters of this book; a few follow that demonstrate the powerful effect that the physical structure of vegetation can have on animal morphology and behavior, and the difficulty of establishing whether such adaptations are primary causes of habitat selection by animals or secondary adjustments to life with particular plants.

Can Adaptations to Plants Determine Distributions?

When animals pass through important developmental stages inside the body of a particular kind of plant, they sometimes become so specialized for exploiting that plant that they are found nowhere else. Some insects develop as larvae inside goldenrod stems, chemically altering the plants' growth physiology to produce a tumorlike gall on the plant that supplies them with both shelter and food. This is a specialized form of herbivory that requires precise chemical correspondence with the plant to alter its allocation of energy in favor of the hypertrophied gall (Hartnett and Abrahamson, 1979). Yucca moths use developing fruits of the yucca plant for brood chambers, and provide apparent compensation by pollinating the flowers (Aker and Udovic, 1981; see also Chap. 2). Fig wasps similarly derive shelter and food from fig flowers while providing pollination services. Males of such wasps can spend their entire lives within the fig flower (Wiebes, 1979). In such cases of extreme specialization, especially the obligate mutualisms of the figs and yuccas, there is evidence of coevolution and it is clear that distribution of the animals will be identical to that of the plants.

In cases of less dramatic specialization, adaptations for life with particular plants still limit the distribution of animals. Some tropical frogs breed exclusively in bromeliads in the forest canopy (Janzen, 1983b) and some mosquitoes breed only inside pitcher plants (Addicott, 1974). Birds' beaks have been extensively studied as mechanical adaptations for the exploitation of particular resources provided by plants, and the following examples show that beak morphologies of a variety of birds have led to resource specialization and limitation of range. Crossbills (*Loxia* spp.) are reddish finches whose mandibles cross like bent scissors at the tips instead of opposing directly like those of other seedeaters. This peculiar bill morphology is the most obviously specialized of temperate songbirds and serves to wedge and lever apart the scales of conifer cones. With a twist of the head, crossbills have access to the seeds inside that are unavailable to other birds unless the cone is fully ripe. Among species of crossbills, larger ones with stouter beaks can use the rigid cones of pines (*Pinus* spp.) while smaller species use the softer cones of spruce (*Picea* spp.) and larch (*Larix* spp.) (Newton, 1967; Lack, 1971; Fig. 8-1). Such morphological specialization carries a price: It is difficult for crossbills to pick up small seeds as other finches do because their bill tips do not meet.

Although they also feed on other plant and animal foods, crossbills depend mainly on the conifers whose seed defenses made their peculiar morphology so useful. Because the seed crops of conifers, especially spruces, are variable in space and time, crossbills are famous in Europe and North America for their nomadic movements and "irruptions" as they track cone production over huge areas (Newton, 1970; Bock and Lepthien, 1976). Their reproductive biology is remarkably flexible; they rear young on conifer seeds at any time of the year depending on cone crops and are not constrained by photoperiodic hormonal responses as are most temperate birds (Tordoff and Dawson, 1965). Pinyon jays exhibit similar flexibility in their close mutualism with pinyon pines (Ligon, 1978).

In contrast to natural systems, human-altered vegetation often favors behavioral and morphological generalization. Successful invaders, like the English sparrow, Norway rat, and German yellow jacket, into North America have been not only tolerant of humans but also are generalized

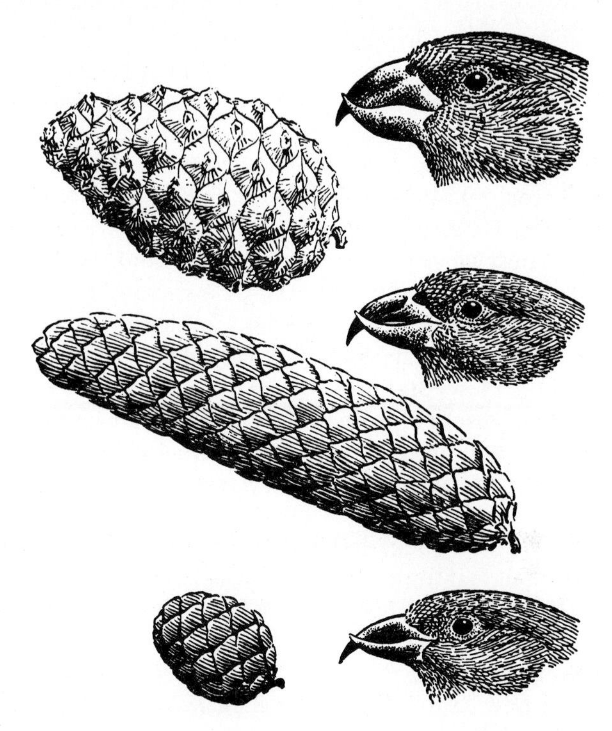

Figure 8–1. The heads of the three crossbill species, and the main cones eaten. *Top:* Parrot crossbill and pine cone; *Center:* Common crossbill and spruce cone; *Bottom:* Two-barred crossbill and larch cone (from Newton, 1972).

omnivores able to thrive in plant communities altered by humans. Native species successful in suburban environments (e.g., cardinals, doves, and house finches) tend to be similarly unspecialized and quick to adapt to horticultural plantings (Emlen, 1974; Jarvinen and Ulfstrand, 1980). Crossbills, however, are conspicuously specialized predators of a few native plant species' seeds, and their way of life, from breeding biology to annual movement patterns, is determined by their intimate relationship with conifers. Their distribution among communities is clearly limited by their adaptations to life with these trees.

Species of the avian genus *Parus* (chickadees and relatives in North America and Europe) using conifers show morphological divergence from those using broadleaf trees; the former have narrower beaks that are more effective at probing for insects hidden in needle clusters (Gibb, 1954; Lack, 1971). Field observations and laboratory studies have shown that these birds possess behavioral adaptations that complement their physical tools. Blue tits live mainly in deciduous woods and prey on insects in galls and dead wood; they excel at tearing and hammering things open. Coal tits use their slender beaks for more delicate probing, often hovering instead of hanging by the feet for access to pine cones surrounded by needles (Partridge, 1976). Finches of various genera assort by the related factors of size, power of the beak, and size of seeds eaten. Distribution and abundance of a particular finch can sometimes be correlated with abundance of seeds of optimal size (Newton, 1972; Grant and Schluter, 1984; Pulliam, 1975, 1985).

Tropical wrens (*Campylorhynchus* spp.) in the savanna of South America choose nesting sites to deter predators and thereby limit their local distributions to particular habitats. One species, the stripe-backed wren, nests preferentially in one species of leguminous tree that is distributed very patchily; this gives the wren a very limited distribution across the savanna. A second species, the bicolored wren, nests in the protective bracts of a much more widely distributed palm; this wren is therefore more cosmopolitan throughout the region. Strong feeding and nesting preferences for particular plants can bring a species to the brink of extinction when the preferred vegetation dwindles. This seems to be the case with the red-cockaded woodpecker's affinity for extensive stands of large longleaf pine that have been made scarce by human lumbering in the southeastern United States (Labisky and Porter, 1984). Giant pandas are endangered in part because their specialized diet of bamboo requires ranging over large areas, and sufficiently large undisturbed areas are now rare (Schaller et al., 1985).

It is clear that behavioral, morphological, and physiological adaptation by animals to life with particular plants can greatly affect the distribution and abundance of animal populations and the composition of animal communities. It is less clear whether biotic interactions like competition among animals have forced these adaptations, or whether apparent ecological divergence of animal species is simply a result of the diverse spectrum of resources provided by plants, requiring a spectrum of specialized exploitative adaptations.

Ecological Segregation Within and Between Habitats

Charles Darwin's thesis that competition for resources, especially food, should be a powerful selective force was based on one of the fundamental observations of natural history brought to full light by the gloomy economist Thomas Malthus: that the reproductive potential of populations far exceeds the capacity of their environments to produce necessary resources, and that powerful destructive forces (disease, predation, starvation) cull every generation. This excess production of young, and the variability among them, provide the ingredients for natural selection. Fifty years after publication of *The Origin of Species*, the American naturalist Joseph Grinnell (1904) wrote: "It is only by adaptations to different sorts of food, or modes of food getting, that more than one species can occupy the same locality. Two species of approximately the same food habits are not likely to remain long evenly balanced in numbers in the same region."

It was another half-century before enough information about the ecological relationships of taxonomically related species had accumulated for the British ornithologist David Lack to summarize evidence bearing on Darwin's idea. The basic question was whether species were distributed across habitat types, or among parts of the same habitat, in ways that minimize contact of related species—especially congeners. Lack detected a fundamental paradox: Although species in a genus tend to share adaptations, especially those for food-gathering, they tend *not* to share habitats. Although congeners are more alike than noncongeners, they diverge in distribution among and within habitat types.

The genus *Parus* is represented by 45 species around the globe at north-temperate latitudes. On reviewing the nine European *Parus* species, Lack (1971) found that, of the 27 pairwise combinations of species whose ranges overlapped, species-pairs were separated ecologically either by occurring in different forest types (12 pairs) or by using the same habitat in different ways (15 pairs). Nowhere did two parid species specialize on the same part of the same habitat. For example, in English oak forests, three *Parus* species regularly coexist: the small blue tit forages high in trees on small twigs, the medium-sized marsh tit forages at middle heights on larger branches, and the large great tit forages low in the vegetation and often on the ground. Although this is a very general summary and there is much overlap among species, substantial differences in foraging height normally exist and are correlated with food types taken (Gibb, 1954; Betts, 1955; Hartley, 1953). Geographic comparisons of populations of the same species also argue that competition can be important. On the Canary Islands, where the blue tit is the only parid, it expands its use of habitat to include pine forest and shows a parallel development of a longer, thinner beak characteristic of other parids that forage on pines (Lack, 1971).

Problems with the Hypothesis that Competition Determines Distribution

Segregation by habitat has been thought to be important for genetic isolation of populations and the development of new species (Mayr, 1970), but its main implications have been as evidence for the importance of interspecific competition in determining distribution (Svardson, 1949; Lack, 1971). In this view, even subtly different adaptations to feeding in a particular plant community would inevitably lead to differences in efficiency; these would result in extirpation of one of a pair of very similar species from a habitat or segregation of the species to different resource bases within a habitat. The variety of distinct resource types provided by plants and their attendant fauna of potential prey would then determine the likelihood that congeners could coexist in a particular habitat by controlling the degree to which divergence between them was possible.

There are serious problems with the above account of forces distributing animal species in and among plant communities. First, it is not surprising that congeneric species should show *some* difference in foraging behavior. Having recognized them as distinct species establishes that morphological differences exist that are likely to be associated with some behavioral differences. Because between- and within-habitat differences in resource use among species occur in many different dimensions (e.g., height in the vegetation, mode of foraging, substrates used, and resulting diet) and quantification of differences including such various dimensions is very difficult, we are lacking any rigorous definition of what a "substantial" difference would be. Testing the possibility of divergence on a geographic scale is likewise not a simple matter. Biologists must first establish what the level of difference would be if the factor in question, competition, were *not* operating. *Some* differences in distribution among species would exist even if each were distributed independently of the others, just because of the morphological differences inherent in their species status, or even if species were just distributed randomly but nonuniversally (Strong et al., 1984b).

Congeners are not expected to be more different in resources used than noncongeners, but the question is whether they are more different than could be expected in the absence of competition. Generating quantitative predictions for such an expectation as a basis of comparison is very difficult. The basic problem is that the hypothesis is an historic one about what the forces *were* that determined *both* morphological and spatial divergence of congeners. There is no resolution to the conundrum that, even though ecological differentiation of species argues that they do not currently compete, it is possible that their differences are the result of past competition.

The best evidence that competition can be a powerful force determining distribution of a species comes from "natural experiments" —especially isolated island populations—and experimental manipulations of species co-occurrence by ecologists. Many examples exist of "ecological release" on islands. Island populations have fewer potential competitors, and often broaden their use of vegetation types, food types, or foraging stations coincident with the absence of putative competitors. Often the direction of change is toward the resource use pattern of the absent species (Schoener, 1975; Diamond, 1978). This result has often been produced by the intentional experimental removal of competitors as well (Connell, 1983; Schoener, 1983). The controversy over the strength of interspecific competition in nature cannot be adequately reviewed here (Schoener, 1982; Connor and Simberloff, 1986), but the degree to which animals actually do divide up the resources provided by plants can be explored to determine whether patterns of such resource partitioning seem to be regular and predictable enough to have been caused by competition.

Studies of Habitat Segregation

Competition and Adaptations to Plants

Environmental gradients like altitude produce sequences of plant communities that often broadly intergrade with one another (Whittaker, 1975). Animal communities normally show parallel change along such a gradient, with species of mammals, for instance, occupying fairly distinct zones along the gradient. In the Sierra Nevada of California, four species of chipmunk (*Eutamias* spp.) inhabit distinct zones of vegetation up an elevational gradient. There is evidence that such habitat segregation is enforced at least in part by overt interspecific aggression (Heller, 1971). Such altitudinal segregation of animals by habitat type is well known, from salamanders in the Great Smoky Mountains to birds in the Peruvian Andes. In many cases, range abutments of similar species are very sharp, suggesting competitive confinement in narrower habitat zones than the animals might be capable of using (e.g., Terborgh, 1971). In some cases, segregation on an environmental gradient, like water depth for fish, seems to result from similarities in foraging that are too great for coexistence in the same habitat (Werner, 1977).

Small oceanic islands in tropical Melanesia are species-poor compared to large areas like New Guinea because of combined effects of isolation from centers of species diversity and small area (Diamond, 1978). Diamond has documented many cases of habitat expansion on islands. For instance, the thrush *Turdus poliocephalus,* a congener of the American robin, occurs on New Guinea only at high altitudes above 2700 m at the edge of forest and alpine grassland. Islands in the New Hebrides and Solomons support as few as one-tenth the 513 species of New Guinean birds. On those islands, the thrush occurs at all elevations and down into sea-level rain forest that is a far cry from its cold alpine haunts on New Guinea. Diamond points out that in spite of the order-of-magnitude change in total species on land masses within the thrush's range, a relatively constant 23 to 36 species coexist in the thrush's habitat, suggesting that increasing competitive pressure in diverse areas forces "retreat" to high elevations. Many cases also exist of dramatic shifts in feeding behavior within habitats on species-poor islands, especially concerning specific foraging behaviors and foraging heights within a forest (Diamond, 1973; Fig. 8-2).

Finches in the genus *Geospiza* live on the dry tropical islands of the Galapagos off the coast of Ecuador. These birds inspired much thought by both Darwin and Lack concerning their adaptive radiation—their historical spread from a single ancestral colonizing form to a group of species in all habitats of the archipelago. The morphology, behavior, and diet of these birds have been very thoroughly studied in an effort to explain both patterns of distribution and morphological variation within and between island populations. Their beaks are clearly adapted to using different sizes of food, especially seeds; the larger species with powerful beaks use large, hard seeds. These large species occur in habitats in which plants produce such seeds, and large finches are conspicuously absent on islands without such plants. When two finch species co-occur on an island, they often segregate into habitats whose plants produce

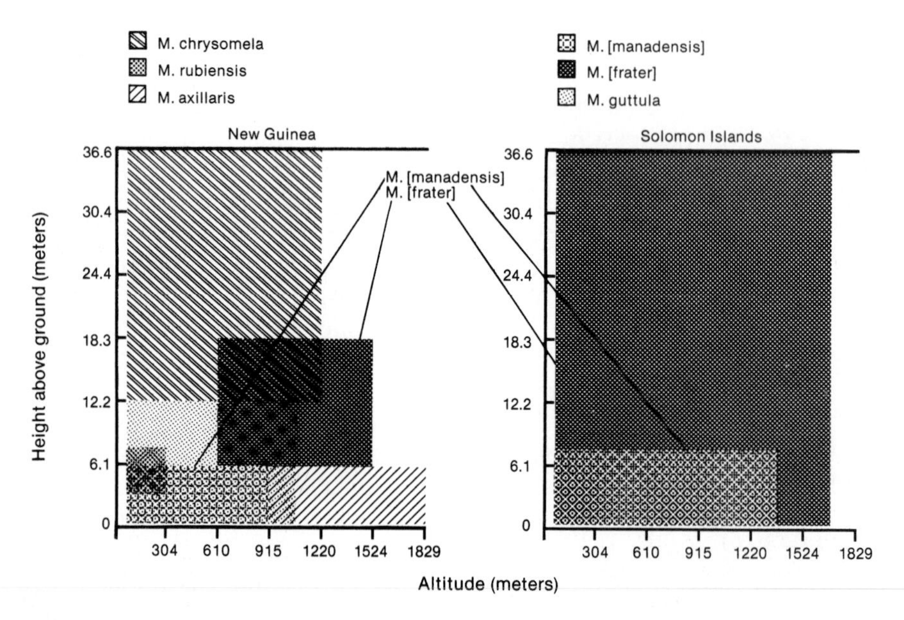

Figure 8–2. Comparison of niche segregation among *Monarcha* flycatchers in New Guinea, where six species occur, and in the Solomon Islands, where only two of these species occur. The preferred altitude *(abscissa)* and height above ground within the rain-forest canopy *(ordinate)* over which each species forages is depicted. In the Solomon Islands, where it is free of most of its competition, M. (frater) greatly expands its altitudinal and vertical ranges. From Diamond (1978).

seeds that match their bill morphology. In fact, such habitat segregation occurs within finch populations: Larger-billed conspecifics forage in habitats producing larger, harder seeds (Grant et al., 1976). In years when plants' production of seeds changes drastically because of climatic fluctuations, directional selection in finch populations becomes apparent; only those finches with bills matched to the existing seed crop survive and the distribution of bill sizes changes (Boag and Grant, 1981). In the dry, nonbreeding season, when both the quality and quantity of food for *Geospiza* finches decline, sympatric species diverge in diet, "retreating" to using seeds that closely match their morphological specializations (Smith et al., 1978).

Finch distributions,. however, are not simply matched to plant distributions, since competitive effects among the finches can limit the number of species found on islands. This is argued by strong patterns of ecological release in which allopatric populations (those on separate islands) are more similar and broad in resource use than sympatric ones, and by seemingly regular patterns of morphological separation in sympatric populations. A given species of finch varies in beak size, habitat choice, and therefore diet among islands, not just in parallel with seed production, but also with the nature of other coexisting finch species (Grant and Schluter, 1984; Grant, 1986b).

Males and females of the same species often differ considerably in resource use within a habitat. Thus, similar reasoning to that outlined above is often invoked to explain sexual divergence as a result of competition (Morse, 1968; Willson et al., 1975). Robert Selander (1966) noted that these differences between the sexes were especially striking for woodpeckers inhabiting islands in the West Indies where no other species of woodpeckers live; this dimorphism is easily as great as between members of different species elsewhere. Males of the species *Centurus striatus,* a species endemic to Hispaniola, are 20 percent heavier than females and have bills that are that much longer. Woodpeckers have long tongues with a hard, barbed tip used to probe holes in wood and extract insects; males of the Hispaniolan woodpecker have 34 percent longer tongues than females. Foraging differences between the sexes correspond to their morphological differences. Along with these sexual differences, the Hispaniolan woodpecker is an extreme habitat generalist: It is abundant in all woodlands on its island from coastal mangrove swamps and moist broad-leaved forests, to coffee plantations, dry thorny woodlands, and pine forests of the interior hills and mountains. By this combination of sexual divergence and habitat generalism, the Hispaniolan woodpecker achieves an abundance and ecological range equaled only by several similar species combined in continental Central America.

Ecological release such as this argues that animals are often behaviorally and genetically capable of exploiting much wider ranges of plant communities and resources than they actually do when sympatric with similarly adapted species. Morphological adaptations like beak shape and size and body size have clearly been historically affected by the nature of resources presented by plants and by interactions with other animals, and both selective forces have altered animal distributions among habitats.

Habitat Segregation Experiments

Experimental approaches by ecologists to the question of factors regulating distribution of animals among habitat types have been very revealing. In some communities, like those in the intertidal zone at the ocean's edge, competitive and predatory pressures have profound effects on animal distribution, but specific adaptations for life with particular plants and partitioning of plant resources by animals are rare. Animals mainly just compete for space (Connell, 1961; Paine, 1984). In Sweden, willow and crested tits (*Parus* spp.) forage in the inner canopy of coniferous forests (mainly pines with some spruce and a few birch trees) while coal tits and goldcrests (relatives of North American kinglets) forage in the needles of the outer parts of the trees. In an experiment removing half of the members of the former two tit species, the unmanipulated species significantly increased their foraging in the inner canopy compared to control plots where no removals were done. This confirms that some species are adapted to using a broader spectrum of plant resources than they normally do, but they are prevented from doing so by similarly adapted competitors (Alatalo et al., 1985a). Of course, there must be limits to this phenomenon; removing the woodpeckers from those forests would not result in goldcrests excavating tree trunks—they lack the equipment to exploit the plants in that way.

Lizards in the genus *Anolis* occur throughout the Caribbean, and various species combinations occur on different islands. Ecological release is often clear for characters like foraging height in the vegetation and even repertoires of color changes to match vegetation (Schoener, 1975). When small areas of Caribbean scrub vegetation are isolated by fencing and one of two species of anole is removed, the remaining lizards show signs of release: Growth increase and use of different strata in the vegetation expands (Pacala and Roughgarden, 1982). Removal experiments with woodpeckers showed ecological release of females: they shifted their foraging activities to parts of trees normally used mainly by males when the latter were removed (Peters and Grubb, 1983).

It is important to remember at this point that many experimental studies, like those with salamanders in the southern Appalachian mountains (Hairston, 1981) have found little evidence of competitive control of distribution among habitats (see review by Connell, 1983). In addition, many coexisting species have nearly identical feeding adaptations and coexist side by side on the same plant resources. Phytophagous insects like leafhoppers (Ross, 1957), aquatic invertebrates (Hutchinson, 1961), and some algae-eating fish (Fryer, 1959) fall into this category. Rathcke (1976) found very high levels of overlap among 13 species of stem-boring insects in a temperate prairie in their use of plant species and parts of the same plants. In spite of probabilities of co-occurrence greater than 70 percent, little evidence of actual competition was found, probably because resources were generally plentiful. Strong (1982, 1984) studied leaf-mining beetles belonging mainly to a single genus that live on tropical *Heliconia* plants. Most of his samples turned up three or more species of beetle in the same stand of plants (sometimes five species in the same rolled leaf), but he found little evidence of competitive effects either in natural distributions or in laboratory experiments. This "harmonious coexistence" is probably the result of heavy parasite and predator pressure on the beetle populations (rather than chemical defenses of the plant) that keep densities low enough to prevent depletion of the plant resources.

Can Distribution and Abundance of Plant Resources Determine Differential Habitat Use?

Lack (1971) reported that *Parus* species in North America generally do not coexist in the same forest type but that two species widely distributed in the eastern deciduous forests, the Carolina chickadee (*P. carolinensis*) and the tufted titmouse (*P. bicolor*), coexist but differ greatly in size and foraging. The titmouse is twice the size of the chickadee, but their foraging is very similar.

Chickadees and titmice are generalists in their use of the forest, searching for insects and seeds from the canopy to the ground and from trunks of trees to branch tips. However, there is still considerable individual variability (Fig. 8-3). This behavioral variability is not closely related to size or sex. The feeding niche of a population can be divided into "between-phenotype" (interindividual) and "within-phenotype" (intraindividual) components (Roughgarden, 1972). Because these birds are such generalists and, relative to their range of behaviors, individual variation is subtle, their populations can be characterized as "monomorphic with generalists": on average, 90 percent of the populations' total niche width is accounted for by the within-phenotype component (individual's average variance) in heights used, parts of trees used at a

© KJPrather 1987

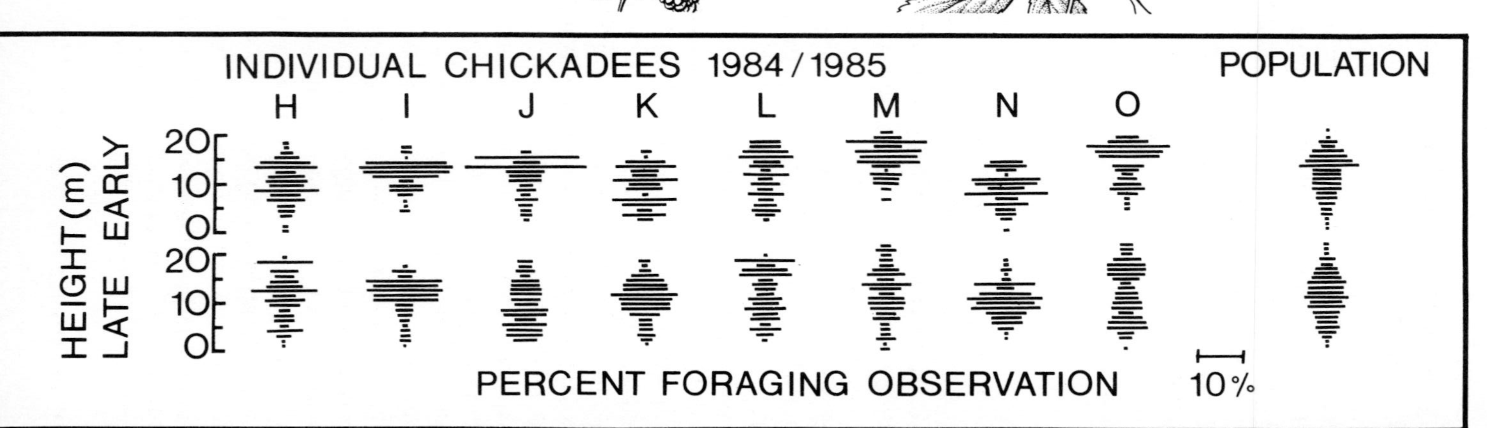
INDIVIDUAL CHICKADEES 1984 / 1985
POPULATION
H
I
J
K
L
M
N
O
HEIGHT (m)
LATE
EARLY
20
10
0
20
10
0
PERCENT FORAGING OBSERVATION
10%

given height, and behavioral techniques of feeding (Thirakhupt and Rabenold, unpublished).

Competition theory would predict that when resources are most limiting, these congeners (and conspecifics) should diverge in feeding characteristics because of competitive effects, as argued earlier for the Galapagos finches. It is very difficult to accurately quantify food availability for omnivores like parids, especially in a forest habitat, but it is certain that in north-temperate winters like those in Indiana where this study was done, food supply dwindles over the course of a winter because there is no primary production to replenish food as it is taken. In late winter, congeners are expected to diverge, but the opposite occurs.

Individual chickadees and titmice were more generalized in their foraging in late winter than in early winter. As the winter progressed and food became more scarce, individuals of both species used a wider variety of foraging strata in the forest, parts of the tree within a stratum, and particular foraging techniques (Fig. 8-4). On the population level, a compilation of thousands of observations for 20 individuals of each species showed that niche width (aggregate foraging) was considerably broader in late winter—so broad that the two populations converged as food became more scarce. Two-dimensional spatial overlap between the species was 20 percent higher when food was most scarce. Competition between and within these populations did not prevent convergence among conspecifics or niche convergence of congeners in spite of a declining food supply. The food supply presented by the forest trees —widely distributed arthropods and seeds requiring little morphological specialization for capture—is intensely seasonal in its production. In this case, these characteristics, and the possibility that it is not energetically feasible for individuals to specialize in foraging, lead to such broad use of the habitat by members of both species that virtually no differentiation between the two species is apparent in habitat use.

Within-Habitat Segregation

Do Plants Determine Limits to Overlap?

Warblers of the genus *Dendroica* are insectivores during the temperate summer breeding season, scouring trees mainly for caterpillars and spiders. Five congeneric species commonly coexist in spruce forests of New England: the Cape May, bay-breasted, black-throated green, blackburnian, and yellow-rumped warblers. These five species are very similar in morphology, much more similar than the five species of *Parus*

Figure 8–3. Foraging height distributions for eight individual chickadees in early and late winter. Each line represents percent foraging in a height category. From Thirakhupt and Rabenold (unpublished).

studied in English oak woods. The warblers are about the size of the smallest parid, but have longer, thinner beaks (likely adaptations to life with conifers). The largest parid is more than twice the size of the smallest (the great tit is 2.15 times the coal tit in weight; Snow, 1953) and has a 40 percent longer beak. The stoutest parid beak has a 40 percent larger depth-to-length ratio. The largest *Dendroica* warbler in MacArthur's (1958) study is less than half again as large as the smallest (yellow-rumped or bay-breasted is 1.42 times the size of the black-throated green; Chapman, 1939; Dunning, 1984) and the longest beak is only 20 percent longer than the shortest. The stoutest warbler beak is only 25 percent thicker for its length than the thinnest.

MacArthur (1958) asked how such similar species could coexist in the same forest and arrived at the following conclusion:

> Thus, of the five species, Cape May warblers and to a lesser degree bay-breasted warblers are dependent upon periods of superabundant food,

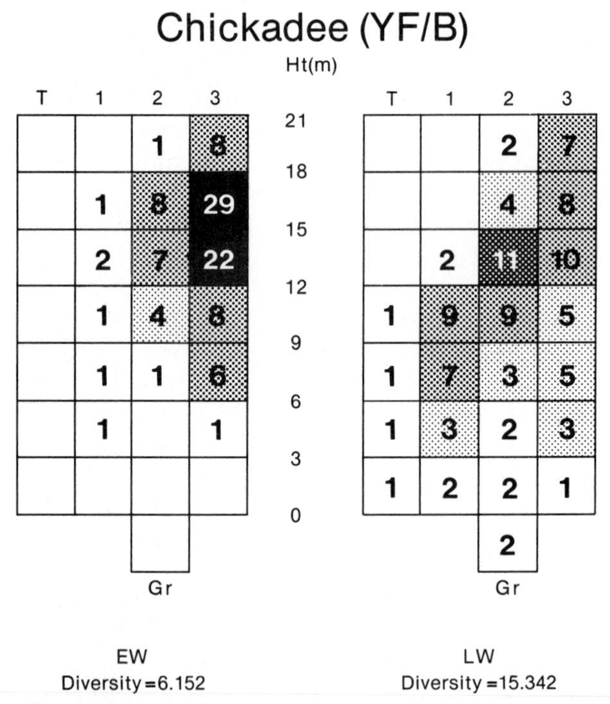

Figure 8–4. The two-dimensional distribution of foraging of an individual chickadee in early and late winter (EW, LW). Percent foraging is shown in each cell. Vertical axis is meters from the ground; horizontal axis is categories of distance from trunk: trunk (T); inner third of branch (1); middle third (2); and outer third (3). Diversity values were calculated using the inverse-Simpson's index (from Thirakhupt and Rabenold, unpublished).

while the remaining species maintain populations roughly proportional to the volume of foliage of the type in which they normally feed. There are differences of feeding position, behavior, and nesting date which reduce competition. These, combined with slight differences in habitats preference and perhaps a tendency for territoriality to have a stronger regulating effect upon the same species than upon others, permit the coexistence of the species.

MacArthur argued that by feeding at slightly different heights, on average, and using subtly different parts of branches and behavioral techniques, the warblers could diverge in diet. He did not, however, directly demonstrate this correlation. He thought it might be more important that foraging-site preferences within a forest could lead to somewhat different distributions among forest types. "Coexistence in one habitat, then, may be the result of each species being limited by the availability of a resource in different habitats" (MacArthur, 1958).

In the Appalachian mountains of the southeastern United States, 1600 km south of MacArthur's study sites, forests dominated by spruce are confined to high elevations. The plant communities of these relict forests share most important species with those in New England, but are somewhat simpler in taxonomic structure; the bird communities are strikingly depauperate (Rabenold, 1978). Old-growth spruce forests in New England commonly support 30 or more species of birds, but equally mature forests in the Great Smoky Mountains of the southern Appalachians normally support only half that number of species. The birds of this high-altitude forest (1500 to 2000 m) are all species characteristic of the northern forest, but they are a very limited subset. Conspicuously absent is the great diversity of warblers found in the north. In fact, coexistence of species of the same genus, or even in the same family, is remarkably absent. Only one species of warbler is common in undisturbed spruce forest, the black-throated green warbler, compared to at least eight species in similar northern forests. Similarly, two species each of the genera *Parus* (chickadees) and *Regulus* (kinglets) occur in the north, but only one species per genus is found in the south. Several finches coexist in the north but only one in the south. In general, the same avian families occur north and south, but the rule is one species per family in the south while the northern forests hold many species per family or even genus. Why is coexistence among similar species lacking in the south in spite of an equal opportunity for the kind of foraging differentiation described by MacArthur?

The first hint of an answer to this question is the fact that nearly half the individuals in the north are highly migratory warblers, and nearly all birds belong to species that characteristically migrate thousands of kilometers in winter to tropical and subtropical climates. In contrast, less than 5 percent of birds in the southern forests are warblers, and the great majority belong to populations that migrate only a few kilometers downslope into the warmer valleys in winter (Rabenold and Rabenold, 1985). These large, nearly resident southern populations arrive for breeding in the spruce forests, where spring begins weeks earlier than in the north, long before the arrival of long-distance migrants whose latitudinal movements are tuned to the shorter northern summer. Evidence suggests that the difference in seasonality of production in the forests, and migratoriness of the birds in response, controls the composition of the bird community rather than the opportunity for resource partitioning or the relative isolation of the southern mountaintops.

Measurements of foraging behavior of 12 species of arboreal insectivores, two decades after MacArthur's study, shed some new light on these forests (Rabenold, 1978). First, what appears to be ecological release is evident in all southern populations: Black-throated green warblers, for instance, greatly expanded their spatial use of the habitat, and their repertoire of foraging behaviors as well (Fig. 8-5a). This might suggest release from competition, but this generalization was just as pronounced for species that are the sole representatives of their families in both northern and southern forests. As seen with the foraging of chickadees and titmice in winter, foraging generalization can result from lower food levels. The longer growing season in the south is associated with lower peak food levels for insectivorous birds. High levels in the north could make foraging differentiation irrelevant to the issue of coexistence.

Similarities of warblers in northern coniferous forests are much more striking than their differences. Bird species "added" to the northern forests, compared to southern communities, are very generalized in their foraging, as exemplified by the yellow-rumped warbler (Fig. 8-5b). If competition were controlling the composition of this community, one might expect a "packing" of species into narrower niches than in depauperate communities; instead, generalists are "stacked" in the north. More extensive behavior sampling than MacArthur's shows that the foraging overlap between warbler species is extreme, far greater than that among the European parids (Rabenold, 1978). Using patterns for the two species illustrated here, the black-throated green warbler's foraging niche is totally included in that of the yellow-rumped warbler's.

Figure 8–5. (a) The two-dimensional distribution of foraging of black-throated green warblers in North Carolina (NC) and Maine (ME). Percent foraging is shown in each cell. Axes are the same as in Fig. 8-4. Diversity values (H) were calculated using the Shannon-Wiener Index (from Rabenold, 1978). (b) Distribution of foraging of yellow-rumped warblers in Maine, for comparison with the distribution of foraging of black-throated green warblers in Maine given in part (a) (from Rabenold, 1978).

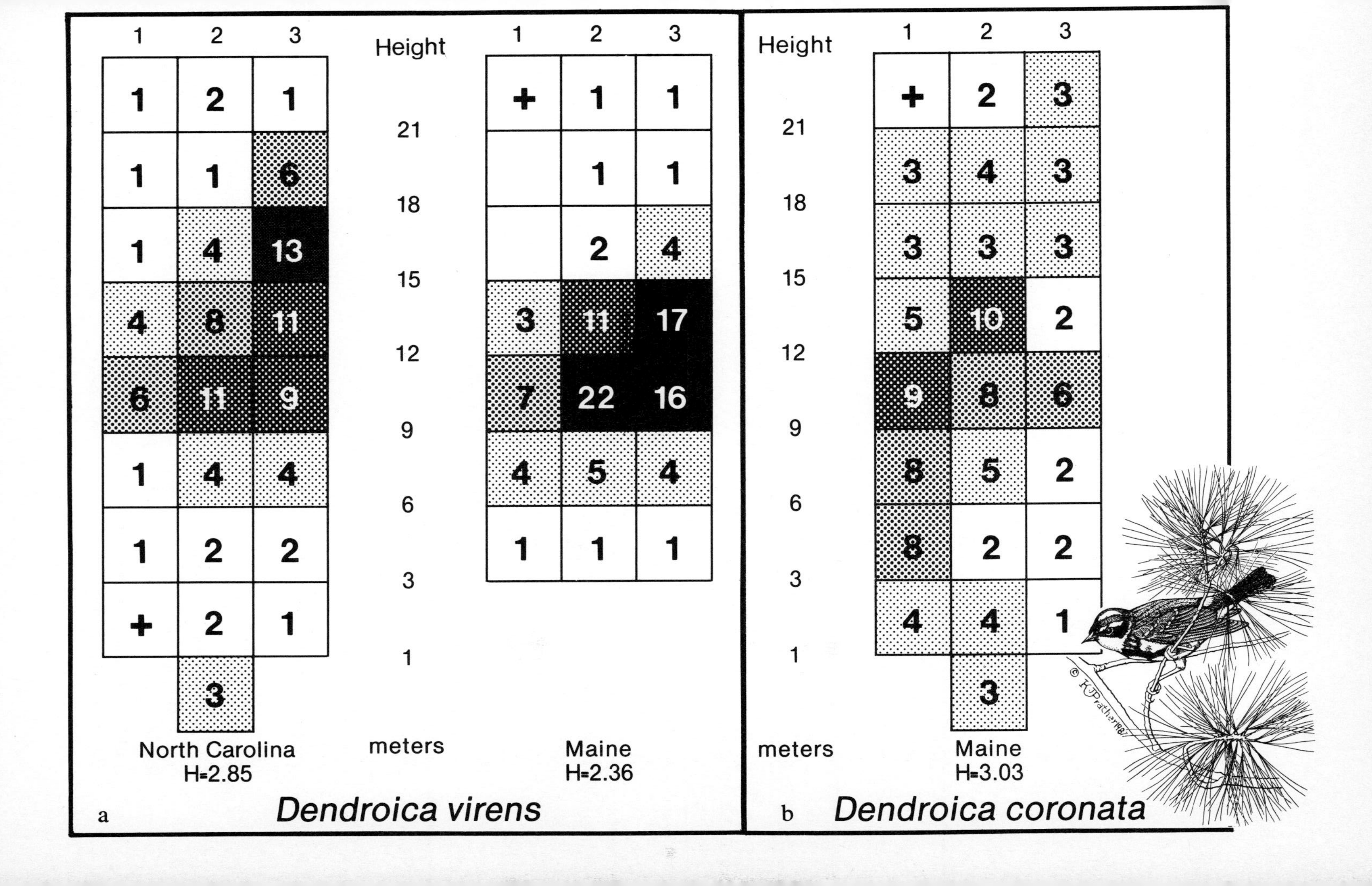

a
Dendroica virens
North Carolina
H=2.85
1 2 3
1 2 1
1 1 6
1 4 13
4 8 11
6 11 9
1 4 4
1 2 2
+ 2 1
3
Height
21
18
15
12
9
6
3
1
meters
Maine
H=2.36
1 2 3
+ 1 1
1 1
2 4
3 11 17
7 22 16
4 5 4
1 1 1
b
Dendroica coronata
Height
21
18
15
12
9
6
3
1
meters
Maine
H=3.03
1 2 3
+ 2 3
3 4 3
3 3 3
5 10 2
9 8 6
8 5 2
8 2 2
4 4 1
3
© KJPrather1987

It seems likely that the stronger resource oscillations produced by the more seasonal productivity of the northern forests have created annual cycles of relaxation of limits to similarity among birds. Competition is probably further reduced by the fact that northern species must migrate long distances to escape the winter and their numbers are probably greatly reduced by the rigors of the trip and limiting conditions on wintering grounds (Keast and Morton, 1980). In southern coniferous forests, the large local populations have an easily accessible refuge in the downslope habitats from which they can get a head start on the latitudinal migrants. Patterns of diversity and coexistence of birds inhabiting deciduous forests in the eastern United States parallel those for the coniferous forests (Rabenold, 1979). In comparison to southern forests, the proliferation of similar bird species in the northern forests is probably due mainly to pulses of productivity by trees in an intensely seasonal environment.

Mechanisms of Habitat Selection

So far, this chapter has examined the ways in which specific adaptations and biotic interactions can determine the distribution of animals within and among plant communities. Animals are generally equipped to efficiently exploit only a subset of plant-generated resources to which they are exposed, and competition can drive animals with inevitably different capabilities to specialize on those resources to which they are best adapted. This view of finite ecological capabilities and limited resources is fundamental to evolutionary ecology and has led to the expectation that animals will generally have very efficient means of assuring settlement in a habitat conducive to survival and reproduction. For plants and sessile animals, habitat selection is mainly passive: Young have limited powers of dispersal and are simply culled if they land in places where their design does not work. Most animals, however, are mobile enough to potentially traverse different habitats in their lifetimes, and many demonstrate distinctly nonrandom choice of environments.

"Hard-wired" Mechanisms of Habitat Choice

Early studies of habitat selection emphasized its "automatic" or invariant nature. In a broad survey of the distribution of British birds among forest types, Lack and Venables (1939) discovered that many species are quite flexible and a few are common in all forest types. However, one-quarter of the species occurred only in deciduous woods, not in pines, and 10 percent lived only in pine forests. Some of these species, like the crossbills mentioned earlier, had obvious adaptations for exploiting a particular kind of forest that limited their range. Some had particular preferences for nest sites, like tree holes that restricted them to forests with trees large enough to have cavities. However, Lack and Venables concluded that habitats are often "instinctively selected" because of their similarity to the "ancestral habitat" of the species. Such preferences could have evolved for a variety of reasons, such as selection pressures created by differential efficiencies or by competitors or predators.

Disadvantages of inappropriate choice of habitats can be sufficiently powerful and immediate to preclude any "shopping" for a workable habitat. Animals that depend on camouflage for their survival may not have the luxury of correcting their mistakes. A grasshopper that resembles a dead leaf may be safe only on a plant with similar dead leaves. A mantis resembling a particular flower will have the greatest success ambushing its prey on a plant normally supporting such flowers. Insects are the masters of camouflage, presumably in response to their sharp-eyed avian predators (Wickler, 1968). Marine polychaete worms in their dispersive larval stage settle preferentially on species of algae to which they can best adhere if the worms are from populations inhabiting turbulent waters (MacKay and Doyle, 1978).

In a classic study of the survival value of substrate matching, Kettlewell (1955a, 1955b) showed that peppered moths actively choose parts of tree trunks where they are least conspicuous to birds. Kettlewell pieced the story together with a prototypical series of experiments: (1) moths actively choose matching backgrounds (pale moths that resemble lichens maneuver onto lichen-covered substrate instead of a dark background when given a choice); (2) birds prey selectively on moths that make mistakes (dark moths on light backgrounds or light on dark are taken first); and (3) marked moths that match the prevailing substrate in the habitat (melanics in sooty woods) survive better than nonmatching moths.

Learning probably plays little role in habitat selection by cryptic insects. When such behavior has a predictably strong effect on survival, and first-time performance is critical, natural selection will operate to produce automatic, instinctive choice. When conditions determining survival and reproduction are less predictable, or the penalty of an unusual choice of habitat less severe, animals can rely more on individual experience and learning (Mayr, 1974). Behavioral ecologists have sought to determine the relative roles of instinct and learning in habitat selection, and found that the contributions vary among animals.

The deer mouse (*Peromyscus maniculatus*) includes races that occupy either forest or prairie in the same region. When individuals of the prairie subspecies are removed from the wild and experimentally provided a choice between forest and prairie vegetation in an enclosure spanning the two habitats, they clearly spend most of their time in the prairie (Wecker, 1963, 1964). Offspring of the field-caught mice similarly preferred the prairie habitat, regardless of whether they were reared in forest vegetation or in the laboratory with no experience with either habitat type. This demonstrates an innate preference, one that is independent of experience. Laboratory lines of this subspecies, kept from natural surroundings for many generations, lost some of this hereditary predisposition to prairie. Laboratory mice reared in the lab or in woods showed no preference, but those reared in prairie showed a later preference more like that of wild mice. In this case, early experience could reinforce a weakened predisposition for a habitat type, but not reverse it. Meadow voles (*Microtus pennsylvanicus*) similarly prefer their usual habitat even when reared in a woodland environment. They ventured into forest vegetation within enclosures only when population densities were high, and their occupation of forest on islands can be attributed to this density response and to absence of forest competitors (Grant, 1971).

Studies with other animals have similarly demonstrated instinctive preference for their normal habitats. Wild-caught and lab-reared chipping sparrows prefer to spend their time near pine needles rather than oak leaves, regardless of early experience (Klopfer, 1963; Klopfer and Hailman, 1965). Birds of the genus *Parus* similarly prefer oak or pine foliage, depending on the normal habitat of the species, even when experimentally deprived of any relevant early experience (Partridge, 1974, 1978). Honeybee workers and queen wasps predictably choose among habitat types in selecting colony sites (Wilson and Hunt, 1966; Seeley, 1977). Some aphids choose cottonwood leaves on the basis of their chemical content (Zucker, 1982); this is undoubtedly an instinctive, "hard-wired" response.

Flexible Mechanisms of Habitat Choice

Development of habitat preferences must be more flexible for animals that regularly inhabit different plant communities. Migratory birds must exercise their choice of habitats over a wide range each year. Dark-eyed juncos (*Junco hyemalis*) in captivity choose projected pictures of their northern breeding habitat when subjected to day lengths appropriate to the breeding season, but choose southern winter habitats when kept in winter day length (Roberts and Weigel, 1984).

In a displacement experiment, Ralph and Mewaldt (1975) moved 800 migratory sparrows away from the site where they first settled in winter. Adult birds did not return to these new locations in the following year; they had apparently established a preference for their previous locations. However, young birds, for which it was the first winter, did return to the new site as long as they were displaced in early winter. Young birds displayed a critical period for developing a site attachment, after which an attachment seemed immutable. Learning that is preprogrammed for a particular context, to occur rapidly and irrevocably during a particular limited time in an animal's life, is called "imprinting." Habitat selection in a variety of animals appears to develop in this way (Thorpe, 1945; Immelmann, 1975). Migratory salmon that spend most of their lives in the ocean far from the streams where they were hatched can return to those streams to spawn because of "imprinting" on the chemical makeup of the home stream's water (Scholz et al., 1976).

Some experiments suggest that specific features of the habitat used in recognition can be quite simple. Tadpoles of some frog species can be affected by early experience to prefer only certain visual patterns. For the red-legged frog, striped patterns are effective in developing a preference, but square-patterned backgrounds are not; the reverse is true for the cascade frog (Wiens, 1972). Both species failed to show any preference when reared in a featureless environment; neither had an innate preference. Red-legged frogs normally live in ponds where shadows of linear branches and plant stems in the water mean relative safety. Cascade frog tadpoles live in lakes where protective vegetation makes more patchy visual patterns on the pond floor. Predators apparently limit the tadpoles to a particular part of the habitat, and the tadpole's learning capacities have evolved so that preference formation can occur only for safe locations near vegetation.

Hooded warblers breed in summer in eastern North America and spend the winter in Central America. In Mexico, males and females both defend territories, but in very different kinds of vegetation. Males are commonest in closed-canopy forests with tall trees, while females tend to inhabit brushier woodlands and old fields (Lynch et al., 1985). In laboratory tests, females preferred artificial visual environments with dense, intersecting horizontal and vertical lines; males preferred visual environments with bold, spaced vertical lines (E. Morton, personal communication). These artificial environments probably present the same basic visual difference as that between shrubby and mature-forest environments, respectively. Generally, little is known of the perceptual or neutral mechanisms underlying habitat selection, but it is clear that it can be an automatic or easily conditioned response, independent of experience or requiring little sampling, based on simple sensory cues.

Strong preferences of animals for particular kinds of vegetation, whether developed by early experience or not, can have broad ecological and evolutionary implications. Gray dogwood grows in open field margins and roadside and its fruits are taken mainly by open-habitat birds like catbirds and thrashers. In contrast, the deep-woods flowering dogwood is visited mainly by forest-dwelling thrushes. Even these flexible migratory birds tend to frequent a particular type of plant community, and the dogwoods are therefore more likely to have their seeds remain in an appropriate habitat as a result (Bromer, unpublished data). The fact that animals have strong inborn preferences, and tendencies to form preferences, argues that powerful selective pressures favor occupation of plant communities matched to animals' resource-exploitation skills.

Habitat Selection Theory

With the vast number of studies of animals inhabiting various plant assemblages, the theory of habitat selection ought to be broadly based and extensively tested with field and laboratory studies. However, the few habitat selection theories are based on optimal foraging theory (Rosenzweig, 1985) and are limited in their applications to empirical data. Habitat selection involves individual variability in behavior, but foraging optimization models cannot accommodate such variation (MacArthur and Pianka, 1966; Maynard Smith, 1974; Stephens and Krebs, 1986). Flexibility in habitat selection is likely to be adaptive in environments where the relative profitability of habitats can change, or when animals' movements cannot carry them to all habitats so that options are limited in an unpredictable way (Levins, 1968).

A habitat can be considered a large patch in which the resources are negligibly or very slowly depleted, but the models of foraging in patchy environments that incorporate the "marginal-value theorem" of Charnov (1976) assume that patches are relatively small and resources quickly exhausted. The marginal-value theorem predicts how long an animal should remain in a patch to maximize the average rate of energy intake given the travel cost between patches and the net energy gain in each patch type (see also Krebs and McCleery, 1986; Stephens and Krebs, 1986). When habitats are substituted for patches in optimal-foraging models, animals should choose habitats with the highest mean benefit (energy gain or fitness) and they should remain in these habitats, totally avoiding less beneficial habitats (Rosenzweig, 1985; Stephens and Krebs, 1986). This model predicts an all-or-none response where all animals will settle in the habitat providing the highest mean gain in energy or fitness, assuming that they can measure rates of gain and that the rates are constant through time. Any change in the benefit ranking of habitats should lead to shifts in habitat selection. This "strawman" model will be falsified by simply showing that some animals occur in suboptimal habitats (Whitham, 1978; Godin and Keenleyside, 1984).

Intraspecific Density Model

The "ideal-free distribution" model of Fretwell and Lucas (1970) and Fretwell (1972) predicts what habitats animals will choose as density increases. Fretwell and Lucas assume that animals are "ideal," meaning that they select the habitat that provides the best chances for survival and reproduction. They also assume that animals are "free" to select habitats so that all individuals will have the same fitness within a habitat regardless of their order of arrival. The model considers several habitats, and in each one, mean fitness declines with increasing population density (Fig. 8-6a). The model makes three predictions:

1. At low densities, only the habitat providing the highest fitness will be used.
2. Animals will choose habitats other than the most-favored only when increasing numbers of individuals in the favored habitat lower the fitness gain to that of the originally less-favored habitats.
3. At high densities, a number of habitats will be occupied at different densities but all yield the same mean individual fitness.

Fretwell and Lucas (1970) also proposed, for comparison, an "ideal-despotic distribution" for territorial animals based on Huxley's (1934) model, where individual fitness is variable both within and between habitats. Territorial behavior reduces the fitness of newcomers in the favored habitat to less than the average of the habitat, resulting in unequal fitness among individuals within a habitat. Furthermore, average fitness is predicted to be greater in the favored habitat.

Whitham (1978, 1980) tested Fretwell's (1972) "ideal-free distribution" model using leaf-galling aphids that colonize the leaves of narrow-leaf cottonwood. The aphid, during the wingless stage of its complex life cycle, forms galls along the midrib of the leaves from which the aphid parthenogenetically produces winged offspring. Cottonwood leaves are the aphid's habitat. Aphids that formed galls on large leaves had higher survivorship and fecundity than those on smaller leaves; large leaves were the preferred habitat. All large leaves were colonized by aphids so that the mean colonized leaf size was 60 percent larger than the mean leaf size of the tree. Many aphids had to choose between colonizing an already occupied large leaf or settling for an open, smaller leaf because of high aphid densities (21.8 aphids per large leaf).

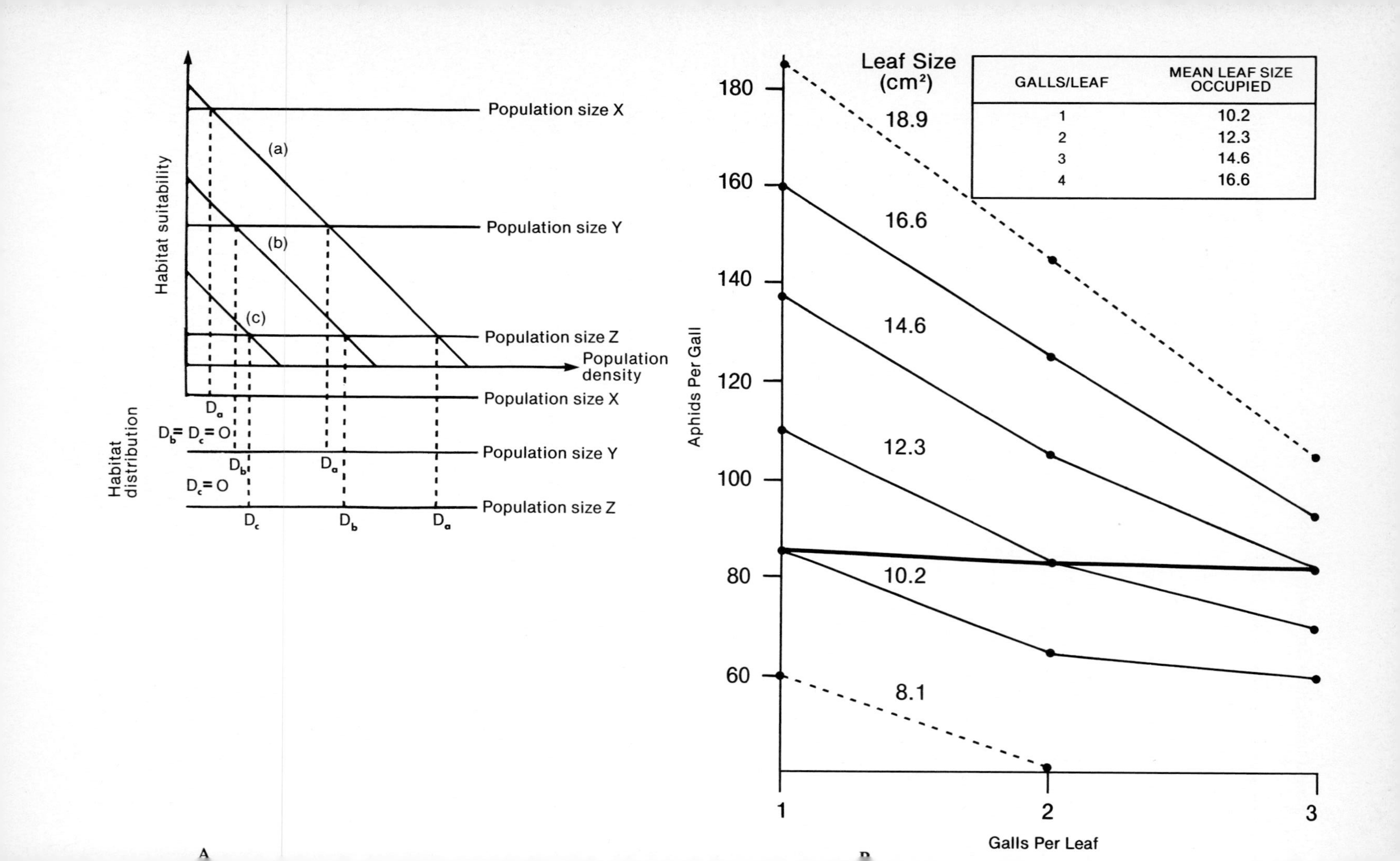

GALLS/LEAF	MEAN LEAF SIZE OCCUPIED
1	10.2
2	12.3
3	14.6
4	16.6

Whitham (1980) found that average fitness (measured as the average number of winged aphids per gall) declined with increasing gall density on a leaf, and the aphids adjusted their densities on varying leaf sizes such that the average fitness was not different for one, two, or three galls per leaf (Fig. 8-6b). Aphids appear to have an ideal, free distribution on cottonwood trees; predictions 2 and 3 are upheld.

With further investigation, Whitham found considerable variation in aphid fitness on any multicolonized leaf. The aphids nearest the base of the leaf produced significantly more offspring (mean = 143.2) than those farther from the base (94.8 and 75.4). This variability violates the "free" assumption of homogeneity within a habitat; subsequent colonizers have a lower fitness than the early colonizers on a leaf. Whitham (1980) attributed the agreement of this aphid data with the ideal, free-distribution model in a heterogeneous (not free) habitat to the territoriality of the aphids. The aphids vigorously defend territories with boundaries determined by kicking and shoving contests (Whitham, 1980), but the inequality of fitness between habitats predicted by the ideal, despotic distribution is not supported by Whitham's (1978, 1980) data. Cichlid fish (Godin and Keenleyside, 1984) and mallard ducks (Harper, 1982) have also been shown to follow the ideal, free distribution. Counter examples or tests that falsify the ideal, free or despotic distributions are difficult to find because the model is usually modified as Whitham (1980) did to account for differences between the species studied and the model assumptions. Some birds behave as Brown (1969) predicted: They simply choose the best habitat available, filling the best habitat first and the less-favorable habitat second.

Figure 8–6. (A) The effect of population density on habitat suitability and hence on habitat distribution. Three habitats are shown, a-with highest suitability, b-with intermediate suitability, and c-with lowest suitability. Horizontal lines representing three total population sizes $X < Y < Z$, are drawn, each intersecting the suitability/density lines at equal suitabilities. Every animal in the population has an equally suitable home area, low habitat suitability being compensated for by an increase in the area per animal. At the three population sizes considered, the density of animals in the three habitats are shown on the habitat distribution axes, below the graph. D_a= density in habitat a, D_b= density in habitat b, and D_c= density in habitat c. Modified after Fretwell and Lucas (1970) (from Partridge, 1978). (B) Family of fitness curves for habitats of varying quality (leaf size) and competitor density (galls per leaf). Points connected by solid narrow lines represent the expected fitness that would be achieved on the mean leaf sizes occupied by the various gall densities per leaf. Points connected by dashed lines represent expected fitnesses that would be achieved on the largest available leaves (18.9 cm^2) and the smallest frequently colonized leaves (8.1 cm^2). The number of aphids per gall not only increases with increased leaf size, but also decreases with increased competitor density. Points connected by the heavy horizontal line are unique because they represent the average fitnesses for one, two, and three stem mothers per leaf (from Whitham, 1980).

Interspecific Density Model

Rosenzweig (1981) developed a graphic model of habitat selection based on some of the ideas of optimal foraging in patches and density-dependence from the ideal, free distribution, but he included the effect of interspecific competition. The simplest form of the model has two species and two habitats. The model assumes that both species can survive in either habitat, but that the individual fitness of each species is highest in one of the habitats. Models incorporating exploitative (Rosenzweig, 1981) and interference competition have been proposed (Pimm and Rosenzweig, 1981), but the latter has been tested (Pimm et al., 1985) so it will be examined here. In this model, both species prefer the same rich habitat but one is dominant over the other when feeding. The model (Fig. 8-7a) plots the density of the dominant and subordinate species along the x and y axes, respectively, with the space dissected by lines of equal habitat choice (isolegs; Rosenzweig, 1981) for each species. Isoleg intercepts and slopes are based on expected fitness values for the species in question depending on the density of the other species and the costs of habitat selection. Given the densities of the dominant and subordinate species, the model predicts what habitat will be chosen by a species depending upon the side of the isoleg on which the point falls.

In this version of Rosenzweig's (1981) habitat selection model, there are two isolegs for the subordinate species (heavy dark lines in Fig. 8-7a). To the left of I_{S1}, the subordinate and dominant species densities are low enough to allow the subordinate to inhabit the rich habitat; to the right of I_{S1}, densities are high enough that the subordinate will use both habitats. To the right of I_{S2}, the dominant density is high enough to completely limit the subordinate to the poor habitat. The isoleg for the dominant species, I_D, is primarily a function of intraspecific density with the dominant choosing the rich habitat to the left of I_D and both habitats on the right side of the isoleg.

Pimm et al. (1985) tested this model with hummingbirds in Arizona feeding at feeders with low (0.35 mol/L) and high (1.2 mol/L) sucrose solutions. Three species were used in the experiment: blue-throated, Rivoli's, and black-chinned hummingbirds. There were no effects between Rivoli's and black-chinneds, but blue-throateds were dominant over both. Densities varied naturally over time with migration and experimentally by mist-netting to remove some hummingbirds. Results of the blue-throated and black-chinned interactions are presented in Fig. 8-7b. At low densities of both species, black-chinneds (subordinate) used primarily the rich habitat, but as densities increased they used both habitats. Once blue-throated density reached about 1000 seconds of

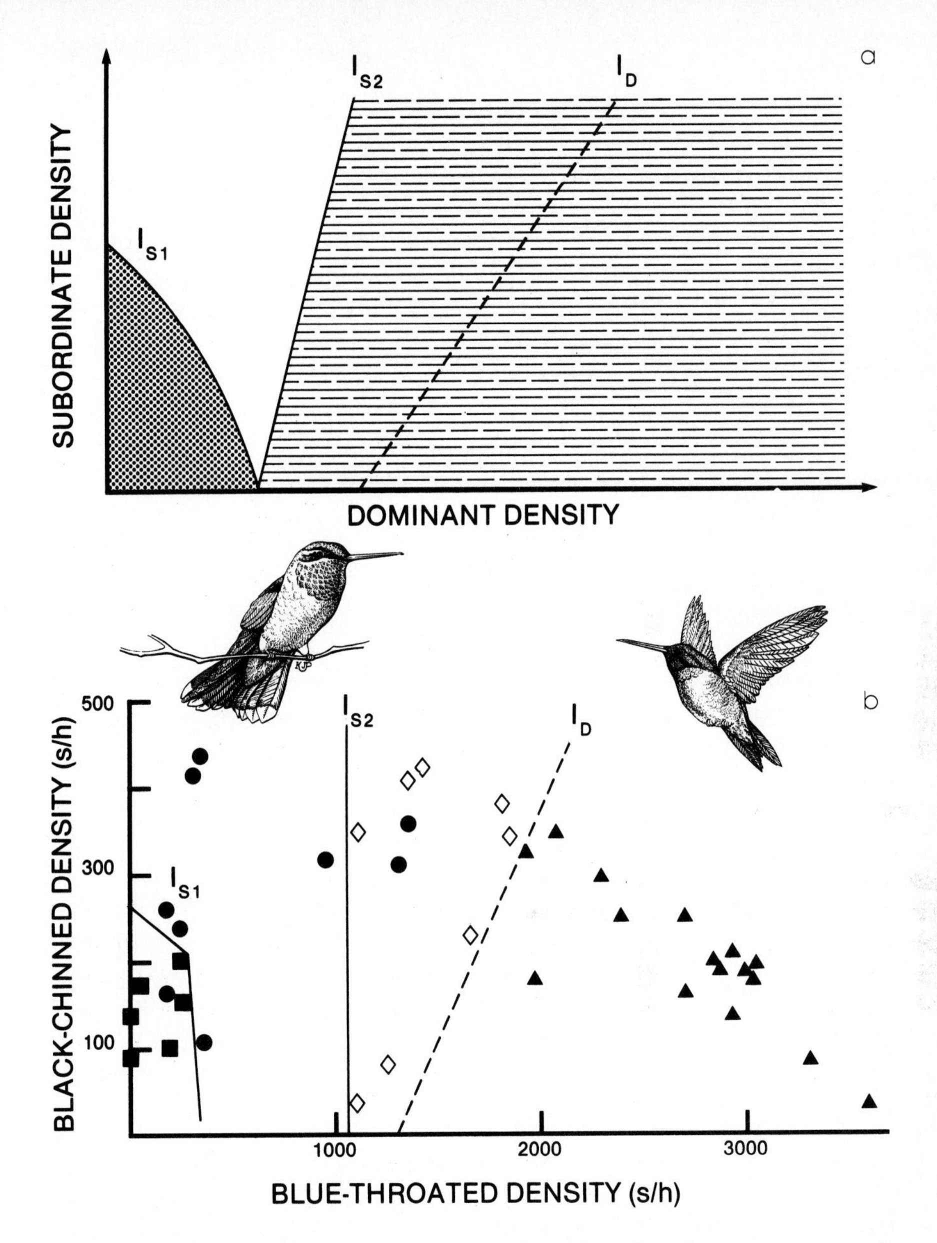

Figure 8–7. (a) Isolegs of a two-patch, cost-free, shared preference system with interference. The subordinate selects the better in the stippled region and the poorer in the shaded region. The subordinate is opportunistic in the clear portion. The dominant may have an isoleg too (dashed line). If so, it selects the better patch to the left of its isoleg. (b) Behavior and isolegs of blue-throated and black-chinned hummingbirds in an experimental system. Code: solid squares, both species select rich; dots: blue-throats select rich, and black-chinneds are opportunistic; diamonds: blue-throats select rich, and black-chinneds select poor; triangles: blue-throats are opportunistic, and black-chinneds select poor (from Rosenzweig, 1985).

territorial or feeding behavior per hour, all black-chinneds were restricted to the poorer feeders. Blue-throateds used only the rich habitats at low and moderate densities, but they started including the poorer feeders once their own density reached high levels. These results indicate that habitat selection can vary with changing balances of interspecific and intraspecific competition. Observational field studies also suggest that high density in a preferred habitat can counteract the fitness gain associated with that habitat to the point that occupation of a less-preferred habitat can be advantageous (Pierotti, 1982).

How Do Characteristics of Plant Resources Affect Animal Population Dynamics?

The habitat preferences and capabilities of many animals may never be fully expressed because of the constraining factors of competition and predation. Interspecific competition has been discussed at some length. This section explores the ways in which productivity and reliability of plant resources can interact with predatory and competitive forces among animals to determine patterns of dispersion, reproduction, and movement in animal populations (Southwood, 1977).

Does Competition Distribute Conspecifics among Habitats?

John Krebs (1971) showed that young birds in English populations of great tits *(Parus major)* are often relegated to inferior hedgerow habitat while older birds enjoy greater reproductive success in mature oak forest. When Krebs removed territorial pairs from the forest, they were quickly replaced by birds from the surrounding hedgerows (Fig. 8-8). Vacancies created in the hedgerows were not filled; competition apparently did not exist for these breeding spots. The spacing effect of territoriality—the active defense of an exclusive area—was limiting the density of breeding birds in the best habitats. However, Krebs argued that territoriality functions more to parcel up prime habitat than to regulate the density of the population. Territories were much smaller in

Figure 8–8. An illustration of the degree of competition for woodland territories in great tits *(Parus major)* of Wytham Woods, England. Woodland habitat occurs inside the solid line; outside is suboptimal hedgerow habitat. A paired hedgerow male (o) had been trying to establish a territory at point x but was repeatedly chased off by male m. When male p dissappeared, male m moved his territory slightly toward the empty territory of p (shown by arrows), leaving a small gap at x. Within a few hours, male o had established a territory at x; he abandoned his hedgerow territory and subsequently bred in the wood. Later, another hedgerow male (u) established a territory within the site formerly occupied by p.

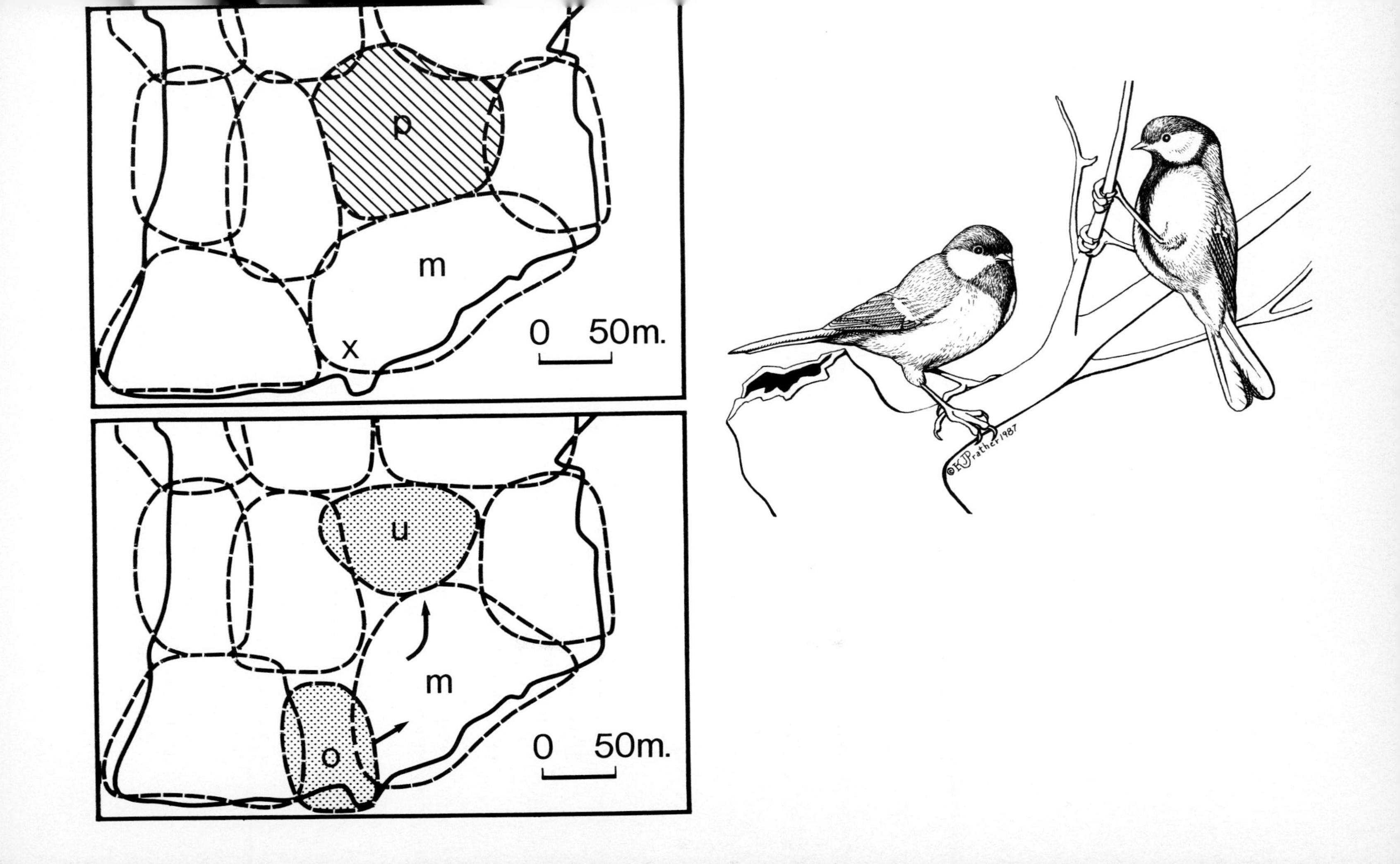
p
m
x
0 50m.
u
m
o
0 50m.
©KJPrather1987

years of good overwinter survival, suggesting that overall density determines territory size more than vice versa. So long as the plant communities intergrade in a gradual way, high population densities can be accommodated by occupation of suboptimal habitat by competitively subordinate animals.

When discontinuities are sharper between plant communities, competition for territories in good habitat can have more drastic implications for individual survival and reproduction. Great tits in hedgerows still managed to produce some young, but studies of red grouse on Scottish moorlands show that young males excluded from prime habitat have a low probability of overwinter survival or of breeding if they do survive (Watson and Miller, 1971). Over a 5-year period, all moorland ground that was covered predominantly with heather was occupied by territorial grouse. Heather, a member of the heath family (Ericaceae) related to the laurels and rhododendrons, is both food and shelter for the grouse. Young grouse that failed to establish territories in heather nearly all died over winter, and vacancies created by removing territory holders were quickly filled. More aggressive males could defend larger territories with more heather and sometimes attracted two mates; less aggressive males with small territories often failed to attract mates at all. Males that died over one winter averaged half the territory size of those that survived.

Watson and Moss (1970, 1972) and Watson (1977) have argued that territorial usurpation of good habitat by aggressive individuals was an effective force in limiting population density because many individuals were evicted from essential resources and prevented from breeding (or surviving the winter). The dependence of these grouse on a single plant population, and the discreteness of the plant resource, undoubtedly contribute to the rewards of aggression and territorial defense and their role in regulation of the breeding population.

Intraspecific competition and dominance behavior probably influence the distribution of individuals in animal populations among plant communities in other vertebrates as well (Brown, 1969; Fretwell, 1969; Howe, 1974; Rohwer, 1977; Gauthreaux, 1978; Spencer and Cameron, 1983; Alatalo et al., 1985a, 1985b; S. Robinson, 1986a). Dominance of male migratory birds over females could result in forcing the females to move farther than the males and occupy very different habitats at different latitudes or altitudes (Rabenold and Rabenold, 1985). In some species, it is proposed that risk of predation is the immediate disadvantage of being excluded from prime habitat, as suggested by the following statement from Errington's (1967) study of mink predation on muskrat: "Muskrats living in suitable habitat and within the limits of crowding that they themselves found tolerable were not much preyed upon by minks." Subordinate muskrats, however, forced to disperse from safe habitat, were in trouble.

Effect of Other Biotic Interactions on Habitat Use

Predation, parasitism, and mutualism are clearly powerful forces limiting the distribution of many animal populations to particular habitats. Predatory crabs can limit mussels to particular habitats (Kitching and Ebling, 1967); dingoes (canine predators in Australia) can greatly depress kangaroo populations (Caughly et al., 1980); lampreys virtually eliminated trout from the Great Lakes (Christie, 1974); mammalian predators can limit nesting birds to islands (S. Robinson, 1985); avian predators can determine the frequency of color morphs in moth populations (Kettlewell, 1955a); and predatory salamanders can have strong effects on the structure of frog communities (Wilbur et al., 1983).

Parasites have perhaps been underestimated in this regard. Malarial parasites have influenced the geographic distribution of monkey species, the rinderpest virus has greatly affected the distribution of hoofed mammals in Africa, and the flagellate protozoans that cause sleeping sickness have greatly limited the ability of humans to colonize areas of Africa (May, 1983; Dobson and May, 1986; Price et al., 1986). Mutualism may have been even more neglected in its potential to limit animal populations' occupation of a variety of habitats (Smith, 1980; Thompson, 1982; Futuyma and Slatkin, 1983; Chap. 1, this volume). Given the power of these interactions among animals to determine population densities and dispersions, how can the nature of plant resources affect their outcomes and contribute to the dynamics of animal populations? Consider now the effects that density, variety, dispersion, and reliability of plant resources can have on the outcomes of predatory and intraspecific competitive interactions.

Plants can sometimes mediate predator–prey relationships among animals by providing a refuge for the prey. In some cases it seems possible that refuge plants are as essential for the continued existence in the community of the prey as are food plants. Two sunfish, bluegills and pumpkinseeds, inhabit many of the same North American lakes and ponds. As adults, they show habitat segregation: Bluegills prey on open-water zooplankton while pumpkinseeds specialize on vegetation-dwelling gastropods. The differences are closely related to functional morphology. However, small fish of both species live in vegetation close to shore where they take very similar prey (up to 70 percent similarity in diet). Predation pressure by piscivorous fish like largemouth bass apparently drives the small bluegills to the refuge of the shallow-water vegetation, where they may experience less-efficient foraging (Mittelbach, 1984). Many animals modify their foraging behavior to take advantage of the cover of vegetation and in doing so compromise their intake efficiency to reduce risk (Lima, 1985; Real and Caraco, 1986; Stephens and Krebs, 1986).

Waterbug populations are confined to low densities in dense vegetation in a British pond when fish are present, but are abundant and widespread when fish are absent. The waterbugs would probably disappear altogether if the dense vegetation were removed (Macan, 1976). Vegetation in marine systems also functions to provide safety to invertebrates from predaceous crabs and fish (Stoner, 1979; Leber, 1985). Some populations of crows show clear preferences for nesting in pine or spruce trees rather than deciduous trees. Losses of eggs and nestlings are lower in the conifers, probably because they offer greater concealment (Loman, 1979). Snowshoe hares prefer forested areas with dense undercover that provides protection, but when populations are dense, many young animals are found in open, high-risk areas. It has been suggested that the relatively constant density of some populations results from a habitat where patches of dense and open vegetation are intermingled. In more completely open habitats, hares are famous for population cycles; the refuge provided by forest patches seems to stabilize the hare population, preventing crashes of both prey and predator (Dolbeer and Clark, 1975). However, it is theoretically possible that predator refuges could actually destabilize predator–prey relationships. Refuges that do not protect a constant number of prey, or only protect certain age-classes (as in the sunfish) could destabilize a prey population (McNair, 1986).

Importance of the Level and Timing of Primary Productivity

Animal populations depend on the productivity of the plant communities they inhabit; this is no surprise and it is especially obvious when a population depends directly on a single plant resource. Red grouse population levels depend not only on the density of heather but its nutrient content, which in turn depends on soil type (Watson and Moss, 1972). Woodrat populations that similarly rely on one species of cactus for food and shelter also vary with cactus productivity (Brown et al., 1972). Acorn woodpeckers and jays that rely on oaks for their livelihood in the southwestern United States are conspicuously affected by the productivity of these trees, and their range limitations are coincident with the distribution of the oaks (Bock and Bock, 1974; Trail, 1980; Edwards, 1986). It is often less obvious how the seasonality or regularity of production by plants can affect animal populations. It was argued earlier that the severity of high-latitude winters and the compression of primary productivity into short growing seasons could sufficiently limit animal survival so that potential competition among animals in the next productive season could be eliminated. Even in tropical systems, there is growing evidence that seasonality of plant productivity, often created by wet and dry seasons, can produce severe annual bottlenecks. These lean seasons could not only limit animal densities, but perhaps preclude certain modes of existence for animals (Smythe, 1986; Terborgh, 1986a).

Plants often introduce their own periodicity to production of biomass used as food by animals. Masting is the synchronous production of fruit in large quantities by a population of plants at longer intervals than dictated by climate; it has profound effects on many characteristics of animal populations, compared to more constant and predictable flowering and fruiting. Perhaps the champions in this category are the bamboos: Many populations flower and fruit in unison at intervals in excess of a century, then die back to be replaced en masse by the next generation (Janzen, 1976). Often the masting interval is longer than the life-span of animals that use the seed as food, so that interflowering intervals are famines, and strong cycles of population density are created, especially in small mammals. More mobile animals can establish very large ranges to encompass plant populations with different masting schedules, like pandas and jungle fowl using bamboo, or wild turkeys using North American beech and oak mast. Jays and nutcrackers using pinyon pine seeds use the strategy of caching seeds below ground in times of plenty, a habit that apparently produces a dispersal benefit for the tree sufficient to encourage cooperation in the form of a nutritious, undefended seed compared to other pines (Balda, 1980).

Tropical forests differ in many ways from temperate-zone and boreal forests but, perhaps most importantly for animal communities, they supply great quantities and varieties of nutritious fleshy fruit. In neotropical lowland humid forests, like those of the Amazon basin, more than half of the tree species produce fruits with edible flesh that appear designed to attract animal dispersers. The fruits are often conspicuously colored and displayed, and often have nutritious pulp (Snow, 1981; Howe and Smallwood, 1982; Baker et al., 1983). Frugivorous birds and primates have probably been best studied for the implications of a fruit-rich diet, and a distinction must be made between fruits relied upon by specialized frugivores and those used by more opportunistic omnivores. The latter are generally small with watery carbohydrate-rich flesh and many small seeds. In the temperate zone, elderberries, raspberries, and grapes are good examples. Fruits adapted for dispersal by full-time frugivores are generally larger, with few large seeds and flesh with high protein and fat content, and they tend to be produced by mature-forest trees. Few examples exist in temperate forests, although flowering dogwoods display most of these attributes (Bromer, unpublished). In tropical trees, fruits can consist of as much as 25 percent protein or 67 percent fat, (Snow, 1981; Herrera, 1985b). There are no exclusively frugivorous birds in the north-temperate region. Even in tropical forests, few species specialize entirely on frugivory; the most speciose family of frugivores is the Cotingidae with fewer than 100 species, while families of insectivores like the tyrant flycatchers have four times that many species. Although frugivore populations can be dense, contributing in some tropical forests more than half of all avian biomass (Terborgh, 1986a), their life histories demonstrate some of the limitations of such a life-style.

Frugivorous birds are perhaps more common in the tropics both because tropical trees have evolved more lucrative enticements to frugivory and because temperate-zone birds have access to a seasonally abundant pulse of high-protein arthropod food (Morton, 1973; Rabenold, 1979). Tropical fruits may be nutritious as fruits go, but they impose important constraints on frugivores. Most frugivores eat a variety of fruit even when single kinds exist in abundance, presumably to compensate for the nutritional imbalance of the fruit of any one plant (Herrera, 1985b). Quetzals in the mountains of Panama must migrate among several habitats in the course of the year to find their preferred fruit of the family Lauraceae (Wheelwright, 1983). Many biologists working in tropical forests have stressed the erratic distribution of fruit in time and space in these habitats and the demands put on the biological rhythms of frugivores (Snow, 1971; Karr, 1976b; Fleming, 1979; Herrera, 1985b). It is rare for even the most specialized frugivorous birds, like the South American cotingas, to use fruit when feeding young. Nestlings fed fruit grow slowly compared to those fed animal protein and they suffer high rates of predation, limiting reproductive potential (Morton, 1973).

Eisenberg (1981) has found that use of a spatially dispersed food source, like fruit or nectar, is associated among the mammals with large brain size. In some groups, like the bats, use of a ubiquitous sparse food, such as flying insects, is associated with smaller brain size than that of species using fruit; the latter generally have a better-developed neocortex. The arboreal habit in groups like marsupials is associated with larger brains and a better-developed cerebellum. Eisenberg has identified an "adaptive syndrome" associated with dispersed food that consists of larger brains, longer life, lower litter size, and longer social association with parents. This suggests that use of spatially and temporally scattered resources like fruit requires greater capacities for information storage and retrieval and longer times devoted to learning and development.

Tropical forests seem more stable and predictable throughout the year than temperate forests with their unproductive winters. However, evidence is accumulating that, because of seasonality of fruiting by trees that is often correlated with patterns of rainfall, lean seasons and off years can be serious bottlenecks for frugivores (Smythe, 1986). In Amazonian Peru, fruit is produced mainly in the wet season and is very scarce in the dry months of May to September (Fig. 8-9) (Terborgh, 1983). During the lean dry season, frugivores must either migrate or modify their diets. Five species of primates studied at Cocha Cashu in Manu National Park are almost completely frugivorous during wet-season production peaks. During the dry season, the primates and other frugivores turn to foods of lower quality: hard palm nuts and high-carbohydrate figs and nectar (Terborgh, 1983, 1986a, 1986b). Even the most abundant of these "keystone" resources, figs, can be unpredictable in their availability on a small spatial scale (Janzen, 1979b).

Terrestrial mammalian frugivores, like agoutis and peccaries, might also be limited by the seasonality and unpredictability in space and time of fruiting by neotropical forest trees (Smythe, 1986). Of course, crop failures in temperate trees, for example, oaks and pinyon pines, can have serious effects on animals, like acorn woodpeckers (MacRoberts and MacRoberts, 1976) and pinyon jays (Ligon, 1978), that specialize on them. Dogwoods in eastern deciduous forests produce large crops of fruit only every other year, so that thrushes that would normally feed heavily on them during fall migration must either be flexible in their diet or keep moving until they come upon a stand that is having a productive year (Bromer, unpublished).

Seasonality of productivity by plants has an obvious effect on animal populations in the temperate zone. North American birds that are

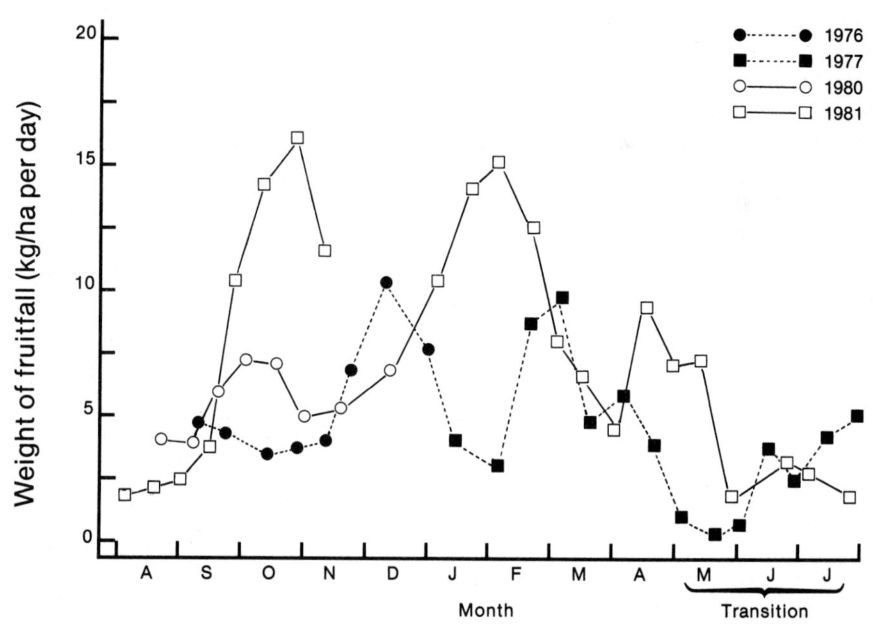

Figure 8–9. Weight of fruitfall in mature forest at Cocha Cashu, Peru in 1976–1977 and in 1980–1981. Measurements are based on biweekly collections from 100 fruit traps. The period from May to July represents the wet to dry season transition, when fruitfall is minimal. (from Terborgh, 1983).

insectivores in summer must eat seeds or migrate in winter; amphibians and reptiles generally retreat underground and estivate; many mammals either hibernate or use stores of plant food cached during better times. These same responses are seen in tropical animals in habitats with harsh dry seasons, like the South American savannas. Even in some of the most equable climates in the world, subtle seasonality of primary plant productivity has striking effects on animal populations.

Fogden (1972) studied one such stable tropical forest in Sarawak, on the island of Borneo, nearly on the equator. Rainfall and temperature there are high all year round. The annual rainfall of 4010 mm at nearby Kuching is more than four times that of Indianapolis, and in every month at least 240 mm fall (2.4 times that of the wettest month in Indianapolis). The range of monthly variation in rainfall between the driest (June to August) and the wettest monsoon months (December to February) is only a twofold increase; this is much less seasonal than in the Peruvian rain forest discussed above, where rainfall in wet months is four times that of dry months (Terborgh, 1983). Temperature and relative humidity are nearly constant year round, especially near the ground in forests where temperature varies only 1 or 2° around 25°C and relative humidity is always near 100 percent. Day length varies less than 10 min/yr. Under these nearly constant conditions, the forest trees still grow with a distinct seasonality of leaf production; the animal populations further magnify that seasonality to the point that birds' breeding seasons are nearly as compressed into the wetter half of the year as north-temperate summer breeding seasons (Fig. 8-10).

Leaf production in the shrub layer of the forest varied nearly 4-fold over the year and nearly 10-fold in the high canopy. Insect densities varied 10-fold over the year for some taxa. Fruiting seasons were not as distinct or predictable as those for insects, but frugivores generally followed the same breeding seasons as insectivores, probably because most of them feed insects to developing young. In general, the birds confined energetically expensive activities, like rearing young and producing new feathers, to the part of the year when the trees' growth was maximal. Some frugivorous species maintained reproductive cycles out of synchrony with the insectivores, depending on the irregular availability of fruit; they were vulnerable to unpredictable periods of fruit scarcity when breeding failed and individuals lost weight. The most specialized frugivores had to range widely to accommodate local failures of fruit crops. Insects, even allowing for their relative scarcity during peak abundances and the seasonal lows, were a more reliable resource; insectivores were more sedentary and had more consistent reproductive success.

Forests dominated by Dipterocarpaceae trees, like those studied by Fogden on Borneo, present a very different regime of resources to

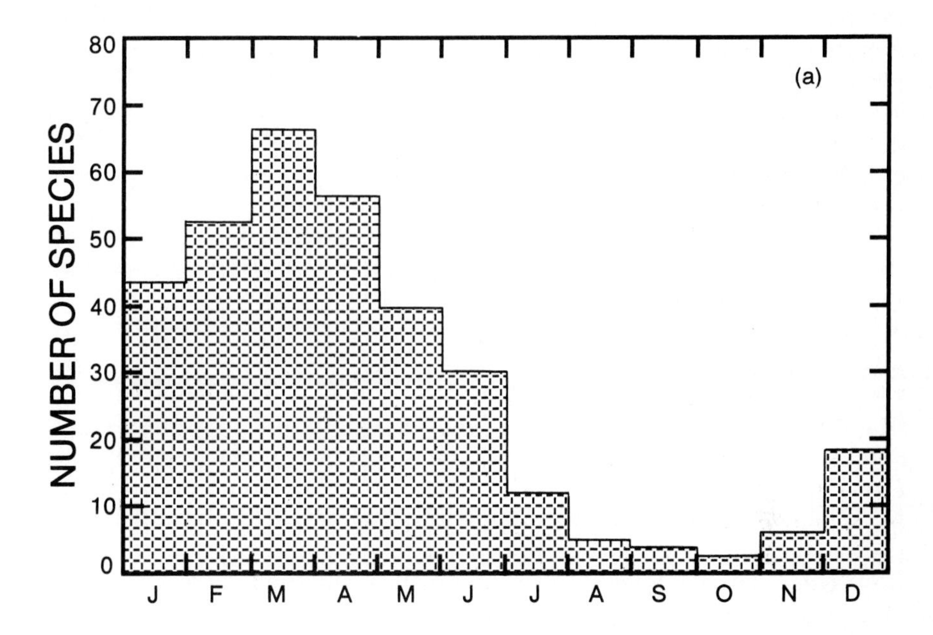

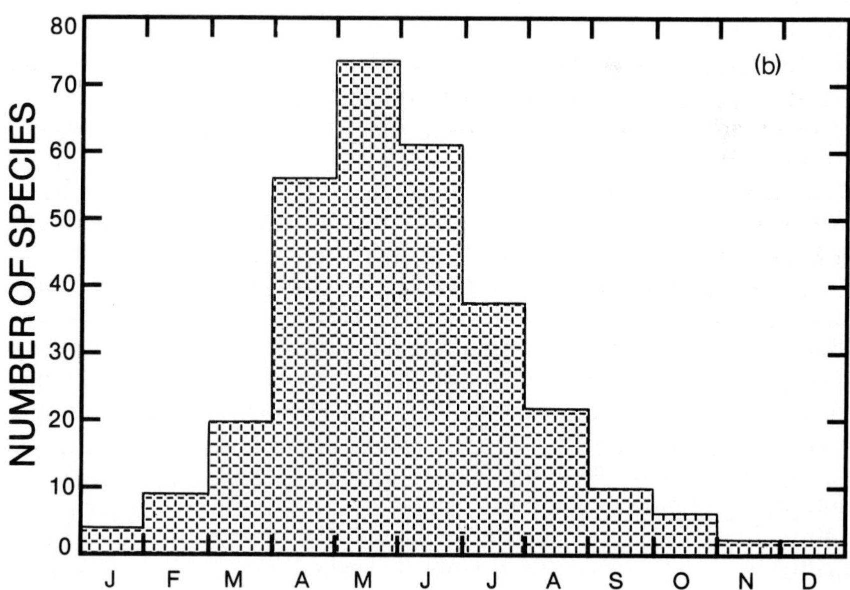

Figure 8–10. The number of passerine species breeding in each month in (a) Sarawak and (b) Britain (from Fogden, 1972).

animals than other lowland tropical forests. Canopies of these forests in Malaysia can consist of 50 to 80 percent dipterocarps, and these trees have well-developed chemical defenses against herbivores and have prodigious cycles of fruit production, with mast crops pulsed at 3- to 11-year intervals. The forests are unproductive as rain forests go, owing to the nutrient-poor sandy soils on which they occur. Masting is interpretable as a mechanism to escape seed predators by starving them for long intervals then swamping the few survivors, but it can also have the effect of making "legitimate" frugivory (that resulting in seed dispersal) equally impossible. It is no accident that seeds of these trees are wind-dispersed (Janzen, 1974a, 1980b). Densities of frugivores are greatly depressed in these forests, undoubtedly because fruit are essentially withheld from them for long intervals. Small mammalian frugivores that are terrestrial and sedentary, like the agoutis and pacas of Central American forests (Smythe, 1986) are absent from Malaysian forests (Janzen, 1980b). Even their scatter-hoarding tactics would probably not be sufficient to carry them through the famines of dipterocarp mast cycles. Large migratory frugivores, like pigeons and pigs, are favored; they can, like the bamboo-using animals mentioned earlier, operate among forests or regions that are unsynchronized.

Animal Movement in Response to Variable Plant Resources

Migratoriness and the distances over which animals range in the course of a year are clearly tied to the seasonality and reliability of food provided by plant communities. Summer primary productivity in the temperate zone is rather predictably controlled by regular changes in temperature produced by the slow rhythmic change of the angle of presentation of the earth's surface to the sun through the orbital cycle. Associated vernal and autumnal bird migrations are correspondingly regular; most often the birds' tendency to move north or south is controlled by simple innate physiological (hormonal) responses to changing day length. However, birds using irregular pulses of production, by masting trees for instance, are different (Ulfstrand, 1963). They have mechanisms like those previously mentioned for crossbills; production of restlessness or breeding condition is tied to the more immediate variable of food availability rather than the indirect index of day length.

Migrants that leave boreal and temperate forests so predictably in autumn may spend the months of the high-latitude winter as itinerants in tropical forests using resources like fruits that are too irregular to be effectively used by tropical residents. Thrushes, like the hermit thrush, that breed in boreal coniferous forest and move through the eastern deciduous forest in autumn eating dogwood berries, grapes, and other fruit, seem to move slowly through Central America exploiting ephemeral fruit crops as they go. Among the roughly 140 species of land birds that move from North America across the Tropic of Cancer and back, it is not known how many are so transient, but this habit is also suggested in some African migrants (Moreau, 1972; Karr, 1980). These very long distance migrants seem to make their living skimming excess productivity from a wide variety of habitats with strongly or regularly pulsing resources. Karr (1976a) has shown that neotropical migrants avoid mature lowland forests whose relative stability might allow residents to effectively track resources.

The itineraries of some migrants carry them to tropical areas just as particular pulses of productivity occur. Some of the more abundant long-distance migratory birds that breed in North America cover enormous areas apparently tracking geographically shifting peaks of fruit production. Eastern kingbirds move quickly through Central America in the fall, on their way to the southern limit of their range south of the equator. From there they follow early dry-season peaks of fruit production northward, even seemingly associating with the fruiting peaks of certain tree species (Morton, 1980). Wildebeest that so dramatically track grassland productivity in their vast migrations across Africa show variation depending on year-to-year changes in rainfall, and they are strikingly sedentary where higher rainfall produces less seasonal productivity, as in the Ngorongoro Crater in Tanzania (Talbot and Talbot, 1963; Dingle, 1980). Hummingbirds arrive in some tropical areas in Mexico from their temperate breeding grounds in time for a temporary abundance of nectar. Resident species, whose numbers are apparently kept down by resource limitation in other seasons, cannot use all of the resource pulse (DesGranges, 1980). Other studies also show that migrants often use ephemeral or irregularly produced resources (Willis, 1966, 1980; Leck, 1972, 1980; Lack, 1976; Pearson, 1980). This is the same scenario of resource peaks permitting the temporary coexistence of migrants and residents presented earlier for migrants in their north-temperate breeding areas (Rabenold, 1978, 1979).

Short-distance migrations, on a regional or altitudinal scale, can be equally responsive to between-habitat variation in productivity peaks. Migration between habitats by tropical animals, even in the "stable" lowland rain forest, is becoming increasingly well known (e.g., Pearson, 1980; Terborgh, 1986a). Hummingbirds migrate altitudinally between habitats while tracking the flowering seasons of a variety of plants. The complex mosaic of vegetation on tropical mountains supports a variety of movement strategies and schedules among hummingbirds. Feinsinger (1976, 1980) studied 14 species in the mountains of Costa Rica and found that during lean seasons of flower scarcity, one or a few socially dominant species can exclude others from a particular habitat, but a richer variety of hummingbirds is accommodated during peaks of flowering. Between flowering peaks of the two major plant species, hummingbird density and diversity was reduced, as some species left for other habitats. Feinsinger (1980) has suggested that "asynchronous migration patterns are instrumental in permitting the regional coexistence of large numbers of short-billed hummingbird species. It follows that regions lacking such habitat mosaics must suffer reduced hummingbird diversity." Coexistence of species in one habitat may depend on population dynamics in another, especially in wide-ranging animals.

Effects of Plants on Animal Spacing and Aggression

Density and dispersion of plant resources can have clear effects on grouping and ranging tendencies in animals. Insectivorous birds, whose prey are distributed thinly but consistently because they depend directly on the abundant and reliable resource of leaves, often are sedentary and have permanent, defended territories. In general, animals should only settle down and space themselves out with territorial aggression when the richness of the resources warrants the effort in defense and when that effort actually improves access to the resource (Brown, 1964; Waser and Wiley, 1980). Several studies have shown that the economics of territoriality is a balance between the richness and concentration of the nectar and the level of competition for it.

Hummingbirds, sunbirds, and honeycreepers (ecological analogs on different continents) show remarkable morphological and physiological adaptations for nectarivory, and their aggressive behavior is tuned to the variability of the resource that they use. Because nectar-producing flowers are often densely clumped, and they often renew nectar that is removed, they can be defensible resources that reward aggressive behavior. Golden-winged sunbirds in Kenya can increase nectar volumes in plants by defending them against other sunbirds and thereby reduce flying time needed for foraging and improve their energy balances. When nectar levels were high and not depleted rapidly by sunbirds, they ceased defense of territories presumably because intruder pressure raised defense costs as likelihood of getting adequate food without defense also rose (Gill and Wolf, 1975). Hawaiian honeycreepers displayed thresholds of territoriality at high and lower flowering densities produced by a tree species with strong year-to-year variation. At high densities, limited data suggested that territoriality was abandoned as it produced diminishing energetic return on investment in aggression, and at low densities, boundaries became too large for defense and resources gained too thin (Carpenter, 1976).

Migrating hummingbirds in North America often set up temporary territories around dense flower patches, and adjust the area defended to maximize their rate of weight gain (Kodric-Brown and Brown, 1978; Carpenter et al., 1983). Hixon et al. (1983) performed experimental reductions and augmentations of flower density for migrant hummingbirds in order to test the idea that variable flower densities dictate a variable territorial response. When flower density was halved, territory size doubled, time spent foraging increased, and resting time decreased. Fruit-eating birds do not often defend their food sources, probably because the fruits are often too ephemeral and too widely scattered. However, when fruiting trees bear concentrated long-lasting crops, they are sometimes defended. Mistle-thrushes in England will defend holly trees if they are isolated and small enough to be defended but large enough to provide a lasting food source (Snow and Snow, 1984).

Effects of Plant Resources on Animal Mating Systems

Plant resources that are sufficiently concentrated in space can favor aggregations of animals into dense breeding colonies. Mating systems in birds are greatly affected by such resource concentrations; polygamy is especially thought to be favored by resources whose concentration produces aggregations of potential mates and relaxation of the need for biparental care. Polygyny (mating of males with several females) has been especially well-studied in birds of the subfamily Icterinae (including New World orioles and blackbirds) (S. Robinson, 1986b). For many blackbirds, it has been proposed that nesting in marshes favors polygyny partly because productivity is high enough to allow females to rear young with little or no help from males. Furthermore, because variation in territory quality is great, some females will do better to mate with a male on a good territory even if she must share him with other females, along with the resources contained in his territory, rather than having sole access to a male on a poor territory (Verner and Willson, 1966; Orians, 1969, 1980). The reasoning behind this "polygyny threshold" hypothesis is similar to that of the ideal, free distribution described earlier: Females occupy good territories until interfemale competition reduces the benefits of further settling on those territories to the level of originally poorer territories; then females should begin to mate monogamously with males on those poorer territories. This process would result in polygyny if some territories are rich enough to compensate for loss of male parental care and interfemale competition; it would also equalize reproductive success of females mated polygynously and monogamously. Polygynous males, however, would do better than monogamous males, putting a premium on defense by males of rich parts of the habitat.

From the male's viewpoint, aggregation of females into relatively small areas of high quality makes it feasible to defend more than one mate and increase reproductive success beyond that possible with monogamy. There is evidence that the quality of a male's territory sometimes depends on the food that it contains (Willson, 1966; Wittenberger, 1980). Ewald and Rowher (1982) supplemented the food available in the territories of male red-winged blackbirds and found that the number of females nesting on those territories, hence the level of polygyny, rose. Other studies have concluded, however, that habitats like marshes afford protection from predators, compared to upland vegetation (Robertson, 1973), and that females aggregate on territories within marshes that are most conducive to safe nesting (Holm, 1973; Searcy, 1979; Lenington, 1980; Picman, 1981). Marsh plant communities appear to encourage polygyny in blackbirds both because their high productivity and physical structure provide highly concentrated food and shelter, and because of the variability of these characteristics within a marsh.

Yellow-hooded blackbirds nest in tropical marshes, probably for similar reasons of abundant food and protection of nests by dense vegetation. In some of these marshes, the seasonal flush of productivity is less pronounced and care of young by both parents is more advantageous than in temperate marshes (Orians, 1980). As a result, these blackbirds are monogamous—the plant community does not supply enough food for females to share the help of males. In more seasonal tropical marshes like those in the Venezuelan savanna, where the flush of food may again be more pronounced and where high predator pressure on nests favors aggregation of females and their nests for group deterrence, yellow-hooded blackbirds are polygynous (Wiley and Wiley, 1980). Yellow-rumped caciques, other members of the Icterinae, also live in the New World tropics but breed in very dense colonies in trees rather than in marshes. Their colonies occur especially in isolated trees or near wasp nests where predation by mammals like monkeys is less severe. In addition, dense synchronous aggregation of females provides the potential for group defense against predators. Once females aggregate into these refuges from predators, polygyny is possible, in part because the forest provides an abundance of fruit and nectar upon which females can feed while searching for insects to feed their young without help from males (S. Robinson, 1986b). Shelter provided by the physical structure of vegetation leads to female aggregation and polygyny in other avian species as well, such as lark buntings (Pleszczynska, 1978) and dickcissels (Zimmerman, 1982).

Several characteristics of fruit provided by tropical trees affect mating systems of frugivorous birds living in tropical forests. Because, as outlined earlier, high-quality fruit can be abundant but erratic in space and time, it can be an easily obtained food but one not readily defended from other frugivores. Some of the more spectacularly polygynous birds are tropical forest frugivores with strikingly elaborate mating displays and display structures. The birds of paradise and bowerbirds of New Guinea and Australia are mostly polygynous; males use gaudy plumage or structures that they have built and decorated to entice females. It has been argued that frugivory has "emancipated" males from parental care duties since females can easily find fruit themselves to feed their young and that males therefore have "leisure time" to devote to competition for mates (Diamond, 1986a). In support of this argument, it is significant that the few monogamous species of these two families specialize on figs of low nutritional value; biparental care is presumably necessary with such a poor food base.

Can Plants Affect Animal Sociality?

Cooperative breeding in birds results from the failure of young mature individuals to disperse from their natal areas. Rather than leaving to breed on their own as is normal for most temperate-zone birds, they remain and help rear the young of other breeding adults (often their parents). It has been proposed that a major factor leading to this kind of social organization is "habitat saturation" (Brown, 1974; Emlen, 1982). When a particular habitat is discrete and productive for an animal population, young birds can "pile up" in their home areas with nowhere to go to establish their own breeding. This impetus for sociality is especially plausible when animals are tied to particular identifiable resources like the large, modified trees used as storage "granaries" by acorn woodpeckers (Stacey, 1979) or when their habitat preferences are particularly limited to very productive vegetation types (Woolfenden and Fitzpatrick, 1984). Cooperative stripe-backed wrens, mentioned earlier, are under sufficient predation pressure that large trees with relatively safe nesting sites limit their habitat distribution. The dispersal options of young birds are further limited by the fact that successful breeding requires not only such a refuge, but a group to help fend off predators and parasites as well (Rabenold, 1984). Territories that support successful breeding over the years are those in which the taxonomic and physical structure of trees is complex, and such groves of trees are in short supply in the Venezuelan savanna.

Studies of herbivorous ungulates further demonstrate how interactions between characteristics of food and shelter provided by plant communities with basic characteristics of animals (like body size) can lead to broad variations in social organization. Jarman (1974) and Jarman and Jarman (1979) have proposed that large species are more likely to form groups than small species, in part because they can better afford to eat low-quality plant food. Plant communities present herbivores with potential foods that vary considerably in quality—especially in protein content. Relatively low-quality foods, like mature leaves and stems with well-developed structural tissue, are more abundant and reliable than more nutritious new growth, seeds, fruits, and storage organs. Small species, because of their higher metabolic rates, will have greater need for these high-quality foods but less need for volume. Small species are therefore more likely to adopt a selective-search, browsing strategy. Large species are more likely to be less discriminating grazers with mouths adapted for volume cropping (McNaughton and Georgiadis, 1986). Grouping interferes with this kind of foraging less, and large ungulates are also more likely to depend on strength, speed, and group deterrence to escape predation. Small species would be less able to form coordinated groups if they were searching for sparse food, and grouping would make them more conspicuous to predators (crypticity is likely to be their main defense against predation).

These ideas were developed during study of the antelope and buffalo of the Serengeti Plain in East Africa, and explain reasonably well the variation in social organization there. Very small species like the dik-dik (4 kg) inhabit woodlands with good cover and live in small nuclear family groups with lifelong pairing of mates on permanent territories large enough to provide all their food requirements. They rarely eat grass, but rather select scattered fruit, seeds, young leaves, and buds from shrubs and herbs.

Larger antelope are not territorial and may range through several vegetation types on an annual or even daily basis. For example, impala (40 to 55 kg) live in more open savanna in larger groups of 10 to 100 individuals. Males defend territories, but mostly just for access to females that roam in groups among males. Impala eat a more mixed diet of grasses and browse (herbs and fruits, flowers and leaves of shrubs), depending on the season (Jarman and Sinclair, 1979; Jarman and Jarman, 1979). At the beginning of the wet season, new plant growth produces an abundance of high-quality food and the impala feed mainly on grass and a few herbs in hilly woodlands. As the dry season approaches, they move into plant communities near rivers where they switch to browsing for seedpods and leaves of shrubs and trees. At this time, herds are smaller and more spread out than in the wet season when nutritious grass is abundant. In drier regions where wet-season grass is less abundant, impala occur in small browsing groups.

Very large eland antelope (700 kg), like the buffalo and wildebeest of the Serengeti, are primarily grazers that occur in large herds of hundreds or even thousands ranging over huge areas. Males cannot defend territories, but compete for estrous females as encountered in the highly mobile herds, and defend them. These large ungulates actively defend herd members against predators. Such general correlations among body size, characteristics of plant material in the diet, group size, and mating systems appear to hold reasonably well for other African ungulates as well (Jarman and Jarman, 1979).

Frugivory, Folivory, and Primate Aggregation

Field studies of primates have been numerous and detailed, and they have best revealed the links between characteristics of the plant community and sociality; this is true in part because primates exhibit a wide range of social organizations, they often utilize plant resources directly through herbivory and frugivory, they are less vulnerable to predation than other social animals like birds, and they are often tolerant of close human observation. As with the antelopes discussed above, there are broad correlations among characteristics of the plant community, body size, group size and social organization; however, in primates, selective pressures internal to their societies can greatly complicate the effort to examine the effects of habitat on population dynamics and sociality.

On a very general level, group sizes tend to be larger, including multiple mature males, in open habitats like savannas than in forests where the structure of vegetation provides protection from predators and impedes communication (Hall, 1965; Crook and Gartlan, 1966; Eisenberg et al., 1972; Rowell, 1972). This is especially true of comparisons across habitats but within genera or families of primates. Open-country baboons sometimes aggregate into very large groups of up to dozens or hundreds at night at safe sleeping sites (trees or cliffs), but break up during the day into smaller groups better suited to searching for scattered food like fruits and tubers (Aldrich-Blake et al., 1971; Kummer, 1971). Small species (generally less than 2 kg) tend to be insectivores and frugivores—like the tarsiers of Malaysia, the galagos, lorises, and bushbabies of Africa, and the night monkeys, tamarins, and marmosets of South America. Larger species are more likely to be folivores using the poorer-quality parts of plant production, like the howler monkeys of South America, langurs of Asia, and the gorillas of Africa. Frugivores tend to live in larger nonterritorial groups but at lower population densities than folivores (Clutton-Brock and Harvey, 1977). These generalizations are especially true for within-habitat comparisons of primate populations.

Several comparative studies have found that, between congeneric species living in the same forest, more frugivorous primates tend to live in larger groups and range more widely (Struhsaker and Oates, 1975; Hladik, 1977; Sussman, 1977; MacKinnon, 1978). This suggests again, as in avian studies, that the temporal and spatial scatter of many fruiting trees in tropical forests, along with the temporary abundance that they provide, might disfavor sedentariness and territoriality and promote aggregation (Snow and Snow, 1984). This can have important implications for mating systems, since aggregations of females encourage male–male competition and polygyny, and since multiple breeding males are accommodated, depending on dominance hierarchies, in the largest of these groups. However, very concentrated food, at trees with abundant fruit crops, for example, can favor aggressive usurpation by dominant males and female choice for such aggressive mates (Janson, 1984).

Does Primate Social Organization Depend on Conflicting Effects of Predators and Fruit Dispersion?

Primate social organization, including the kind of sexual competition among males and the importance of female kinship, is likely to be fundamentally determined by group size. In turn, environmental factors like predation pressure and the distribution and abundance of plant resources, especially fruits and photosynthetic parts, probably determine optimal group size (Waser and Wiley, 1980; Van Schaik and van Hooff, 1983; Terborgh and Janson, 1986). Basic morphological attributes of primates, like body size, dentition, and digestive abilities, will determine vulnerability to particular predators and ability to utilize particular plant resources. For an average-sized primate, larger groups provide more safety but also generate more competition for all but the richest resources. Small groups should be better able to use resources that occur in small patches, but they would be vulnerable to predation. Group size is probably a compromise between these conflicting pressures (Fig. 8-11). In support of these general ideas, many of the smallest primates are nocturnal, possibly because predators—especially avian —can best be avoided then, and many of the largest live in small groups. Comparisons of conspecific populations in different habitats, or of different species in the same habitat, argue that the size of trees used for food is generally correlated with group size and that larger groups of a particular species travel more and visit more trees in a day than smaller

Figure 8–11. Model for the optimization of group size in primates via evolutionary compromise between benefits derived from enhanced safety from predators and costs in decreased access to preferred feeding sites as a function of increasing group size. The dotted lines labeled S, M, and L correspond, respectively, to group sizes that maximize the differences between benefits and costs of sociality (technically, in fitness units) for species feeding on small, medium, and large resources (from Terborgh and Janson, 1986).

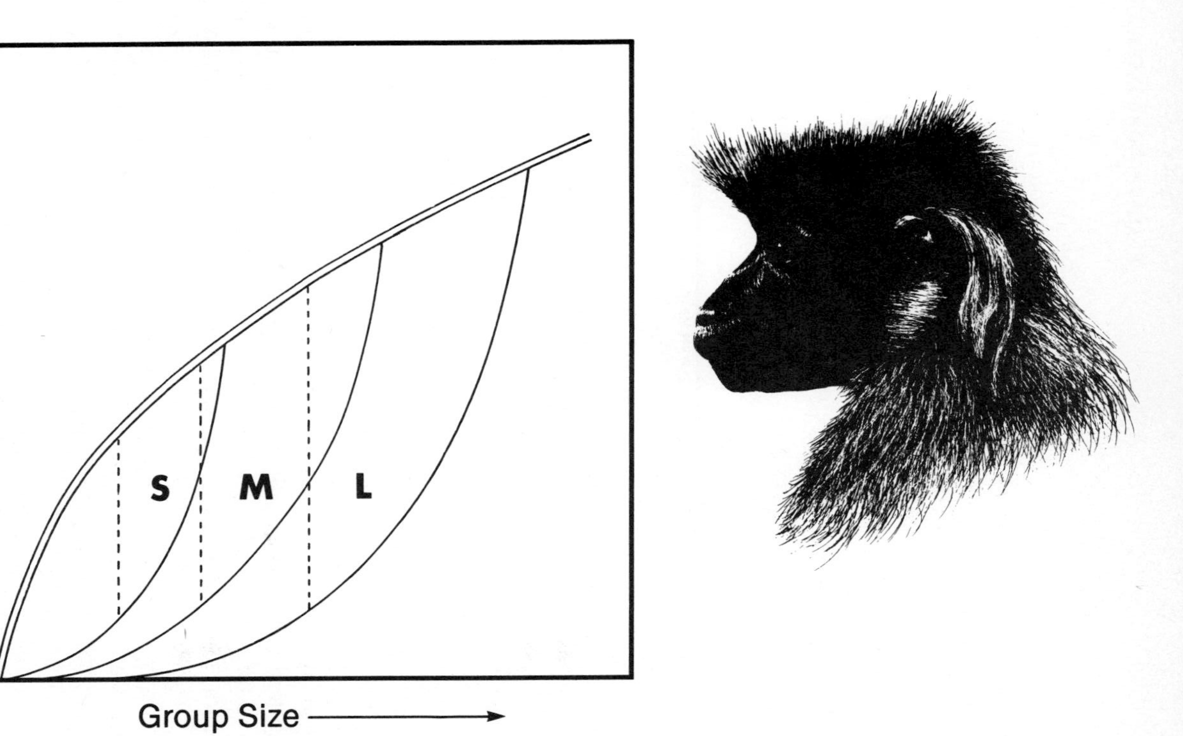
Per capita cost in lost feeding success (———)
Per capita benefit in increased predator protection (═══)
S
M
L
Group Size

groups. Furthermore, because increasing risk and cost of travel with increasing group size must reach some maximum tolerable level, species using fruit that occurs in small scattered patches must live in small groups and rely on crypticity to escape predation (Terborgh, 1983; Terborgh and Janson, 1986).

Mangabeys live in mature forests of East Africa and are omnivorous, eating mainly fruits along with new leaves, buds, shoots, and seeds of a variety of plants as well as invertebrate prey (Waser, 1977). Nonetheless, they do not use the 70 species of trees in their forest indiscriminately. Some of the commonest trees are proportionally well represented in the mangabey's diet, but some rare trees are heavily used while some common trees are avoided. The ratio of the representation of a tree species in the diet to the representation in the forest overall varied 1000-fold among the 25 trees most often used. Some of the trees favored for foraging by insects have characteristics that seem to encourage that activity: large size with extensive branch surface area carrying heavy loads of moss and epiphytes that harbor insects. Fruits of most species that are heavily preferred are borne in dense and asynchronous crops. It is not understood what nutritional constraints might explain why the monkeys eat only particular fruits, or why they tend to select only one plant part—fruit, flower, *or* leaves—from each tree species. It is possible that only certain potentially toxic compounds often found in plant foods can be handled by their digestive systems, and that other consumers in the forest have different capabilities (Waser, 1977).

Mangabey ranging patterns are tied to the complex schedules of growth and reproduction of their food trees. "The rarity, large size (or spatial clumping), and asynchronous fruiting of many preferred mangabey food trees, combined with the fact that fruit persists on many of them for short periods and may be edible even more briefly, results in sequentially available and concentrated food sources which are likely to be separated by considerable distances" (Waser, 1977). These mangabeys have large annual ranges on the order of 4 km^2 that could contain 100,000 trees more than 10 m tall. Operating in such a spatially and temporally complex environment of food sources with greatly varying profitability would seem to be quite a challenge. It is not clear whether grouping in the mangabeys helps them find food, but it is possible that experience and information shared among a dozen or so individuals could be in their communal interest. It does seem clear that the mangabey's opportunism could not be pursued within an area defendable as a territory. Their activities are often centered around single, large, fruiting trees, but they must move on after a few days. Larger groups must travel more, possibly putting an upper limit on group size. By comparing movements of different-sized groups, Waser estimated that for groups larger than 12 to 15 individuals, the energetic costs of adding new members quickly becomes prohibitive given the richness and dispersion of food. Levels of aggression within a 28-member group prohibited feeding by the whole group in even the largest fruiting tree. "Mangabeys specialize on large but temporary food concentrations, which present them with an abundant but moving resource base" (Waser, 1977).

Capuchin monkeys, like the mangabeys, are omnivorous opportunists that eat mostly ripe fruit along with new plant shoots and buds, invertebrates, and occasional small vertebrates. J. Robinson (1986) studied several groups of these monkeys in Venezuela and has found that a group of 20 animals ranges over 275 ha. Capuchins in one group ate fruits from 50 species of plants in 30 families and fruit constituted the largest category of food items (46 percent). In general, the capuchins might take animal food to compensate for nutritional deficiencies of fruit. This seems plausible, since they take time to laboriously forage for invertebrates even when ensconced in a large fruiting tree with an abundance of food. Invertebrates comprise only 33 percent of the items in the capuchin diet annually, but capuchins spend 55 percent of their time foraging for these items (J. Robinson, 1986).

Fruiting trees in the capuchins' habitat, as in other tropical forests, are likely to be hundreds of meters apart, so that the monkeys' use of space is more likely determined by these plant resources than by animal food that is relatively uniform in its dispersion. Robinson argues that the foraging behavior required by the dispersion and abundance of fruit crops affects the social characteristics of capuchins through its effect on group size. Large groups could have an advantage in searching for fruiting trees in a large complex range, especially if the experience of old, postreproductive individuals could direct the group to particular trees. Most important, however, is the ability of large groups to displace smaller groups with overlapping ranges from lucrative fruit crops. This could explain why the more frugivorous capuchin species live in larger groups (Wrangham, 1980; Robinson, 1986; Rubenstein and Wrangham, 1986).

A quick review of the social tendencies of our closest primate relatives, the great apes, can serve to summarize the potential effects of plant resource distribution and abundance on sociality. Because of their size and strength, gorillas, chimpanzees, and orangutans are relatively safe from predators; this focuses the effects of resources on sociality. Chimpanzees are omnivores that concentrate on fruit, and they display a very fluid and complex "fission–fusion" system of group membership and a correspondingly sophisticated system of social bonds (Wrangham, 1977, 1986). Orangutans are often virtually solitary frugivores that also show flexibility in group size (MacKinnon, 1974; Rodman, 1984). Their great size probably makes sharing a fruit crop less feasible and finding the next crop less urgent than for small monkeys like capuchins. Gorillas live in larger groups (10 to 15) than orangutans probably, in part, because their herbivorous diet makes intragroup competition for food less of a problem (Schaller, 1963; Fossey and Harcourt, 1977; Watts, 1985). The flexible-membership social system of chimpanzees probably directly reflects the spatial and temporal variability of fruit as a staple food. The more cohesive groups of gorillas probably reflect the greater stability of leaves and stems as food. The flexible but generally small group sizes of orangutans, and their comparatively asocial existence, probably reflect the result of frugivory relatively unconstrained by predation or metabolic demands, and the effects of patchy plant food distributed in relatively small parcels.

How do Characteristics of Plant Resources Affect Animal Diversity?

The last section showed that the effects of the distribution and abundance of plant resources on animal populations can be profound. It seems obvious that these effects could also determine the composition of animal communities and especially the number of animal species inhabiting a particular plant community. This section returns to issues raised early in the chapter: Given the tendency of adaptations for exploitation of plant resources to become specialized, perhaps because of interspecific and intraspecific competition, do the abundance and variety of resources provided by plants determine animal diversity by providing potential for differentiation and coexistence? Examination of tropical plant communities, themselves strikingly species-rich, contributes considerably to understanding biogeography of animals.

Primate Communities

Returning to habitats already touched upon, the patchy "gallery" forests along rivers in the savanna region of Venezuela support only two species of primate: the capuchin monkeys studied by Robinson (1986) and more folivorous red howler monkeys. Savannas that surround that gallery forest, with scattered groves of trees separated by open grassland, support only the howler monkeys. In comparison, forests in Uganda (Waser, 1977; Waser and Case, 1981) and in the Amazon basin of Peru (Terborgh, 1983, 1986a, 1986b) support communities of 11 and 13 primate species, respectively, in areas of similar size. It is undoubtedly important that the gallery forests exist as narrow strips along rivers (several kilometers wide) in much larger areas of open savanna, thus limiting migration and population sizes. However, the seasonally dry savanna forest has several other important ecological characteristics in comparison to forests that are richer in primate species: (1) plant resources like fruits are less dependable (more seasonal); (2) diversity of those resources is not great; (3) structurally, the vertical and horizontal complexity of the forest is low; and (4) productivity is relatively low.

Considering the above factors in order, Robinson (1986) found that the dry-season bottleneck for the frugivorous capuchins in Venezuela (9 degrees north latitude) was severe enough that only two species of fruit were available and the monkeys had to use a greater variety of foods than in the productive wet season. Waser (1977) found no such severe seasonal bottleneck in the moist evergreen Ugandan forest situated nearly on the equator, although seasonal and annual variation in diet was pronounced. Terborgh (1983, 1986a, 1986b) has stressed that seasonal bottlenecks can be important, even in climates that seem very benign and constant to temperate-zone observers, as that of his evergreen Peruvian rain-forest study site at 11 degrees south latitude. As mentioned earlier, primates including capuchins, howlers, tamarins, and spider monkeys must turn to a few "keystone resources" like palm nuts, figs, and nectar to tide them over. Palms and figs were also critical dry-season resources for Robinson's capuchins, but the severity of the dry season for them must be much greater; no rain falls for 3 to 4 months and many of the trees drop their leaves in the dry season, in marked contrast to the other two sites.

The species diversity of trees in the capuchins' forest is lower than in the other two forests. In a 7.2-ha area, Robinson identified 80 species of trees more than 4 m tall; the five most abundant species accounted for nearly 40 percent of individuals sampled. If a 10-m height criterion were used, the total number would drop to approximately 60 to 65 species. In the surrounding savanna, a similar sample would yield only 30 to 40 species. The capuchins ate fruit from at least 50 species of plants, but more than half of their fruit diet came from two families of plants comprising eight species; one-third of the observations of capuchins eating ripe fruit were of two species of fig, in spite of the fact that only 3 percent of the trees in the forest were figs. The Ugandan forest is somewhat more diverse. Waser found 70 species of trees more than 10 m tall in a 5.8-ha sample, and the five most abundant species accounted for 50 percent of the individuals. The mangabeys ate the fruits of 54 plant species, and the equitability of these species in the diet was greater than for the capuchins: The favored fruit accounted for less than a quarter of the mangabeys' diet. The Amazonian forest studied by Terborgh has much higher diversity of trees than the other two sites; even in a small patch of upland forest, more than 200 species of trees could be found in just a few hectares! A small sample of 50 fruit traps of 0.1 m^2 yielded 40 species of fruit in 1 month. Using this rich collection of fruit, primates segregate to some degree by sizes of fruits eaten; large monkeys eat larger and harder fruit (Terborgh, 1983). The diet of capuchin monkeys (a different species) was probably also more diverse than in Venezuela, although comparable data are not available; they ate up to 20 species of fruits in 4-month periods and ate fruits and other parts of more than 90 species of plants in a year (Janson, 1985). As in Venezuela, they ranged over large areas in search of figs in the dry season.

The physical structure of the Venezuelan forest is less well-developed than the other two sites; trees only occasionally reach a height of 20 m and the canopy is not closed. In the savanna where the capuchins are rare, large trees are uncommon and groves of large trees with touching canopies are still more rare. The Ugandan forest has a closed canopy 25 to 30 m high with some emergent trees reaching 50 m. Similarly, the Peruvian forest canopy is at 30 to 35 m with emergents up to 60 m tall. This structural complexity may not be important in fostering primate diversity directly by allowing spatial segregation, but it is probably associated with a more diverse and abundant resource base. The horizontal structure of the latter two forests could be more complex as well, since they are more a mix of successional stages than the Venezuelan forest.

The much greater mass of photosynthetic vegetation in the primate-rich forests, and their greater annual rainfall (208 cm in Peru; 166 cm in Uganda; 145 cm in Venezuela), suggest that productivity is higher there. The Peruvian rain forest produces prodigious amounts of fruit: 2000 kg/ha/yr. High productivity by plants might be expected to be correlated with high consumer biomass. The Venezuelan savanna forest and more open woodland support 330 to 645 kg/km^2 of the two species of monkey that live there (Eisenberg et al., 1979). Four forests in Malaysia, Borneo, Uganda, and Peru that have more than five common species support 335 to 2650 kg/km^2 of primate biomass (Waser, 1987). It seems likely that primate-rich forests are productive, but not enough data exist to judge conclusively whether primary productivity itself is important in determining primate diversity. High productivity would increase both the usable diversity of food (since even rare species could produce significant quantities) and the patchiness of the food source (because single trees could be major resources). That energy resources for primates might ultimately be limiting is suggested by the fact that, in Uganda, biomass per group is fairly constant across four species (72–134 kg) in spite of order-of-magnitude variation in both individual body size and group size (Watson and Case, 1981).

Waser (1977, 1987; Waser and Case, 1981) has argued that the dispersion of food, especially fruit, into rich patches scattered in space and time favors primate coexistence and "in particular, the complex seasonal and interseasonal variation in plant phenology that characterizes many tropical forests provides precisely the conditions argued to relegate competition to an occasional and minor role" (Waser, 1987).

Large, rich patches of fruit attract large competitively dominant species like chimpanzees. The evanescence and unpredictability of these patches make them occasionally available to smaller mobile species, and the existence of small patches provides a refuge for small species from potential competitive exclusion. This temporal and spatial complexity of plant resources, which is exaggerated by high productivity and absence of severe seasonal bottlenecks, is probably the key to primate diversity.

The Generality of Effects of Stability, Diversity, and Productivity

The variety of animal species in a particular habitat can be affected by many factors other than the features of plant resources. The history of the speciation process in the region and patterns of migration over millenia will affect the pool of species from which an interacting set in a particular local site could be drawn. The climate of a habitat could affect animal diversity directly, but the stronger effects are likely to be indirect through characteristics of the plant community dictated by climate. Interactions among animals themselves could certainly affect diversity but, as this chapter has shown, plants can mediate these interactions in important ways. The most basic definition of animal diversity is the number of species found in a particular area, but it is also important to consider the evenness of representation of each species. A 20-species community is not really much more diverse than a 10-species community if the additional 10 species are very rare. For simplicity, however, "diversity" in this section refers to number of species (species richness).

Even considering only a few characteristics of plant communities, it quickly becomes obvious that those that seem likely to be causally related to animal diversity are also closely interrelated with each other. For instance, if a few square kilometers of equatorial rain forest in Borneo hold 135 species of reptiles and amphibians while the norm for temperate forests is near 40 (Lloyd et al., 1968), which aspects of the plant resources are most likely to be responsible? The rain forest is less seasonal, more productive, more taxonomically diverse, and more physically complex than temperate forests; furthermore, it is more productive, partly because it is more constant and more physically complex and partly because of its taxonomic diversity. It is difficult to rank interdependent factors by their importance in producing a common result; nonetheless, as with the primate communities, it is possible to show that all can be important.

Stability or reliability of resources could foster animal diversity by allowing specialization and by reducing the competitive impact of overlap in resource use (Klopfer and MacArthur, 1961; Connell and Orias, 1964; Orians, 1969; MacArthur, 1972). With a stable resource base, it is easier to imagine new species moving into a habitat to exploit a narrowly defined resource. Vegetation can buffer the seasonality of climate, and animal communities could be more diverse in complex vegetation for this reason. The lower, more buffered strata of forest vegetation support a more diverse and a more seasonally constant fauna than the upper strata that provides more seasonally variable resources in some systems (Karr, 1976b). Desert communities show great month-to-month and year-to-year variation in productivity because rainfall is often erratic and clearly controls primary productivity. Desert rodents (Brown, 1975; Davidson et al., 1980) are more diverse where rainfall, hence production of plant food, is more predictable. The highly seasonal Venezuelan savanna supports few small mammals compared to other tropical or temperate habitats (August, 1983). The high number of mammalian species in tropical forests compared to temperate ones is mainly due to an increase in the number of bat species, and Fleming (1973) suggested that this is in part due to the year-round availability of fruit and insects that depend on plants in tropical forests. As shown earlier, temporal unpredictability of the fruiting phenology of masting species can ultimately depress numbers of mammalian species (Janzen, 1980b).

It is possible that in many systems there are moderate levels of variability in resources that could maximize animal diversity. If resources were absolutely constant, competitively dominant species could more readily crowd out other species. Periodic change in the resource base, as in moderately seasonal habitats and those whose resources vary somewhat from year to year, could produce changes in competitive advantage from one species to the other or just keep all species below densities at which competitive dominance could result in extirpation of some species. Under these conditions, animal diversity would be increased (Hutchinson, 1961; Pianka, 1975; Connell, 1978; Rabenold, 1979). Wiens (1977) has suggested that severe "bottlenecks"—poor years for primary production—in semiarid habitats could enhance species diversity of birds by depressing numbers below levels at which competition would be important. In general, however, strong fluctuations or marked unpredictability of resources probably depress consumer diversity unless a diversity of alternative resources exists. In a comparison of six tropical lowland forests, Pearson (1977) has argued that extreme dry seasons and their often unpredictable timing of resource depression are associated with depauperate avifaunas.

Variety of plant resources could promote animal diversity by providing potential for a wider range of specialization and by providing alternative resources in bottleneck periods. A community of plants with diverse physiologies and phenologies could stabilize productivity through compensating responses to environmental fluctuation (McNaughton, 1977). The more diverse the flora, the more likely divergence among plants in flowering and fruiting times will provide animals with some food source to turn to in unproductive seasons. Pulliam (1975) has found that the variety of seeds produced in semiarid grassland and woodland habitats is closely associated with the number, size, and relative density of sparrows supported. Beak sizes of sparrows are known to be related to the sizes of seeds eaten, and sparrow population densities are known to be related to abundances of seeds in winter. Using a model combining these relationships, Pulliam could measure abundances and size distributions of seeds and predict the relative densities of sparrows living in different habitats. Grassland and dry, oak woodland produced mainly medium-sized seeds and supported only one medium-sized sparrow, as predicted. A riparian woodland plant community produced a wider variety of food—more small and large seeds—and supported two sparrow species: a small and a large species with a predicted numerical bias toward the former. However, in years of low seed abundance, only one species of sparrow occupied even the riparian habitat. When food is scarce, sparrows converge in their seed-size preferences. For resource diversity to affect animal diversity, productivity across the resource spectrum must be adequate. Pulliam (1985) has concluded that "seed-size partitioning is unlikely to play any major role in the coexistence of sparrow species" because coexistence in habitats with varieties of seeds is temporary and sparrows sort themselves by habitat type during lean times when numbers are actually limited.

Studies of finches on the Galapagos Islands off the coast of Ecuador, mentioned earlier, have also shown that the variety of seeds available is a good predictor of the number and kind of consumers. The distributions of seeds by size and hardness were described for 15 islands on which the number of granivorous finches varied from one to three. The islands support variable vegetation types from moist forest to desert scrub, and the kind of seed available varies greatly as well. On some islands, small hard seeds were most abundant, while on others large, soft seeds were the norm. Given known relationships between beak morphology and ability to crack seeds of varying size and hardness, and proposing limits to similarity in diet among finches, it has been possible to predict the composition of bird communities on a particular island based on the kinds of seeds available (Grant and Schluter, 1984; Grant, 1986a). Islands with multiple peaks of seed production along a seed size/hardness gradient support two or three species of finch; islands with a simpler variety of seeds support only one finch species.

Relationships between the variety of resources provided by plants and animal diversity are clear for other animals as well. Insectivorous birds that have foraging or nesting preferences for certain types of trees are more diverse in forests with many species of trees. In New Hampshire, northern hardwood forests are dominated by beech, maple, and birch, with a mixture of spruce and hemlock. Some species of birds, like blackburnian warblers, prefer to forage in conifers while others, like least flycatchers, prefer to forage in deciduous trees with more open canopies (Robinson and Holmes, 1982; Sherry and Holmes, 1985). A forest with a mixture of coniferous and deciduous trees is likely, therefore, to support a diversity of birds, especially when the evenness of their representation is considered. Avian species diversity increases with diversity of plant forms in desert scrub communities because the birds are selective in the placement of their nests (Tomoff, 1974).

Animal diversity on the larger scale of tens or hundreds of square kilometers would also be enhanced by a mixture of kinds of plant communities, including successional stages. This is especially apparent in studies of animal diversity on islands (Dueser and Brown, 1980). Of course, the degree to which this would be true would depend largely on the degree of specificity of animals' preferences for particular plant communities or particular plant foods. Animals in some low-diversity regions can be very broad in their habitat tolerances (Cody, 1986), while many species in high-diversity regions have narrow habitat tolerances (Lovejoy, 1974; Diamond, 1986b). The degree to which diversity of primary resources within and among plant communities is associated with animal diversity is of great importance to the design of nature reserves (Foster, 1980).

The structural complexity of plant communities has often been correlated with animal diversity, and animals often segregate by strata in physically complex habitats (MacArthur and MacArthur, 1981; Pianka, 1967; Schoener, 1968; Rotenberry and Wiens, 1980; Askins, 1983; August, 1983). In many cases, this physical complexity is probably correlated with diversity of resources, so separation of the effect of physical structure itself on animal diversity is difficult. In desert rodent associations, however, the physical structure of vegetation provides cover for species more vulnerable to predation. This probably promotes coexistence with species having better-developed escape tactics that typically forage in the open for the wind-blown seeds caught in depressions. Even though the two sets of rodents may use seeds from the same species of plants, they use different fractions of the crop defined by the structure of the woody plants (not necessarily the seed producers) around them (Rosenzweig, 1973; Price, 1978; Kotler, 1984). Horizontal structural complexity, or patchiness, of vegetation can also affect animal communities. Treefall gaps in forests can contribute to animal diversity (Schemske and Brokaw, 1981), but again it is unclear whether the effect is due to an increase in the variety of food types or in variety of structure itself.

Primary productivity is broadly related to animal diversity, and could underlie some of the above correlations (Brown, 1981). For instance, more diverse and physically complex plant communities are often more productive as well, as in some studies of fish in lakes (Tonn and Magnuson, 1982). Habitats that are consistently very dry are unproductive and depauperate world wide. Clear correlations between animal diversity and primary productivity exist for desert rodents and ants (Brown and Davidson, 1977; Davidson et al., 1980) and grassland arthropods (Kirchner, 1977). Studies of avian communities in the same habitats in productive and unproductive years produce strong correlations in some studies (Smith, 1982) but not in others (Dunning and Brown, 1982).

Plant Communities and the Increase in Avian Diversity at Lower Latitudes

The global pattern of diversification of animal communities in the tropics puts the effects of plant resources on animal diversity in the broadest perspective. All of the factors considered above—the reliability, variety, structure, and abundance of resources provided by plants—are implicated in explaining tropical diversity because many tropical habitats exceed any temperate analog in all of these attributes. Bird communities provide the best comparative base since they have been extensively studied at both tropical and temperate latitudes. Forests have received the most attention from biologists. Broad historical explanations are based on proposals of high rates of speciation in the tropics (Mayr, 1969; Prance, 1982) or freedom from interruption of the speciation process (Wallace, 1878; MacArthur, 1969), making tropical habitats "older" than north-temperate ones that have relatively recently emerged from beneath glaciers. These explanations remain plausible, but a full understanding of tropical diversification of taxa such as *Aves* requires further consideration of the ecological effects of plant resource characteristics.

Some tropical rain forests have prodigious physical structure, as mentioned earlier, and this factor alone undoubtedly contributes to avian diversity (MacArthur et al., 1966; Karr and Roth, 1971; Terborgh, 1980a, 1980b). However, structurally simple vegetation, like that of the savanna woodland of Venezuela, still supports several times the number of bird species contained in the best-developed North American forest. Studies of Amazonian forests show that their great physical structure cannot account for most of the Amazonian diversification among birds (Lovejoy, 1974; Pearson, 1975; Terborgh, 1985). Primary productivity is very high in some tropical habitats (Whittaker, 1975). However, Karr (1971, 1975) has shown that the energetic requirements of some tropical avifaunas are not greater than their temperate counterparts. Even though tropical avian biomass was higher in this study, smaller body sizes and more benign temperatures combined to reduce energy demands per unit mass. In addition, many tropical habitats that support diverse avifaunas are probably not more productive than temperate analogs. The year-round *variety* of abundant foods provided directly and indirectly by plants probably explains much of the striking bird diversity.

The diversity of animal life in some tropical habitats is staggering. The trend at both the local and regional levels for species diversity to increase with decreasing latitude is very general for a variety of animals on a variety of taxonomic levels (Dobzhansky, 1950; Fischer, 1960; Simpson, 1964; Pianka, 1966; Stehli et al., 1969; Whittaker, 1975). Of course, some tropical habitats support a low diversity of animals, such as the arid woodland on the northern coast of Peru that consists of two common tree species and a few shrubs. Alpine vegetation in the northern Andes supports an avifauna that is only slightly more diverse than temperate alpine habitats. In the extremes, glaciers on the slopes of Chimborazo in Ecuador or Mt. Kenya in Africa, astride the equator, are as barren of life as any in Greenland. However, the variety of animals in productive lowland habitats, and bird life in particular, can be overwhelming, in part because visitors from the temperate zone see animals that bear little resemblance to familiar categories.

The country of Colombia, with half the land area of the eastern United States, has more than 1550 species of birds (DeSchauensee, 1964), twice as many as America north of Mexico on one-twentieth the area. On a larger scale, South America is inhabited by 2936 species of birds; this is nearly one-third the global total and it is nearly four times the total for the larger area of North America.

Purely tropical morphological and taxonomic varieties of birds suggest a broadening of the resource spectrum in tropical habitats. Among families of birds that are represented at both high and low latitudes, many show broader morphological variety in addition to a proliferation of species. For example, flycatchers inhabiting a single hectare in the Venezuelan savanna range from the tiny tody flycatcher (8 cm long) to the boat-billed flycatcher (22 cm); this is greater than the size range of flycatchers for all of eastern North America, and the range of bill morphology and foraging behavior is even more distinct. It has been suggested for insectivorous birds that greater insect diversity, especially in size, has contributed to such morphological range and increased numbers of species (Schoener, 1971) and wider variation in behavioral foraging tactics (Orians, 1969). Similar generalizations could be made about avian families that use plant resources directly, like the fruit-and-seed-eating doves, finches, and tanagers. Sixteen species of hummingbirds inhabit North America, while hundreds take plant nectar in South America and display a much wider range of beak morphology.

Mainly tropical families show modes of exploitation of plant resources that simply do not exist in the temperate zone. Specialized frugivorous families, like the cotingas, manakins, barbets, and toucans of South America and the birds of paradise and bowerbirds of New Guinea suggest, again, that production of soft, nutritious fruits by tropical forest trees is a major factor contributing to tropical avian diversity.

Comparison of bird communities in temperate and tropical forests (in areas on the order of a few square kilometers) has shown generally that a large part of the diversity increase is accounted for by the addition of "new" resources to the tropical forests: fruits and nectar (Karr, 1971, 1975). These resources exist in temperate forests, but they are used very little by birds. It was the great proliferation of hummingbirds in South America that most excited the imagination of one of the fathers of the theory of evolution, Alfred Russell Wallace. Snow (1981) has surveyed the diets of tropical frugivores and compiled the plant genera most responsible for sustaining full-time frugivory. Three families are especially associated with specialized frugivores (birds with diets consisting mainly of nutritious fruits): Lauraceae, Palmae, and Burseraceae. "Lauraceous fruits seem the archetype of fruits adapted for dispersal by specialized frugivorous birds, with a comparatively thin layer of very nutritious flesh enclosing a single large seed" (Snow, 1981). They offer high levels of protein (up to 14 percent) and fat (up to 44 percent). Snow points out that the tropical forests of Africa are poor in plants of the above families, compared to the American tropics and Australasia, and that the number of specialized frugivores in the African forests is low as well. In fact, African rainforests have substantially fewer species of birds generally than neotropical forests, although historical considerations might help explain this as well (Karr, 1976c).

Terborgh and colleagues have studied the bird communities of a wet lowland forest in Amazonian Peru where more than 500 species reside in a 10-km^2 area (Terborgh et al., 1984; Terborgh, 1985; Terborgh and Robinson, 1986). In the main study plot of 110 ha, at least 330 species of birds were found, compared to the maximum of about 40 in temperate forests (Terborgh, 1985). This high diversity is partly produced by a mosaic of successional forest types created by river meanders. However, the bird associations in the different vegetation types were not very distinct; the "between-habitat" component of diversity did not seem strong. The "guild structure," or variety of ways of making a living for birds, of the forest compared to temperate forests is instructive. Nearly half of the excess tropical diversity could be accounted for by tighter "niche packing" within guilds represented in both temperate and tropical forests. This especially comprises small insectivores; many more species in the tropical forest were searching for small insects in roughly the same way temperate birds do. For instance, in the list of 526 species for the vicinity of the biological station, 68 were flycatchers in the family Tyrannidae (compared to the three to four species expected in a temperate forest) and 39 were tanangers in the family Thraupidae (compared to one to two species in temperate forests). Several exclusively tropical families were added to these shared guilds, like the 28 species of ovenbirds (Furnariidae).

Exclusively tropical guilds that used fruit and nectar accounted for a significant part of the tropical increase in species diversity. Along with the 19 species of hummingbirds (Trochilidae), conspicuous are the frugivorous cotingas (15 species, Cotingidae), manakins (8 species, Pipridae), toucans (8 species, Ramphastidae), and parrots (18 species, Psittacidae). Antbirds (53 species, Formicariidae), specialized in following ant swarms to take other insects that are flushed, are another conspicuous tropical family pursuing a way of life not found in temperate forests. Increased physical structure of the forest would undoubtedly help explain the high diversity of woodpeckers (16 species, Picidae) and woodcreepers (17 species, Dendrocolaptidae).

Obviously, the variety, reliability, and productivity of plant resources are all important in explaining tropical bird diversity. The balance of relative importance among these factors undoubtedly varies among guilds of animals. Insectivorous bird species differentiate mainly on the basis of foraging behavior that is sensitive to the physical structure of the habitat while nectarivores and frugivores differentiate mainly by resource type, so that plant species diversity will be more important to them. Along a tropical elevational gradient, the number of hummingbirds does not differ much between high-elevation elfin forests only 3 m tall and the 50-m-tall lowland forests, but the number of insectivores, like flycatchers, differs greatly (Terborgh, 1977, 1985). In all, Terborgh (1985) estimates that 34 percent of the additional tropical diversity is accounted for by unique resources provided by the tropical forest (approximately 20 percent by frugivory and nectarivory). It seems clear that the great variety of abundant resources provided by this forest, with its hundreds of tree species and great physical structure, promotes avian diversity. This includes various processes of plant–animal interactions discussed throughout this chapter: increased specialization of adaptations to common resources, mediation of competition and predation, and specialization to uniquely tropical plant resources.

Selected References

Cody, M. L. 1985. An introduction to habitat selection in birds. pp. 2–56. In: M. L. Cody, ed. Habitat selection in birds. Academic Press, NY.

Diamond, J., and T. J. Case, eds., 1986. Community ecology. Harper & Row Publishers, NY.

Grant, P. R. 1986b. Ecology and evolution of Darwin's finches. Princeton University Press, NJ.

Terborgh, J. 1986b. Community aspects of frugivory in tropical forests. pp. 371–384. In: A. Estrada and T. H. Fleming, eds. Frugivores and seed dispersal. Dr. Junk, Dordrecht, Netherlands.

Wrangham, R. W., and D. I. Rubenstein. 1986. Social evolution in birds and mammals. pp. 452. In: D. I. Rubenstein and R. W. Wrangham, eds. Ecological aspects of social evolution—birds and mammals. Princeton University Press, Princeton, NJ.

9

Plant–Animal Interactions in Agricultural Ecosystems

BENJAMIN R. STINNER AND DEBORAH H. STINNER

Department of Entomology
The Ohio State University
Ohio Agricultural Research and Development Center
Wooster, OH 44691

Introduction

Agricultural ecosystems are areas of land in which physical, chemical, and biological processes are manipulated by humans for production of food and/or fiber (including forestry and clothing) (Fig. 9-1). Agriculture encompasses the set of practices or manipulations required to produce desired products for human use. The structure and function of natural ecosystems over much of the world have been dramatically changed by agriculture. This transformation, involving about 190 million ha of land in the United States (Clawson, 1973; Chapin, 1983), is historical and ongoing and affects ecological processes at the population, community, and ecosystem levels as well as being one of the most powerful anthropogenic forces shaping landscapes (Cox and Atkins, 1979; Chapin, 1983;

The authors thank G. J. House, S. C. Rabatin, and O. G. Hall for reading and commenting on this chapter.

Forman and Godron, 1986). When a forest or prairie is cleared and plowed, the most striking change is dramatic alteration in vegetation structure, which in turn has a profound impact on aboveground plant and animal distribution and interactions. Not so obvious, but equally important in affecting distribution and interactions among soil-inhabiting organisms, are the concomitant changes in physical and chemical properties of soil (Coleman, 1985). Modifications of microclimatic conditions, both above and below ground, also have important influences on plant and animal distributions and interactions. This chapter provides an overview of how agriculture alters plant–animal interactions in comparison to those in natural ecosystems, including a discussion of the evolutionary history of some key plant–animal interactions in agroecosystems.

History of Agriculture

Origins of Domesticated Plants and Animals

Agriculture is about 8,000 to 10,000 years old (De Candolle, 1959; Sauer, 1969). There are several theories to explain the origins of plant and animal culture, and while authors differ on specifics, there is general agreement that animal husbandry predated plant culture by several thousand years (Reed, 1977). These early, animal-based cultures were largely nomadic because of the need to move grazing animals (mostly sheep and goats) to new grasslands as old areas became overgrazed. The nomadic people supplemented their diets by gathering wild plant material. Seeds of these plants were often thrown on refuse piles in disturbed sites around the nomadic villages. It has been hypothesized that plant domestication developed out of these trash heaps as humans gradually recognized that some weedy, opportunistic plant species could provide food on a sustainable basis (Baker, 1970). Eventually, more plants were grown and selected to produce food and fiber materials. Examples of some of these archetypal crop plants and major centers for domestication of plant and animal species are listed in Table 9-1. Interaction between plants and animals increased under human control, and this domestication process exerted strong ecological and evolutionary influences on the weeds, insects, and microorganisms associated with the crop and livestock species. The results of these influences on

Figure 9–1. Typical farming systems occurring in midwestern United States. The photograph on the top shows diverse cropping involving grain, forage, and orchard crops. The lower photo illustrates animal agriculture and the pathway that much of the vegetable matter produced in North America and Europe follows: direct consumption by animals rather than humans.

plant–animal interactions in modern agricultural ecosystems is discussed in subsequent sections of this chapter.

It is important to emphasize the lasting effects these domestication processes have had on human social systems. In particular, the development of sustained plant culture permitted the establishment of large, complex, and stable societies (Sauer, 1969). Plant material for human and animal consumption could be grown on the same land year after year, and people no longer had to move in order to replenish their resources. Thus, the domestication process (i.e., the development of agriculture) and development of human societies are linked in a biological and cultural evolution.

Development of Modern Agriculture

During most of the history of agriculture, farming was labor intensive. Human and animal power provided, and still do in many areas of the world, most of the energy input to maintain domesticated plant production (Bourne et al., 1980). In this way, humans and animals serve to

Table 9-1 Major geographical areas of plant domestication. Examples are representative rather than all-inclusive.

Area	Food Type	Examples
Western Asia	Cereals and grain legumes	Wheat (*Triticum aestivum*)
		Rye (*Secale cereale*)
		Oats (*Avena sativa*)
		Barley (*Hordeum vulgare*)
		Chick pea (*Cicer arietinum*)
		Garden pea (*Pisum sativum*)
		Broad bean (*Vicia faba*)
		Lentil (*Lens esculenta*)
	Root crops	Beet (*Beta vulgaris*)
		Turnip (*Brassica rapa*)
		Carrot (*Daucus carota*)
		Radish (*Raphanus sativus*)
	Vegetables	Onion (*Allium cepa*)
		Cabbage (*Brassica oleracea*)
		Parsley (*Petroselinum sativum*)
		Lettuce (*Lactuca sativa*)
	Fruits and nuts	Melon (*Cucumis* spp.)
		Quince (*Cydonia oblonga*)
		Fig (*Ficus carica*)
		Date palm (*Phoenix dactylifera*)
		Almond (*Prunus amygdalus*)
		Cherry (*Prunus avium*)

Table 9-1 (*continued*)

Area	Food Type	Examples
		Pear (*Pyrus communis*)
		Apple (*Pyrus malus*)
		Grape (*Vitis vinifera*)
	Forage crops	Alfalfa (*Medicago sativa*)
		Orchard grass (*Dactylis glomerata*)
		Timothy (*Phleum pratense*)
		Clovers (*Trifolium* spp.)
		Bromegrass (*Bromus inermis*)
		Fescue grass (*Festuca* spp.)
		Bent grasses (*Agrostis* spp.)
		Vetches (*Vicia* spp.)
		Rye grasses (*Lolium* spp.)
Africa	Cereals and grain legumes	Millet (*Brachiaria deflexa*)
		Sorghum (*Sorghum bicolor*)
		African rice (*Oryza glaberrima*)
		Pearl millet (*Pennisetum americanum*)
		Cowpea (*Vigna unguiculata*)
	Root crops	Yam (*Dioscorea rotundata*)
	Oil crops	Oil palm (*Elaeis guineenis*)
		Castor oil (*Ricinus communis*)
Eastern Asia	Cereals and grain legumes	Rice (*Oryza sativa*)
		Buckwheat (*Fagopyrum esculentum*)
		Japanese millet (*Echinochloa frumentacea*)
		Soybean (*Glycine max*)
	Oil crops	Tung oil (*Aleurites fordii*)
		Rapeseed (*Brassica compestris*)
		Mustard seed (*Brassica juncea*)
	Fruits and nuts	Chinese chestnut (*Castanea henryi*)
		Loquat (*Eriobotrya japonica*)
		Apricot (*Prunus armeniaca*)
		Peach (*Prunus persica*)
		Water chestnut (*Trapa natans*)
	Spices	Chinese shallot (*Allium bakeri*)
		Cinnamon (*Cinnamonum cassia*)
		Chinese pepper (*Zanthoxylum bungei*)
		Horseradish (*Wasabia japonica*)
		Ginger (*Zingiber officinale*)
Americas	Cereals and grain legumes	Maize (*Zea mays*)
		Amaranth (*Amaranthus cruenthus*)
		Peanut (*Arachis hypogaea*)
		Common bean (*Phaseolus vulgaris*)
		Lima bean (*Phaseolus lunatus*)
	Root crops	Potato (*Solanum tuberosum*)
		Sweet potato (*Ipomoea batatus*)
		Arrow root (*Maranta arundinaceae*)
	Vegetables	Tomato (*Lycopersicon esulentum*)
		Squash (*Cucurbita* spp.)
		Pepper (*Capsicum* spp.)

subsidize plant production through soil preparation, cultivation, harvest, and other cultural operations.

During the last 100 years, a revolution has occurred in agriculture. Farming became very mechanized in the United States and Europe and rapidly is becoming so in other areas of the world. In the mid-1800s, the McCormick grain harvester and other machines were developed, allowing one person to do the work of many. Gradually, fewer people were needed to produce crops in industrialized countries and the farm population declined. This trend has increased dramatically since World War II, until now, in the United States, less than 3 percent of the population produces all the food consumed (USDA, 1986). These statistics are frequently used to support the boast that industrialized agriculture is very productive and extremely efficient. Modern agriculture is very productive on a per area or per animal basis (e.g., more grain per acre, more milk per cow) and is extremely efficient in terms of output per unit of labor input (Fig. 9-2). Yet, the efficiency of output per unit of material and energy input is appreciably lower than it was 40 or 50 years ago. This latter point is illustrated by data showing that the average ratio of energy (kcal) output per kilocalorie of input for mechanized maize production in the United States is 2:9 compared to the ratio 3:4 for horse-powered maize production (Pimentel, 1984). That is, fossil fuel energy has been substituted for human and animal labor.

In many areas of the world, agriculture is still practiced in a labor-intensive manner with technology that is not appreciably different from what it was hundreds or even thousands of years ago. This labor-intensive agriculture is carried out within the context of village communities in relatively closed economies where energy, materials, and currency are exchanged within a localized area without much participation in the world market (Mitchell, 1984). The mechanized form of agriculture is characteristic of higher latitudes, whereas the labor-intensive agriculture is practiced predominantly in equatorial regions. This chapter focuses primarily on plant–animal interactions in agricultural ecosystems created by industrialized societies. However, it should be pointed out that differences between industrialized and nonindustrialized agriculture can have important consequences on plant–animal associations.

Comparison of Modern Agroecosystems and Natural Ecosystems

Although various agricultural practices effect plant–animal interactions differently, most modern agricultural ecosystems share certain characteristics with natural ecosystems (Fig. 9-3). For example, agriculture almost always entails subsidizing naturally occurring levels of material

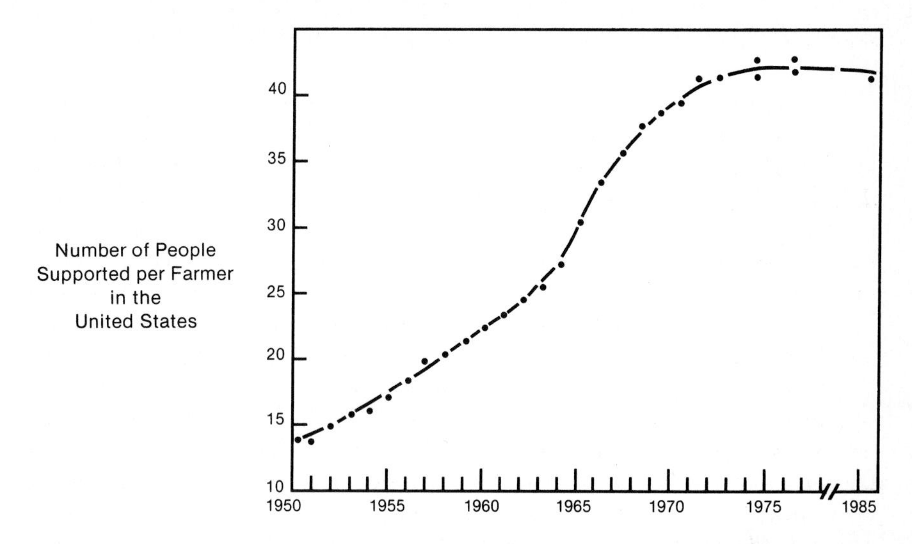

Figure 9–2. Changing patterns in resource use and labor in agriculture during the past 30 to 40 years. Note that after 1970 there has been no appreciable increase in the number of people supported per farmer.

and energy inputs (i.e., fertilizers, pesticides, and mechanical disturbances) to achieve desired amounts of output or production. As a result, material and energy inputs and outputs are usually much larger in agricultural than in natural ecosystems. Large material and energy inputs can greatly modify the environment in which plants and animals interact. In general, the larger the inputs, the greater the modification of naturally occurring interactions (Stinner and House, in press).

Furthermore, agricultural management of a land area often produces lower species diversity than occurs in the natural state (Edwards, 1977; Price and Waldbauer, 1982). Therefore, the number of potential interactions between plants and animals is reduced. Another major difference between agricultural and natural ecosystems is that anthropogenic selection dominates natural selection in agricultural ecosystems.

Finally, ecosystem processes in agricultural systems are controlled by external forces (i.e., human activities) in contrast to the ecological and evolutionary control in natural ecosystems. These differences between agricultural and natural ecosystems are the main factors that determine the impact of agriculture on plant–animal interactions. A list comparing specific ecological characteristics of agricultural and natural ecosystems is presented in Table 9-2. The contrasts shown represent extremes (i.e., conventional, mechanized agriculture with tillage vs virgin or mature second-growth natural ecosystems), whereas in reality most of the parameters vary by degree, depending on what specific management scheme is considered. The remainder of this chapter deals with the characteristics in Table 9-2 that pertain specifically to effects of agriculture on plant–animal interactions.

Life History and Genetic Characteristics of Plants and Animals in Agricultural vs Natural Ecosystems

Selection and Fitness

The definition and concept of fitness presented in Chap. 1 of this book and developed in subsequent chapters must be modified to explain genetic and life history characteristics of plants and animals in agricultural systems. In contrast to natural ecosystems, humans are the major selective force determining the distribution of plants and animals and their interactions in agricultural ecosystems. Fitness in wild populations is a function of adaptations to a wide gamut of factors leading to reproductive success over many generations. Fitness as applied to domesticated plants and animals can be defined as some combination of traits that convert energy and nutrients into products for human con-

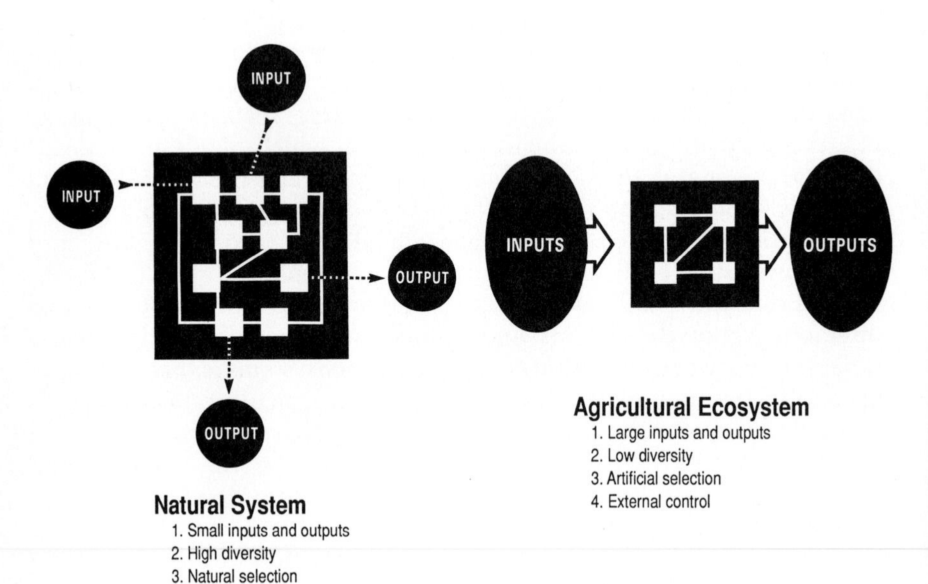

Figure 9–3. Major conceptual differences between natural and agricultural ecosystems. Small boxes represent groups of organisms ("components") and lines connecting boxes indicate flows of materials and energy among components. Circles and ellipses represent input and output of material and energy.

sumption. Depending on what characteristics are selected, many domesticated varieties of plants and animals would have difficulty in surviving in natural ecosystems. For example, forage plants have been bred for increased palatability over their wild congeners (e.g., more digestible nutrients per unit biomass, less tough-fibered tissue). However, greater susceptibility to disease and insect pests often accompanies increased palatability (Hanson, 1972). In plants, there is frequently a loss in competitive ability (Cox and Atkins, 1979). Similar patterns are observed with animals. Cattle that are bred for high meat or milk production in domesticated situations are often less able to withstand suboptimal conditions (e.g., food material with low-nutrient content, harsh environmental conditions) and are less resistant to diseases and parasites (Cole and Garrett, 1980).

Another key difference between selection in agricultural and natural ecosystems is the rate at which genetic changes occur in populations of plants and animals. Although agriculture is a relatively recent phenomenon in evolutionary time, the domestication process has caused very rapid genetic changes. Moreover, recombinant genetics and biotechnology have the potential to increase the rate of genetic change above the rates obtained through traditional breeding techniques (Austin, 1986).

In natural ecosystems, changes in one species as a result of selection pressures can produce subsequent changes in other species that interact with it. As emphasized in Chap. 1, these interactions can involve both reciprocal and diffuse coevolution. These processes also occur in agricultural ecosystems, but via anthropogenic selection pressures. For example, as plants were domesticated and human societies became more persistent in a given location, there was a concomitant selection and evolution in livestock toward types that were more sedentary, larger, and more productive on a per-animal basis (Clutton-Brock, 1981). Accordingly, forage and grain crops were developed for livestock consumption that were less fibrous, easier to digest, and that could be produced on a consistent basis (Table 9-3). The net result is that the genetic and life-history characteristics of plant and animal populations change through time, albeit through human intervention.

Genetic Variability

Wild populations typically have more genetic variation than their domesticated counterparts in agricultural ecosystems (Allard, 1960; Mason, 1984). The genetic composition of most wild plant and animal populations is comprised of a range of genotypes with characteristically wide frequency distribution (Stebbins, 1971). This genetic diversity

Table 9-2 Comparison of ecological characteristics of organisms and ecological processes in natural and agricultural ecosystems*

Characteristics	Agricultural Ecosystems	Natural Ecosystems
Life history and genetic characteristics of plants and animals:		
†Life-history strategies	*r*-selected	K-selected
†Life cycles	Short (annual)	Long (perennial)
Niche breadth	Broad	Narrow
†Reproductive allocation	High	Low
†Genetic variability	Low	High
†Defensive allocation	Low	High
†Mortality	Density independent	Density dependent
Population and community characteristics:		
Total organic material	Low	High
†Plant architecture	Simple	Complex
†Species diversity	Low	High
†Symbiotic associations	Few	Many
Chemical diversity	Low	High
†Food webs	Simple	Complex
†Population dynamics	Fluctuating	Constant
Energetics:		
†Net production (yield) per unit biomass	High	Low
Standing crop of biomass	Low	High
†Biomass supported per unit energy flow	Low	High
Food chains	Simple, short	Complex, long
Nutrient flux:		
Element cycles	Open (leaky)	Closed (conservative)
Pathways for chemical transformations	Simple	Complex
Exchange rates	Rapid between soil and plant biomass	Slow, efficient
†Control	Abiotic, anthropogenic	Biological

*The table presents characteristics as either-or comparisons. In reality these parameters vary by degree depending on specific situations being contrasted.

†Discussed in some detail in the text.

Source: Modified from Odum, 1969.

permits adaptation to a range of environmental conditions so that local populations or demes can be distinguished. If environmental conditions change, this genetic diversity confers an ability to adapt and persist (Ford, 1975). In contrast, the genetic characteristics of domesticated plants and animals have relatively narrow frequency distributions as a result of selection for specific purposes. If environmental conditions shift in agricultural ecosystems, then either new genetic varieties must be developed or the crops or animals must be subsidized with energy or material inputs.

The majority of modern varieties of hybrid maize or corn (*Zea mays*) that are grown in the United States are represented by a limited number of inbred lines or races of genetic material (Jugenheimer, 1976). By comparison, the wild analog of maize, teosinte (*Zea mexicana*), is much more genetically varied with many diverse, local races occurring in Mexico and Central America. Even older, domesticated varieties of maize were more genetically varied than the present genetic stock (Jugenheimer, 1976). Phenotypically, teosinte and maize look very different. The exaggerated size of the indehiscent fruiting bodies (corn ears) in modern maize contrasts with small reproductive organs in teosinte that retain their dispersal abilities (Fig. 9-4). Modern maize now requires high levels of energy and material inputs for successful growth (Pimentel, 1984).

The consequences of this trend toward genetic homogeneity in domesticated plants and animals are very important in determining the husbandry requirements of these organisms. One benefit is the uniform requirements of crop and livestock varieties in concert with their high yield potential. A negative aspect of this homogeneity is that diseases and pests can often very rapidly overcome defenses in genetically uniform populations. Also, as domesticated organisms become less variable through intensive, directed breeding programs, the potential to lose valuable genetic resources increases markedly (Schery, 1972; Cole and Garrett, 1980). This latter concern is somewhat allayed by several germ-plasm "banks" located around the world, where, for sound ecological reasons, diverse cultures of plant varieties are maintained.

Life-History Strategies and Life Cycles

The plants and animals that have been domesticated for agricultural purposes had life-history traits that predisposed them for domestication. For example, most of the major grain and vegetable crops, such as maize, soybeans (*Glycine max*), wheat (*Triticum aestivum*), and rye (*Secale cereale*) have evolved from "weedy" types of wild ancestors that were opportunistic, with annual life cycles as well as high growth and reproductive rates (Sauer, 1969). These ephemeral and rapidly growing plants allocate large portions of their energy and nutrients to reproductive organs, producing nutritious fruits and seeds. Essentially, agricultural breeding practices (artificial selection) over thousands of years have exaggerated these characteristics so that increasingly, larger portions of plant resources are allocated to those organs utilized by human consumers (see the maize example, Fig. 9-4).

Table 9-3 Brief survey of forage plant species and their application in grazing systems

Plant	Description
Alfalfa (*Medicago sativa*)	Originated in southwest Asia and is thought to have spread throughout the Mediterranean area by 500 B.C. where it was held in high regard as forage crop. Perennial legume with very high capacity for nitrogen-fixation activity. Has worldwide distribution as forage crop; although spp. limited to moderate to high soil fertility conditions. In terms of protein and vitamin content and palatability, alfalfa has no superior. Especially valued as very productive dairy forage in the U.S. and Europe. Does well in polyculture with cool-season C3 grasses, and contributes to soil improvement when used in rotation with other crop spp.
Clovers (*Trifolium* spp.)	At least 30 spp. of clovers are used in grazing systems world wide. Like alfalfa, clover is a legume and contributes to soil fertility via nitrogen fixation. Because of high diversity in the genus, clovers are adapted to a wide range of climatic conditions—arid vs humid, high-fertility vs low-fertility soils. Perennial, biennual, and annual habits are all included within the genus. Clovers are palatable, nutritious forage crops and important source of protein for livestock. Most do well in mixed plantings with grass spp.
Other legumes	Although clovers and alfalfa are the major forage legumes in North America and many other areas of the world, there is great diversity of other legume genera and species that are very important conponents in grazing systems. Representative list of legumes includes birdsfoot trefoil (*Lotus corniculatus*), lespedeza (*Lespideza* spp.), vetch (*Vicia* spp.), sainfoin (*Onobrychis vicifolia*), lupine (*Lupinus* spp.), cowpea (*Vigna sinensis*).
Bluegrass (*Poa pratensis*)	Thought to be native to North America and Eurasia and adapted to a wide variety of soil and climatic types. Grows best in cooler regions having moderate to high soil fertility. Long-lived perennial, propagating vegetatively with rhizomes. One of the most dominant pasture grasses in humid areas because it withstands heavy grazing pressure so well. In natural grasslands, does not do well in the absence of grazing by large herbivores. Grows well in mixtures with legumes having a short growth habit (such as white clover, *Trifolium repens*).
Bromegrass (*Bromus* spp.)	*Bromus* comprises about 60 spp. in grass subfamily Festucoidae (originated in Middle East). Resembles oats and is adapted to cool climates. Has aggressive reproductive strategies both in terms of vegetative spread (rhizomes) and self-seeding. Rugged, persistant forage crop in North America and Europe, being both palatable and nutritious. More sensitive to heavy grazing pressure than in bluegrass.

Table 9-3 *(continued)*

Plant	Description
Timothy (*Phleum pratense*) and orchardgrass (*Dactylis glomerata*)	Native Eurasian cool-season, perennial bunch grasses providing high-quality grazing and hay crops. Do particularly well in mixed plantings with tall-growing legumes such as alfalfa (*Medicago sativa*) and red clover (*Trifolium pratense*).
Fescue grass (*Festuca* spp.)	More than 100 spp. belong to *Festuca,* most of which are adapted to wider growing conditions than strictly cool season grasses (such as timothy) and orchardgrass. Tolerant of both drought and poorly drained conditions. There are both tall and short varieties of fescue so that genus has application as pasture and hay crop.
TROPICAL FORAGES:	Forage species suitable for tropical and subtropical regions often have characteristics and requirements that are quite different from spp. grown in temperate zones. Pasturelands in tropics often replace rain-forest climax vegetation. Therefore, because of long growing seasons and aggressive nature of this natural, weedy vegetation, grasslands in tropical climates form unstable subclimaxes. Rather than discuss individual species in tropical grasslands, pasture associations are listed below and characterized according to different types of habitats.
Grasslands derived from broad-leaved woodlands	Broad-leaved woodlands are characterized by deciduous and evergreen trees with a continuous groundcover dominated by grasses. This type of vegetation common in East Africa is typified by trees in genera *Brachystegia, Julbernardia* and *Isoberlinia* and tussock grasses mostly in genus *Hypanhenia.* These understory grasses are most prevalent in arid subclimates and least common in more humid areas. In these types of ecosystems, pasterage commonly known as *sour veld* is common, meaning that young grass is palatable to cattle and sheep but soon becomes rank and low in nutritive value. These areas also contain some legumes and more palatable grasses which can be encouraged with controlled burning practices.
Grasslands derived from montane evergreen forests.	Grasslands resulting from montane forests usually occur at elevations between 1000 and 3000 m with rainfall from 100–400 cm/yr, that is, a fairly moderate, moist climate. Areas like this are found in Africa, India, and South America. Grass spp. in these areas are short to medium height and, for example, in Kenya are dominated by the Kikuyer grass (*Pennisetum clandestinum*), which makes a productive association with the clover *Trifolium semipilosum.* Montane-derived grasslands are considerably more productive and can support more livestock per unit land area than pasturelands derived from broad-leaved forests.
Cultivated pastures	Sown pastures are only practical in regions having moderate to high rainfall, where climax vegetation is highland forest or wetter types of broad-leaved woodland. In cultivated pastures, grasses and legumes are sown into tilled soil and fertilized with either animal manure or chemical fertilizer. Some grass species used in cultivated tropical pastures include: *Brachiaria brizantha, Cenchrus*

Table 9-3 *(continued)*

Plant	Description
	ciliaris, Chloris gayana, Cynodon dactylon, Digitaria decumbens, and *Setaria sphacelata.* Legume species also play a major role in these cultivated pastures and some of the more common spp. are *Trifolium subterraneum, Trifolium repens, Lotononis bainesii, Piseraria phaseoloides,* and *Stylosanthes guianesis.* Mixed grass–legume pastures typically yield more than comparable stands of grass alone and have better nutritive value in polyculture.

Similar evolutionary trends have occurred in animal livestock species where production of particular qualities have been exaggerated through artificial selection and management practices. Wild counterparts of grazing mammals such as goats, bison, and sheep (members of the family Bovidae) are preadapted for domestication. These animals efficiently convert high-cellulose fibrous plant material (e.g., mostly grasses and legumes) into high-protein meat and milk products. Sheep and cattle digestive tracts have evolved to take advantage of the primary production occurring in grasslands and steppes. The animals have ruminant digestive systems with multichambered stomachs containing symbiotic microorganisms that help break down cellulose (see Chap. 5). Apart from the digestive systems that convert plant fiber to animal protein, most of these grazing animals possess very gregarious behavioral characteristics allowing the animals to be kept in large groups or herds with minimal effort. In addition to being a source of meat, these animals produce milk that can be converted to easily stored milk products (e.g., cheese) that contribute to the stability of food resources. Reed (1969) pointed out that the automatic elimination of aggressive animals in bovine herds leads to submissiveness, and those animals that breed best in captivity contribute most to future gene pools. In addition to modifying behavioral characteristics, selective breeding in captivity increases the traits most directly beneficial to humans (e.g., rapid muscle and flesh development, high milk production). However, these traits were gained with an accompanying loss in other traits. Domesticated animals can no longer effectively defend themselves against predators or adjust to harsh environmental conditions.

The life-history characteristics of most pest species also predispose them to success in agricultural ecosystems. A pest can be defined as any organism that causes loss in quantity or quality of agricultural products. Just as is the case for many crop species, weed and insect pests often can be characterized as *r*-selected (i.e., opportunistic, with well-developed dispersal abilities, and rapid intrinsic rates of increases) (Radosevich and Holt, 1984). Grime (1979) classifies most of the common weeds as *ruderals,* a term that characterizes plants found in highly disturbed but potentially productive environments. These plants are usually herbs with characteristically short life-spans and high seed production. They typically occupy the earliest phases of succession.

Effects of Agriculture on Population and Community Characteristics

Distribution of Plants and Animals

Humans intentionally and inadvertently control the distribution of many plants and animals through agricultural practices. The spread of agriculture throughout the world has so profoundly influenced the distribution of many plants and animals, that it can be compared to major geological forces such as continental drift in determining the distribution of species. This is true for both the species intentionally selected for agricultural production and the pest species, including weeds, insect pests, and pathogens (Cox and Atkins, 1979). When an exotic crop species is introduced to an area, associated species often are inadvertently introduced. For example, the European corn borer (*Ostrinia nubilalis*), a noctuid moth, gained worldwide distribution as a result of the spread of maize agriculture. The insect originated in eastern Europe, and the first record of its presence in North America was in the Boston

Figure 9–4. Maize *(Zea mays)*, left, and teosinte *(Zea mays mexicana)*, right. Note the extreme differences in size of reproductive organs and general appearance of the two types.

area in 1917 (Caffrey and Worthley, 1927). Since that time, the insect has spread over North America to become one of the most serious pests of maize. Moreover, there is substantial evidence that the European corn borer has, over the past 50 to 60 years, divergently evolved into several ecotypes, each locally adapted to different geographic and farming conditions (Showers et al., 1975). Agriculture has allowed the European corn borer not only to expand its distribution, but also to increasingly specialize and exploit a single host-plant, maize.

Species Composition and Diversity

When a virgin forest is cut, cleared, plowed, and planted with a crop, the species composition is greatly changed. The dominant species change from endemic species that evolved reciprocally with other species to species that in many cases evolved in different geographic regions and types of ecosystems. Maize and soybeans, both of which are exotic species to the United States, have replaced much of the native forest and prairie vegetation. These crop species often bring associated pests with them, as discussed in the maize–European corn borer example above. Similarly, the Western corn rootworm (*Diabrotica vigifera*) is believed to have evolved in Mexico and is now one of the most serious pests of maize. In addition to associated insect species, introduction of exotic crops can also bring with them associated plant species. *Setaria*, or foxtail, an ubiquitous weed in North America, is thought to have originated in Mexico and Central America in association with wild maize (Branson and Krysan, 1981).

Besides the direct effects on species composition associated with replacement of dominant species, there are important indirect effects. Indeed, these indirect effects may have more impact on the total species composition and diversity of agricultural ecosystems than the direct effects. Changes in plant architecture and landscape diversity as a result of agricultural practices will strongly influence the presence and distribution of many animal species (Risch et al., 1983; Forman and Godron, 1986). Lawton and Schroder (1977), in examining a gradient of habitats from forest to annual plant communities, showed that insect numbers and species richness became increasingly depauperate in the sequence: forest > shrubs > perennial herbs > weeds and other annuals > monocotyledons. Plowing and cultivation appreciably change the soil environment physically and stimulate the mineralization of nutrients from a relatively immobile to a mobile form. This process then affects nutrient content of plant tissues (Stinner et al., 1984b), which in turn, may influence herbivory levels (Scriber, 1984). In addition, plowing and cultivation significantly influence populations of soil-inhabiting predators that also may influence herbivore populations (House and Stinner, 1983; Brust et al., 1986).

Types of Interactions

Agriculture affects the quantity and types of interactions among organisms. Because agriculture generally reduces species composition and diversity, the repertory of interactions occurring among plants and animals in agricultural ecosystems is often much more restricted than those in natural ecosystems. For example, cattle on a midwestern farm consume a limited variety of plant species, for example, maize and alfalfa (*Medicago sativa*), compared to white-tailed deer (*Odocoileus virginianus*) feeding on a wide variety of vegetation in the same region. Moreover, in agricultural ecosystems, each cultivar or livestock variety is genetically quite uniform, resulting in reduced behavioral variation among individuals. Plowing and insecticide applications decrease higher trophic levels, such as predator and parasite guilds that help regulate herbivore populations (Price, 1984; Brust et al., 1985).

Although there is little direct evidence for how agriculture affects the relative proportions of different kinds of interactions (e.g., antagonism, mutualism, commensalism), it has been hypothesized that there is a shift toward antagonistic interactions (e.g., crop–weeds, plant–pathogens, plant–herbivores) in agricultural versus natural ecosystems (Crossley et al., 1984; Stinner and House, in press). It is these antagonistic interactions in particular that humans spend a great deal of energy (mechanical and chemical) trying to control.

Large-Mammal Herbivory in Grazing Systems

In natural ecosystems, herbivores have major impacts on the characteristics of plants and plant communities, including physiology, morphology, population dynamics, species diversity, and production (see Chaps. 4 and 5) and in regulating rates of energy and nutrient transfers (Chew, 1974; Mattson and Addy, 1975; and Chap. 5). Herbivory is strongly controlled by human intervention in agricultural ecosystems, either by maximization as in large-herbivore grazing systems (e.g., cattle and sheep), or minimization, as in grain and vegetable systems. In grazing systems, pasture lands are often fertilized with inorganic nutrients (mainly nitrogen, phosphorous, and potassium) to increase the growth and nutritional value of forage plants. This fertilizer subsidy functions to increase the quantity of grazers per unit of land area (e.g., carrying capacity) that a grassland can handle without incurring the adverse effects of overgrazing. Plant breeders select for characteristics in forage plants that respond positively to anthropogenic subsidies (e.g., fertilizer, cultivation) to maximize biomass production (Heath et al., 1973). In turn, livestock varieties that utilize these forage cultivars are selected to convert these forage plants into products for human consumption (Reed, 1969).

In both natural and agricultural grazing systems, there are optimal densities of grazer animals per unit of land area. In natural grasslands, the mechanisms that regulate this density are complex functions of the biology and ecology of both the plants and animals, while in agricultural grazing systems, humans exert this control. In both systems, if there is too much grazing, adverse effects on plants ensue with eventual shifts in plant community composition (see Chap. 5).

Induced regression and *induced progression* are terms used to denote a continuum of negative and positive effects of consumers on plant communities and land quality (e.g., fertility, erosion) (Lewis, 1972). Induced regression in pasturelands by overgrazing is typified by reduced mulch cover, high fluctuations in the soil microclimates, soil compaction (Reynolds and Packer, 1963), disruption of normal phenology of plant production, and alteration of biogeochemical cycles (Vogel, 1966; Rauzi et al., 1968). To promote positive effects or progression in agricultural grazing ecosystems, humans must regulate the distribution and density of grazing animals. Also, maintaining a mixture and diversity of plant species is critical to maintaining plant productivity. Legume–grass mixtures, for example, produce more biomass than single-species cultures of either plant type (De Wit, 1960; Donald, 1963).

Insect Herbivory in Agricultural Systems

Insect herbivores play a major role in the ecology of cropping systems. It is estimated that 25 percent of all crops in the United States are lost annually to insects and other arthropods (USDA, 1985). As Weis and Berenbaum point out in Chap. 4, 21 percent of all known insects are herbivores, distributed among eight orders. Agriculturally important insect herbivores are found in each of these orders. Insects damage nearly all crop species. These phytophages can be divided into tissue feeders and sap (phloem and xylem) feeders. The former encompass the majority of herbivorous arthropods, but the latter group is especially important as vectors of plant pathogens (Nault et al., 1976). Table 9-4 shows a brief list of arthropod taxa and their involvement as crop pests. From this list it should be apparent that there are certain taxa, such as Lepidoptera (moths), that are major crop pests.

Annual cropping systems (e.g., maize, soybeans, cotton) are "open" plant communities with more or less constant migration of insects back and forth from adjacent environments. Price and Waldbauer (1982) point out that individual crop fields should be considered as islands in a matrix composed of a mosaic of habitat types (e.g., woods, fallow field, and land planted in crops). Insects (both herbivores and predator–parasite guilds) invade and reinvade crop fields from the surrounding matrix so that there is likely to be a gradient in population density from field edge to center. As a consequence, herbivory levels and insect damage to crops are often patchy or follow gradients from field edges to centers. Lepidopteran larvae, such as the stalk borer (*Paipapema nebris*), are harbored in the perennial grasses surrounding maize fields. Stalk borer larvae migrate from field edges into maize fields to feed on young maize plants. Damage to maize plants declines with distance away from field edges (Stinner et al., 1984a). Price (1976) found that in an annual soybean system, most herbivores never reached an equilibrium in terms of community composition because of migration from adjacent habitats.

Two major aspects of agriculture contribute to the high levels of insect herbivory often observed in crop systems. First, most conventional cropping systems used in large-scale mechanized agriculture are planted in monoculture. Second, crop plants have reduced chemical defenses as a result of human selection to increase palatability and nutritional value. The ancestors of crop plants were short-lived colonizing species that occurred in successionally immature environments in mixed stands. In monoculture, crop plants are uniformly distributed in high densities, therefore they are readily apparent to potential herbivores. If herbivores find one crop plant, it is likely that they will find more. Most insect herbivores can find their hosts more efficiently when other plant species are not present to interfere. This pattern especially holds true for insects that have a single host species (Tahvanainen and Root, 1977; Rausher, 1981). Also, there are data to suggest insects are less inclined to stay on their host plants in polyculture versus monoculture (Bach, 1980). Therefore, many herbivorous insects are more common per plant in monocultures than in polycultures. A formalization of this concept is termed the *resource concentration hypothesis* that states: "Herbivores are more likely to find and remain on hosts that are growing in dense or nearly pure stands; and that species which are more specialized in feeding habits frequently attain higher relative densities in simple environments" (Root, 1973).

Concentrations of alkaloid, terpene, and tannin compounds that offer resistance to herbivore attack in wild plants have been purposely reduced in crop plants. In natural systems, the likelihood of herbivores finding their hosts has been analyzed in terms of plant apparency. Rhoades and Cates (1976) recognized two plant strategies for defense

Table 9-4 Major crop species and their associated insect pests

Crop	Associated Insect Pest *Common Name (Order: Family)*	Description
Corn or maize	European corn borer (Lepidoptera: Pyralidae)	Larvae damage maize by feeding on leaves of young maize plants and later burrow into the stalks causing a structural weakening of the plant. Entrance holes serve as entry sites for both saprophytic and pathogenic fungi and bacteria. Species has worldwide distribution.
	Black cutworm (Lepidoptera: Noctuidae)	Larvae feed on and cut off young maize plants at soil level. Has worldwide distribution.
	Corn earworm (Lepidoptera: Noctuidae)	Larvae damage developing grain; insect's feeding areas serve as sites for mold to develop. Earworm feeding can also interfere with pollination; causes economically important damage to cotton and soybeans. Endemic in North America but has spread to other continents.
	Corn rootworm (Coleoptera: Chrysomelidae)	Larvae feed on maize roots which reduces nutrient and water absorption and can result in the plants falling over in high winds. Several spp. of rootworms or *Diabrotica* exist in U.S., although genus is most intensively speciated in Mexico and Central America.
Cotton	Boll weevil (Coleoptera: Curculionidae)	Is one of the most destructive insects of cotton. Feeds on developing cotton seed pods or bolls and can completely destroy a cotton harvest when beetles are especially abundant. Pest is a native of Mexico and Central America although it has spread rapidly to most of cotton-growing regions of the U.S.
	Pink bollworm (Lepidoptera: Noctuidae)	Larvae of the insect feed extensively on cotton bolls. Regarded as the most serious pest of cotton on worldwide basis.
	Lygus bugs (Hemiptera: Miridae)	These sucking insects damage cotton and other plants by withdrawing sap from leaves and reproductive organs. Have worldwide distribution.
	Tobacco budworm (Lepidoptera: Noctuidae)	Pest is in same genus as corn earworm (*Heliothis*) and feeds on developing cotton bolls. Is a serious pest of cotton in southeastern U.S.
Soybean	Bean leaf beetle (Coleoptera: Chrysomelidae)	Feed on soybean leaves and can cause up to 40% defoliation of leaf biomass.

Table 9-4 *(continued)*

Crop	Associated Insect Pest *Common Name (Order: Family)*	Description
		Also distribute pathogenic viruses to soybeans. Ranges over most of North America.
	Mexican bean beetle (Coleoptera: Coccinellidae)	Larvae damage soybeans by "scraping" off leaf epidermal tissue producing veined appearance on soybean foliage. Is a serious pest of soybeans primarily in Atlantic Coast area and Ohio Valley.
	Soybean looper (Lepidoptera: Noctuidae)	Larvae damage both foliage and developing pods of soybeans. Although species occurs from South America to Canada, economic losses in the northern portion of its range are minimal.
	Green cloverworm (Lepidoptera: Noctuidae)	Damages soybeans by eating leaf tissue between main veins of leaflets. Cloverworm is a particularly serious pest because maximum defoliation usually occurs during blossom and pod development times. Has a wide distribution in most soybean growing areas but causes most severe damage in the midwestern U.S.
Small grains (wheat, oats, rye)	Hessian fly (Diptera: Cecidomyiidae)	Larvae damage wheat by tunneling into plants, either killing young plants or reducing grain yield by up to 35–40%. Introduced from Europe; has spread over most of North America.
	Cereal leaf beetle (Coleoptera: Chrysomelidae)	Relatively recent pest of wheat, oats, and rye in the U.S. Was first identified from collections made in 1962. Chews on leaves and sheaths of the grass family and is particularly damaging to young grain crops.
	Green bug (Homoptera: Aphididae)	This aphid is considered the most destructive pest of small grains in central and south-central U.S. Damages plants by sucking plant sap and by injecting toxic substances that results in necrosis of plant tissue.
	Grasshoppers (Orthoptera: Cyrtacanthacridinae)	Consume young plants completely, strip leaves from older plants, and feed on maturing grain. Some spp. are migratory (locusts) and most damaging in arid parts of the world.

Note: The examples presented here are intended to be representive rather than exhaustive.

against herbivores that appear to be characteristic of different stages of succession. One strategy employs defenses that are effective against many species of plant feeders that predominate in mature communities where plants are long-lived, occur in relatively pure stands (e.g., oak or maple trees) and are, therefore, *apparent* or easily found. Apparent plants appear to have evolved defenses that are not directed toward specific herbivores. The mechanisms of this defense strategy include tough leaves, low available nutrient content, and large amounts of unspecific chemicals such as tannins.

In contrast, short-lived, colonizing plants in less successionally mature environments usually occur in mixed species stands and are therefore more difficult to find and are *unapparent* to herbivores. This latter type of plant is protected by defenses directed toward herbivores that have specialized feeding habits on particular plant species. These defenses include diverse toxic chemicals that provide barriers to specific groups of herbivores. Crop plants have evolved from these short-lived species and artificial selection has reduced the protective chemicals. Moreover, domesticated plants are usually grown in monoculture so that their apparency is high. In essence, agriculture has changed both the defense capabilities and apparency of plants.

Weis and Berenbaum (Chap. 4) point out that the nutritional status of host plants also affects herbivory. Protein nitrogen is a component of plant tissue that is critical for insect growth and reproduction (Scriber, 1984). As nitrogen content of plant foliage increases, assimilation efficiency, growth, and fecundity of herbivores often improve (Onuf, 1978; Mattson, 1980). Since high protein content is usually one of the traits selected in agricultural plants, it is little wonder that herbivory levels are high in crop habitats. Not only is it easier for insect herbivores to find crop plants, but once found, crop plants provide an easily assimilated and high-quality resource.

To partially compensate for the palatability of most crop plants, resistant varieties of crops are bred to decrease herbivory by certain insects. There are many traits that determine resistance, and they often involve more than nutrient status or toxicity of plants to insects. Morphological (e.g., leaf and stem pubescence) and phenological (e.g., crops maturing before or after peak herbivore densities) characteristics are also important mechanisms conferring resistance to plants (Day, 1972; Jones, 1977; Gallun and Khush, 1980). It is important to emphasize, however, that resistant varieties are rarely immune from attack. Rather, resistant varieties support smaller populations of herbivores than the susceptible varieties (Rufener et al., 1986).

A well-documented example of breeding for host-plant resistance is the work on genetic resistance in maize against the European corn borer (Beck, 1965; Scott and Guthrie, 1966). Certain genetic strains of young maize plants are highly resistant to corn borer survival. A complex biochemical apparently responsible for the resistance, 2,4-dihydroxy-7-methoxybenzoxazine-3-one or DIMBOA, was isolated from these resistant maize varieties. Subsequent breeding experiments demonstrated that the presence and concentration of DIMBOA had a genetic component that could be introduced into commercial varieties of maize. The introduction of genetic resistance against herbivores into crop plants is one of the most environmentally sound means of controlling pests.

Predation and Parasitism

In natural ecosystems, interactions between plants and animals are strongly influenced by higher trophic levels—predators and parasites—a point developed in earlier chapters of this book. The third or secondary consumer trophic level is extremely important, because it often plays a regulator role in limiting herbivore population growth. In agricultural ecosystems, predator and parasite guilds frequently are severely depressed by cultural practices (Foster and Ruesink, 1984; Brust et al., 1986), and therefore the effects of higher trophic levels in controlling herbivory are substantially diminished. Agricultural practices often destroy habitat for predators and insecticides kill both pests and predators indiscriminately (Gholson et al., 1978; Brust et al., 1985).

Two major groups of entomophagous insects are distinguished: parasites and predators. Parasites and parasitoids are usually more specialized while mixed groups of predators are relatively generalized in their feeding habits. The former group plays a key role in classic biological control programs and involves identification, collection, mass rearing, and release of specific parasitoids or predators of a particular pest. Examples include the release of predatory mites to control plant-feeding mites (Oatman et al., 1968) and parasitic wasps to control Mexican bean beetles (*Epilachna varivestis* Mulsant) on soybeans (Stevens et al., 1975). Notably, most of the successes with introduction and mass release of parasites and predators have occurred in warm, fairly stable climates. Attempts at biocontrol through introduction and release have been less successful in more harsh climates. In temperate zones, conditions are often such that once a particular prey population is reduced, alternative food resources are in short supply so that the entomophages become locally extinct before prey numbers increase (Stehr, 1982).

Predatory entomophagous insects include such taxonomically diverse taxa as ladybird beetles (Coccinellidae), ground beetles, (Carabidae), spiders (Arachnids), lacewings (Neuroptera: Chrysopidae), and assassin bugs (Hemiptera: Anthocoridae). These predators, with generalized feeding habits, differ ecologically from their more specialized counterparts in that they cannot usually reproduce nor can they track pest populations (numerically respond) as rapidly as can the more specialized parasitoids (Stehr, 1982). Nevertheless, these generalist predators can play an important role in keeping herbivore populations below economic thresholds in agricultural systems, if they are provided with suitable habitat and are not killed by insecticides (Brust et al., 1986). However, even under the best conditions for the predators, once prey numbers exceed a certain level, predators may no longer be able to limit prey numbers. Figure 9-5 illustrates the distinction between predatory and parasitic entomophages and their effects on the hypothetical population dynamics of insect herbivores.

Multitrophic Interactions

As stressed in earlier chapters, plant–animal interactions should be examined within the context of at least three trophic levels to be fully understood. In agricultural ecosystems, multitrophic interactions involving plants and insects have been addressed in terms of how plant nutritional status and resistance (defense) interacts with the dynamics of predators and parasitoids. Price (1986) points out that many mechanisms of plant defense may positively or negatively influence higher trophic levels. Painter (1951) found that resistant plant varieties could affect predators and parasites (natural enemies) in two fundamental ways by: (1) decreasing insect herbivory to levels that reduce successful host finding, and (2) altering prey quality due to the prey feeding upon resistant plant varieties. Bergman and Tingey (1979) and Price (1986) expanded Painter's dichotomy to encompass the following mechanisms of interactions:

1. *Semiochemically mediated interactions* are those in which a chemical produced by a plant attracts or repels natural enemies to or from their prey. For example, the aphid parasitoid, *Diaeretiella rapae*, is attracted to mustard oil (allyl isothiocyanate) in crucifer plants that harbor aphids (Read et al., 1970).
2. *Chemically mediated interactions* are those in which nutritional or defensive chemical components of plant material are sequestered from plant to herbivore to predator or parasite. Thurston and Fox (1972) showed that the larval survivorship of the parasite *Apanteles congregaters* was lower when its host, the herbivore larvae *Manduca sexta*, fed on a high-nicotine (from tobacco) diet compared to a low-nicotine diet.

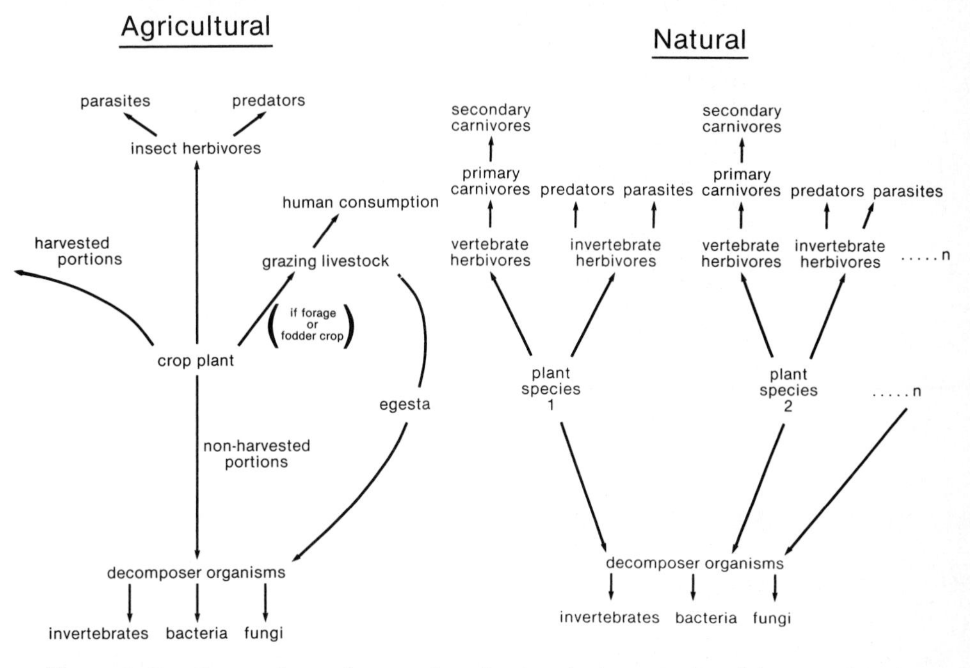

Figure 9–5. Comparison of exemplary food webs in agricultural (monoculture) and natural habitats. The potential trophic complexity in the latter is much greater than in simplified agricultural systems.

3. *Physically mediated interactions* involve physical characteristics of plants that either facilitate or impede natural enemy utilization of herbivorous prey. Pimentel (1961) reported that within the cabbage genus *Brassica* parasitism is higher on caterpillars feeding on open-leaved varieties like broccoli than on tight-leaved cabbage.

Duffey et al. (1986) offer the caveat that at present no body of knowledge permits us to completely interrelate all the mechanisms affecting tri-trophic interactions involving crop plants, herbivores, and natural enemies. However, this area of plant–animal interactions is gaining a good deal of interest and enthusiasm among researchers (Boethel and Eikenbary, 1986).

Influence of Insecticides on Interactions and Pest Management

Human control of insect herbivory in cropping systems is substantial. At one time it was thought that insect pests could be eradicated with chemical toxicants (Hunter and Hinds, 1904). At first, inorganic poisons such as arsenic, copper, and cyanide compounds were used to kill insects

(Smith, 1978). Later, in the 1940s, DDT (dichlorodiphenyltrichloroethane), an environmentally persistent, chlorinated hydrocarbon, emerged as a major insect-control measure. DDT did kill insects by the millions and was used extensively (Luckmann and Metcalf, 1982). However, within 10 years of its introduction certain species of insects, like the common housefly (*Musca domestica*) evolved genetic resistance to what had promised to be a panacea for insect control. This loss in efficacy of DDT coupled with the environmentally damaging effects of chlorinated hydrocarbon type pesticides, caused toxicologists to search for new chemistry to control insect pests.

Presently, there are three classes of insecticides used in modern agriculture: organophosphates, carbamate, and pyrethroid compounds. None of these toxicants are environmentally persistent, but they still can cause serious pollution problems (Brown, 1978). In addition, many insects and other arthropods have evolved genetic resistance against these toxicants (Harding and Dyar, 1970; Georghiou and Taylor, 1976). During the 1960s, it became apparent that eradication or complete control of insect pests would probably never be accomplished—in point of fact, no insect-pest species has ever been made extinct through human efforts (however, we may have caused extinction of nonpest species).

Many insecticides (along with herbicides and fungicides) are relatively unspecific in what organisms they affect. Therefore, the indirect effects of a particular compound may be more important than its direct effects (Ingham, 1985). Before a chemical can be marketed as an insecticide, the manufacturer must be able to demonstrate that the chemical does in fact control a pest or a suite of pests on a specific crop. However, many of the pesticides are also toxic to other organisms. For example, toxaphene was developed as an insecticide, but at high rates of application it is used as a herbicide in cotton agricultural systems (Reynolds et al., 1982). An herbicide that removes weeds may significantly impact insect herbivores because the toxicant removes habitat for parasites and predators (Alterei and Whitcomb, 1979). For many other examples of nontarget effects of insecticides, herbicides and fungicides on plants, invertebrates, and microorganisms, see Pimentel and Edwards (1982).

As an alternative to complete chemical control, economic entomologists have developed the integrated pest management (IPM) concept in which crop pests are managed with a variety of biological (e.g., predators and parasites, host-plant resistance), chemical, and cultural (e.g., varying planting dates, tillage) procedures. Under the IPM concept, certain population densities of pests are tolerated and chemicals are applied only when it is economically worthwhile to do so. IPM involves application of some key ecological phenomena observed in natural ecosystems, such as: (1) maintaining diverse trophic webs, (2) relying on several rather than one source of mortality against pests and diseases, (3) tolerating low levels of herbivore populations to dampen oscillations of parasite and predator populations, and (4) incorporating genetic resistance into crop varieties. Although IPM has permeated the agricultural world, not all cropping systems utilize IPM (Pedigo et al., 1986). For example, perishable, directly consumed crops such as vegetables and fruits are often treated with insecticides on a regular basis, regardless of pest densities. These commodities must be near-cosmetic perfection because consumers demand blemish-free produce. Still, IPM is a powerful concept shaping agricultural practice. It is an example of how application of knowledge from natural ecosystems can help produce more stable and environmentally safe agricultural ecosystems.

Interaction Networks

Figure 9-6 compares networks of interactions occurring in a representative farming system that includes both plant and animal crop species and a representative natural ecosystem. It should be obvious that agriculture decreases the complexity of interactions. In addition, some interactions between plants and animals are direct and rather straightforward, such as plant biomass channeled into cattle or insect-herbivore biomass, while others (e.g., three- and four-trophic interactions among crops, weeds, insect-herbivores, predators, and parasitoids) exert their influence on crops mostly via indirect mechanisms. For example, weeds can be important indirect mediators of herbivory interactions since few cropping systems are weed-free. Weeds harbor many arthropod species that can contribute either negatively or positively to crop production (Tripathi, 1977; Crossley et al., 1984). For example, for the stalk borer (*Papaipema nebris*), weeds provide protected places for the adult insects to lay eggs (Stinner et al., 1984a). In contrast, for the black cutworm *Agrotis ipsilon,* weeds serve as alternate food plants and removing weeds can increase damage to maize plants (Levine et al., 1981). In addition, some weed species serve as reservoirs for diseases such as maize chlorotic dwarf virus that is spread by leafhoppers (Cicadellidae) from Johnson grass (*Sorghum halepense*) to maize (Nault et al., 1976).

Many of the crop–insect herbivore interactions in agriculture can be thought of as a three-way interaction among insects, crops, and weeds. The influence of weeds on crop–insect interactions can become much more complex when the focus is shifted from two- to three-trophic levels. Noncrop plant species in or adjacent to agricultural fields can provide appreciable habitat space for entomophagous arthropods or predators

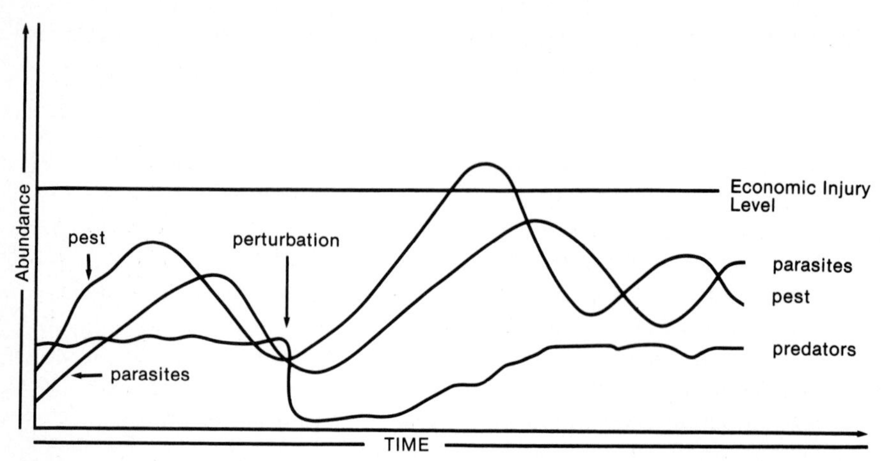

Figure 9–6. Generalized relationships among arthropod pest, predator, and parasite population dynamics. The economic injury level indicates the density at which pest numbers significantly reduce crop yield quantity or quality. Predator numbers remain at low but relatively stable densities until some disturbance like insecticide application lowers their numbers. After predator numbers are lowered, parasite–pest population cycles increase in amplitude until predator numbers recover. In essence, parasite and predators operate synergistically to affect pest population dynamics.

and parasites of pest species and in this indirect way contribute a positive influence on crop production.

Interactions in Soils

Another important group of interactions that are indirect determinants of plant growth and primary production occur in soil systems. Studies on belowground biotic interactions have depended largely on functional interpretations (e.g., roles of organisms in energy flow and nutrient cycling) as explanations for observed phenomena (Coleman, 1985; Ingham et al., 1985). Little evolutionary interpretation has been applied to data collected on plant–animal interactions occurring in soil systems. Yet, a rich variety of biological interactions exist in soil environments of natural and agricultural ecosystems. Major groups of soil-inhabiting animals that can influence plant physiology and ultimately fitness parameters of plants include arthropods, earthworms, nematodes, isopods, and mollusks (e.g., slugs and snails). While some interactions between invertebrates and plants in soils have direct effects, such as root feeding, there is growing evidence that many, perhaps most, soil-inhabiting insects and other invertebrates exert their influence on plants indirectly through processes mediated by microorganisms (Coleman, 1985).

The diagram in Fig. 9-7 indicates the kinds of organisms found in soil systems. The indicated biota and their interactions with other organisms can have both negative, neutral, or positive influences on plant growth and reproduction. For example, root-feeding nematodes and insects can appreciably reduce the capacity of plants to absorb water and nutrients (Wallwork, 1970; Branson et al., 1980). The net result is antagonism toward the plants. Alternatively, many groups of soil invertebrates alone, and in concert with microorganisms, appear to function in ways that benefit plant processes. Soil-inhabiting insects and arthropods play an important role in breaking down decaying plant material into nutrients available for uptake by growing plants (Coleman et al., 1983). Soil animals, by grazing on bacteria and fungi, stimulate mineralization of nitrogen and phosphorus which increases the rates of nutrient absorption into plant biomass (Ingham et al., 1985).

The complexity of many interactions between soil animals and plants makes them difficult to analyze according to traditional evolutionary paradigms. However, because many of these interactions involve nutrient transformations, examination of them in context of fitness relationships and ecosystem processes (i.e., energy and nutrient dynam-

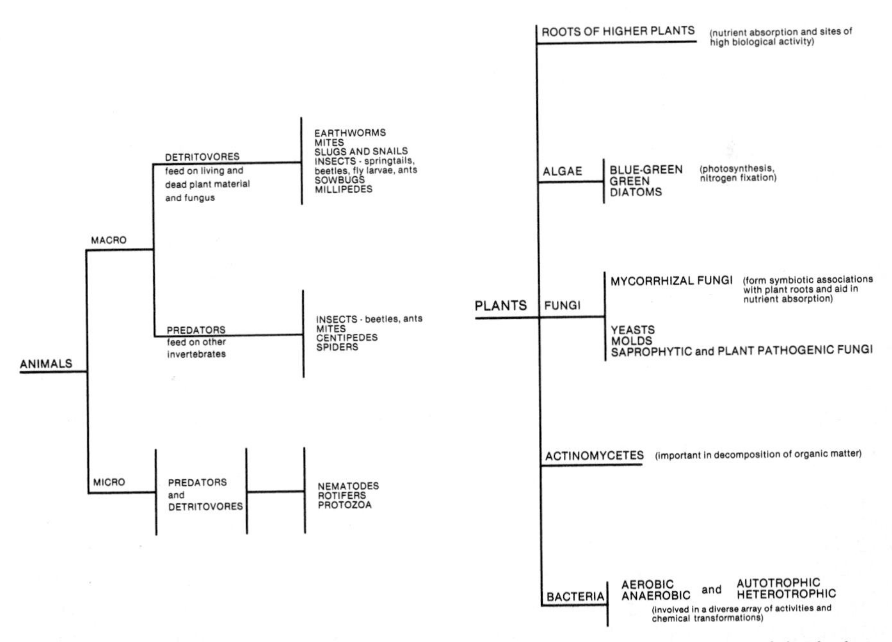

Figure 9–7. Taxonomic and functional classification of soil biota (modified after Brady, 1974).

ics) concomitantly may provide a more comprehensive understanding. Most taxa of higher plants, for example, form symbiotic associations with mycorrhizal fungi. The fungus aids plant roots in nutrient absorption and in return receives photosynthetically fixed carbon compounds from the plants. The relationship has a net mutual benefit for both plant and fungus (Harley and Smith, 1983). In turn, the fungus may be consumed by certain groups of soil invertebrates including earthworms, nematodes, and arthropods (Coleman, 1985). It has been hypothesized that these animals, by grazing on fungal biomass, can alter nutrient absorption characteristics that will ultimately affect the plants.

Agriculture can greatly influence these complex soil interactions via its impact on soil biota, physical structure, and chemical characteristics. Agricultural practices can be placed into three major groupings: (1) soil tillage and cultivations, (2) application of organic and inorganic fertilizers, and (3) application of xenobiotics (primarily fossil fuel-derived pesticides). Each of these practices has an intended purpose of ultimately increasing crop yield, but as previously discussed for insecticides, there are also nontarget effects that affect plant growth. For example, apart from removing weeds and loosening soil to enhance seedling establishment, soil tillage changes the nutrient status of soil that in turn affects decomposer organisms. Chemical fertilizer application also can have appreciable impact on soil macrofauna and microfauna. Fertilizer added in organic form (manure and leguminous plants) instead of inorganic form, increases soil microbial biomass and enzyme activity (Brady, 1974). As pointed out earlier, perhaps the most complex effects of agricultural manipulations on soil ecology are observed with the nontarget effects of pesticides. Few insecticides and herbicides are without significant impact on at least some groups of soil microorganisms and animals (Edwards and Thompson, 1973; Ingham et al., 1985).

Relationships between aboveground and belowground plant–animal interactions can be important in both natural and agricultural ecosystems, although there have been few empirical studies on this topic. Janzen (1973b), in a paper putting forth the idea that plants can be viewed as islands for insect herbivores, argued that two or more plant herbivores are potential competitors even if they exploit different plant parts. Foliage and root feeders, for example, can affect each other, because they are connected via a plant's vascular system. There is evidence from ecosystem processes suggesting that herbivore (both mammal and insect) defoliation causes changes in microbial populations and nutrient absorption patterns in plants in forest ecosystems (Schowalter et al., 1986). Furthermore, Seastedt et al., (1983) showed that insect herbivory in forest ecosystems increased rates of nutrient cycling, which may have significant impact on nutrient uptake by trees and shrubs. Although there are few if any data sets for agricultural ecosystems, one can speculate that similar processes operate, especially since herbivory levels are often appreciably greater in agricultural than in forested ecosystems.

General Discussion and Future Considerations

Several areas of innovation are impacting agriculture on a worldwide basis. Four areas are described here: (1) conservation tillage, (2) multiple cropping, (3) new crop species, and (4) recombinant genetics and biotechnology. These topics were chosen because they will change the agroecological arenas in which plant–animal interactions occur. Also, the topics are important because they have the potential to contribute appreciably to lowering the increasing economic and environmental costs of material and energy inputs to agricultural ecosystems.

Conservation Tillage

The use of reduced or conservation tillage practices for the production of food and fiber crops has increased dramatically in recent years (Allmaras and Dowdy, 1985). Traditionally, crop production entailed sufficient soil tillage to provide a seed bed and to reduce pests. In contrast, specific conservation tillage practices range from minimal plowing (for example, using only a disk or harrow) to the complete lack of tillage, most commonly termed no-tillage. Conservation tillage practices not only conserve energy and soil, but also reduce machinery-intensive operations and save on fuel and machinery costs. For these and other reasons, farmers and agricultural scientists are enthusiastic about conservation tillage. Yet, conservation tillage is not a panacea. Reduced tillage practices are not well-suited to poorly drained soil types (Napier et al., 1984), and rely heavily on herbicides for weed control (Hayes, 1982). Ecologists also have taken a particular interest in reduced tillage systems, in terms of both basic and applied research. Complete lack of tillage, as in no-tillage, can markedly alter ecological phenomena in agricultural systems, especially soil processes (Selles et al., 1984). Because the physical, chemical, and biological characteristics of soil are so different in tilled vs nontilled agricultural systems, there is great potential for plant–animal interactions to change appreciably as more land is placed in conservation tillage management.

Figure 9-8 diagrams some of the major differences between plowed and unplowed soil systems. In general, numbers and diversity of soil organisms and the interactions among the biota increase with reduced tillage. There are more diverse assemblages of fungi, bacteria, nematodes, and arthropods in unplowed than in plowed soils. Plowing mixes soil and rapidly incorporates organic materials (e.g., manure, crop, and weed debris) into the soil matrix. In contrast, no-tillage production practices result in organic materials being left undisturbed and stratified in the top soil layers. The degree of disturbance and position of organic materials in the various kinds of reduced tillage lie somewhere between these two extremes of plowed and unplowed soils. Effects on biological, chemical, and physical characteristics change accordingly from plowed to the unplowed or no-tillage condition (Carter and Rennie, 1984).

In many ways, no-tillage soils are very similar to those found in natural ecosystems like forests and prairies (Doran, 1980; House et al., 1984). Some biota associated with surface-maintained residues have a negative impact on crops. These include a few species of crop-damaging arthropods and nematodes (All et al., 1984). However, beneficial interactions such as biological control of pests (Brust et al., 1985) and biotic regulation of organic matter decomposition increase under reduced tilled conditions (Coleman et al., 1984). More specifically, a gradient of decomposition products from fresh surface crop residues to humified organic matter within the upper soil strata occurs under continuous, reduced, and especially no-tillage management. These residues promote environmental conditions conducive to the proliferation of an abundant and diverse soil fauna and microfloral communities (House and Stinner, 1983; Doran, 1980). A significant feature of continuous no-tillage cropping practices is that they enhance the predatory and saprophagous soil arthropod community. Predaceous carabid beetles (House and All, 1981) and spiders (Blumberg and Crossley, 1983), microarthropods, and other decomposer fauna, such as earthworms (Edwards and Lofty, 1973), occur in higher numbers under no-tillage than in conventionally plowed systems. Aboveground interactions between plants and insects may also be altered by conservation tillage practices. Foster and Ruesink (1984) reported that changes in weed flora resulting from conservation tillage agriculture can affect both the levels of herbivory by the black cutworm on maize and the behavior of parasites of the black cutworms.

It does appear as though the introduction of conservation tillage agriculture has the net effect of increasing the kind and number of interactions among plants, invertebrates, and microorganisms. Therefore, conservation tillage helps to restore at least some natural ecological characteristics (e.g., increased biological complexity) to agricultural ecosystems. This should help create more efficient (maximizing yield per unit of input), less environmentally degrading, and more sustainable agricultural ecosystems.

Figure 9–8. Generalized diagram of soil profiles in plowed and reduced tillage systems.

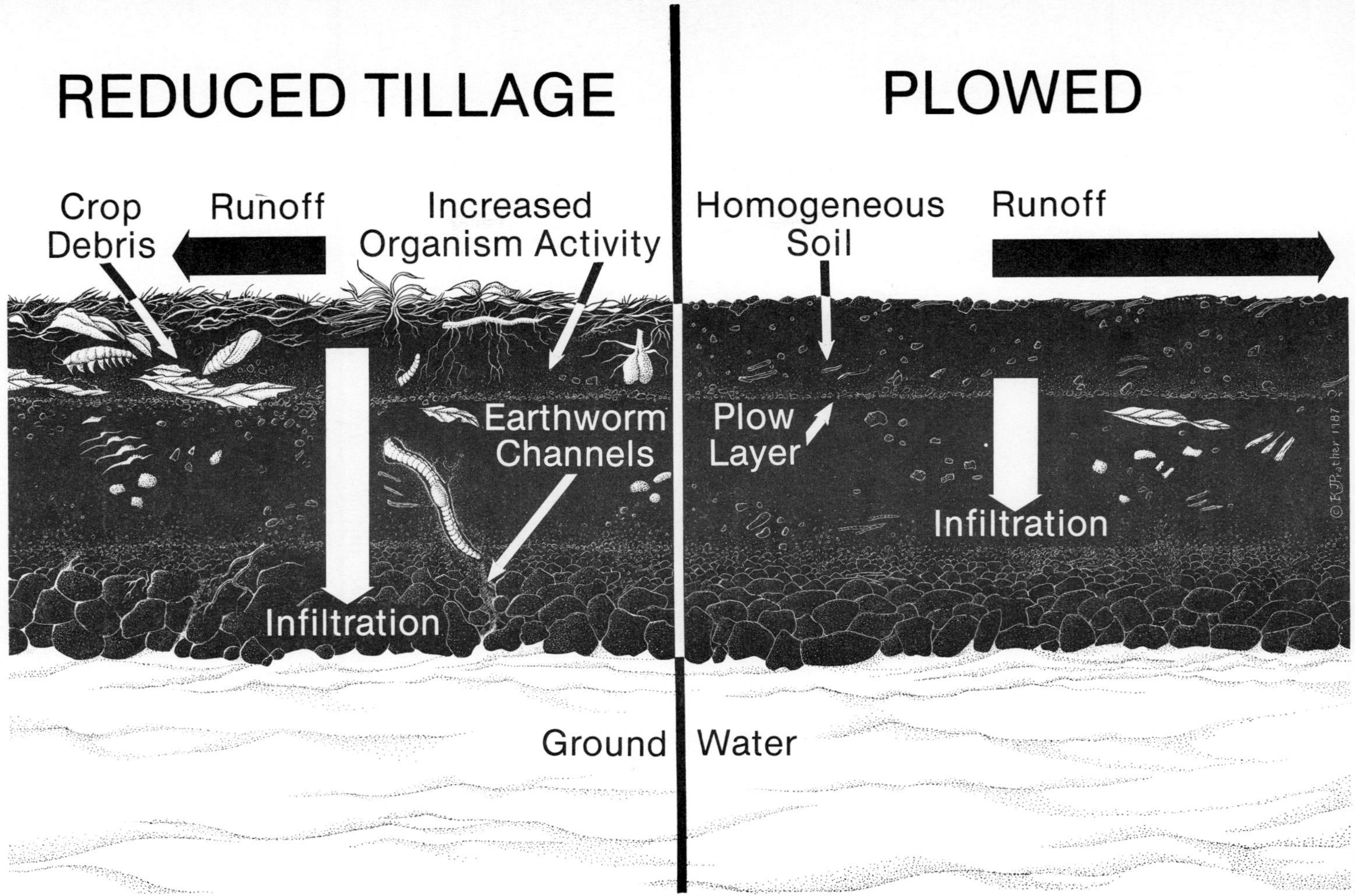

REDUCED TILLAGE

PLOWED

Crop Debris

Runoff

Increased Organism Activity

Homogeneous Soil

Runoff

Earthworm Channels

Plow Layer

Infiltration

Infiltration

Ground Water

© RJPrather 1987

Multiple Cropping

Growing two or more crops together or within the same year on the same area of land is not a new concept but rather an old practice receiving new interest (Fordham, 1983; Horwith, 1985). Rotation or sequential cropping encompasses growing two crops in one year on the same land, but not simultaneously. Interplanting or intercropping (also termed companion planting and polyculture) is the practice of growing two or more crop species together on the same land at the same time. Table 9-5 lists diverse examples of multiple cropping schemes, representing both mechanized and labor-intensive agriculture. The adoption of conservation tillage practices has increased the potential for new kinds of multiple cropping practice. The agronomic reasons for this interest in growing two or more crop species on the same piece of land are varied and include more nutritional diversity, reduced soil erosion, and greater profitability (Jeffers et al., 1977; Horwith, 1985).

Polyculture presents a more architecturally and chemically diverse

Table 9-5 Examples of multiple cropping (interplanting and rotation)

Type	Location
Interplanting	
*Interplanted maize (*Zea mays*) + soybean (*Glycine max*), maize + ground nut (*Arachis hypogaea*)	India
*Interplanted cassava (*Manihot esculenta*) + groundnut, cassava + cowpea (*Vigna unguiculata*)	Columbia
*Interplanted maize + greengram (*Phaseolus aureus*)	Nigeria
*Interplanted maize and soybean	United States (Iowa)
Maize interplanted with sorghum (*Sorghum bicolor*)	Central America
*Intercropped wheat (*Triticum aestivum*) and soybeans	United States (Ohio)
*Maize + white clover (*Trifolium repens*) interplanted as 'live mulch'	United States (Pennsylvania)
Rotation	
*Maize, soybeans, or sorghum following winter cover crops of rye (*Secale cereale* or wheat)	United States (Eastern portion)
*Maize following vetch (*Vicia* spp.) using no-tillage methods	United States (Kentucky)
*Maize following alfalfa (*Medicago sativa*) using conventional and no-tillage methods	United States (Ohio)

*Indicates a legume is present in the crop combinations.

habitat for insect herbivores, predators, and parasites than monoculture. Therefore, the number and kinds of interactions among plants and animals should be greater in polyculture than in monoculture. Risch et al. (1983) summarized numerous intercropping studies and reported that with 198 arthropod species, 53 percent of the herbivore species were less abundant in interplanted systems, 18 percent were more abundant, 9 percent showed little or no difference, and 20 percent demonstrated varying responses. In a recent study, Brust et al. (1986b) found significantly higher densities and species richness of macroarthropod predators that inhabit the soil (primarily ground beetles and spiders) in interplanted maize and alfalfa than in monoculture maize. This latter study indicates that maize–alfalfa polyculture supports more complex predator–prey food webs than monoculture systems.

As evidenced by the survey in Table 9-5, legumes play a dominant role in multiple cropping schemes worldwide, serving mainly as a substitute for commercial nitrogen for fertilizer. Legumes form symbiotic associations with bacteria of the genus, *Rhizobium*. These microorganisms have the ability to "fix" atmospheric nitrogen which subsequently contributes to the nutrition of the legume plant. When legumes are grown in rotation or in polyculture with other crop plants such as maize, wheat, or sorghum, the fixed nitrogen can be transferred to these crops either directly through root secretions or indirectly through decomposition of the legume plant tissue. The latter mechanism is usually responsible for most of the nitrogen transfer (Ladd et al., 1983; Power et al., 1983). However, nitrogen accretion and soil exudates associated with legumes also provide resources for soil organisms at several trophic levels. Microorganism activity has been reported to increase when legumes are included in cropping systems (Dick, 1984; Wahua, 1984). Indirectly, other taxa such as nematodes and soil mites that feed on the bacteria and fungi are also affected (Stinner, unpublished data). In turn, these organisms are involved in many of the nutrient transformations that influence soil fertility (Coleman, 1985; Chlarholm, 1985). Multiple cropping with legumes not only increases nitrogen input through biological fixation, but also changes patterns of nitrogen transformation within soil systems and, ultimately, mechanisms of nutrient retention and loss (Chujo and Daimon, 1984; Ebelhar et al., 1984). In addition, using legumes with a prostrate growth habit as a "living mulch" between row crops has the potential to suppress weeds and reduce herbicide application (Elkins et al., 1984). Multiple cropping, especially with legumes, has the potential to contribute toward reducing expensive and potentially environmentally damaging chemical fertilizer and pesticide inputs to agricultural ecosystems. Multiple cropping with legumes is another way to help restore more natural ecological characteristics to agricultural ecosystems.

New Crop Species

Domesticated plant species represent a very small fraction (less than 1 percent) of all seed plant species on earth. During the past 10 to 15 years, there has been increased emphasis placed on developing new crop species for production of food and oil for human consumption. Undoubtedly, these introductions will have important effects on plant–animal interactions for both mammalian and insect herbivores. Felger (1979) has estimated there are at least 100 plant species that are candidates for domestication and incorporation into production systems. Major representative species of this group and their attributes are summarized in Table 9-6. Some species are intended primarily for direct human consumption, some for animal fodder, and others for energy purposes.

Utilization of these species as "new" crops will have beneficial and detrimental effects on agricultural ecosystems around the world. As with present domesticated plant species, each new crop will have its complement of pests, production problems (i.e., soil fertility needs, machinery, and labor requirements), and marketing limitations. However, there are important reasons why new crop species should be developed. Newly introduced species would increase possibilities for polyculture and add appreciably more genetic variability to farming systems (Bemis et al., 1979; Feine et al., 1979). This variability could be exploited to increase resistance to insect pests and pathogens, and to rapidly breed cultivars adapted to a wide range of geographic conditions.

One of the major impetuses for the domestication and development of new crop species is the more efficient use of resources. As shown in Table 9-6, some of the more promising candidates for continued domestication are legumes that are capable of fixing atmospheric nitrogen. This capability is especially important in nutrient-poor regions or where fertilizer costs are prohibitive. One of the most important considerations in developing new crop species for agriculture is the production of food and other materials in regions in which present crops do not grow well unless expensive material and energy inputs are added. For example, maize and cotton require extensive use of irrigation, fertilizer, and pesticides in order to produce acceptable yields in the southwestern United States. Similarly, when crops and management systems developed for temperate climates are exported to subtropical and equatorial regions, failure to obtain sustainable yields and pest problems are often encountered (Webster and Wilson, 1980). Some of the species listed in Table 9-6 are preadapted to the environments for which they are intended so that many obstacles to their successful culture in these areas should be minimized.

It is also important to consider the kinds of technology and management systems that will be tied to new crop species. In the equatorial regions of China, Africa, South America, and Central America, the exportation of modern crop varieties, such as hybrid maize and wheat, and the accompanying large-scale agricultural production system (what has been termed the "green revolution"), have not met with long-term success (Webster and Wilson, 1980). The reasons for this lack of success vary with cropping systems, geographic regions, and economic systems. However, it is safe to say that generally it is an uphill effort to try to export large-scale, energy-intense monoculture farming to equatorial, developing countries. These new species are promising in that they should offer more flexibility in management so that appropriate levels of technology can be matched to specific needs of people as determined by both physical and socioeconomic factors. It will be exceedingly important to develop these new species for compatibility in economically viable and sustainable farming systems through conservation tillage and/or multiple cropping practices that do not necessarily rely on high-powered technology. However, the social factor cannot be underestimated in getting a new crop accepted. If the local people and their culture do not appreciate a new crop because of taste or social reasons, it does not matter how appropriate it and its associated technology are for the region.

Recombinant Genetics and Biotechnology

Recombinant DNA technology has been lauded as a new revolution in biology, especially in agriculture (Bliss, 1984). New developments and the expanded potential for creating varieties of plants and animals are occurring very rapidly as indicated by the burgeoning biotechnological literature. The impact of genetic engineering on the ecology of plant–animal interactions at this time is largely speculative because few biotechnological products have been released in field situations. Essentially, genetic engineering compresses the time frame needed to breed new plant and animal varieties. The technology also enables the genetic combination of unrelated groups of taxa so that "new" organisms can be created. This latter application is an aspect of biotechnology that often causes deep dissension on social and environmental issues. Genetic engineering could influence plant–animal interactions in several major areas: changing the nutritional and photosynthetic characteristics of crop plants, developing insect-pest and plant pathogen-resistant varieties, and engineering herbicide-resistant crops (Comai and Stalker, 1984). The former two areas would influence herbivores directly and the latter indirectly through changing weed–crop community structure.

It is now recognized that in order for genetically engineered organisms to be successfully introduced into agricultural systems, there

Table 9-6 Examples of new crop species and their anticipated roles in agriculture

Crop	Description
Winged bean (*Psophocarpus tetragonolobus*)	This large-seeded legume is a climbing perennial that is well adapted to tropical and subtropical regions. Seeds are high in oil and protein and tubers have high protein content. Winged bean has great potential for use in high rainfall areas for village agriculture which typically depends on low-protein yams and cassava. Winged bean grows well in poor soils, because it is very active in fixing atmospheric nitrogen. Compatible with polyculture procedures; presently hosts few pathogens.
Amaranth (*Amaranthus* spp.)	Although amaranth has been cultivated since 5000 BC in Mexico and Central America, the plant has not received much attention for modern, large-scale production until recently. Grain or seed from this genus has high nutritional value with 12–15% protein and high lysine levels. Young leaves can be utilized as very palatable and vitamin-rich vegetable. Has a C_4 metabolic pathway with characteristic high photosynthetic efficiency. One of the few dicotyledons that can be utilized as a grain crop. Amaranth has the advantage of wide genetic diversity and few pests compared to the crops commonly grown today.
Buffalo gourd (*Cucurbita foetidissima*)	Indigenous to western North America with potential as a major arid lands crop. This perennial member of the squash genus is capable of producing acceptable yields of vegetable oil, protein, and starch. Produces extremely large roots with exceptionally high starch content. Major advantage of this crop is large yield potential in desert-like conditions.
Mesquite (*Prosopis glandulosa*)	Common, woody leguminous shrub or small tree growing in arid lands of southwestern U.S. Seed pods contain as much as 13% protein and 30% sucrose. At one time, mesquite formed the staple food for Indians in southern California deserts. Culture of sp. in arid areas would obviate need for irrigation, nitrogen fertilizer, and tillage. Reduction of tillage is especially important for reducing wind- and water-induced soil erosion in these environments.
Saltbush (*Atriplex* spp.)	Consists of approx. 200 species, many which have been recognized for potential forage plants. These species, commonly termed "saltbushes" are characterized by high biomass production, high protein content and mineral concentrations sufficient for animal nutrition. Most spp. are halophytes, can be irrigated with saline or blackish water, and will grow in environments hostile to other types of agricultural plants. Characteristic of *Atriplex* to concentrate salt in its tissues could be exploited to reclaim saline contaminated land.

Table 9-6 *(continued)*

Crop	Description
Cassia (*Cinnamonum cassia*)	Native to southern China and northern Indochina. Potential as new, oil-bearing crop in humid tropics. Cassia oil is particularly valued for its flavor and fragrance for both food and perfume preparations. It is another old crop receiving new attention. Primarily useful as a cash crop in village agricultural systems. Has treelike growth form; bark, buds, and leaves are all valuable as oil-bearing organs.
Guayule (*Parthenium argentatum*)	Guayule is native to Mexico and southwestern U.S.; produces natural rubber in large quantities. Grows bushlike and is seldom more than 0.6 m high. Is a desert plant, able to survive up to several years without appreciable rainfall. Drought or cool weather switches guayule plants from production of biomass to synthesis of rubber. Large-scale guayule culture was attempted at turn of century and, largely for political reasons, was not successful. Wild guayule populations have large genetic diversity so that it is possible to breed strains of the species adapted to a wide variety of climatic conditions.

must be some comprehension of how the genetically altered plants or animals will respond ecologically to effect both positive and negative influences on crop production (Hauptli, 1985). At present, there is very little empirical or conceptual information to make an accurate assessment of these ecological responses. It seems reasonable that research and application of recombinant genetic techniques should be done by teams of scientists possessing expertise ranging from molecular genetics to agricultural ecology and economics. In this way, long-range consequences can be evaluated with comprehensive management objectives in mind.

Selected References

Boethel, D. J., and R. D. Eikenbary, eds. 1986. Interactions of plant resistance and parasitoids and predators of insects. Ellis Harwood: John Wiley and Sons, NY.

Cox, G. W., and M. D. Atkins. 1979. Agricultural ecology. W. H. Freeman and Company, San Francisco, CA.

De Candolle, A. L. 1959. Origin of cultivated plants. Hafner Co., NY.

Lowrance, R., B. R. Stinner, and G. J. House, eds.. 1984. Agricultural ecosystems. John Wiley and Sons, NY.

Metcalf, R., and W. Luckmann, eds. 1982. Introduction to insect pest management. John Wiley and Sons, NY.

Price, P. W. 1984. Insect ecology. John Wiley and Sons, NY.

Literature Cited

Abbott, H. G., and T. F. Quink. 1970. Ecology of eastern white pine seed caches made by small forest mammals. *Ecology* 51:271–278.

Abrahamson, W. G. 1975. Reproductive strategies in dewberries. *Ecology* 56:721–726.

Abrahamson, W. G. 1980. Demography and vegetative reproduction. pp. 89–106. In: O. T. Solbrig, ed. Demography and the evolution of plant populations. Blackwell Scientific Publications, Oxford.

Abrahamson, W. G., and K. D. McCrea. 1986. The impacts of galls and gallmakers on plants. *Proc. Entomol. Soc. Wash.* 88:364–367.

Abrahamson, W. G., and A. E. Weis. 1987. The nutritional ecology of Arthropod gall-makers. pp. 235–258. In: F. Slansky, Jr., and J. G. Rodriquez, eds. Nutritional ecology of insects, mites and spiders. John Wiley and Sons, NY.

Abrahamson, W. G., P. O. Armbruster, and G. D. Maddox. 1983. Numerical relationships of the *Solidago altissima* stem gall insect-parasitoid guild food chain. *Oecologia (Berl.)* 58:351–357.

Ackerman, J. D., M. R. Mesler, K. L. Lu, and A. M. Montalvo. 1982. Food-foraging behavior of male Euglossini (Hymenoptera: Apidae): vagabonds or trapliners? *Biotropica* 14:241–248.

Adams, R. 1972. Watership down. Rex Collings Publishers, London.

Adams, R. M., II, and G. W. Smith. 1977. An S.E.M. survey of the five carnivorous pitcher plant genera. *Am. J. Bot.* 64:265–272.

Addicott, J. F. 1974. Predation and prey community structure: an experimental study of the effect of mosquito larvae on the protozoan communities of pitcher plants. *Ecology* 55:475–492.

Agnew, A. D. Q., and J. E. C. Flux. 1970. Plant dispersal by hares (*Lepus capensis* L.) in Kenya. *Ecology* 51:735–737.

Aker, C. L., and D. Udovic. 1981. Oviposition and pollination behavior of the yucca moth, *Tegeticula maculata* (Lepidoptera: Prodoxidae), and its relation to the reproductive biology of *Yucca whipplei* (Agavaceae). *Oecologia (Berl.)* 49:96–101.

Alatalo, R. V., L. Gustafsson, M. Linden, and A. Lundberg. 1985a. Interspecific competition and niche shifts in tits and the goldcrest: an experiment. *J. Anim. Ecol.* 54:977–984.

Alatalo, R. V., A. Lundberg, and S. Ulfstrand. 1985b. Habitat selection in the pied flycatcher *Ficedula hypoleuca*. pp. 59–83. In: M. L. Cody, ed. Habitat selection in birds. Academic Press, NY.

Aldenius, J., B. Carlsson, and S. Karlsson. 1983. Effects of insect trapping on growth and nutrient content of *Pinguicula vulgaris* in relation to the nutrient content of the substrate. *New Phytol.* 93:53–60.
Aldrich-Blake, F. P. G., T. K. Dunn, R. I. M. Dunbar, and P. M. Headley. 1971. Observations on baboons, *Papio anubis,* in an arid region in Ethiopia. *Folia Primatol.* 15:1–35.
All, J. N., R. L. Hussey, and D. G. Cummins. 1984. Southern corn billbug (Coleoptera: Curculionidae) and plant-parasitic nematodes: influence of no-tillage, coulter-in-row-chiseling, and insecticides on damage to corn. *J. Econ. Entomol.* 77:178–182.
Allard, R. W. 1960. Principles of plant breeding. John Wiley and Sons, NY.
Allmaras, R. R., and R. H. Dowdy. 1985. Conservation tillage systems and their adoption in the United States. *Soil Tillage Res.* 5:197–222.
Alterei, M., and W. H. Whitcomb. 1979. The potential use of weeds in the manipulation of beneficial insects. *HortScience.* 14(1):12–18.
Anderson, R. C., and O. L. Loucks. 1979. White-tail deer (*Odocoileus virginianus*) influence on structure and composition of *Tsuga canadensis* forests. *J. Appl. Ecol.* 16:855–861.
Anderson, R. C., and S. Schelfhout. 1980. Phenological patterns among tall grass prairie plants and their implications for pollinator competition. *Am. Midl. Natur.* 104:253–263.
Armbruster, W. S. 1984. The role of resin in angiosperm pollination: ecological and chemical considerations. *Am. J. Bot.* 71:1149–1160.
Archer, S., and L. L. Tieszen. 1980. Growth and physiological responses of tundra plants to defoliation. *Arct. Alp. Res.* 12:531–552.
Arms, K., P. Feeny, and R. C. Lederhouse. 1974. Sodium: stimulus for puddling behavior by tiger swallowtail butterflies, *Papilia glaucus. Science* 185:372–374.
Aronow, L., and F. Kerdel-Vegas. 1965. Seleno-cystathionine, a pharmacologically active factor in the seeds of *Lecythis ollaria. Nature* 205:1185–1186.
Askew, R. R. 1961. A study of the biology of species of the genus *Mesopobolus* Westwood (Hymenoptera: Pteromalidae) associated with cynipid galls on oak. *Trans Ray. Entomol. Soc. Lond.* 113:155–173.
Askins, R. A. 1983. Foraging ecology of temperate-zone and tropical woodpeckers. *Ecology* 64:945–956.
Augspurger, C. K. 1981. Reproductive synchrony of a tropical shrub: experimental studies on effects of pollinators and seed predators on *Hybanthus prunifolius* (Violaceae). *Ecology* 62:775–788.
August, P. V. 1983. The role of habitat complexity and heterogeneity in structuring tropical mammal communities. *Ecology* 64:1495–1507.
Austin, R. B. 1986. Molecular biology and crop improvement. Cambridge University Press, Cambridge, UK.
Ayensu, E. S. 1974. Plant and bat interactions in West Africa. *Ann. Mo. Bot. Gard.* 61:702–747.
Bach, C. E. 1980. Effect of plant density and diversity on the population dynamics of a specialist herbivore, the striped cucumber beetle, *Acalymma vittata* (Fab.). *Ecology* 61:1515–1530.
Baird, J. W. 1980. The selection and use of fruit by birds in an eastern forest. *Wilson Bull.* 92:63–73.
Baker, H. G. 1970. Plants and civilization. Wadsworth, Belmont, CA.
Baker, H. G. 1975. Sugar concentrations in nectars from hummingbird flowers. *Biotropica* 7:37–41.
Baker, H. G. 1976. "Mistake" pollination as a reproductive system with special reference to the Caricaceae. *Linn. Soc. Symp. Ser.* 2:161–169.
Baker, H. G., and I. Baker. 1975. Studies of nectar-constitution and pollinator–plant coevolution. pp. 100–140. In: L. E. Gilbert, and P. H. Raven, eds. Coevolution of animals and plants. University of Texas Press, Austin, TX.
Baker, H. G., and I. Baker. 1983a. A brief historical review of the chemistry of floral nectar. pp. 126–152. In: B. Bentley, and T. Elias, eds. The biology of nectaries. Columbia University Press, NY.
Baker, H. G., and I. Baker. 1983b. Some evolutionary and taxonomic implications of

variation in the chemical reserves of pollen. pp. 43–52. In: D. L. Mulcahy, and E. Ottaviano, eds. Pollen: biology and implications for plant breeding. Elsevier, Amsterdam.

Baker, H. G., and I. Baker. 1983c. Floral nectar sugar constituents in relation to pollinator type. pp. 117–141. In: C. E. Jones, and R. J. Little, eds. Handbook of experimental pollination biology. Van Nostrand Reinhold, NY.

Baker, H. G., and P. H. Hurd, Jr. 1968. Intrafloral ecology. *Ann. Rev. Entomol.* 13:385–414.

Baker, H. G., P. A. Opler, and I. Baker. 1978. A comparison of the amino acid complements of floral and extrafloral nectaries. *Bot. Gaz.* 139:322–332.

Baker, H. G., K. S. Bawa, G. W. Frankie, and P. A. Opler. 1983. Reproductive biology of plants in tropical forests. pp. 183–215. In: F. B. Golley, ed. Tropical rainforest ecosystems. A. Structure and function. Elsevier, Amsterdam.

Balda, R. P. 1980. Are seed caching systems co-evolved? *Acta XVIII Int. Ornithol. Congr.*, pp. 1185–1191.

Baldwin, I. T., and J. C. Schultz. 1983. Rapid changes in tree leaf chemistry induced by damage: evidence for communication between plants. *Science* 221:277–279.

Barber, J. T. 1978. *Capsella bursa-pastoris* seeds: are they carnivorous? *Carniv. Plant News.* 7:39–42.

Barlow, B. A., and D. Wiens. 1977. Host–parasite resemblance in Australian mistletoes: the case for cryptic mimicry. *Evolution* 31:69–84.

Barrett, S. C. H. 1987. The evolution, maintenance and loss of self-incompatibility systems. In: J. Lovett Doust, and L. Lovett Doust, eds. Plant reproductive strategies. Oxford Press, NY.

Barrett, S. C. H., and D. E. Glover. 1985. On the Darwinian hypothesis of the adaptive significance of tristyly. *Evolution* 39:766–774.

Barth, F. G. 1985. Insects and flowers. Princeton University Press, Princeton, NJ.

Barton, A. M. 1986. Spatial variation in the effect of ants on an extrafloral nectary plant. *Ecology* 67:495–504.

Bate-Smith, E. C., I. K. Ferguson, K. Hutson, S. R. Jensen, B. J. Nielsen, and T. Swain. 1975. Phytochemical interrelationships in the Cornaceae. *Biochem. Syst. Ecol.* 3:79–89.

Batzli, G. O. 1975. The influence of grazers on tundra vegetation and soils. *Proc. Circumpolar Conf. North. Ecol.* 1:215–225.

Batzli, G. O. 1978. The role of herbivores in mineral cycling. pp. 95–112. In: D. C. Adriano, and I. L. Brisbin, eds. Environmental chemistry and cycling processes. ERDA Symp. Ser., U. S. Dep. of Energy, Washington, D.C.

Batzli, G. O. 1983. Responses of arctic rodent populations to nutritional factors. *Oikos* 40:396–406.

Batzli, G. O. 1985. Nutrition. pp. 779–811. In: R. H. Tamarin, ed. Biology of the new world *Microtus. Spec. Publ. No. 8,* Am. Soc. Mammalogists.

Batzli, G. O., and F. R. Cole. 1977. Nutritional ecology of microtine rodents: digestibility of forage. *J. Mammal.* 60:740–750.

Batzli, G. O., and H. G. Jung. 1980. Nutritional ecology of microtine rodents. Resource utilization near Atkasook, Alaska. *Arct. Alp. Res.* 12:483–499.

Batzli, G. O., and F. A. Pitelka. 1970. Influence of meadow mouse populations on California grassland. *Ecology* 51:1027–1039.

Batzli, G. O., R. G. White, S. F. MacLean, Jr., F. A. Pitelka, and B. Collier. 1980. The herbivore-based trophic system. pp. 335–410. In: J. Brown, P. C. Miller, L. L. Tieszen, and F. L. Bunnell, eds. An arctic ecosystem: the coastal tundra at Barrow, Alaska. Dowden, Hutchinson and Ross, Stroudsburg, PA.

Bawa, K. S. 1980a. Evolution of dioecy in flowering plants. *Ann. Rev. Ecol. Syst.* 11:15–39.

Bawa, K. S. 1980b. Mimicry of male by female flowers and intrasexual competition for pollinators in *Jacaratia delichaula* (D. Smith) Woodson (Caricaceae). *Evolution* 34:467–474.

Bawa, K. S., and J. H. Beach. 1981. Evolution of sexual systems in flowering plants. *Ann. Mo. Bot. Gard.* 68:254–274.

Bazzaz, F. A., R. W. Carlson, and J. L. Harper. 1979. Contribution to reproductive effort by photosynthesis of flowers and fruits. *Nature* 279:554–555.
Beattie, A. J. 1983. The distribution of ant-dispersed plants. *Sunderbd. Naturwiss. Verh. Hamburg* 7:249–270.
Beattie, A. J. 1985. The evolutionary ecology of ant-plant mutualisms. Cambridge studies in ecology. Cambridge University Press, Cambridge, UK.
Beattie, A. J., C. Turnbull, R. B. Knox, and E. G. Williams. 1984. Ant inhibition of pollen function: a possible reason why ant pollination is rare. *Am. J. Bot.* 71:421–426.
Beaver, R. A. 1983. The communities living in *Nepenthes* pitcher plants: fauna and food webs. pp. 129–159. In: J. H. Frank, and L. P. Lounibos, eds. Phytotelmata: terrestrial plants as hosts for aquatic insect communities. Plexus Publishing, Medford, NJ.
Beaver, R. A. 1985. Geographical variation in food web structure in *Nepenthes* pitcher plants. *Ecol. Entomol.* 10:241–248.
Beck, E. W. 1947. Some studies on the *Solidago* gall caused by *Eurosta solidaginis* Fitch. Ph.D. thesis, University of Michigan, Ann Arbor, MI.
Beck, S. D. 1965. Resistance of plants to insects. *Ann. Rev. Entomol.* 10:207–222.
Beckmann, R. L., Jr., and J. M. Stucky. 1981. Extrafloral nectaries and plant guarding in *Ipomoea pandurata* (L.) G. F. W. Mey (Convolvulaceae). *Am. J. Bot.* 68:72–79.
Begon, M., J. L. Harper, and C. R. Townsend. 1986. Ecology: individuals, populations, and communities. Sinauer Associates, Sunderland, MA, 876 pp.
Bell, E. A., and D. H. Janzen. 1971. Medical and ecological considerations of L-dopa and 5-HTP in seeds. *Nature* 229:136–137.
Bell, G. 1985. On the function of flowers. *Proc. Roy. Soc. London Ser. B* 224:223–265.
Bell, R. H. V. 1971. A grazing system in the Serengeti. *Sci. Am.* 225:86–93.
Belovsky, G. E. 1984. Moose and snowshoe hare competition and a mechanistic explanation from foraging theory. *Oecologia (Berl.)* 61:150–159.
Belovsky, G. E., and P. A. Jordan. 1981. Sodium dynamics and adaptations of a moose population. *J. Mammal.* 62:613–621.
Belsky, A. J. 1986. Does herbivory benefit plants? A review of the evidence. *Am. Natur.* 127:870–892.
Belt, T. 1874. The naturalist in Nicaragua. J. Murray, London.
Bemis, W. P., J. W. Berry, and C. W. Weber. 1979. The buffalo gourd. pp. 65–87. In: G. A. Ritchie, ed. New agricultural crops. Am. Assoc. Advan. Sci. Washington, DC.
Bentley, B. L. 1977a. Extrafloral nectaries and protection by pugnacious bodyguards. *Ann. Rev. Ecol. Syst.* 8:407–427.
Bentley, B. L. 1977b. The protective function of ants visiting the extrafloral nectaries of *Bixa orellana* (Bixaceae). *J. Ecol.* 56:17–38.
Bentley, B. L. 1981. Ants, extrafloral nectaries and the vine life-form: an interaction. *Trop. Ecol.* 22:127–133.
Bentley, S., and J. B. Whittaker. 1979. Effects of grazing by a chrysomelid beetle, *Gastrophysa viridula*, on competition between *Rumex obtusifolia* and *Rumex crispus*. *J. Ecol.* 67:79–90.
Benzing, D. H. 1970. An investigation of two bromeliad myrmecophytes: *Tillandsia butzii* Mez, *T. caput-medusae* E. Morren, and their ants. *Bull. Torrey Bot. Club* 97:109–115.
Benzing, D. H. 1973. Mineral nutrition and related phenomena in Bromeliaceae and Orchidaceae. *Q. Rev. Biol.* 48:277–290.
Benzing, D. H. 1976. Bromeliad trichomes: structure, function, and ecological significance. *Selbyana* 1:330–348.
Benzing, D. H. 1980. Biology of the bromeliads. Mad River Press, Eureka, CA.
Benzing, D. H., and A. Renfrew. 1971. The biology of the epiphytic bromeliad *Tillandsia circinata* Schlecht. I. The nutritional status of populations in south Florida. *Am. J. Bot.* 58:867–873.
Benzing, D. H., K. Henderson, B. Kessel, and V. Sulak. 1976. The absorptive capacities of bromelian trichomes. *Am. J. Bot.* 63:1009–1014.
Benzing, D. H., T. J. Givnish, and D. L. Bermudez. 1985. Absorptive trichomes in *Brocchinia reducta* (Bromeliaceae) and their evolutionary significance. *Syst. Bot.* 10:81–91.

Bequaert, J. 1922. Ants in their diverse relations to the plant world. *Bull. Am. Mus. Natur. Hist.* 45:333–583.
Berenbaum, M. R. 1980. Adaptive significance of midgut pH in Lepidoptera. *Am. Natur.* 115:138–146.
Berenbaum, M. R. 1983. Coumarins and caterpillars: a case for coevolution. *Evolution* 37:163–179.
Berenbaum, M. R., and E. Miliczky. 1984. Mantids and milkweed bugs: efficacy of aposematic coloration against invertebrate predators. *Am. Midl. Natur.* 111:64–68.
Berenbaum, M. R., A. R. Zangerl, and J. K. Nitao. 1986. Constraints on chemical coevolution: wild parsnips and the parsnip webworm. *Evolution* 40:1215–1228.
Berg, R. ·Y. 1972. Dispersal ecology of *Vancouveria* (Berberidaceae). *Am. J. Bot.* 59:109–112.
Berger, P. J., E. H. Sanders, P. D. Gardner, and N. C. Negus. 1977. Phenolic plant compounds functioning as reproductive inhibitors in *Microtus montanus. Science* 195:575–577.
Berger, P. J., N. C. Negus, E. H. Sanders, and P. D. Gardner. 1981. Chemical triggering of reproduction in *Microtus montanus. Science* 214:69–70.
Bergman, J. M., and W. W. Tingey. 1979. Aspects of the interaction between plant genotypes and biological control. *Bull. Entomol. Soc. Am.* 25:275–299.
Bernays, E. A. 1982. The insect on the plant—a closer look. *Proc. 5th Int. Symp. Insect–Plant Relationships.* Pudoc, Wagenenger.
Bernays, E. A. 1986. Diet-induced head allometry among foliage-chewing insects and its importance for graminovores. *Science* 231:495–497.
Bernays, E. A., and R. Chapman. 1977. Deterrent chemicals as a basis of oligophagy in *Locust migratoria* (L.). *Ecol. Entomol.* 2:1–18.
Bernays, E. A., and S. Woodhead. 1982. Plant phenols utilized as nutrients by a phytophagous insect. *Science* 216:201–203.
Berthold, P. 1976a. Der Seidenschwanz *Bombycilla garrulus* als frugivorer Ernahrungsspezialist. *Experimentia* 32:1445.
Berthold, P. 1976b. The control and significance of animal and vegetable nutrition in omnivorous songbirds. *Ardea* 64:140–154.
Berthold, P. 1977. Proteinmangel als Ursache der schaedigenden Wirkung rein vegetabilischen Ernahrung omnivorer Singvogel Arten. *J. Ornithol.* 118:202–203.
Bertin, R. I. 1982a. Floral biology, hummingbird pollination and fruit production of trumpet creeper (*Campsis radicans,* Bignoniaceae). *Am. J. Bot.* 69:122–134.
Bertin, R. I. 1982b. The evolution and maintenance of andromonoecy. *Evol. Theory* 6:25–32.
Bertin, R. I. 1982c. The ecology of sex expression in red buckeye. *Ecology* 63:445–456.
Bertin, R. I. 1987. Paternity in plants. In: J. Lovett Doust and L. Lovett Doust, eds. Plant reproductive strategies. Oxford Press, NY.
Bertin, R. I., and M. F. Willson. 1980. Effectiveness of diurnal and nocturnal pollination of two milkweeds. *Can. J. Bot.* 58:1744–1746.
Bertsch, A. 1984. Foraging in male bumblebees (*Bombus lucorum* L.): maximizing energy or minimizing water load? *Oecologia (Berl.)* 62:325–336.
Best, L. S. 1981. The effect of specific fruit and plant characteristics on seed dispersal. Ph.D. thesis, University of Washington, Seattle, WA.
Betts, M. 1955. The food of titmice in oak woodland. *J. Anim. Ecol.* 24:282–323.
Bierzychudek, P. 1981. Pollinator limitation of plant reproductive effort. *Am. Natur.* 117:838–842.
Birk, Y. 1969. Saponins. pp. 169–210. In: I. E. Liener, ed. Toxic constituents of plant foodstuffs. Academic Press, NY.
Blau, P. A., P. Feeny, L. Contordo, and D. S. Robson. 1978. Allyglucosinolate and herbivorous caterpillars: a contrast in toxicity and tolerance. *Science* 200:1296–1298.
Bliss, F. A. 1984. The application of new plant biotechnology to crop improvement. *HortScience* 19:43.
Blom, P. E., and W. H. Clark. 1980. Observations of ants (Hymenoptera: Formicidae) visiting the extrafloral nectaries of the barrel cactus *Ferocactus gracilis* (Cactaceae) in Baja California, Mexico. *Southwest. Natur.* 25:181–196.

Blumberg, A. Y., and D. A. Crossley, Jr. 1983. Comparison of soil surface arthropod populations in conventional tillage, no-tillage and old-field systems. *Agro-Ecosystems* 8:247–253.
Boag, P. T., and P. R. Grant. 1981. Intense natural selection on a population of Darwin's finches (Geospizinae) in the Galapagos. *Science* 214:82–85.
Bock, C. E., and J. H. Bock. 1974. Geographical ecology of the acorn woodpecker: abundance vs. diversity of resources. *Am. Natur.* 108:694–698.
Bock, C. E., and L. W. Lepthien. 1976. Synchronous eruptions of boreal seed-eating birds. *Am. Natur.* 110:559–571.
Boecklen, W. J. 1984. The role of extrafloral nectaries in the herbivore defense of *Cassia fasciculata. Ecol. Entomol.* 9:243–249.
Boethel, D. J., and D. Eikenbary, eds. 1986. Interactions of plant resistance and parasitoids and predators of insects. John Wiley and Sons, NY.
Boettcher, J. 1979. Ant–plant relationships: defense of *Yucca glauca* L. (Liliaceae) by ants. M.A. thesis, University of Nebraska, Omaha, NB.
Bond, W., and P. Slingsby. 1984. Collapse of an ant–plant mutualism: the Argentine ant *(Iridomyrmex humilis)* and myrmecochorous Proteaceae. *Ecology* 65:1031–1037.
Bopp, M., and E. Weiler. 1985. Leaf blade movement of *Drosera* and auxin distribution. *Naturwissenchaften* 72:434.
Borrer, D. J., D. M. DeLong, and C. A. Triplehorn. 1976. An introduction to the study of insects, 4th ed. Holt, Reinhart and Winston, NY.
Boston, H. L., and M. S. Adams. 1986. The contribution of crassulacean acid metabolism to the annual productivity of two aquatic vascular plants. *Oecologia (Berl.)* 68:615–622.
Boucher, D. H., S. James, and K. H. Keeler. 1982. The ecology of mutualism. *Ann. Rev. Ecol. Syst.* 13:315–347.
Bourne, C., E. R. LeFranc, and F. Nunes. 1980. Small farming in the less developed countries. Caribbean Development Bank, San Juan, PR.
Boyden, T. C. 1982. The pollination biology of *Calypso bulbosa* var. *americana* (Orchidaceae): initial deception of bumblebee visitors. *Oecologia (Berl.)* 55:178–184.
Bradshaw, W. E. 1983. Interaction between the mosquito *Wyeomyia smithii*, the midge *Metriocnemus knabi*, and their carnivorous host *Sarracenia purpurea*. pp. 161–189. In: J. H. Frank, and L. P. Lounibos, eds. Phytotelmata: terrestrial plants as hosts for aquatic insect communities. Plexus Publishing, Medford, NJ.
Bradshaw, W. E., and R. A. Creelman. 1984. Mutualism between the carnivorous purple pitcher plant *Sarracenia purpurea* and its inhabitants. *Am. Midl. Natur.* 112:294–304.
Brady, N. C. 1974. The nature and properties of soils. Macmillan, NY.
Branson, T. F., and J. L. Krysan. 1981. Feeding and oviposition behavior and life cycle strategies of *Diabrotica:* an evolutionary view with implications for pest management. *Environ. Entomol.* 10:826–831.
Branson, T. F., G. R. Sutter, and J. R. Fisher. 1980. Plant response to stress induced by artificial infestations of western corn rootworm. *Environ. Entomol.* 9:253–257.
Brattsten, L. B. 1979a. Ecological significance of mixed function oxidases. *Drug Metab. Rev.* 10:35–58.
Brattsten, L. B. 1979b. Biochemical defense mechanisms in herbivores against plant allelochemicals. pp. 199–271. In: G. A. Rosenthal, and D. H. Janzen, eds. Herbivores their interaction with secondary plant metabolites. Academic Press, NY.
Brattsten, L. B., C. F. Wilkinson, and J. Eisner. 1977. Herbivore-plant interactions: mixed-function oxidases and secondary plant substances. *Science* 196:1349–1352.
Breitwisch, R., P. G. Merritt, and G. H. Whitesides. 1984. Why do northern mockingbirds feed fruit to their nestlings? *Condor* 86:281–287.
Brewer Carias, C. 1976. La vegetacion del mundo perdido. Fundacion Eugenio Mendoza, Caracas.
Bridwell, J. C. 1968. *Speculamus erythrinae* a new bruchid affecting the seeds of *Erythrina* (Coleoptera). *J. Wash. Acad. Sci.* 28:69–76.
Briese, D. T. 1982a. Partitioning of resources amongst seed-harvesters in an ant community in semi-arid Australia. *Aust. J. Ecol.* 7:299–307.

Briese, D. T. 1982b. Relationship between the seed-harvesting ants and the plant community in the semi-arid environment. pp. 11–24. In: R. C. Buckley, ed. Ant–plant interactions in Australia. Dr. W. Junk, Publisher, The Hague.

Brower, L. P., J. V. Z. Brower, and J. M. Corvine. 1967. Plant poisons in the terrestrial food chain. *Proc. Natl. Acad. Sci.* 57:893–898.

Brown, A. W. 1978. Ecology of pesticides. John Wiley and Sons, NY.

Brown, J. H. 1975. Geographical ecology of desert rodents. pp. 315–341. In: M. L. Cody, and J. M. Diamond, eds. Ecology and evolution of communities. Belknap, Cambridge, MA.

Brown, J. H. 1981. Two decades of homage to Santa Rosalia: toward a general theory of diversity. *Am. Zool.* 21:877–888.

Brown, J. H., and D. W. Davidson. 1977. Competition between seed-eating rodents and ants in desert ecosystems. *Science* 196:877–888.

Brown, J. H., and A. Kodric-Brown. 1979. Convergence, competition, and mimicry in a temperate community of hummingbird-pollinated flowers. *Ecology* 60:1022–1035.

Brown, J. H., and G. A. Lieberman. 1973. Resource utilization and coexistence of seed-eating desert rodents in sand dune habitats. *Ecology* 54:788–797.

Brown, J. H., G. A. Lieberman, and W. F. Dengler. 1972. Woodrats and cholla: Dependence of a small mammal population on the density of cacti. *Ecology,* 53:310–313.

Brown, J. H., O. J. Reichman, and D. W. Davidson. 1979. Granivory in desert ecosystems. *Ann. Rev. Ecol. Syst.* 10:201–227.

Brown, J. L. 1964. The evolution of diversity in avian territorial systems. *Wilson Bull.* 6:160–169.

Brown, J. L. 1969. Territorial behavior and population regulation in birds. *Wilson Bull.* 81:293–329.

Brown, J. L. 1974. Alternate routes to sociality in jays—with a theory for the evolution of altruism and communal breeding. *Am. Zool.* 14:63–80.

Brown, M., and R. G. Can. 1984. Interactions between *Azotabacter chroococcum* and vesicular—arbuscular mycorrhiza and their effects on plant growth. *J. Appl. Bacteriol.* 56:429–438.

Brown, W. L. 1960. Ants, acacias and browsing mammals. *Ecology* 41:587–592.

Brown, W. L., and R. W. Taylor. 1970. Hymenoptera superfamily Formicoidae. pp. 951–959. In: CSIRO. The insects of Australia. Melbourne University Press, Melbourne.

Brunner, F., and D. E. Fairbrothers. 1978. A comparative serological investigation within the Cornales. *Serol. Mus. Bull.* 53:2–5.

Brust, G. E., B. R. Stinner, and D. A. McCartney. 1985. Tillage and soil insecticide effects on predators—black cutworm (Lepidoptera: Noctuidae) interactions in corn agroecosystems. *J. Econ. Entomol.* 78:1389–1392.

Brust, G. E., B. R. Stinner, and D. A. McCartney. 1986a. Predator activity and predation in corn agroecosystems. *Environ. Entomol.* 15:1017–1021.

Brust, G. E., B. R. Stinner, and D. A. McCartney. 1986b. Predation by soil inhabiting arthropods in intercropped and monoculture agroecosystems. Agric. Ecosystems Environ. 18:145–154.

Bryant, J. P. 1981. Phytochemical deterrence of snowshoe hare browsing by adventitious shoots of four Alaskan trees. *Science* 213:889–890.

Bryant, J. P., and P. J. Kuropat. 1980. Selection of winter forage by subarctic browsing vertebrates: the role of plant chemistry. *Ann. Rev. Ecol. Syst.* 11:261–285.

Bryant, J. P., F. S. Chapin III, and D. R. Klein. 1983. Carbon/nutrient balance of boreal plants in relation to vertebrate herbivory. *Oikos* 40:357–368.

Buckley, R. C. 1982. Ant-plant interactions: a world review. pp. 111–142. In: R. C. Buckley, ed. Ant–plant interactions in Australia. Dr. W. Junk, Publisher, The Hague.

Burbridge, F. W. T. 1880. *Gardeners Chronicle* 180:201.

Burtt, B. D. 1929. A record of fruits and seeds dispersed by mammals and birds from the Singida district of Tanganyika territory. *J. Ecol.* 17:351–355.

Büsgen, M. 1883. Die Bedeutung des Insektenfanges fur *Drosera rotundifolia. Bot. Z.* 41:569–577.

Buttrose, M. S., W. J. R. Grant, and J. N. A. Lott. 1977. Reversible curvature of style branches of *Hibiscus trionum* L., a pollination mechanism. *Aust. J. Bot.* 25:567–570.

Caffrey, D. J., and L. H. Worthley. 1927. A progress report on the investigations of the European corn borer. pp. 155. *U.S. Dep. Agric. Bull.* 1476.

Cahalane, V. H. 1942. Catching and recovery of food by the western fox squirrel. *J. Wildl. Manage.* 6:338–352.

Cameron, C. J., G. L. Donald, and C. G. Paterson. 1977. Oxygen-fauna relationships in the pitcher plant, *Sarracenia purpurea* L. with special reference to the chironomid *Metriocnemus knabi* Coq. *Can. J. Zool.* 55:2018–2023.

Campbell, B. C., and S. S. Duffy. 1979. Tomatine and parasitic wasps. Potential incompatibility of plant antibiosis with biological control. *Science* 205:700–702.

Campbell, D. R. 1985. Pollinator sharing and seed set of *Stellaria pubera:* competition for pollination. *Ecology* 66:544–553.

Carlquist, S. 1976. Wood anatomy of Byblidaceae. *Bot. Gaz.* 137:35–38.

Carlquist, S. 1982. Wood anatomy of Daphniphyllaceae: ecological and phylogenetic considerations, review of Pittosporalean families. *Brittonia* 34:252–266.

Carpenter, F. L. 1976. A threshold model of feeding territoriality and a test with a Hawaiian honeycreeper. *Science* 194:639–642.

Carpenter, F. L. 1978. Hooks for mammal pollination? *Oecologia (Berl.)* 35:123–132.

Carpenter, F. L., D. C. Paton, and M. A. Hixon. 1983. Weight gain and adjustment of feeding territory size in migrant hummingbirds. *Proc. Natl. Acad. Sci.* 80:7259–7273.

Carroll, C. R., and D. H. Janzen. 1973. Ecology of foraging by ants. *Ann. Rev. Ecol. Syst.* 4:231–257.

Carter, M. R., and D. A. Rennie. 1984. Nitrogen transformations under zero and shallow tillage. *Soil Sci. Soc. Am. J.* 48:1077–1081.

Caspar, J. 1966. Monograpahie der Gattung *Pinguicula* L. *Bibliotheca Bot.* 127/128:1–209.

Casper, B. B., and T. R. LaPine. 1984. Changes in corolla color and other floral characteristics in *Cryptantha humilis* (Boraginaceae): cues to discourage pollinators? *Evolution* 38:128–141.

Caughley, G., G. C. Grigg, J. Caughley, and G. J. E. Hill. 1980. Does dingo predation control the densities of kangaroos and emus? *Austr. Wildl. Res.* 7:1–12.

Caughley, G., and J. H. Lawton. 1981. Plant-herbivore systems. pp. 132–166. In: R. M. May, ed. Theoretical ecology. Principles and applications, 2nd ed. Sinauer Associates, Sunderland, MA.

Chandler, G. E., and J. W. Anderson. 1976a. Studies on the nutrition and growth of *Drosera* species with reference to the carnivorous habit. *New Phytol.* 76:129–141.

Chandler, G. E., and J. W. Anderson. 1976b. Studies on the origin of some hydrolytic enzymes associated with the leaves and tentacles of *Drosera* species and their role in heterotrophic nutrition. *New Phytol.* 77:51–62.

Chandler, G. E., and J. W. Anderson. 1976c. Uptake and metabolism of insect metabolites by leaves and tentacles of *Drosera* species. *New Phytol.* 77:625–634.

Chapin, F. S., III. 1983. Patterns of nutrient absorption and use by plants from natural and man-modified environments. pp. 175–187. In: H. A. Mooney, and M. Godron, eds. Disturbance and ecosystems. Springer-Verlag, NY.

Chapman, F. M. 1939. Handbook of the birds of eastern North America. D. Appleton-Century Co., reprinted in 1966 by Dover Publications, NY.

Charles-Dominique, P. 1977. Ecology and behavior of nocturnal primates. Columbia University Press, NY.

Charles-Dominique, P. 1986. Feeding strategies of frugivorous bats (Phyllostomidae) and their relation to the evolution of pioneer plants. In: A. Estrada, T. H. Fleming, C. Vazquez-Yanes, and R. Dirzo, eds. Frugivores and seed dispersal. Dr. W. Junk, Publisher, The Hague.

Charnov, E. L. 1976. Optimal foraging. The marginal value theorem. *Theor. Pop. Biol.* 9:129–136.

Charnov, E. L. 1979. Simultaneous hermaphroditism and sexual selection. *Proc. Natl. Acad. Sci.* 76:2480–2484.

Charnov, E. L., J. P. Maynard Smith, and J. J. Bull. 1976. Why be an hermaphrodite? *Nature* 263:125–126.
Chase, V. C., and P. H. Raven. 1975. Evolutionary and ecological relationships between *Aquilegia formosa* and *A. pubescens* (Ranunculacea), two perennial plants. *Evolution* 29:474–486.
Chew, R. 1974. Consumers as regulators of ecosystems: an alternative to energetics. *Ohio J. Sci.* 6:359–370.
Chiang, H. C. 1973. Bionomics of the northern and western corn rootworms. *Ann. Rev. Entomol.* 18:47–72.
Chlarholm, M. 1985. Possible roles for roots, bacteria, protozoa and fungi in supplying nitrogen to plants. pp. 355–366. In: A. H. Fitter, ed. Ecological interactions in soil. Blackwell Scientific Publications, London.
Christiansen, N. L. 1976. The role of carnivory in *Sarracenia flava* L. with regard to specific nutrient deficiencies. *J. Elisha Mitchell Sci. Soc.* 92:144–147.
Christie, W. J. 1974. Changes in the fish species composition of the Great Lakes. *J. Fish. Res. B Can.* 31:827–854.
Chujo, H., and H. Daimon. 1984. Plant growth and fate of nitrogen in mixed croppings, intercropping and crop rotation. *Japan. J. Crop Sci.* 53:213–221.
Cibula, D. A., and M. Zimmerman. 1984. The effect of plant density on departure decisions: testing the marginal value theorem using bumblebees and *Delphinium nelsonii. Oikos* 43:154–158.
Clancy, F. G., and M. D. Coffey. 1977. Acid phosphatase and protease release by the insectivorous plant *Drosera rotundifolia. Can. J. Bot.* 55:480–488.
Clawson, M. 1973. America's lands and its uses. Johns Hopkins Press, Baltimore, MD.
Clemens, W. A., and Z. Kielan-Jaworowska. 1979. Multituberculata. pp. 99–149. In: J. A. Lillegraven, Z. Kielan-Jaworowska, and W. A. Clemens, eds. Mesozoic mammals. The first two-thirds of mammalian history. University of California Press, Berkeley, CA.
Clutton-Brock, B. J. 1981. Domesticated animals. University of Texas Press, Austin, TX.
Clutton-Brock, T. H., and P. H. Harvey. 1977. Primate ecology and social organization. *J. Zool.* 183:1–39.
Cody, M. L. 1986. Diversity, rarity, and conservation in Mediterranean-climate regions. pp. 123–152. In: M. E. Soule, ed. Conservation biology. Sinauer Associates, I, Sunderland, MA.
Cole, F. R., and G. O. Batzli. 1979. Nutrition and population dynamics of the prairie vole, *Microtus ochrogaster*, in central Illinois. *J. Anim. Ecol.* 48:455–470.
Cole, H. H., and Garrett, eds. 1980. Animal agriculture. W. H. Freeman, San Francisco, CA.
Cole, S., F. R. Hainsworth, A. C. Kamil, T. Mercier, and L. L. Wolf. 1982. Spatial learning as an adaptation in hummingbirds. *Science* 217:655–657.
Coleman, D. C. 1985. Through a ped darkly: an ecological assessment of root-soil-microbial-faunal interactions. pp. 1–22. In: A. H. Fitter, ed. Ecological interactions in soil. Blackwell Scientific Publications, London.
Coleman, D. C., C. P. P. Reid, and C. V. Cole. 1983. Biological strategies of nutrient cycling in soil systems. *Adven. Ecol. Res.* 13:1–55.
Coleman, D. C., C. V. Cole, and E. T. Elliott. 1984 Decomposition organic matter turnover and nutrient dynamics in agroecosystems. pp. 83–104. In: R. Lowrance, B. R. Stinner, and G. J. House, eds. Agricultural ecosystems. John Wiley and Sons, NY.
Coley, P. D., J. P. Bryant, and F. S. Chapin III. 1985. Resource availability and plant antiherbivore defense. *Science* 230:895–899.
Comai, L., and D. M. Stalker. 1984. Impact of genetic engineering on crop protection. *Crop Prot.* 3:399–408.
Conn, E. E. 1979. Cyanide and cyanogenic glycosides. pp. 387–412. In: G. A. Rosenthal, and D. H. Janzen, eds. Herbivores: their interaction with secondary plant metabolites. Academic Press, NY.
Conn, E. E. 1981. Secondary plant products, vol. 7: The biochemistry of plants. Academic Press, NY.
Connell, J. H. 1961. The influence of interspecific competition and other factors on the distribution of the barnacle *Chthamalmus stellatus. Ecology* 42:710–723.

Connell, J. H. 1978. Diversity in tropical rain forests and coral reefs. *Science* 199:1302–1310.
Connell, J. H. 1983. On the prevalence and relative importance of interspecific competition: evidence from field experiments. *Am. Natur.* 122:661–696.
Connell, J. H., and E. Orias. 1964. The ecological regulation of species diversity. *Am. Natur.* 98:399–414.
Connor, E. F., and D. Simberloff. 1986. Competition, scientific method, and null models in ecology. *Am. Sci.* 74:155–162.
Cook, A. D., P. R. Atsatt, and C. A. Simon. 1971. Doves and dove weed: multiple defense against avian predation. *Bioscience* 21:277–281.
Cooper, S. M., and N. Owen-Smith. 1985. Condensed tannins deter feeding by browsing ruminants in a South African savanna. *Oecologia (Berl.)* 67:142–146.
Coppock, D. L., J. K. Detling, J. E. Ellis, and M. I. Dyer. 1983a. Plant-herbivore interactions in a North American mixed grass prairie. I. Effects of black-tailed prairie dogs on intraseasonal aboveground plant biomass and nutrient dynamics and plant species diversity. *Oecologia (Berl.)* 56:1–9.
Coppock, D. L., J. E. Ellis, J. K. Detling, and M. I. Dyer. 1983b. Plant-herbivore interactions in a North American mixed grass prairie. II. Responses of bison to modification of vegetation by prairie dogs. *Oecologia (Berl.)* 56:10–15.
Corbet, S. A., I. Cuthill, M. Fallows, T. Harrison, and G. Hartley. 1981. Why do nectar-foraging bees and wasps work upward on inflorescences? *Oecologia (Berl.)* 51:79–83.
Cornell, H. V. 1982. The secondary chemistry and complex morphology of galls formed by the Cynipinae (Hymenoptera): why and how? *Am. Midl. Natur.* 110:225–232.
Cornell, H. V., and J. O. Washburn. 1979. Evolution of the richness area correlation for cynipid gall wasps on oak trees: a comparison of two geographic areas. *Evolution* 33:257–274.
Courtney, S. P. 1983. Models of host plant location by butterflies—the effect of search images and search efficiencies. *Oecologia (Berl.)* 59:317–321.
Covich, A. P. 1976. Analyzing shapes of foraging areas: some ecological and economic theories. *Ann. Rev. Ecol. Syst.* 7:235–257.
Cox, G. W., and M. D. Atkins. 1979. Agricultural ecology. W. H. Freeman and Company, San Francisco, CA.
Cox, P. A. 1983. Search theory, random notion, and the convergent evolution of pollen and spore morphology in aquatic plants. *Am. Natur.* 121:9–31.
Cramer, C. L., T. B. Ryder, J. N. Bell, and C. J. Lamb. 1985. Rapid switching of plant gene expression induced by fungal elicitor. *Science* 227:1240–1243.
Crawley, M. J. 1983. Herbivory: the dynamics of animal-plant interactions. University of California Press, Berkeley, CA.
Crawley, M. L. 1985. Reduction of oak fecundity by low-density herbivore populations. *Nature* 314:163–164.
Crepet, W. L. 1979. Insect pollination: a paleontological perspective. *BioScience* 29:102–108.
Cronquist, A. 1981. An integrated system of classification of flowering plants. Columbia University Press, NY.
Crook, J. H., and J. S. Gartlan. 1966. Evolution of primate societies. *Nature* 210:1200–1203.
Crossley, D. A., Jr., G. J. House, R. M. Snider, R. J. Snider, and B. R. Stinner. 1984. Positive interactions in agroecosystems. pp. 73–82. In: R. Lowrence, B. R. Stinner, and G. J. House, eds. Agricultural ecosystems. John Wiley and Sons, NY.
Cruden, R. W. 1977. Pollen-ovule ratios: a conservative indicator of breeding systems in flowering plants. *Evolution* 31:32–46.
Cruden, R. W., and S. M. Hermann-Parker. 1979. Butterfly pollination of *Caesalpinia pulcherrima,* with observations on a psychophilous syndrome. *J. Ecol.* 67:155–168.
Cruden, R. W., S. M. Hermann, and S. Peterson. 1983. Patterns of nectar production and plant-pollinator coevolution. pp. 80–125. In: B. Bentley and T. Elias, eds. The biology of nectaries. Columbia University Press, NY.
Culver, D. C., and A. J. Beattie. 1978. Myrmecochory in *Viola:* dynamics of seed-ant interactions in some West Virginia species. *J. Ecol.* 66:53–72.

Czyhrinciw, N. 1969. Typical fruit technology. *Advan. Food Res.* 16:153–214.
Danell, K., and K. Huss-Danell. 1985. Feeding by insects and hares on birches earlier affected by moose browsing. *Oikos* 44:75–81.
Darwin, C. 1859. On the origin of species. Murray, London.
Darwin, C. 1874. Insectivorous plants. Murray, London.
Darwin, C. 1877. The different forms of flowers on plants of the same species. Murray, London.
Darwin, C. 1890. The various contrivances by which orchids are fertilized by insects. Murray, London.
Darwin, C. 1878. Experiments on the nutrition of *Drosera rotundifolia*. *J. Linn. Soc. Bot.* 17:17–32.
Davidar, P., and E. S. Morton. 1986. The relationship between fruit crop sizes and fruit removal rates by birds. *Ecology* 67:262–265.
Davidson, D. W. 1977. Seed diversity and community organization in desert seed-eating ants. *Ecology* 58:711–724.
Davidson, D. W., J. H. Brown, and R. S. Inouye. 1980. Competition and the structure of granivore communities. *Bioscience* 30:233–238.
Day, P. R. 1972. Crop resistance to pests and pathogens. pp. 257–271. In: Pest control: strategies for the future. Natl. Acad. Sci., Washington, DC.
DeAngelis, D. L., E. W. Stiles, W. C. Johnson, D. M. Sharpe, and P. K. Schreiber. 1977. A model for the dispersal of seeds by animals. Oak Ridge National Laboratory Environmental Sciences Division Publication No. 1053, Oak Ridge, TN.
DeBuhr, L. 1975a. Phylogenetic relationships of the Sarraceniaceae. *Taxon* 24:297–306.
DeBuhr, L. 1975b. Observations of *Byblis gigantea* in southwestern Australia. *Carniv. Plant News.* 4:60–63.
DeBuhr, L. 1977. Sectional reclassification of *Drosera* subgenus *Eragaleium* (Droseraceae). *Aust. J. Bot.* 25:209–218.
De Candolle, A. L. 1959. Origin of cultivated plants. Hafner Co., NY.
DeLong, D. W. 1971. The bionomics of leafhoppers. *Ann. Rev. Entomol.* 16:179–210.
Denslow, J. S., and T. C. Moermond. 1982. The effect of accessibility on rates of fruit removal from tropical shrubs: an experimental study. *Oecologia (Berl.)* 54:170–176.
DeSchauensee, R. Meyer. 1964. The birds of Colombia. Livingston, Wyneewood, PA.
DesGranges, J. L. 1980. Migrant hummingbird's accommodation into tropical communities. pp. 395–410. In: A. Keast, and E. S. Morton, eds. Migrant birds in the neotropics. Smithsonian Institution Press, Washington, DC.
Detling, J. K., M. I. Dyer, C. Procter-Gregg, and D. T. Winn. 1980. Plant-herbivore interactions: examination of potential effects of bison saliva on regrowth of *Bouteloua gracilis* (H.B.K.) Lag. *Oecologia (Berl.)* 45:26–31.
Deuth, D. 1977. The function of extrafloral nectaries in *Aphelandra deppeana*. Schl. and Cham. (Acanthaceae). *Brenesia* 10/11:135–145.
Deuth, D. 1980. The protection of *Cassia fruticosa* (Leguminosae) from herbivores by ants foraging at the foliar nectaries. M.A. thesis, University of Colorado, Boulder, CO.
Devlin, B., and A. G. Stephenson. 1985. Sex differential floral longevity, nectar secretion, and pollinator foraging in a protandrous species. *Am. J. Bot.* 72:303–310.
De Wit, C. T. 1960. On competition. *Versl. Landouk. Onderz.* 66:1–82.
Diamond, J. M. 1973. Distributional ecology of New Guinea birds. *Science* 179:759–769.
Diamond, J. M. 1978. Niche shifts and the rediscovery of interspecific competition. *Am. Sci.* 66:322–331.
Diamond, J. M. 1986a. Biology of birds of paradise and bowerbirds. *Ann. Rev. Ecol. Syst.* 17:17–38.
Diamond, J. M. 1986b. The design of a nature reserve system for Indonesian New Guinea. pp. 485–503. In M. E. Soule, ed. Conservation biology. Sinauer Associates, Sunderland, MA.
Dick, W. A. 1984. Influence of long-term tillage and crop rotation combinations on soil enzyme activities. *Soil Sci. Soc. Am. J.* 48:569–574.
Diels, L. 1906. Droseraceae. Das Pflanzenreich, vol. 27, Engelmann, Leipzig.
Dinesman, L. G. 1967. Influence of vertebrates on primary production of terrestrial

communities. pp. 261–266. In: K. Petrusewicz, ed. Secondary productivity of terrestrial ecosystems. Institute of Ecology, Polish Academy of Sciences, Warsaw.
Dingle, H. 1980. Ecology and evolution of migration. pp. 1–101. In: S. A. Gauthreaux, ed. Animal migration, orientation and navigation. Academic Press, NY.
Dixon, K. W., and J. S. Pate. 1978. Phenology, morphology, and reproductive biology of the tuberous sundew, *Drosera erythrorhiza* Lindl. *Aust. J. Bot.* 26:441–454.
Dixon, K. W., J. S. Pate, and W. J. Bailey. 1980. Nitrogen nutrition of the tuberous sundew, *Drosera erythrorhiza,* with special reference to catch of arthropod fauna by its glandular leaves. *Aust. J. Bot.* 28:283–297.
Dobson, A. P., and R. M. May. 1986. Disease and conservation. pp. 345–366. In: M. E. Soule, ed. Conservation biology. Sinauer Associates, Sunderland, MA.
Dobzhansky, T. 1950. Evolution in the tropics. *Am. Sci.* 38:209–221.
Dobzhansky, T. H. 1968. On some fundamental concepts of Darwinian biology. *Evol. Biol.* 2:1–34.
Dodson, C. H. 1975. Coevolution of orchids and bees. pp. 91–99. In: L. E. Gilbert and P. H. Raven, eds. Coevolution of animals and plants. University of Texas Press, Austin, TX.
Dodson, C. H., and G. P. Frymire. 1961. Natural pollination of orchids. *Ann. Mo. Bot. Gard.* 48:137–171.
Dodson, C. H., R. L. Dressler, H. G. Hills, R. M. Adams, and N. H. Williams. 1969. Biologically active compounds in orchid fragrances. *Science* 164:1243–1249.
Dolbeer, R. A., and W. R. Clark. 1975. Population ecology of snowshoe hares in the central Rocky Mountains. *J. Wildl. Manage.* 39:535–549.
Donald, C. M. 1963. Competition between crop and pasture plants. *Advan. Agron.* 15:1–118.
Doran, J. W. 1980. Soil microbial and biochemical changes associated with reduced tillage. *Soil Sci. Soc. Am. J.* 44:765–771.
Doyle, J. A. 1978. Origin of angiosperms. *Ann. Rev. Ecol. Syst.* 9:365–392.
Dubost, G. 1979. The size of African forest artiodactyls as determined by the vegetation structure. *Afr. J. Ecol.* 17:1–17.
Dueser, R. D., and W. C. Brown. 1980. Ecological correlates of insular rodent diversity. *Ecology* 61:50–56.
Duffey, S. S. 1980. Sequestration of plant natural products by insects. *Ann. Rev. Entomol.* 25:447–477.
Duffey, S. S., K. A. Bloem, and B. C. Campbell. 1986. Consequences of sequestration of plant natural production on plant-insect-parasitoid interactions. pp. 31–60. In: D. J. Boethel, and D. Eikenbary, eds. Interactions of plant resistance and parasitoids and predators of insects. John Wiley and Sons, NY.
Dunning, J. B. 1984. Body weights of 686 species of North American birds. Western Bird-Banding Association Monogr. No. 1, 38 pp.
Dunning, J. B., Jr., and J. H. Brown. 1982. Summer rainfall and winter sparrow densities: a test of the food limitation hypothesis. *Auk* 99:123–129.
Ebelhar, S. A., W. W. Frye, and R. L. Blevins. 1984. Nitrogen from legume cover crops for no-tillage corn. *Agron. J.* 76:51–55.
Edwards, C. A., 1977. Investigations into the influence of agricultural practice on soil invertebrates. *Ann. Appl. Biol.* 87:515–520.
Edwards, C. A., and J. R. Lofty. 1973. The influence of cultivations on soil animal populations. pp. 349–407. In: J. Vaneck, ed. Progress in soil zoology. Academia, Czechoslovakia.
Edwards, C. A., and J. R. Lofty. 1980. Effects of earthworm upon the root growth of direct drilled cereals. *J. Appl. Ecol.* 17:533–543.
Edwards, C. A., and A. R. Thompson. 1973. Pesticides and the soil fauna. *Res. Rev.* 45:1–79.
Edwards, J. 1985. Effects of herbivory by moose on flower and fruit production of *Aralia nudicaulis. J. Ecol.* 73:861–868.
Edwards, T. C., Jr. 1986. Ecological distribution of the gray-breasted jay: the role of habitat. *Condor* 88:456–460.
Ehrlich, P. R., and P. H. Raven. 1964. Butterflies and plants: a study in coevolution. *Evolution* 18:586–608.

Eisenberg, J. F. 1981. The mammalian radiations. An analysis of trends in evolution, adaptation, and behavior. University of Chicago Press, Chicago, IL.
Eisenberg, J. F., N. A. Muckenhirn, and R. Rudran. 1972. The relation between ecology and social structure in primates. *Science* 176:863–874.
Eisenberg, J. F., M. A. O'Connell, and P. V. August. 1979. Density, productivity and distribution of mammals in two Venezuelan habitats. pp. 187–207. In: J. F. Eisenberg, ed. Vertebrate ecology in the northern neotropics. Smithsonian Institution Press, Washington, DC.
Eisner, T., and J. Shepperd. 1965. Caterpillar feeding on a sundew plant. *Science* 150:1608–1609.
Eleutarius, L. N., and S. B. Jones. 1969. A floristic and ecological study of pitcher plant bogs in south Mississippi. *Rhodora* 71:29–34.
Elias, T. S. 1983. Extrafloral nectaries: their structure and distribution. pp. 174–203. In: B. Bentley, and T. Elias, eds. The biology of nectaries. Columbia University Press, NY.
Elias, T. S., and H. Gelband. 1975. Nectar: its production and function in trumpet creeper. *Science* 189:289–290.
Elias, T. S., W. Rosich, and L. Newcombe. 1975. The foliar and floral nectaries of *Turnera ulmifolia* L. *Am. J. Bot.* 62:570–576.
Elkins, D., D. Fredesking, R. Marashi, and B. McVay. 1984. Living mulch for no-till corn and soybeans. *J. Soil Water Cons.* 38:431–436.
Ellison, L. 1960. Influence of grazing on plant succession of rangelands. *Bot. Rev.* 26:1–78.
Emberger, L. 1968. Les plantes fossiles dan leurs rapports avec les vegetaux vivants. Masson, Paris.
Emlen, J. T. 1974. An urban bird community in Tucson, Arizona: derivation, structure, regulation. *Condor* 76:184–197.
Emlen, S. T. 1982. The evolution of helping. I. An ecological constraints model. *Am. Natur.* 119:29–39.
Engelbrecht, L., U. Orban, and W. Heese. 1969. Leaf-miner caterpillars and cytokinins in the "green islands" of autumn leaves. *Nature* 223:319–321.
Erickson, E. H., and S. L. Buchman. 1983. Electrostatics and pollination. pp. 173–184. In: C. E. Jones, and R. J. Little, eds. Handbook of experimental pollination biology. Van Nostrand Reinhold, NY.
Erickson, R. 1978. Plants of prey in Australia. Lamb Publications, Perth.
Ericson, L. 1977. The influence of voles and lemmings on the vegetation in a coniferous forest during a 4-year period in northern Sweden. *Wahlenbergia* 4:1–113.
Errington, P. L. 1967. Of predation and life. Iowa State University Press, Ames, IA.
Esau, K. 1977. Anatomy of seed plants, 2nd ed. John Wiley and Sons, NY.
Estes, J. A., and R. W. Thorp. 1975. Pollination ecology of *Pyrrhopappus carolinianus* (Compositae). *Am. J. Bot.* 62:148–159.
Ewald, P. W., and S. Rohwer. 1982. Effects of supplemental feeding on timing of breeding, clutch-size, and polygyny in red-winged blackbirds *(Agelaius phoeniceus). J. Anim. Ecol.* 51:429–450.
Eyde, R. H. 1985. The case for monkey-mediated evolution in big-bracted dogwoods. *Arnoldia* 45:2–9.
Faegri, K., and L. van der Pijl. 1979. The principles of pollination ecology. Pergamon Press, NY.
Faeth, S. H. 1985. Host leaf selection by leaf miners: interactions among three trophic levels. *Ecology* 66:870–875.
Faeth, S. H., E. F. Connor, and D. Simberloff. 1981. Early leaf abscission: a neglected source of mortality for foliovores. *Am. Natur.* 117:409–415.
Falconer, D. S. 1981. Introduction to quantitative genetics, 2nd ed. Longman Group, Harlow, Essex, UK.
Feeny, P. 1970. Seasonal changes in oak leaf tannins and nutrients as a cause of spring feeding by winter moth caterpillars. *Ecology* 51:565–581.
Feeny, P. 1976. Plant apparency and chemical defense. *Rec. Advan. Phytochem.* 10:1–40.
Feeny, P. 1977. Defensive ecology of the Cruciferae. *Ann. M. Bot. Gard.* 64:221–234.
Feeny, P. 1982. Coevolution of plants and insects. Chapt. II. In: T. R. Odhiambo, ed.

Current themes in tropical sciences 2: natural products for innovative pest management. Pergammon Press, Oxford.
Feine, L. B., R. R. Harwood, C. S. Kauffman, and J. P. Senft. 1979. Amaranth. pp. 41–63. In: G. Ritchie, ed. New agricultural crops. Am. Assoc. Advan. Sci., Washington, DC.
Feinsinger, P. 1976. Organization of a tropical guild of nectarivorous birds. *Ecol. Monogr.* 46:257–291.
Feinsinger, P. 1980. Asynchronous migration patterns and the coexistence of tropical hummingbirds. pp. 411–420. In: A. Keast, and E. S. Morton, eds. Migrant birds in the neotropics. Smithsonian Institution Press, Washington, DC.
Feinsinger, P. 1983. Variable nectar secretion in a *Heliconia* species pollinated by hermit hummingbirds. *Biotropica* 15:48–52.
Felger, R. S. 1979. Ancient crops for the twenty-first century. pp. 4–19. In: G. A. Ritchie, ed. New agricultural crops. Am. Assoc. Advan. Sci., Washington, DC.
Field, C., and H. A. Mooney. 1986. The photosynthesis-nitrogen relationship in wild plants. pp. 24–25. In: T. J. Givnish, ed. On the economy of plant form and function. Cambridge University Press, Cambridge, UK.
Fischer, A. G. 1960. Latitudinal variation in organic diversity. *Evolution* 14:64–81.
Fish, D. 1976a. Structure and composition of the aquatic invertebrate community inhabiting epiphytic bromeliads in south Florida and the discovery of an insectivorous bromeliad. Ph.D. thesis, University of Florida, Gainesville, FL.
Fish, D. 1976b. Insect-plant relationships of the insectivorous pitcher plant *Sarracenia minor. Fla. Entomol.* 59:199–203.
Fish, D., and D. W. Hall. 1978. Succession and stratification of aquatic insects inhabiting the leaves of the insectivorous pitcher plant, *Sarracenia purpurea. Am. Midl. Natur.* 99:172–183.
Fleming, T. H. 1973. Numbers of mammal species in North and Central American forest communities. *Ecology* 54:555–563.
Fleming, T. H. 1979. Do tropical frugivores compete for food? *Am. Zool.* 19:1157–1172.
Fleming, T. H. 1981. Fecundity, fruiting pattern, and seed dispersal in *Piper amalago* (Piperaceae), a bat-dispersed tropical shrub. *Oecologia (Berl.)* 51:42–46.
Fleming, T. H., and E. R. Heithaus. 1981. Frugivorous bats, seed shadows, and the structure of tropical forests. *Biotropica (suppl.)* 13:45–53.
Flemion, F., M. C. Ledbetter, and E. S. Kelley. 1954. Penetration and damage of plant tissues during feeding by the tarnished plant bug. *Contrib. Boyce Thompson Inst.* 17:950–954.
Fogden, M. P. L. 1972. The seasonality and population dynamics of equatorial birds in Sarawak. *Ibis* 114:307–343.
Folkerts, G. W. 1982. The Gulf Coast pitcher plant bogs. *Am. Sci.* 79:260–267.
Ford, E. B. 1975. Ecological genetics. Chapman and Hall, London.
Fordham, R. 1983. Intercropping-what are the advantages? *Outlook Agric.* 12:142–146.
Forman, R. T. T., and M. Godron. 1986. Landscape ecology. John Wiley and Sons, NY.
Fossey, D., and A. H. Harcourt. 1977. Feeding ecology of free-ranging mountain gorilla *(Gorilla gorilla beringei)*. pp. 415–447. In: T. H. Clutton-Brock, ed. Primate ecology: feeding and ranging behaviour in lemurs, monkeys and apes. Academic Press, London.
Foster, M. A., and W. G. Ruesink. 1984. Modeling black cutworm–parasitoid–weed interactions in reduced tillage corn. *Agric. Ecosys. Environ.* 16:13–28.
Foster, M. S. 1977. Ecological and nutritional effects of food scarcity on a tropical frugivorous bird and its fruit source. *Ecology* 58:73–85.
Foster, M. S. 1978. Total frigivory in tropical passerines: a reappraisal. *Tropical Ecol.* 19:131–154.
Foster, M. S., and R. W. McDiarmid. 1983. Nutritional value of the aril of *Trichilia cuneata* a bird-dispersed fruit. *Biotropica* 15:26–31.
Foster, R. B. 1980. Heterogeneity and disturbance in tropical vegetation. pp. 75–92. In: M. E. Soule, and B. A. Wildox, ed. Conservation ecology. Sinauer Associates, Sunderland, MA.
Foster, R. B. 1982. The seasonal rhythm of fruit fall on Barro Colorado Island. In: E. G. Leigh, Jr., A. S. Rand, and D. S. Windsor, eds. The ecology of a tropical forest:

seasonal rhythms and long-term changes. Smithsonian Institution Press, Washington, D.C.
Fowden, L., D. Lewis, and H. Tristram. 1967. Toxic amino acids: their action as antimetabolites. *Advan. Enzymol.* 29:89–163.
Fowler, S. V., and J. H. Lawton. 1985. Rapidly induced defenses and talking trees: the devil's advocate position. *Am. Natur.* 126:181–195.
Fox, L. R. 1981. Defense and dynamics in plant-herbivore systems. *Am. Zool.* 21:853–864.
Fox, L. R., and P. A. Morrow. 1981. Specialization: species property or local phenomenon? *Science* 211:887–893.
Fraenkel, G. S. 1959. The raison d'etre of secondary plant substances. *Science* 129:1466–1470.
Francis, G. R., and A. B. Stephenson. 1972. Marten ranges and food habits in Algonquin Provincial Park, Ontario. Research Report (Wildlife) No. 91. Ministry of Natural Resources, Ontario, Canada.
Frank, D. H. 1976. The morphological interpretation of epiascidiate leaves. *Bot. Rev.* 42:345–388.
Frank, J. H., and G. F. O'Meara. 1984. The bromeliad *Catopsis bertieroniana* traps terrestrial arthropods but harbors *Wyeomyia* larvae (Diptera: Culicidae). *Fla. Entomol.* 67:418–424.
Freeman, D. C., L. G. Klikoff, and K. T. Harper. 1976. Differential resource utilization by the sexes of dioecious plants. *Science* 193:597–599.
Fretwell, S. 1969. Dominance behaviour and winter habitat distribution in juncos *(Junco hyemalis). Bird Banding* 40:1–25.
Fretwell, S. D. 1972. Populations in a seasonal environment. Monographs in Population Biology Vol. 5, Princeton University Press, Princeton, NJ.
Fretwell, S. D., and H. L. Lucas. 1970. On territorial behavior and other factors influencing habitat distribution in birds. I. Theoretical development. *Acta Biotheor.* 19:16–36.
Frisch, K. von. 1967. The dance language and orientation of bees. Harvard University Press, Cambridge, MA.
Fritz, R. S., and D. H. Morse. 1981. Nectar parasitism of *Asclepias syriaca* by ants: effect on nectar levels, pollinia insertion, pollinaria removal and pod production. *Oecologia (Berl.)* 50:316–319.
Frost, P. G. H. 1980. Fruit-frugivore interactions in a South African coastal dune forest. Acta. XVII Congr. Int. Ornithol., Deutsch Ornithologien-Gesellschest, Berlin.
Fryer, G. 1959. Some aspects of evolution in Lake Nyasa. *Evolution* 13:440–451.
Futuyma, D. J. 1976. Food plant specialization and environmental predictability. *Am. Natur.* 110:285–292.
Futuyma, D. J. 1983. Evolutionary interactions among herbivorous insects and plants. pp. 207–231. In: D. J. Futuyma, and M. Slatkin, eds. Coevolution. Sinauer Associates, Sunderland, MA.
Futuyma, D. J., and M. Slatkin. 1983. Introduction. pp. 1–13. In: D. J. Futuyma and M. Slatkin, eds. Coevolution. Sinauer Associates, Sunderland, MA.
Galen, C., and R. C. Plowright. 1985. The effects of nectar level and flower development on pollen carry-over in inflorescences of fireweed *(Epilobium angustifolium)* (Onagraceae). *Can. J. Bot.* 63:488–491.
Galil, J., and D. Eisikovitch. 1968. On the pollination ecology of *Ficus sycomorus* in East Africa. *Ecology* 49:259–269.
Galil, J., and D. Eisikovitch. 1974. Further studies on pollination ecology in *Ficus sycomorus.* II. Pocket filling and emptying by *Ceratosolen arabicus* Magr. *New Phytol.* 73:515–528.
Gallun, R. L., and G. S. Khush. 1980. Genetic factors affecting expression and stability of resistance. pp. 64–85. In: F. G. Maxwell, and P. R. Jennings, eds. Breeding plants resistant to insects. John Wiley and Sons, NY.
Gauthreaux, S. A., Jr. 1978. The ecological significance of behavioral dominance. pp. 17–54. In: P. P. G. Bateson, and P. H. Klopper, eds. Perspectives in ethology. Plenum Press, NY.

Gentry, A. H. 1978. Anti-pollinators for mass-flowering plants? *Biotropica* 10:68–69.
Georghiou, G. P., and C. E. Taylor. 1976. Pesticide resistance as an evolutionary phenomenon. pp. 31–37. In: Proc. 15th Int. Congr. Entomol., Washington, DC.
Gholson, L. E., C. C. Beegle, R. L. Best, and J. C. Owens. 1978. Effects of several commonly used insecticides on cornfield carabids in Iowa. *J. Econ. Entomol.* 71:416–418.
Gibb, J. 1954. Feeding ecology of tits, with notes on treecreeper and goldcrest. *Ibis* 96:513–543.
Gibson, T. C. 1983. Competition, disturbance, and the carnivorous plant community in the southeastern United States. Ph.D. dissertation, University of Utah, UT.
Gilbert, L. E. 1971. Butterfly-plant coevolution: has *Passiflora adenopoda* won the selectional race with heliconiine butterflies? *Science* 172:585–586.
Gilbert, L. E. 1972. Pollen feeding and reproductive biology of *Heliconius* butterflies. *Proc Natl. Acad. Sci.* 69:1403–1407.
Gilbert, L. E. 1975. Ecological consequences of a coevolved mutualism between butterflies and plants. pp. 210–240. In: L. E. Gilbert, and P. H. Raven, eds. Coevolution of animals and plants. University of Texas Press, Austin, TX.
Gilbert, L. E., and P. H. Raven, eds. 1975. Coevolution of animals and plants. University of Texas Press, Austin, TX.
Gill, F. B., and L. L. Wolf. 1975. Economics of feeding territoriality in the golden-winged sunbird. *Ecology* 56:333–345.
Gill, F. B., and L. L. Wolf. 1977. Nonrandom foraging by sunbirds in a patchy environment. *Ecology* 58:1284–1296.
Givnish, T. J., ed. 1972. Ecology of New Jersey Pine Barrens cedar swamps. Misc. Publ. Dep. of Geology, Princeton University, Princeton, NJ.
Givnish, T. J., E. L. Burkhardt, R. E. Happel, and J. D. Weintraub. 1984. Carnivory in the bromeliad *Brocchinia reducta*, with a cost-benefit model for the general restriction of carnivorous plants to sunny, moist, nutrient-poor habitats. *Am. Natur.* 124:479–497.
Givnish, T. J., R. W. McDiarmid, and W. R. Buck. 1986. Fire adaptation in *Neblinaria celiae* (Theaceae), a high-elevation rosette shrub endemic to a wet equatorial tepui. *Oecologia (Berl.)* 70:481–485.
Godin, J-G. J., and M. H. A. Keenleyside. 1984. Foraging on patchily distributed prey by a cichlid fish (Teleostei, Cichlidae): A test of the ideal free distribution theory. *Anim. Behav.* 32:120–131.
Goldstein, J. L., and T. Swain. 1963. Changes in tannins in ripening fruits. *Phytochemistry* 2:371–383.
Goldsmith, T. H. 1980. Hummingbirds see near ultraviolet light. *Science* 207:786–788.
Golley, F. B., and J. B. Gentry. 1964. Bioenergetics of the southern harvester ant *Pogonomyrmex badius. Ecology* 45:217–245.
Gori. D. F. 1983. Post-pollination phenomena and adaptive floral changes. pp. 31–49. In: C. E. Jones, and R. J. Little, eds. Handbook of experimental pollination biology. Van Nostrand Reinhold, NY.
Gottsberger, G. 1978. Seed dispersal by fish in the inundated regions of Humatia, Amazonia. *Biotropica* 10:170–183.
Gottsberger, G., J. Schrauwen, and H. F. Linskens. 1984. Amino acids and sugars in nectar, and their putative evolutionary significance. *Plant Syst. Evol.* 145:55–77.
Gould, S. J. 1980. The panda's thumb: more reflections in natural history. W. W. Norton, NY.
Goulding, M. 1980. The fishes and the forest. University of California Press, Berkeley.
Grant, P., and D. Schluter. 1984. Interspecific competition inferred from patterns of guild structure. pp. 201–233. In: D. R. Strong, D. Simberloff, L. G. Abele, and A. B. Thistle, eds. Ecological communities. Princeton University Press, Princeton, NJ.
Grant, P. R. 1971. The habitat preference of *Microtus pennsylvanicus* and its relevance to the distribution of this species on islands. *J. Mamm.* 52:351–361.
Grant, P. R. 1986a. Interspecific competition in fluctuating environments. pp. 173–191. In: J. Diamond, and T. J. Case, eds. Community ecology. Harper & Row, NY.
Grant, P. R. 1986b. Ecology and evolution of Darwin's finches. Princeton University Press, Princeton, NJ.

Grant, P. R., B. R. Smith, J. N. M. Smith, I. J. Abbott, and L. K. Abbott. 1976. Darwin's finches: population variation and natural selection. *Proc. Na. Acad. Sci.* 73:257–261.
Grant, S. 1984. Beauty and the beast: the coevolution of plants and animals. Charles Scribner's Sons, NY.
Grant, V., and K. A. Grant. 1965. Pollination in the *Phlox* family. Columbia University Press, NY.
Green, S., T. L. Green, and Y. Heslop-Harrison. 1979. Seasonal heterophylly and leaf gland features in *Triphyophyllum* (Dioncophyllaceae), a new carnivorous plant genus. *Bot. J. Linn. Soc.* 78:99–116.
Greenberg, R. 1981. Frugivory in some migrant tropical forest wood warblers. *Biotropica* 13:215–223.
Greig-Smith, P. W. 1986. Bicolored fruit displays and frugivorous birds: the importance of fruit quality to dispersers and seed predators. *Am. Natur.* 127:246–251.
Grime, J. P. 1979. Plant strategies and vegetation processes. John Wiley and Sons, NY.
Grinnell, J. 1904. The origin and distribution of the chestnut-backed chickadee. *Auk* 21:364–382.
Gross, J. E., Z. Wang, and B. A. Wunder. 1985. Effects of food quality and energy needs: changes in gut morphology and capacity of *Microtus ochrogaster*. *J. Mammal.* 66:661–667.
Gulmon, S. L., and C. C. Chu. 1981. The effect of light and nitrogen on photosynthesis, leaf characteristics, and dry matter allocation in the chaparral shrub, *Diplacus aurantiacus*. *Oecologia (Berl.)* 49:207–212.
Gwynne, M. D. 1969. Notes on the nutritive values of *Acacia* pods in relation to *Acacia* seed distribution by ungulates. *East Afr. Wildl. J.* 7:176–178.
Hainsworth, F. R., and L. L. Wolf. 1976. Nectar characteristics and food selection by hummingbirds. *Oecologia (Berl.)* 25:101–113.
Hairston, N. G. 1981. An experimental test of a guild: salamander competition. *Ecology* 62:65–72.
Hairston, N. G., F. E. Smith, and L. B. Slobodkin. 1960. Community structure, population control and competition. *Am. Natur.* 94:421–425.
Hall, K. R. L. 1965. Social organization of the old-world monkeys and apes. *Symp. Zool. Soc. Lond.* 14:265–289.
Hamilton, W. D. 1964. The genetic evolution of social behavior I and II. *J. Theor. Biol.* 7:1–52.
Handel, S. N. 1976. Dispersal ecology of *Carex pendunculata* (Cyperaceae), a new North American myrmecochore. *Am. J. Bot.* 63:1971–1979.
Handel, S. N. 1978. The competitive relationship of three woodland sedges and its bearing on the evolution of ant dispersal of *Carex pendunculata*. *Evolution* 32:151–162.
Handel, S. N., and J. P. Kochmer. 1986. Phylogenetic constraints and the evolution of flowering time of temperate animal-pollinated angiosperms. Program IV Int. *Congr. Ecol.* pp. 170–171 [Abstract].
Hanson, C. H. 1972. Alfalfa science and technology. Am. Soc. Agron., Madison, WI.
Harborne, J. B. 1982. Introduction to ecological biochemistry, 2nd ed. Academic Press, NY.
Hardee, D. D., and T. B. Davich. 1966. A feeding deterrent for the boll weevil, *Anthonomus grandis*, from tung meal. *J. Econ. Entomol.* 59:1267–1270.
Harder, L. D. 1983. Flower handling efficiency of bumble bees: morphological aspects of probing time. *Oecologia (Berl.)* 57:274–280.
Harder, R. 1963. Blutenbildung durch tierisch Zusatznahrung and andere Faktoren bei *Utricularia exoleta* R. Braun. *Planta* 59:459–471.
Harder, R. 1970. *Utricularia* als Objekt fur Heteroptrophieuntersuchungen (Wechselwirkung von Saccharose and Acetat). *Zeitschr. Pflanzenphysiol.* 63:181–184.
Harder, R., and I. Zemlin. 1967. Forderung der Entwicklung und des Bluhens von *Pinguicula lusitanica* durch futterung in axenischer kultur. *Planta* 73:181–193.
Harding, J. A., and R. C. Dyar. 1970. Resistance induced in European corn borers in the laboratory by exposing successive generations to DDT, Diazonon, or Carbaryl. *J. Econ. Entomol.* 63:250–253.

Hare, J. D., and T. G. Andreadis. 1983. Variation in the susceptibility of *Leptinotarsa decemlineala* (Coleoptera: Chrysomelidae) when reared on different host plants to the fungal pathogen, *Beauveria bassiana* in the field and laboratory. *Environ. Entomol.* 12:1891–1896.
Harley, J. L., and S. E. Smith. 1983. Mycorrhizal symbiosis. Academic Press, NY.
Harper, D. G. C. 1982. Competitive foraging in mallards: ideal free ducks. *Anim. Behav.* 30:575–584.
Harper, J. L. 1969. The role of predation in vegetational diversity. In: Diversity and stability in ecological systems. *Brookhaven Symp. Biol.* 22:48–62.
Harper, J. L., P. H. Lovell, and K. G. Moore. 1970. The shapes and sizes of seeds. *Ann. Rev. Ecol. Syst.* 1:327–356.
Hartley, P. H. T. 1953. An ecological study of the feeding habits of the English titmice. *J. Anim. Ecol.* 33:261–288.
Hartling, L. K., and R. C. Plowright. 1979. Foraging by bumble bees on patches of artificial flowers: a laboratory study. *Can. J. Zool.* 57:1866–1870.
Hartnett, D. C., and W. G. Abrahamson. 1979. The effects of stem gall insects on life history patterns in *Solidago canadensis. Ecology* 60:910–917.
Haukioja, E. 1980. On the role of plant defenses in the fluctuation of herbivore populations. *Oikos* 35:202–213.
Haukioja, E., and P. Niemela. 1979. Birch leaves as a resource for herbivores: seasonal occurrence of increased resistance in foliage after mechanical damage of adjacent leaves. *Oecologia (Berl.)* 39:151–160.
Hauptli, H. 1985. Genetically, engineered plants: environmental issues. *BioTechnology* 3:28–37.
Hawkins, R. P. 1961. Observations on the pollination of red clover by bees. I. The yield of seed in relation to the numbers and kinds of pollinators. *Ann. Appl. Biol.* 49:55–65.
Hay, M. E., and P. J. Fuller. 1981. Seed escape from heteromyid rodents: the importance of microhabitat and seed preference. *Ecology* 62:1395–1399.
Hayes, W. A. 1982. Minimum tillage farming. p. 202. No-tillage Farmer, Inc., Brookfield, WI.
Heads, P. A., and J. H. Lawton. 1985. Bracken, ants and extrafloral nectaries. II. The effect of ants on the insect herbivores of bracken. *J. Anim. Ecol.* 53:1015–1031.
Heath, M. E., D. S. Metcalf, and R. F. Barnes. 1973. Forages. Iowa State University Press, Ames, IA.
Heinrich, B. 1975. Energetics of pollination. *Ann. Rev. Ecol. Syst.* 6:139–170.
Heinrich, B. 1976. The foraging specializations of individual bumblebees. *Ecol. Monogr.* 46:105–128.
Heinrich, B. 1979a. Bumblebee economics. Harvard University Press, Cambridge, MA.
Heinrich, B. 1979b. Resource heterogeneity and patterns of movement in foraging bumblebees. *Oecologia (Berl.)* 40:235–245.
Heinrich, B. 1983. Insect foraging energetics. pp. 187–214. In: C. E. Jones, and R. J. Little, eds. Handbook of experimental pollination biology. Van Nostrand Reinhold, NY.
Heinrich, B., and S. L. Collins. 1983. Caterpillar leaf damage, and the game of hide-and-seek with birds. *Ecology* 64:592–602.
Heinrich, B., and P. H. Raven. 1972. Energetics and pollination ecology. *Science* 176:597–602.
Heithaus, E. R. 1981. Seed predation by rodents on three ant-dispersed plants. *Ecology* 62:136–145.
Heller, H. C. 1971. Altitudinal zonation of chipmunks (*Euramias*): interspecific aggression. *Ecology* 52:424–433.
Hendrix, S., and E. J. Trapp. 1981. Plant-herbivore interactions: insect induced changes in host plant sex expression and fecundity. *Oecologia (Berl.)* 49:119–122.
Hepburn, J. S., E. Q. St. John, and F. M. Jones. 1920. The absorption of nutrients and allied phenomena in the pitchers of the Sarraceniaceae. *J. Franklin Inst.* 189:147–184.
Herrera, C. M. 1981. Are tropical fruits more rewarding to dispersers than temperate ones? *Am. Natur.* 118:896–907.

Herrera, C. M. 1982a. Defense of ripe fruit from pests: its significance in relation to plant-disperser interactions. *Am. Natur.* 120:219–241.
Herrera, C. M. 1982. Seasonal variation in the quality of fruits and diffuse coevolution between plants and avian dispersers. *Ecology* 63:773–785.
Herrera, C. M. 1984. Adaptation to frugivory of Mediterranean avian seed dispersers. *Ecology* 65:609–617.
Herrera, C. M. 1985a. Determinants of plant–animal coevolution: the case of mutualistic dispersal of seeds by vertebrates. *Oikos* 44:132–141.
Herrera, C. M. 1985b. Habitat-consumer interactions in frugivorous birds. pp. 341–367. In: M. L. Cody, ed. Habitat selection in birds. Academic Press, NY.
Herrera, C. M., and P. Jordano. 1981. *Prunus mahaleb* and birds: the high-efficiency seed dispersal system of a temperate fruiting tree. *Ecol. Monogr.* 51:203–218.
Herring, E. M. 1951. Biology of leaf miners. Dr. W. Junk, Publisher, The Hague.
Heslop-Harrison, Y. 1975. Enzyme release in carnivorous plants. pp. 525–578. In: J. T. Dingle, and R. T. Dean, eds. Lysosomes in biology and pathology. North Holland Publishing, Amsterdam.
Heslop-Harrison, Y. 1976. Enzyme secretion and digest uptake in carnivorous plants. pp. 463–476. In: N. Sunderland, ed. Perspectives in experimental biology. Pergamon Press, Oxford.
Heslop-Harrison, Y. 1978. Carnivorous plants. *Sci. Am.* 238:104–115.
Heslop-Harrison, Y., and J. Heslop-Harrison. 1980. Chloride ion movement and enzyme secretion from the digestive glands of *Pinguicula*. *Ann. Bot.* 45:7219–731.
Heslop-Harrison, Y., and R. B. Knox. 1971. A cytochemical study of the leaf gland enzymes of insectivorous plants of the genus *Pinguicula*. *Planta* 96:183–211.
Heyneman, A. J. 1983. Optimal sugar concentrations of floral nectars—dependence on sugar intake efficiency and foraging costs. *Oecologia (Berl.)* 60:198–213.
Heywood, V. H., ed. 1978. Flowering plants of the world. Mayflower Books, London.
Hickey, W. C. 1961. Growth form of crested wheatgrass as affected by site and grazing. *Ecology* 42:173–176.
Hickman, J. C. 1974. Pollination by ants: a low-energy system. *Science* 184:1290–1292.
Hicks, K. L., and J. O. Tahvanainen. 1974. Niche differentiation by crucifer-feeding flea beetles (Coleoptera: Chrysomelidae). *Am. Midl. Natur.* 91:406–424.
Hilbert, D. W., D. M. Smith, J. K. Detling, and M. I. Dyer. 1981. Relative growth rates and the grazing optimization process. *Oecologia (Berl.)* 51:14–18.
Hilder, E. J. 1966. Distribution of excreta by sheep at pasture. *Proc. 10th Int. Grassl. Congr.* pp. 977–981. Helsinki, Finland.
Hixon, M. A., F. L. Carpenter, and D. C. Paton. 1983. Territory area, flower density, and time budgeting in hummingbirds: an experimental and theoretical analysis. *Am. Natur.* 122:366–391.
Hladik, C. M. 1977. A comparative study of the feeding strategies of two sympatric species of leaf monkeys: *Presbytis senex* and *Presbytis entellus*. pp. 324–353. In: T. H. Clutton-Brock, ed. Primate ecology: feeding and ranging behaviour in lemurs, monkeys and apes. Academic Press, London.
Hladik, C. M., A. Hladik, J. Bousset, P. Valdebouze, G. Viroben. and J. Delort-Laval. 1971. Le regime alimentaire des primates de l'ile de Barro Colorado (Panama). *Folia Primat.* 16:85–122.
Hobbs, R. J., and H. A. Mooney. 1985. Community and population dynamics of serpentine grassland annuals in relation to gopher disturbance. *Oecologia (Berl.)* 67:342–351.
Hocking, B. 1970. Insect association with the swollen thorn acacias. *Trans. Roy. Entomol. Soc. Lond.* 122:211–255.
Hocking, B. 1975. Ant–plant mutualism: evolution and energy. pp. 78–90. In: L. E. Gilbert, and P. H. Raven, eds. Coevolution of animals and plants. University of Texas Press, Austin, TX.
Hodgson, B. E. 1928. The host plants of the European corn borer in New England. pp. 64. U. S. Dep. Agric. Tech. Bull. 77, Washington, D.C.
Holm, C. H. 1973. Breeding sex ratios, territoriality, and reproductive success in the red-winged blackbird (*Agelaius phoeniceus*). *Ecology* 54:356–365.

Holthuijzen, A. M. A., and C. S. Adkisson. 1984. Passage rate, energetics, and utilization efficiency of the cedar waxwing. *Wilson Bull.* 96:680–684.
Horsfield, P. 1978. Evidence for xylem feeding by *Philaenus spumonus* (L). (Homoptera: Cercopidae). *Entomol. Exp. Appl.* 24:95–99.
Horvitz, C. C. 1980. Seed dispersal and seedling demography of *Calathea microcephala* and *C. ovandensis.* Dissertation, Northwestern University, Evanston, IL.
Horvitz, C. C. 1981. Analysis of how ant behaviors affect germination in a tropical myrmecochore *Calathea microcephala* (Marantaceae): microsite selection and aril removal by neotropical ants *Odontomachus, Pachycondyla,* and *Solenopsis* (Formicidae). *Oecologia (Berl.)* 51:47–52.
Horvitz, C. C., and A. J. Beattie. 1980. Ant dispersal *Calathea* (Marantaceae) seeds by carnivorous ponerines (Formicidae) in a tropical rain forest. *Am. J. Bot.* 67:321–326.
Horvitz, C. C., and D. W. Schemske. 1984. Effects of ants and an ant-tended herbivore on seed production of a neotropical herb. *Ecology* 65:1369–1378.
Horvitz, C. C., and D. W. Schemske. 1986. Seed dispersal and environmental heterogeneity in a neotropical herb: a model of population and patch dynamics. pp. 169–186. In: A. Estrada, and T. H. Fleming, eds. Frugivores and seed dispersal. Dr. W. Junk, Publisher, Dordrecht, Netherlands.
Horwith, B. 1985. A role for intercropping in modern agriculture. *BioScience* 35:286–291.
House, G. J., and J. N. All. 1981. Carabid beetles in soybean agroecosystems. *Environ. Entomol.* 10:194–196.
House, G. J., and B. R. Stinner. 1983. Arthropods in no-tillage soybean agroecosystems: community composition and ecosystem interactions. *Environ. Manage.* 7:23–28.
House, G. J., B. R. Stinner, D. A. Crossley, and E. P. Odum. 1984. Nitrogen cycling in conventional and no-tillage agroecosystem analysis of pathways and processes. *J. Appl. Ecol.* 21:991–1012.
Howe, H. F. 1974. Age-specific differences in habitat selection by the American redstart. *Auk* 91:161–162.
Howe, H. F. 1977. Bird activity and seed dispersal of a tropical wet forest tree. *Ecology* 58:539–550.
Howe, H. F. 1979. Fear and frugivory. *Am. Natur.* 114:925–931.
Howe, H. F. 1980. Monkey dispersal and waste of a neotropical fruit. *Ecology* 61:944–959.
Howe, H. F., and D. De Steven. 1979. Fruit production, migrant bird visitation, and seed dispersal of *Guarea glabra* in Panama. *Oecologia (Berl.)* 39:185–196.
Howe, H. F., and G. F. Estabrook. 1977. On intraspecific competition for avian dispersers in tropical trees. *Am. Natur.* 111:817–832.
Howe, H. F., and R. B. Primack. 1975. Differential seed dispersal by birds of the tree *Casearia nitida* (Flacourtiaceae). *Biotropica* 7:278–283.
Howe, H. F., and J. Smallwood. 1982. Ecology of seed dispersal. *Ann. Rev. Ecol. Syst.* 13:201–228.
Howe, H. F., and G. A. Vande Kerckhove. 1979. Fecundity and seed dispersal of a tropical tree. *Ecology* 60:180–189.
Howe, H. F., and G. A. Vande Kerckhove. 1980. Nutmeg dispersal by tropical birds. *Science* 210:925–927.
Howe, J. G., W. E. Grant, and L. J. Folse. 1982. Effects of grazing by *Sigmodon hispidus* on the regrowth of annual rye-grass *(Lolium perenne)*. *J. Mammal.* 63:176–179.
Hubbell, S. P., J. H. Howard, and D. F. Weimer. 1984. Chemical leaf repellency to an attine ant: seasonal distribution among potential host plant species. *Ecology* 65:1067–1076.
Huffaker, C. B., and C. E. Kennett. 1959. A ten-year study of vegetational changes associated with biological control of Kalmath weed. *J. Range Manage.* 12:69–82.
Hulme, A. C. 1971. The biochemistry of fruits and their products, vol. 2. Academic Press, NY.
Hulspas-Jordan, P. M., and J. C. van Lenteren. 1978. The relationship between host-plant leaf structure and parasitization efficiency of the parasitic wasp *Encarsaria formossa* Gahan (Hymenoptera: Aphelinidae) *Med. Fac. Landouww. Ryksuniv. Gent.* 43:341–440.

Hunter, P. E., and C. A. Hunter. 1964. A new *Aneotus* mite from pitcher plants. *Proc. Entomol. Soc. Wash.* 66:39–46.
Hunter, W. D., and W. E. Hinds. 1904. The Mexican cotton boll weevil. *U.S. Dep. Agric.*, Div. Entomol. Bull. No. 45, Washington, DC.
Hutchinson, G. E. 1961. The paradox of the plankton. *Am. Natur.* 95:137.
Huxley, C. R. 1978. The ant-plants *Myrmecodia* and *Hydnophytum* (Rubiaceae) and the relationships between their morphology, ant occupants, physiology and ecology. *New Phytol.* 80:231–268.
Huxley, C. R. 1980. Symbiosis between ants and epiphytes. *Biol. Rev.* 55:321–340.
Huxley, C. R. 1982. Ant-epiphytes of Australia. pp. 63–74. In: R. C. Buckley, ed. Ant–plant interactions in Australia. Dr. W. Junk, Publisher, The Hague.
Huxley, J. S. 1934. A natural experiment on the territorial instinct. *Brit. Birds* 27:270–277.
Immelmann, K. 1975. Ecological significance of imprinting and early learning. *Ann. Rev. Ecol. Syst.* 6:15–38.
Ingham, E. R. 1985. Review of the effects of 12 selected biocides on target and non-target soil organisms. *Crop Prot.* 4:3–32.
Ingham, R. E., J. A. Trofymow, E. R. Ingham, and D. C. Coleman. 1985. Interactions of bacteria, fungi and this nematode grazer: effects on nutrient cycling and plant growth. *Ecol. Monogr.* 55:119–140.
Ingram, R., and L. Taylor. 1982. The genetic control of a non-radiate condition in *Senecio squalidus* L. and some observations on the role of ray florets in the Compositae. *New Phytol.* 91:749–756.
Inouye, D. W. 1978. Resource partitioning in bumblebees: experimental studies of foraging behavior. *Ecology* 59:672–678.
Inouye, D. W. 1980a. The effect of proboscis and corolla tube lengths on patterns and rates of flower visitation by bumblebees. *Oecologia (Berl.)* 45:197–201.
Inouye, D. W. 1980b. The terminology of floral larceny. *Ecology* 61:1251–1253.
Inouye, D. W., and O. R. Taylor, Jr. 1979. A temperate region plant-ant-seed predator system: consequences of extra floral nectar secretion by *Helianthella quinquenervis*. *Ecology* 60:1–7.
Inouye, D. W., and G. D. Waller. 1984. Responses of honey bees (*Apis mellifera*) to amino acid solutions mimicking floral nectars. *Ecology* 65:618–625.
Isley, F. B. 1944. Correlations between mandibular morphology and food specificity in grasshoppers. *Ann. Entomol. Soc. Am.* 37:47–62.
Ivie, G. W., D. Bull, R. Beier, N. Pryor, and E. Oertli. 1983. Metabolic detoxification; mechanism of insect resistance to plant psoralens. *Science* 221:59–61.
Jacob, F. 1977. Evolution and tinkering. *Science* 196:1161–1166.
Jameson, D. A. 1963. Responses of individual plants to harvesting. *Bot. Rev.* 29:532–594.
Janson, C. 1985. Aggressive competition and individual food consumption in wild brown capuchin monkeys (*Cebus apella*). *Behav. Ecol. Sociobiol.* 18:125–138.
Janson, C. H. 1983. Adaptation of fruit morphology to dispersal agents in a neotropical forest. *Science* 219:187–189.
Janson, C. H. 1984. Female choice and mating system of the brown capuchin monkey *Cebus apella* (primates: Cebidae). *Z. Tierpsychol.* 65:177–200.
Janson, C. H. 1986. Capuchin counterpart. *Natur. Hist.* 95:45–57.
Janzen, D. H. 1966. Coevolution of mutualism between ants and acacias in Central America. *Evolution* 20:249–275.
Janzen, D. H. 1967a. Fire, vegetation structure and the ant x acacia interaction in Central America. *Ecology* 48:26–35.
Janzen, D. H. 1967b. Interaction of the bull's horn acacia (*A. cornigera* L.) with one of its ant inhabitants (*Pseudomyrmex ferruginea* F. Smith) in eastern Mexico. *Kansas Univ. Sci. Bull.* 47:315–558.
Janzen, D. H. 1969a. Seed-eaters versus seed size, number, toxicity, and dispersal. *Evolution* 23:1–27.
Janzen, D. H. 1969b. Allelopathy by myrmecophytes: the ant *Azteca* as an allelopathic agent of *Cecropia*. *Ecology* 50:147–152.
Janzen, D. H. 1970. Herbivores and the number of tree species in tropical forests. *Am. Natur.* 104:501–528.

Janzen, D. H. 1971a. Euglossine bees as long-distance pollinators of tropical plants. *Science* 171:203–205.
Janzen, D. H. 1971b. Seed predation by animals. *Ann. Rev. Ecol. Syst.* 2:465–492.
Janzen, D. H. 1972. Protection of *Barteria* (Passifloraceae) by *Pacysima* ants (Pseudomyrmicinae) in a Nigerian rain forest. *Ecology* 53:885–892.
Janzen, D. H. 1973a. Dissolution of mutualism between *Cecropia* and its *Azteca* ants. *Biotropica* 5:15–28.
Janzen, D. H. 1973b. Host plants as islands. II. Competition in evolutionary and contemporary time. *Am. Natur.* 107:786–790.
Janzen, D. H. 1974a. Tropical blackwater rivers, animals, and mast fruiting by Dipterocarpaceae. *Biotropica* 6:69–103.
Janzen, D. H. 1974b. Epiphytic myrmecophytes in Sarawak: mutualism through the feeding of plants by ants. *Biotropica* 6:237–259.
Janzen, D. H. 1974c. Swollen-thorn acacias of Central America. *Smithsonian Contrib. Bot.* 13:1–131.
Janzen, D. H. 1975. Behavior of *Hymenaea courbaril* when its predispersal seed predator is absent. *Science* 189:145–147.
Janzen, D. H. 1976. Why bamboos wait so long to flower. *Ann. Rev. Ecol. Syst.* 7:347–391.
Janzen, D. H. 1977a. A note on optimal mate selection by plants. *Am. Natur.* 111:365–371.
Janzen, D. H. 1977b. Why fruits rot, seeds mold, and meat spoils. *Am. Natur.* 111:691–713.
Janzen, D. H. 1978. The size of a local peak in a seed shadow. *Biotropica* 10:78.
Janzen, D. H. 1979a. Why food rots. *Natur. Hist.* 88:60–66.
Janzen, D. H. 1979b. How to be a fig. *Ann. Rev. Ecol. Syst.* 10:13–51.
Janzen, D. H. 1980a. When is it coevolution? *Evolution* 34:611–612.
Janzen, D. H. 1980b. Heterogeneity of potential food abundance for tropical small land birds. pp. 545–555. In: A. Keast, and E. S. Morton, eds. Migrant birds in the neotropics. Smithsonian Institution Press, Washington, DC.
Janzen, D. H. 1981a. Guanacaste tree seed-swallowing by Costa Rican range horses. *Ecology* 62:587–592.
Janzen, D. H. 1981b. *Enterolobium cyclocarpum* seed passage rate and survival in horses, Costa Rican Pleistocene seed dispersal agents. *Ecology* 62:593–601.
Janzen, D. H. 1981c. Digestive seed predation by a Costa Rican Baird's tapir. *Biotropica (suppl.)* 13:59–63.
Janzen, D. H. 1982a. Differential seed survival and passage rates in cows and horses, surrogate Pleistocene dispersal agents. *Oikos* 38:150–156.
Janzen, D. H. 1982b. Dispersal of small seeds by big herbivores: foliage is the fruit. *Am. Natur.* 123:338–353.
Janzen, D. H. 1982c. Removal of seeds from dung by tropical rodents: influence of habitat and amount of dung. *Ecology* 63:1887–1900.
Janzen, D. H. 1983a. Dispersal of seeds by vertebrate guts. In: D. J. Futuyma, and M. Slatkin, eds. Coevolution. Sinauer Associates, Sunderland, MA.
Janzen, D. H., ed. 1983b. Costa Rican natural history. University of Chicago Press, Chicago, IL.
Janzen, D. H. 1985. On ecological fitting. *Oikos* 45:308–310.
Janzen, D. H., and P. S. Martin. 1982. Neotropical anachronisms: the fruits the gomphotheres ate. *Science* 215:19–27.
Janzen, D. H., G. A. Miller, J. Hackforth-Jones, C. M. Pond, K. Hooper, and D. Janos. 1976. Two Costa Rican bat-generated seed shadows of *Andira inermis* (Leguminosae). *Ecology* 57:1068–1075.
Jarman, P. J. 1974. The social organization of antelope in relation to their ecology. *Behaviour* 48:215–266.
Jarman, P. J., and M. V. Jarman. 1979. The dynamics of ungulate social organization. pp. 185–220. In: A. R. E. Sinclair, and M. Morton-Griffiths, eds. Serengeti-dynamics of an ecosystem. University of Chicago Press, Chicago, IL.
Jarman, P. J., and A. R. E. Sinclair. 1979. Feeding strategy and the pattern of resource partitioning in ungulates. pp. 130–163. In: A. R. E. Sinclair, and M. Norton-

Griffiths, eds. Serengeti-dynamics of an ecosystem. University of Chicago Press, Chicago, IL.
Jarvinen, O., and S. Ulfstrand. 1980. Species turnovers of a continental bird fauna: Northern Europe 1850–1970. *Oecologia (Berl.)* 46:186–195.
Jay, M., and P. Lebreton. 1972. Chemotaxonomic research on vascular plants. XXVI. The flavenoids of the Sarraceniaceae, Nepenthaceae, Droseraceae, and Cephalotacaceae: a critical study of the Sarraceniales. *Natur. Can.* 99:607–613.
Jeffers, D. L., G. B. Triplett, Jr., and H. N. Lafever. 1977. Relay intercropping wheat and soybeans. p. 11. Ohio Agric. Res. Dev. Cent. Res. Circ. 233.
Joel, D. M., and S. Gepstein. 1985. Chloroplasts in the epidermis of *Sarracenia purpurea* ssp. *purpurea* (the American pitcher plant) and their possible role in carnivory: an immunocytochemical approach. *Physiol. Plant.* 63:71–75.
Joel, D. M., B. E. Juniper, and A. Dafni. 1985. Ultraviolet patterns in the traps of carnivorous plants. *New Phytol.* 101:585–593.
Johnson, F. M., J. Stubbs, and R. A. Klawitter. 1964. Rodent repellent value of arasan-endrin mixtures applied to acorns. *J. Wildl. Manage.* 28:15–19.
Johnson, R. A., M. F. Willson, J. N. Thompson, and R. I. Bertin. 1985. Nutritional values of wild fruits and consumption by migrant frugivorous birds. *Ecology* 66:819–827.
Jolivet, P. 1986. Les fourmis et les plantes. Un exemple de coevolution. Societé Nouvelle des Editions Boubee, Paris.
Jones, C. E. 1978. Pollinator constancy as a pre-pollination isolating mechanism between sympatric species of *Cercidium*. *Evolution* 32:189–198.
Jones, C. E., and R. J. Little. 1983. Handbook of experimental pollination of biology. Van Nostrand Reinhold, NY.
Jones, D. A. 1966. On the polymorphism of cyanogenesis in *Lotus corniculatus*. I. Selection by animals. *Can. J. Genet. Cytol.* 8:556–567.
Jones, F. G. W. 1977. Pests, resistance and fertilizers. 12th Colloquium of the International Potash Institute. pp. 111–135, Schweiz, Switzerland.
Jones, F. M. 1921. Pitcher plants and their moths. *Natur. Hist.* 21:296–316.
Jones, F. M. 1935. Pitcherplants and their insect associates. pp. 25–33. In: M. V. Walcott, ed. Illustrations of North American pitcherplants. Smithsonian Institution, Washington, DC.
Jordano, P. 1984. Seed weight variation and differential avian dispersal in blackberries *Rubus ulmifolius*. *Oikos* 43:149–153.
Judd, W. W. 1959. Studies on Byron Bog in southwestern Ontario. X. Inquilines and victims of the pitcher plant, *Sarracenia purpurea* L. *Can. Entomol.* 91:171–180.
Jugenheimer, R. W. 1976. Corn improvement, seed production and uses. John Wiley and Sons, NY.
Juniper, B. E., and K. Buras. 1962. How pitcher plants trap insects. *New Sci.* 13:75–77.
Kahn, D. M., and H. V. Cornell. 1983. Early leaf abscission and foliovores: comments and considerations. *Am. Natur.* 122:428–432.
Karban, R. 1980. Periodical cicada nymphs impose periodical oak tree wood accumulation. *Nature* 287:326–327.
Karban, R. 1982. Experimental removal of 17-year cicada nymphs and growth of host apple trees. *J. N.Y. Entomol. Soc.* 90:74–81.
Karlsson, P. S., and B. Carlsson. 1984. Why does *Pinguicula vulgaris* trap insects? *New Phytol.* 97:25–30.
Karr, J. R. 1971. Structure of avian communities in selected Panama and Illinois habitats. *Ecol. Monogr.* 41:207–233.
Karr, J. R. 1975. Production, energy pathways, and community diversity in forest birds. pp. 161–177. In: F. B. Golley, and E. Medina, eds. Tropical ecological systems. Springer-Verlag, Berlin.
Karr, J. R. 1976a. On the relative abundances of migrants from the north temperate zone in tropical habitats. *Wilson Bull.* 88:433–458.
Karr, J. R. 1976b. Seasonality, resource availability, and community diversity in tropical bird communities. *Am. Natur.* 110:973–994.
Karr, J. R. 1976c. Within- and between-habitat avian diversity in African and neotropical lowland habitats. *Ecol. Monogr.* 46:457–481.
Karr, J. R. 1980. Patterns in the migration systems between the north temperate zone and

the tropics. pp. 529–544. In: A. Keast, and E. S. Morton, eds. Migratory birds in the neotropics. Smithsonian Institution Press, Washington, DC.
Karr, J. R., and R. R. Roth. 1971. Vegetation structure and avian diversity in several New World areas. *Am. Natur.* 105:423–435.
Kaufmann, W., H. Hagemeister, and G. Dirksen. 1980. Adaptation to changes in dietary composition, level and frequency of feeding. pp. 587–602. In: Y. Ruckebusch and P. Thivend, eds. Digestive physiology and metabolism in ruminants. AVI Publishing, Westport, CT.
Keast, A., and E. S. Morton, eds. 1980. Migrant birds in the neotropics. Smithsonian Institution Press, Washington, DC.
Keeler, K. H. 1977. The extrafloral nectaries of *Ipomoea carnea* (Convolvulaceae). *Am. J. Bot.* 64:1182–1188.
Keeler, K. H. 1980. Function of extrafloral nectaries in *Ipomoea leptophylla* (Convolvulaceae). *Am. J. Bot.* 67:216–222.
Keeler, K. H. 1981a. Function of *Mentzelia nuda* (Loasaceae) postfloral nectaries in seed defense. *Am. J. Bot.* 68:295–299.
Keeler, K. H. 1981b. A model of selection for facultative, non-symbiotic mutualism. *Am. Natur.* 118:488–498.
Keeler, K. H., and R. B. Kaul. 1984. Distribution of defense nectaries in *Ipomoea* (Convolvulaceae). *Am. J. Bot. 71:1364–1372.*
Keeley, S. E., S. C. Keeley, C. C. Swift, and J. Lee. 1984. Seed predation due to the Yucca-moth symbiosis. *Am. Midl. Natur.* 112:187–191.
Keith, L. B. 1983. Role of food in hare population cycles. *Oikos* 40:385–395.
Kellerman, C., and E. von Raumer. 1878. Vegatationsversuche an *Drosera rotundifolia,* mit und ohne Fleischfutterung. *Bot. Z.* 36:209–218, 225–229.
Kerner von Marilaun, A. 1878. Flowers and their unbidden guests. Kegan Paul, London.
Kettlewell, H. B. D. 1955a. Selection experiments on industrial melanism in the lepidoptera. *Heredity* 9:323–342.
Kettlewell, H. B. D. 1955b. Recognition of appropriate background by the pale and black phases of lepidoptera. *Nature* 175:943–944.
Kevan, P. G. 1975. Sun-tracking solar furnaces in high Arctic flowers: significance for pollination and insects. *Science* 189:723–726.
Kevan, P. G., and M. A. Lane. 1985. Flower petal microtexture is a tactile cue for bees. *Proc. Natl. Acad. Sci.* 82:4750–4752.
Kirchner, T. B. 1977. The effects of resource enrichment on the diversity of plants and arthropods in a shortgrass prairie. *Ecology* 58:1334–1344.
Kitching, J. A., and F. J. Ebling. 1967. Ecological studies at Lough INE. *Advan. Ecol. Res.* 4:197–291.
Kleinfeldt, S. 1978. Ant-gardens: the interaction of *Codonanthe crassifolia* (Gesneriaceae) and *Crematogaster longispina* (Formicidae). *Ecology* 59:449–456.
Klimstra, W. D., and F. Newsome. 1960. Some observations on the food coactions of the common box turtle *Terrapene carolina. Ecology* 41:639–647.
Klocke, J. A., and B. Chan. 1982. Effects of cotton condenses tannin on feeding and digestion in the cotton pest, *Meliothis zea. J. Insect Physiol.* 23:158–166.
Klopfer, P. 1963. Behavioral aspects of habitat selection: the role of early experience. *Wilson Bull.* 75:15–22.
Klopfer, P., and R. H. MacArthur. 1960. Niche size and faunal diversity. *Am. Natur.* 94:293–300.
Klopfer, P., and R. H. MacArthur. 1961. On the causes of tropical species diversity: niche overlap. *Am. Natur.* 95:223–226.
Klopfer, P. H., and J. P. Hailman. 1965. Habitat selection in birds. *Advan. Study Behav.* 1:279–303.
Knight, R. S., and W. R. Siegfried. 1983. Inter-relationships between type, size and colour of fruits and dispersal in southern African trees. *Oecologia (Berl.)* 56:405–412.
Knox, R. B., and J. Kenrick. 1983. Polyad function in relation to the breeding system of *Acacia.* pp. 411–417. In: D. L. Mulcahy, and E. Ottaviano, eds. Pollen: biology and implication for plant breeding. Elsevier, NY.
Knox, R. B., J. Kenrick, P. Bernhardt, R. Marginson, G. Beresford, I. Baker, and H. G.

Baker. 1985. Extrafloral nectaries as adaptations for bird pollination in *Acacia terminalis*. *Am. J. Bot.* 72:1185–1196.

Kochmer, J. P., and S. N. Handel. 1986. Constraints and competition in the evolution of flowering phenology. *Ecol. Monogr.* 56:303–325.

Kodric-Brown, A., and J. H. Brown. 1978. Influence of economics, interspecific competition and sexual dimorphism on territoriality of migrant rufous hummingbirds. *Ecology* 59:285–296.

Kondo, K., and P. S. Lavarack. 1984. A cytotaxonomic study of some Australian species of *Drosera* L. (Droseraceae). *Bot. J. Linn. Soc.* 88:317–333.

Koptur, S. 1979. Facultative mutualism between weedy vetches bearing extrafloral nectaries and weedy ants in California. *Am. J. Bot.* 66:1016–1020.

Koptur, S. 1984. Experimental evidence for defense of *Inga* saplings (Mimosoideae) by ants. *Ecology* 25:1787–1793.

Koptur, S., A. R. Smith, and I. Baker. 1982. Nectaries in some neotropical species of *Polypodium* (Polypodiaceae). Preliminary observations and analyses. *Biotropical* 14:108–113.

Kotler, B. P. 1984. Risk of predation and the structure of desert rodent communities. *Ecology* 65:689–701.

Krebs, J. R. 1971. Territory and breeding density in the great tit, *Parus major* L. *Ecology* 52:2–22.

Krebs, J. R., and R. H. McCleery. 1986. Optimization in behavioral ecology. pp. 91–121. In: J. R. Krebs, and N. B. Davies, eds. Behavioral ecology: an evolutionary approach. 2nd ed. Blackwell Scientific Publications, Oxford.

Krefting, L. W., and E. I. Roe. 1949. The role of some birds and mammals in seed germination. *Ecol. Monogr.* 19:269–286.

Kuhlman, H. M. 1971. Effects of insect defoliation on trees. *Ann. Rev. Entomol.* 16:289–324.

Kullenberg, B., G. Bergstrom, B. G. Svensson, J. Tengo, and L. Agren. 1984. The ecological station of Uppsala University on Oland 1963–1983. Almquist and Wiksell International, Stockholm.

Kummer, H. 1972. Primate societies, group techniques of ecological adaptation. Aldine, Chicago, IL.

Kuroda, N. 1962. Comparative growth rates in two gray starling chicks, artificially raised with animal and plant foods. *Misc. Rep. Yamashina's Inst. Ornithol. Zool.* 3:174–184.

Labarca, C., and F. Loewus. 1973. The nutritional role of pistil exudate in pollen tube wall formation in *Lilium longiflorum*. *Plant Physiol.* 52:87–92.

Labisky, R. F., and M. L. Porter. 1984. Home range and foraging habitat of red-cockaded woodpeckers in north Florida. Department of Wildlife and Range Sciences, School of Forest Resources and Conservation, University of Florida, Gainesville, FL. 49 pp.

Lack, D. 1971. Ecological isolation in birds. Blackwell Scientific Publications, Oxford.

Lack, D. 1976. Island biology. Studies in Ecology, vol. 3. University of California Press, Berkeley, CA.

Lack, D., and L. S. V. Venables. 1939. The habitat distribution of British woodland birds. *J. Anim. Ecol.* 8:39–71.

Ladd, J. N., M. Amato, R. B. Jackson, and J. H. A. Butler. 1983. Utilization by wheat crops of nitrogen from legume residues decomposing in soils in the field. *Soil Biol. Biochem.* 15:231–238.

Lamont, B. 1985. The significance of flower colour change in eight co-occurring shrub species. *Bot. J. Linnean Soc.* 90:145–155.

Laverty, T. M., and R. C. Plowright. 1985. Competition between hummingbirds and bumble bees for nectar in flowers of *Impatiens biflora*. *Oecologia (Berl.)* 66:25–32.

Lawton, J. H., and P. A. Heads. 1985. Bracken, ants and extrafloral nectaries. I. The components of the system. *J. Anim. Ecol.* 53:995–1014.

Lawton, J. H., and D. Schroder. 1977. Effect of plant type, size of geographical range and taxonomic isolation on number of insect species associated with British plants. *Nature* 265:137–140.

Leber, K. M. 1985. The influence of predatory decapods, refuge and microhabitat selection on seagrass communities. *Ecology* 66:1951–1964.

Leck, C. F. 1971. Overlap in the diet of some neotropical birds. *Living Bird* 10:89–106.

Leck, C. F. 1972. Seasonal changes in feeding pressures of fruit- and nectar-eating birds in Panama. *Condor* 74:54–60.
Leck, C. F. 1980. The impact of some North American migrants at fruiting trees in Panama. *Auk* 89:842–850.
Leighton, M., and D. R. Leighton. 1982. The relationship of size of feeding aggregate to size of food patch: howler monkeys (*Allouattal palliata*) feeding in *Trichilia cipo* fruit trees on Barro Colorado Island. *Biotropica* 14:81–90.
Lenington, S. 1980. Female choice and polygyny in red-winged blackbirds. *Anim. Behav.* 28:347–361.
Leroi, B., and M. Jarry. 1981. Relationships of the bruchid *Acanthoscelides obtectus* and various species of *Phaseolus*: influence on fecundity and possibility of larval development. *Entomol. Exp. Appl.* 30:73–82.
Lersten, N. R., and R. W. Pohl. 1985. Extrafloral nectaries in *Cipadessa* (Meliaceae). *Ann. Bot.* 56:363–366.
Levey, D. J., T. C. Moermond, and J. S. Denslow. 1984. Fruit choice in neotropical birds: the effect of distance between fruits on preference patterns. *Ecology* 65:844–850.
Levin, D. A. 1972. The role of phenolics in plant defense. *Univ. Ark. Mus. Occ. Pap. No. 4*, pp. 165–190.
Levin, D. A. 1973. The role of trichomes in plant defense. *Quart. Rev. Biol.* 48:3–15.
Levin, D. A. 1975. Pest pressure and recombination systems in plants. *Am. Natur.* 109:437–451.
Levin, D. A. 1984. Inbreeding depression and proximity-dependent crossing success in *Phlox drummondii. Evolution* 38:116–127.
Levine, E., S. L. Clement, and D. A. McCartney. 1981. Agronomic practices influence black cutworm damage on corn. *Ohio Rep.* 66:29–30.
Levins, R. 1968. Evolution in changing environments. Princeton University Press, Princeton, NJ.
Lewis, J. K. 1972. Range management viewed in the ecosystem framework. pp. 97–187. In: G. Van Dyne, ed. The ecosystem concept in natural resource management. Academic Press, NY.
Lewis, T. 1973. Thrips: their biology, ecology and economic importance. Academic Press, London.
Lewis, T., G. V. Pollard, and G. C. Dibley. 1974. Rhythmic foraging in the leaf cutting ant *Atta cephalotes* (Formicidae: Attini). *J. Anim. Ecol.* 43:143–153.
Lewis, W. H., and M. P. F. Elvin-Lewis. 1977. Medical botany: plants affecting man's health. John Wiley and Sons, NY.
Lieberman, D., J. B. Hall, M. D. Swaine, and M. Lieberman. 1979. Seed dispersal by baboons in the Shai Hills, Ghana. *Ecology* 60:65–75.
Ligon, J. D. 1978. Reproductive interdependence of pinyon jays and pinyon pines. *Ecol. Monogr.* 48:111–126.
Lill, W. J., L. J. Gut, P. H. Westigaid, and C. E. Warren. 1986. Perspectives on arthropod community structure, organization, and development in agricultural crops. *Ann. Rev. Entomol.* 31:455–478.
Lima, S. L. 1985. Maximizing feeding efficiency and minimizing time exposed to predators: a trade-off in the black-capped chickadee. *Oecologia* (Berl.) 66:60–67.
Lincoln, D. E., D. Couvet, and N. Sionit. 1986. Response of an insect herbivore to host plants grown in carbon dioxide enriched atmospheres. *Oecologia* (Berl.) 69:556–560.
Lindroth, R. L. 1979. Diet optimization by generalist mammalian herbivores. *Biologist* 61:41–58.
Lindroth, R. L. 1987. Adaptations of mammalian herbivores to plant chemical defenses. In: K. C. Spencer, ed. Chemical mediation of coevolution. American Institute of Biological Sciences (in press).
Lindroth, R. L., and G. O. Batzli. 1984a. Food habits of the meadow vole (*Microtus pennsylvanicus*) in bluegrass and prairie habitats. *J. Mammal.* 65:600–606.
Lindroth, R. L., and G. O. Batzli. 1984b. Plant phenolics as chemical defenses: effects of natural phenolics on survival and growth of prairie voles (*Microtus ochrogaster*). J. Chem. Ecol. 10:229–244.

Lindroth, R. L., and G. O. Batzli. 1986. Inducible plant chemical defenses: a cause of vole population cycles? *J. Anim. Ecol.* 55:431–449.
Lindroth, R. L., G. O. Batzli, and D. S. Seigler. 1986. Patterns in the phytochemistry of three prairie plants. *Biochem. Syst. Ecol.* 14:597–602.
Linhart, Y. B. 1973. Ecological and behavioral determinants of pollen dispersal in hummingbird-pollinated *Heliconia*. *Am. Natur.* 107:511–523.
Linzey, A. V. 1984. Patterns of coexistence in *Synaptomys cooperi* and *Microtus pennsylvanicus*. *Ecology* 65:382–393.
Lloyd, D. G. 1979. Parental strategies of angiosperms. *N. Z. J. Bot.* 17:595–606.
Lloyd, D. G. 1980. Sexual strategies in plants. III. A quantitative method for describing the gender of plants. *N. Z. J. Bot.* 18:103–108.
Lloyd, D. G. 1982. Selection of combined versus separate sexes in seed plants. *Am. Natur.* 120:571–585.
Lloyd, D. G., and C. J. Webb. 1986. The avoidance of interference between the presentation of pollen stigmas in angiosperms. I. Dichogamy. *N. Z. J. Bot.* 24:135–162.
Lloyd, D. G., and J. M. A. Yates. 1982. Intrasexual selection and the segregation of pollen and stigmas in hermaphrodite plants, exemplified by *Wahlenbergia albomarginata* (Campanulaceae). *Evolution* 36:903–913.
Lloyd, F. E. 1934. Is *Roridula* a carnivorous plant? *Can. J. Res.* 10:780–786.
Lloyd, F. E. 1942. The carnivorous plants. Chronica Botanica, Waltham, MA.
Lloyd, M., R. F. Inger, and F. W. King. 1968. On the diversity of reptile and amphibian species in a Bornean rainforest. *Am. Natur.* 102:497–515.
Lock, J. M. 1972. The effects of hippopotamus grazing on grasslands. *J. Appl. Ecol.* 60:445–467.
Lockley, R. M. 1974. The private life of the rabbit. Macmillan, NY.
Loman, J. 1979. Nest tree selection and vulnerability to predation among hooded crows *Corvus corone cornix*. *Ibis* 121:204–207.
Longstreth, D. J., and P. S. Nobel. 1980. Nutrient influences on leaf photosynthesis. *Plant Physiol.* 65:541–543.
Louda, S. M. 1982. Limitations of the recruitment of the shrub *Haplopappus squarrosus* (Asteraceae) by flower- and seed-feeding insects. *J. Ecol.* 70:43–53.
Lovejoy, T. E. 1974. Bird diversity and abundance in Amazon forest communities. *Living Bird* 13:127–192.
Luckmann, W. H., R. L. Metcalf. 1982. The pest management concept. pp. 1–32. In: R. L. Metcalf and W. H. Luckmann, eds. Introduction to insect pest management. John Wiley and Sons, NY.
Luckner, M. 1984. Secondary metabolism in microorganisms, plants, and animals. Springer-Verlag, NY.
Lukefahr, M. J., L. W. Nobel, and J. E. Houghtaling. 1966. Growth and infestation of bollworms and other insects on glanded and glandless strains of cotton. *J. Econ. Entomol.* 59:817–820.
Lüttge, U. 1964. Untersuchungen zur Physiologie der Carnivoren-Drusen. I. Mitteilung. Die an den Verdauungsvorgangen beteiligten Enzyme. *Planta* 63:103–117.
Lüttge, U. 1965. Untersuchungen zur Physiologie der Carnivoren-Drusen. II. Mitteilung. Uber die Resorption verschiedener Substanzen. *Planta* 66:331–334.
Lüttge, U. 1983. Ecophysiology of carnivorous plants. pp. 489–517. In: O. L. Lange, P. S. Nobel, C. B. Osmond, and H. Ziegler, eds. Encyclopedia of plant physiology, vol. 12C. Springer-Verlag, Heidelberg.
Lynch, J. F., E. S. Morton, and M. van der Voort. 1985. Habitat segregation between the sexes of wintering hooded warblers (*Wilsonia citrina*). *Auk* 102:714–721.
Macan, T. T. 1976. A twenty-one-year study of water bugs in a moorland fishpond. *J. Anim. Ecol.* 45:913–922.
MacArthur, R. H. 1958. Population ecology of some warblers of northeastern coniferous forest. *Ecology* 39:599–619.
MacArthur, R. H. 1969. Patterns of communities in the tropics. *Biol. J. Linn. Soc.* 1:19–30.
MacArthur, R. H. 1972. Geographical ecology. Princeton University Press, Princeton, NJ.

MacArthur, R. H., and J. MacArthur. 1961. On bird species diversity. *Ecology* 42:594–598.
MacArthur, R. H., and E. R. Pianka. 1966. On optimal use of a patchy environment. *Am. Natur.* 100:603–609.
MacArthur, R. H., H. Recher, and M. Cody. 1966. On the relation between habitat selection and species diversity. *Am. Natur.* 100:319–332.
Macfarlane, J. M. 1908. Nepenthaceae. Das Pflanzenreich, vol. 36. Engelmann, Leipzig.
Mack, R. N., and J. N. Thompson. 1982. Evolution in steppe with few large, hooved mammals. *Am. Natur.* 119:757–773.
MacKay, T. F. C., and R. W. Doyle. 1978. An ecological genetic analysis of the settling behaviour of a marine plychaete. I. Probability of settlement and gregarious behaviour. *Heredity* 40:1–12.
MacKinnon, J. R. 1974. The behaviour and ecology of wild orangutans *(Pongo pygmaaeus)*. *Anim. Behav.* 22:3–74.
MacKinnon, J. R. 1978. Comparative feeding ecology of six sympatric primate species in West Malaysdia. *Rec. Advan. Primatol.* 1:305–321.
MacRoberts, M. H., and B. R. MacRoberts. 1976. Social organization and herbavior of the acorn woodpecker in central coastal California. Ornithological Monographs No. 21, American Ornithologists' Union, 115 pp.
Madison, M. 1977. Vascular epiphytes: their systematic distribution and salient features. *Selbyana* 2:1–13.
Malyshev, S. I. 1968. Genesis of the Hymenoptera and the phases of their evolution. Methuen, London.
Mani, M. S. 1964. Ecology of plant galls. Dr. W. Junk, Publisher, The Hague.
Mann, J. 1978. Secondary metabolism. Clarendon Press, Oxford.
Manzur, M. I., and S. P. Courtney. 1984. Influence of insect damage in fruits of hawthorn on bird foraging and seed dispersal. *Oikos* 43:265–270.
Marchant, N. G., and A. S. George. 1982. *Drosera*. *Flora Australia* 8:9–64.
Marcus, C., and E. P. Lichtenstein. 1979. J. Agric. Food Chem. 27:1217–1223.
Marden, J. H. 1984. Remote perception of floral nectar by bumblebees. *Oecologia (Berl.)* 64:232–240.
Martin, A. C., H. S. Zim, and A. L. Nelson. 1951. American wildlife and plants. McGraw-Hill, NY.
Martin, M. M., and J. Martin. 1984. Surfactants: their role in preventing the precipitation of proteins by tannins in insect guts. *Oecologia (Berl.)* 61:342–345.
Mason, I. L. 1984. Evolution of domesticated animals. Longman, NY.
Mattson, W. J., Jr. 1980. Herbivory in relation to plant nitrogen content. *Ann. Rev. Ecol. Syst.* 11:119–161.
Mattson, W. J., and N. D. Addy. 1975. Phytophagous insects as regulators of forest primary production. *Science* 180:515–522.
May, R. M. 1983. Parasitic infections as regulators of animal populations. *Am. Sci.* 71:36–45.
Maynard Smith, J. 1974. Models in ecology. Cambridge University Press, Cambridge, MA.
Mayr, E. 1969. Bird speciation in the tropics. *Biol. J. Linn. Soc.* 1:1–17.
Mayr, E. 1970. Populations, species, and evolution. Harvard University Press, Cambridge, MA.
Mayr, E. 1974. Behavior programs and evolutionary strategies. *Am. Sci.* 62:650–659.
McCrea, K. D., and W. G. Abrahamson. 1985. Evolutionary impacts of the goldenrod ball gallmaker on *Solidago altissima* clones. *Oecologia (Berl.)* 68:20–22.
McCrea, K. D., W. G. Abrahamson, and A. E. Weis. 1985. Goldenrod ball gall effects on *Solidago altissima:* 14C translocation and growth. *Ecology* 66:1902–1907.
McDade, L. A., and S. Kinsman. 1980. The impact of floral parasitism in two neotropical hummingbird-pollinated plant species. *Evolution* 34:944–958.
McDaniel, S. 1971. The genus *Sarracenia*. *Bull. Tall Timbers Res. Sta.* 9:1–36.
McDonnell, M. J., and E. W. Stiles. 1983. The structural complexity of old field vegetation and the recruitment of bird-dispersed plant species. *Oecologia (Berl.)* 56:109–116.

McDonnell, M. J., E. W. Stiles, G. P. Cheplick, and J. J. Armesto. 1984. Bird-dispersal of *Phytolacca americana* L. and the influence of fruit removal on subsequent fruit development. *Am. J. Bot.* 71:895–901.
McKendrick, J. D., G. O. Batzli, K. R. Everett, and J. C. Swanson. 1980. Some effects of mammalian herbivores and fertilization on tundra soils and vegetation. *Arct. Alp. Res.* 12:565–578.
McKey, D. 1975. The ecology of coevolved seed dispersal systems. pp. 159–191. In: L. E. Gilbert, and P. H. Raven, eds. Coevolution of animals and plants. University of Texas Press, Austin, TX.
McKey, D. 1984. Interaction of the ant-plant *Leonardoxa africana* (Caesalpinaceae) with its obligate inhabitants in a rainforest in Cameroon. *Biotropica* 16:81–99.
McLain, D. I. 1983. Ants, extrafloral nectaries and herbivory on the passion vine *Passiflora incarnata. Am. Midl. Natur.* 110:433–439.
McNab, B. K. 1983. Ecological and behavioral consequences of adaptation to various food resources. pp. 664–697. In: J. F. Eisenberg, and D. G. Kleiman, eds. Advances in the study of mammalian behavior. Spec. Publ. No. 7, Am. Soc. Mammologists.
McNab, B. K. 1986. The influence of food habits on the energetics of eutherian mammals. *Ecol. Monogr.* 56:1–19.
McNair, J. N. 1986. The effects of refuges on predator-prey interactions: a reconsideration. *Theor. Pop. Biol.* 29:38–63.
McNaughton, S. J. 1976. Serengeti migratory wildebeest: facilitation of energy flow by grazing. *Science* 191:92–94.
McNaughton, S. J. 1977. Diversity and stability of ecological communities: a comment on the role of empiricism in ecology. *Am. Natur.* 111:515–525.
McNaughton, S. J. 1979. Grazing as an optimization process: grass-ungulate relationships in the Serengeti. *Am. Natur.* 113:691–703.
McNaughton, S. J. 1983a. Serengeti grassland ecology: the role of composite environmental factors and contingency in community organization. *Ecol. Monogr.* 53:291–320.
McNaughton, S. J. 1983b. Physiological and ecological implications of herbivory. pp. 657–677. In: O. L. Lange, P. S. Nobel, C. B. Osmond, and H. Ziegler, eds. Physiological plant ecology. III. Responses to the chemical and biological environment. Encyclopedia of Plant Physiology; New Series, vol. 12C. Springer-Verlag, NY.
McNaughton, S. J. 1983c. Compensatory plant growth as a response to herbivory. *Oikos* 40:329–336.
McNaughton, S. J. 1985a. Ecology of a grazing ecosystem. The Serengeti. *Ecol. Monogr.* 55:259–294.
McNaughton, S. J. 1985b. Interactive regulation of grass yield and chemical properties by defoliation, a salivary chemical, and inorganic nutrition. *Oecologia (Berl.)* 65:478–486.
McNaughton, S. J., and N. J. Georgiadis. 1986. Ecology of African grazing and browsing mammals. *Ann. Rev. Ecol. Syst.* 17:39–66.
McNaughton, S. J., J. L. Tarrants, M. M. McNaughton, and R. H. Davis. 1985. Silica as a defense against herbivory and a growth promoter in African grasses. *Ecology* 66:528–535.
McNeil, S., and T. R. E. Southwood. 1978. The role of nitrogen in the development of insect/plant relationships. pp. 77–98. In: J. Harborne, ed. Biochemical aspects of plant and animal coevolution. Academic Press, London.
McVaugh, R., and T. J. Mickel. 1963. Notes on *Pinguicula* sect. *Orcheosanthus. Brittonia* 15:134–140.
Mead, R. J., A. J. Oliver, D. R. King, and P. H. Hubach. 1985. The co-evolutionary role of fluoroacetate in plant–animal interactions in Australia. *Oikos* 44:55–60.
Medina, E. 1970. Relationships between nitrogen level, photosynthetic activity, and carboxydismutase activity in *Atriples patula* leaves. *Carnegie Inst. Wash. Year Book* 69:655–662.
Medina, E. 1971. Effect of nitrogen supply and light intensity during growth on the photosynthetic capacity and carboxydismutase activity of *Atriplex patula* ssp. *hastata. Carnegie Inst. Wash. Year Book* 70:551–559.

Meeuse, B., and S. Morris. 1984. The sex life of flowers. Facts on File, NY.

Messina, F. J. 1982. Plant protection as a consequence of an ant-membracid mutualism: interactions on Goldenrod (*Solidago* sp.). *Ecology* 62:1433–1440.

Metcalfe, C. R., and L. Chalk. 1950. Anatomy of the dicotyledons, vol. 1. Clarendon Press, London.

Meyer, D. G., and J. R. Strickler. 1979. Capture enhancement in a carnivorous aquatic plant: function of antennae and bristles in *Utricularia vulgaris. Science* 203:1022–1025.

Michaels, H. J., and F. A. Bazzaz. 1986. Resource allocation and demography of sexual and apomictic *Antennaria parlinii. Ecology* 67:27–36.

Miles, D., M. Howard, and V. Naresh. 1974. A draught from the poison pitcher. *Sci. News* 106:286.

Milewski, A. V., and W. J. Bond. 1982. Convergence of myrmecochory in Mediterranean Australia and South Africa. pp. 89–98. In: R. C. Buckley, ed. Ant–plant interactions in Australia. Dr. W. Junk, Publisher, The Hague.

Miller, P. F. 1973. The biology of some *Phyllonorycter* species (Lepidoptera: Gracillaridae) mining leaves of oak and beech. *J. Natur. Hist.* 7:391–409.

Milton, K. 1981. Food choice and digestive strategies of two sympatric primate species. *Am. Natur.* 117:496–505.

Mitchell, R. 1984. The ecological basis for comparative primary production. pp. 13–53. In: R. Lowrance, B. R. Stinner, and G. J. House, eds. Agricultural ecosystems. John Wiley and Sons, NY.

Mittelbach, G. G. 1984. Predation and resource partitioning in two sunfishes (Centrarchidae). *Ecology* 65:499–513.

Mitter, C., D. Futuyma, J. Schneider, and J. Hare. 1979. Genetic variation in host plant relations in a parthenogenetic moth. *Evolution* 33:777–790.

Mody, N. V., R. Henson, P. A. Hedin, U. Kokpol, and D. H. Miles. 1976. Isolation of the insect paralyzing agent coniine from *Sarracenia flava. Specialia* 13:829.

Moermond, T. C., and J. S. Denslow. 1983. Fruit choice in neotropical birds: effects of fruit type and accessibility on selectivity. *J. Anim. Ecol.* 52:407–420.

Moermond, T. C., and J. S. Denslow. 1985. Neotropical frugivores: patterns of behavior, morphology and nutrition with consequences for fruit selection. In: P. A. Buckley, M. S. Foster, E. S. Morton, R. S. Ridgely, and N. G. Buckley, eds. Neotropical Ornithology. Am. Ornithologists Union Monographs. Allen Press, Laurence, KS.

Mooney, H. A. 1972. The carbon balance of plants. *Ann. Rev. Ecol. Syst.* 3:315–346.

Moran, N., and W. D. Hamilton. 1980. Low nutritive quality as defense against herbivores. *J. Theor. Biol.* 86:247–254.

Moreau, R. E. 1972. The Palearctic-African bird migration systems. Academic Press, NY.

Morris, J. G., and Q. R. Rogers. 1982. Nutritionally related metabolic adaptations of carnivores and ruminants. pp. 165–180. In: N. S. Margaris, M. Arianoutsou-Faraggitaki, and R. J. Reiter, eds. Adaptations to terrestrial environments. Plenum Press, NY.

Morrison, D. 1978. Foraging ecology and energetics of the frugivorous bat *Artebeus jamaicensis. Ecology* 59:716–723.

Morse, D. H. 1968. A quantitative study of foraging of male and female spruce-woods warblers. *Ecology* 49:779–784.

Morse, D. H. 1978. Size-related foraging differences of bumble bee workers. *Ecol. Entomol.* 3:189–192.

Morse, D. H., and R. S. Fritz. 1983. Contributions of diurnal and nocturnal insects to the pollination of common milkweed (*Asclepias syriaca* L.) in a pollen-limited system. *Oecologia (Berl.)* 60:190–197.

Morton, E. S. 1971. Food and migration habits of the Eastern Kingbird in Panama. *Auk* 88:925–926.

Morton, E. S. 1973. On the evolutionary advantages and disadvantages of fruit eating in tropical birds. *Am. Natur.* 107:8–22.

Morton, E. S. 1980. Adaptations to seasonal changes by migrant land birds in the Panama Canal Zone. pp. 437–456. In: A. Keast, and E. S. Morton, eds. Migrant birds in the neotropics. Smithsonian Institution Press, Washington, DC.

Mosquin, T. 1971. Competition for pollinators as a stimulus for the evolution of flowering time. *Oikos* 22:398–402.
Moss, R., D. Welch, and P. Rothery. 1981. Effects of grazing by mountain hares and red deer on the production and chemical composition of heather. *J. Appl. Ecol.* 18:487–496.
Motten, A. F., D. R. Campbell, D. E. Alexander, and H. L. Miller. 1981. Pollination effectiveness of specialist and generalist visitors to a North Carolina population of *Claytonia virginica*. *Ecology* 62:1278–1287.
Mulcahy, D. L. 1979. The rise of the angiosperms: a genecological factor. *Science* 206:20–23.
Napier, T. L., C. S. Thraen, A. Gore, and W. R. Gol. 1984. Factors affecting adoption of conventional and conservation tillage practices in Ohio. *J. Soil Water Conserv.* 39:205–209.
Natr, L. 1975. Influence of mineral nutrition on photosynthesis and the use of assimilates. pp. 537–555. In: J. W. Cooper, ed. Photosynthesis and productivity in different environments. Cambridge University Press, Cambridge, MA.
Nault, L. R., D. T. Gordon, D. C. Robertson, and D. E. Bradfute. 1976. Host range of maize chlorotic dwarf virus. *Plant Dis. Rep.* 60:374–377.
Nettancourt, D. de. 1977. Incompatibility in angiosperms. Springer-Verlag, Berlin.
Newton, I. 1967. The adaptive radiation and feeding ecology of some British finches. *Ibis* 109:33–98.
Newton, I. 1970. Irruptions of crossbills in Europe. pp. 337–357. In: A. Watson, ed. Animal populations in relation to their food resources. Blackwell Scientific Publications, Oxford.
Newton, I. 1972. Finches. Collins, London.
Nicholls, K. W., B. A. Bohm, and R. Ornduff. 1985. Flavonoids and affinities of the Cephalotacaceae. *Biochem. Syst. Ecol.* 13:262–264.
Niklas, K. J., and S. L. Buchmann. 1985. Aerodynamics of wind pollination in *Simmondsia chinensis* (Link) Schneider. *Am. J. Bot.* 72:530–539.
Nilsson, L. A. 1983. Mimesis of bellflower *(Campanula)* by the red helleborine orchid *Cephalanthera rubra*. *Nature* 305:799–800.
Nilsson, L. A., L. Jonsson, L. Rason, and E. Randrianjohany. 1985. Monophily and pollination mechanisms in *Angraecum arachnites* Schltr. (Orchidaceae) in a guild of long-tongued hawk-moths (Sphingidae) in Madagascar. *Biol. J. Linnean Soc.* 26:1–19.
Noy-Meir, I. 1975. Stability of grazing systems: an application of predator–prey graphs. *J. Ecol.* 63:459–481.
Oatman, E. R., J. A. McMurty, and V. Voth. 1968. Suppression of the two-spotted spider mite on strawberry with mass releases of *Phytoseiulus persimilis*. *J. Econ. Entomol.* 61:1517–1521.
O'Dowd, J. D. 1979. Foliar nectar production and ant activity on a neotropical tree, *Ochroma pyramidale*. *Oecologia (Berl.)* 45:233–248.
O'Dowd, D. J. 1980. Pearl bodies of a neotropical tree *Ochroma pyramidale*: ecological implications. *Am. J. Bot.* 67:543–549.
O'Dowd, D. J. 1982. Pearl bodies as ant-food: an ecological role for some leaf emergences of tropical plants. *Biotropica* 14:40–49.
O'Dowd, D. J., and E. A. Catchpole. 1983. Ants and extrafloral nectaries: no evidence for plant protection in *Helichrysum* spp.-ant interactions. *Oecologia (Berl.)* 59:191–200.
O'Dowd, D. J., and M. E. Hay. 1980. Mutualism between harvester ants and a desert ephemeral: seed escape from rodents. *Ecology* 61:531–540.
Odum, E. P. 1969. The strategy of ecosystem development. *Science* 164:262–270.
Ohmart, C. P., L. Stewart, and J. Thomas. 1983. Leaf consumption by insects in three eucalyptus forest types in southeastern Australia and their role in short-term nutrient cycling. *Oecologia (Berl.)* 51:379–384.
Oksanen, L., and T. Oksanen. 1981. Lemmings *(Lemmus lemmus)* and gray-sided voles *(Clethrionomys rufus)* in interaction with their resources and predators on Finnmarks vidda, northern Norway. *Rep. Kevo Subarct. Res. Stat.* 17:7–31.

Onuf, C. P. 1978. Nutritive value as a factor in plant-insect interactions with an emphasis on field studies. pp. 85–96. In: G. G. Montgomery, ed. The ecology of arboreal folivores. Smithsonian Institution Press, Washington, DC.
Orians, G. H. 1969. The number of bird species in some tropical forests. *Ecology* 50:783–801.
Orians, G. H. 1980. Some adaptations of marsh-nesting blackbirds. Monographs in Population Biology No. 14. Princeton University Press, Princeton, NJ.
Orians, G. H., and N. P. Pearson. 1979. On the theory of central place foraging. In: D. J. Horn, ed. Analysis of ecological systems. Ohio State University Biosciences Colloquia. Ohio State University Press, Columbus, OH.
Ostler, W. K., and K. T. Harper. 1978. Floral ecology in relation to plant species diversity in the Wasatch Mountains of Utah and Idaho. *Ecology* 59:848–861.
Ott, J. R., L. A. Real, and E. M. Silverfine. 1985. The effect of nectar variance on bumblebee patterns of movement and potential gene dispersal. *Oikos* 45:333–340.
Owen, D. F. 1980. How plants may benefit from the animals that eat them. *Oikos* 35:230–235.
Owen, D. F., and R. G. Wiegert. 1981. Mutualism between grasses and grazers: an evolutionary hypothesis. *Oikos* 36:376–378.
Pacala, S., and J. Roughgarden. 1982. Resource partitioning and interspecific competition in two two-species insular anolis lizard communities. *Science* 217:444–446.
Paine, R. T. 1984. Ecological determinism in the competition for space: the first MacArthur lecture. *Ecology* 65:1339–1348.
Painter, R. H. 1951. Insect resistance in crop plants. Macmillan, NY.
Parker, M. A. 1985. Size-dependent herbivore attack and the demography of an arid grassland shrub. *Ecology* 66:850–860.
Parker, M. A., and R. B. Root. 1981. Insect herbivores limit habitat distribution of a native composite. *Machaeranthera canescens. Ecology* 62:1390–1392.
Parkes, D. M., and N. D. Hallam. 1984. Adaptation for carnivory in the West Australian pitcher plant *Cephalotus follicularis. Aust. J. Bot.* 32:595–604.
Partridge, L. 1974. Habitat selection in titmice. *Nature* 247:573–574.
Partridge, L. 1976. Field and lab observations on the foraging and feeding techniques of blue tits *(Parus caeruleus)* and coal tits *(Parus ater)* in relation to their habitat. *Anim. Behav.* 24:534–544.
Partridge, L. 1978. Habitat selection. pp. 351–376. In: J. R. Krebs, and N. B. Davies, eds. Behavioural ecology, an evolutionary approach.
Pate, J. S. 1986. Economy of symbiotic nitrogen fixation. pp. 299–325. In: T. J. Givnish, ed. On the economy of plant form and function. Cambridge University Press, NY.
Pate, J. S., and K. W. Dixon. 1978. Mineral nutrition of *Drosera erythrohiza* Lindl. with special reference to its tuberous habit. *Aust. J. Bot.* 26:455–464.
Pearson, D. L. 1975. The relation of foliage complexity to ecological diversity of three Amazonian bird communities. *Condor* 77:453–466.
Pearson, D. L. 1977. A pantropical comparison of bird community structure on six lowland forest sites. *Condor* 79:232–244.
Pearson, D. L. 1980. Bird migration in Amazonian Ecuador, Peru, and Bolivia. pp. 273–284. In: A. Keast, and E. Morton, eds. Migrant birds in the neotropics. Smithsonian Institution Press, Washington, DC.
Pease, J. L., R. H. Vowles, and L. B. Keith. 1979. Interaction of snowshoe hares and woody vegetation. *J. Wildl. Manage.* 43:43–60.
Pedigo, L. P., S. H. Hutchins, and L. G. Higley. 1986. Economic injury level in theory and practice. *Ann. Rev. Entomol.* 31:341–368.
Pellmyr, O., and J. M. Patt. 1986. Function of olfactory and visual stimuli in pollination of *Lysichiton americanum* (Araceae) by a staphylinid beetle. *Madrono* 33:47–54.
Percival, M. S. 1961. Types of nectar in angiosperms. *New Phytol.* 60:235–281.
Petal, J. 1978. The role of ants in ecosystems. pp. 293–325. In: M. V. Brian, ed. Production ecology of ants and termites. Cambridge University Press, NY.
Peters, W. D., and T. C. Grubb. 1983. An experimental analysis of sex-specific foraging in the downy woodpecker, *Picoides pubescens. Ecology* 64:1437–1443.
Petrusewicz, K. 1967., ed. Secondary production of terrestrial ecosystems. Polish Academy of Science, Institute of Ecology, Warsaw, 2 vols.

Philbrick, C. T. 1986. Hydroautogamy: a unique form of self pollination in angiosperms. *Am. J. Bot.* 73:780–781.
Pianka, E. R. 1966. Latitudinal gradients in species diversity: a review of concepts. *Am. Natur.* 100:33–46.
Pianka, E. R. 1967. On lizard species diversity: North American flatland deserts. *Ecology* 48:333–351.
Pianka, E. R. 1975. Niche relations of desert lizards. pp. 292–314. In: M. L. Cody, and J. M. Diamond, eds. Ecology and evolution of communities. Belknap, Cambridge, MA.
Picado, C. 1913. Les Bromeliacees epiphytes considerees comme milieu biologique. *Bull. Sci. France Belg. Ser. 7,* 47:216–360.
Pickett, C. H., and D. W. Clark. 1979. The function of extrafloral nectaries in *Opuntia acanthocarpa* (Cactaceae). *Am. J. Bot.* 66:618–625.
Pickett, S. T. A., and P. S. White, eds. 1985. The ecology of natural disturbance and patch dynamics. Academic Press, Orlando, FL.
Picman, J. 1981. The adaptive value of polygyny in marsh-nesting red-winged blackbirds: renesting, territory tenacity, and mate fidelity of females. *Can. J. Zool.* 59:2284–2296.
Pierotti, R. 1982. Habitat selection and its effect on reproductive output in the herring gull in Newfoundland. *Ecology* 63:854–868.
Pijl, L. van der. 1957. The dispersal of plants by bats. *Acta Bot. Neerl.* 6:291–315.
Pijl, L. van der. 1982. Principles of dispersal in higher plants, 3rd ed. Springer-Verlag, NY.
Pillemer, E. A., and W. M. Tingey. 1976. Hooked trichomes: a physical plant barrier to a major agricultural pest. *Science* 193:482–484.
Pimentel, D. 1961. An evaluation of insect resistance in broccoli, brussels sprouts, cabbage, collards, and kale. *J. Econ. Entomol.* 54:156–158.
Pimentel, D., and C. A. Edwards. 1982. Pesticides and ecosystems. *BioScience* 34:595–600.
Pimentel, D. 1984. Energy flow in agroecosystems. pp. 121–132. In: R. Lowrance, B. R. Stinner, and G. J. House, eds. Agricultural ecosystems. John Wiley and Sons, NY.
Pimm, S. L., and M. L. Rosenzweig. 1981. Competitors and habitat use. *Oikos* 37:1–6.
Pimm, S. L., M. L. Rosenzweig, and W. Mitchell. 1985. Competition and food selection: field tests of a theory. *Ecology* 66:798–807.
Pisarski, B. 1978. Comparison of various biomes. pp. 326–341. In: M. V. Brian, ed. Production ecology of ants and termites. Cambridge University Press, NY.
Pleszczynska, W. K. 1978. Microgeographic prediction of polygyny in the lark bunting. *Science* 201:935–937.
Plummer, G. L. 1963. Soils of the pitcher plant habitats in the Georgia coastal plain. *Ecology* 44:727–734.
Plummer, G. L., and T. H. Jackson. 1963. Bacterial activities within the sarcophagus of the insectivorous plant, *Sarracenia flava*. *Am. Midl. Natur.* 69:462–469.
Plummer, G. L., and J. B. Kethley. 1964. Foliar absorption of amino acids, peptides, and other nutrients by the pitcher plant, *Sarracenia flava*. *Bot. Gaz.* 125:245–260.
Pojar, J. 1974. Reproductive dynamics of four plant communities of southwestern British Columbia. *Can. J. Bot.* 52:1819–1834.
Poole, R. W., and B. J. Rathcke. 1979. Regularity, randomness, and aggregation in flowering phenologies. *Science* 203:470–471.
Potter, B. The tale of Peter rabbit. Frederick Warne, NY.
Powell, J. A. 1980. Evolution of larval food preferences in microlepidoptera. *Ann. Rev. Entomol.* 25:133–159.
Power, J. F., R. F. Follett, and G. E. Carlson. 1983. Legumes in conservation tillage systems: a research perspective. *J. Soil Water Conserv.* 38:217–218.
Prance, G. T. (ed.) 1982. Biological diversification in the tropics. Columbia University Press, NY.
Pratt, T. K. 1984. Examples of tropical frugivores defending fruit-bearing plants. *Condor* 86:123–129.
Pratt, T. K., and E. W. Stiles. 1983. How long fruit-eating birds stay in the plants where they feed: implications for seed dispersal. *Am. Natur.* 122:797–805.

Price, M. V. 1978. The role of microhabitat in structuring desert rodent communities. *Ecology* 59:910–921.
Price, P. W. 1976. Colonization of crops by arthropods: non-equilibrium communities in soybean fields. *Environ. Entomol.* 5:605–611.
Price, P. W. 1977. General concepts on the evolutionary biology of parasites. *Evolution* 31:405–420.
Price, P. W. 1980. Evolutionary biology of parasites. Princeton University Press, Princeton, NJ.
Price, P. W. 1984. Insect ecology, 2nd ed. John Wiley and Sons, NY.
Price, P. W. 1986. Ecological aspects of host plant resistance and biological control: interactions among three trophic levels. pp. 111–127. In: D. J. Boethel, and R. D. Eikenbary, eds. Interactions of plant resistance and parasitoids and predators of insects. John Wiley and Sons, NY.
Price, P. W., and G. P. Waldbauer. 1982. Ecological aspects of pest management. pp. 33–68. In: R. L. Metcalf, and W. H. Luckmann, eds. Introduction to insect pest management. John Wiley and Sons, NY.
Price, P. W., C. E. Booton, P. Gross, B. A. McPheron, J. N. Thompson, and A. E. Weis. 1980. Interactions among three trophic levels: influence of plants on interactions between insect herbivores and natural enemies. *Ann. Rev. Ecol. Syst.* 11:41–65.
Price, P. W., M. Westoby, B. Rice, P. R. Atsatt, R. S. Fritz, J. N. Thompson, and K. Mobley. 1986. Parasite mediation in ecological interactions. *Ann. Rev. Ecol. Syst.* 17:487–506.
Pringsheim, E., and O. Pringsheim. 1962. Axenic culture of *Utricularia*. *Am. J. Bot.* 49:989–901.
Proctor, M., and P. Yeo. 1972. The pollination of flowers. Taplinger, NY.
Proctor, V. W. 1968. Long distance dispersal of seeds by retention in digestive tracts of birds. *Science* 166:321–322.
Pryce-Jones, J. 1944. Some problems associated with nectar, pollen and honey. *Proc. Linn. Soc. Lond.* 1944:129–174.
Pulliam, H. R. 1975. Coexistence of sparrows: a test of community theory. *Science* 189:474–476.
Pulliam, H. R. 1985. Foraging efficiency, resource partitioning, and the coexistence of sparrow species. *Ecology* 66:1829–1836.
Pyke, G. H. 1978. Optimal foraging: movement patterns of bumblebees between inflorescences. *Theor. Popul. Biol.* 13:72–98.
Pyke, G. H., H. R. Pulliam, and E. L. Charnov. 1977. Optimal foraging: a selective review of theory and tests. *Q. Rev. Biol.* 52:137–154.
Rabb, R. L., and F. E. Guthrie. 1964. Resistance of tobacco hornworms to certain insecticides in North Carolina. *J. Econ. Entomol.* 57:995–996.
Rabenold, K. N. 1978. Foraging strategies, diversity, and seasonality in bird communities of Appalachian spruce-fir forests. *Ecol. Monogr.* 48:397–424.
Rabenold, K. N. 1979. A reversed latitudinal diversity gradient in avian communities of eastern deciduous forests. *Am. Natur.* 114:275–286.
Rabenold, K. N. 1984. Cooperative enhancement of reproductive success in tropical wren societies. *Ecology* 65:871–885.
Rabenold, K. N., and P. P. Rabenold. 1985. Variation in altitudinal migration, winter segregation, and site tenacity in two species of dark-eyed juncos in the southern Appalachians. *Auk* 102:805–819.
Racine, C. H., and J. F. Downhower. 1974. Vegetative and reproductive strategies of *Opuntia* (Cactaceae) in the Galapagos Islands. *Biotropica* 6:175–186.
Radosevich, S. R., and J. S. Holt. 1984. Weed ecology. John Wiley and Sons, NY.
Ralph, C. J., and L. R. Mewaldt. 1975. Timing of site fixation upon the wintering grounds in sparrows. *Auk* 92:698–705.
Ranta, E. 1984. Proboscis length and the coexistence of bumblebee species. *Oikos* 43:189–196.
Raposo, J. 1970. Bein' green. Sesame Street Inc., ASCAP, NY.
Rathcke, B. J. 1976. Competition and coexistence within a guild of herbivorous insects. *Ecology* 57:76–87.

Rathcke, B. 1983. Competition and facilitation among plants for pollination. pp. 305–329. In: L. Real, ed. Pollination biology. Academic Press, NY.
Rathcke, B., and E. P. Lacey. 1985. Phenological patterns of terrestrial plants. *Annu. Rev. Ecol. Syst.* 16:179–214.
Rathcke, B. J., and R. W. Poole. 1975. Coevolutionary race continues: butterfly larval adaptation to plant trichomes. *Science* 187:175–176.
Rausher, M. D. 1978. Search image for leaf shape in a butterfly. *Science* 200:1071–1073.
Rausher, M. D. 1981. The effect of native vegetation on the susceptibility of *Aristolachia reticulata* (Aristolochiaceae) to herbivore attack. *Ecology* 62:1187–1195.
Rausher, M. D., and P. Feeny. 1980. Herbivory, plant density, and plant reproduction success: the effect of *Battus philenor* on *Aristolochia reticulata*. *Ecology* 61:905–917.
Rauzi, F., R. L. Lang, and L. I. Painter. 1968. Effects of nitrogen fertilization on native rangeland. *J. Range Manage.* 21:287–290.
Raven, P. H. 1972. Why are bird-visited flowers predominately red? *Evolution* 26:674.
Rayor, L. S. 1985. Effects of habitat quality on growth, age of first reproduction, and dispersal in Gunnison's prairie dogs *(Cynomys gunnisoni)*. *Can. J. Zool.* 63:2835–2840.
Read, D. P., P. P. Feeny, and R. B. Root. 1970. Habitat selection by the aphid parasite *Diaeretiella rapae* (Hymenoptera: Braconidae) and hyperparasite *Charips brassicae* (Hymenoptera: Cynipidae). *Can. Entomol.* 102:1567–1578.
Real, L. A. 1981. Uncertainty and pollinator–plant interactions: the foraging behavior of bees and wasps on artificial flowers. *Ecology* 62:20–26.
Real, L. 1983. Pollination biology. Academic Press, NY.
Real, L., and T. Caraco. 1986. Risk and foraging in stochastic environments. *Annu. Rev. Ecol. Syst.* 17:371–390.
Reardon, P. O., C. L. Leinweber, and L. B. Merrill. 1972. The effect of bovine saliva on grasses. *J. Anim. Sci.* 34:897–898.
Reardon, P. O., C. L. Leinweber, and L. B. Merrill. 1974. Response of sideoats grama to animal saliva and thiamine. *J. Range Manage.* 27:400–401.
Reed, C. A. 1969. The pattern of animal domestication in the prehistoric near East. pp. 343–389. In: P. J. Ucko, and G. W. Dimbley, eds. The domestication and exploitation of plants and animals. Duckworth, London.
Reed, C. A., ed. 1977. Origins of agriculture. Mouton, The Hague.
Rees, W. E., and N. A. Roe. 1980. *Puya raimondii* (Pitcairnoideae, Bromeliaceae) and birds: a hypothesis on nutrient relationships. *Can. J. Bot.* 58:1262–1268.
Regal, P. J. 1976. Ecology and evolution of flowering plant dominance. *Science* 196:622–629.
Rehr, S. S., P. P. Feeny, and D. H. Janzen. 1973. Chemical defense in Central American non-ant-acacias. *J. Anim. Ecol.* 42:405–416.
Reichle, D. E., R. A. Goldstein, R. I. Hook, and G. J. Dodson. 1973. Analysis of insect consumption in a forest canopy. *Ecology* 54:1076–1084.
Reynolds, H. G., and P. E. Packer. 1963. Effects of trampling on soil and vegetation. *U.S. Dep. Agric. Misc. Publ.* 940:117–122.
Reynolds, H. T., P. L. Adkisson, R. F. Smith, and R. E. Frisbie. 1982. Cotton insect pest management. pp. 375–441. In: R. L. Metcalf, and W. H. Luckmann, eds. Introduction to insect pest management. John Wiley and Sons, NY.
Rhoades, D. F. 1979. Evolution of plant chemical defense against herbivores. pp. 3–54. In: G. A. Rosenthal, and D. H. Janzen, eds. Herbivores: their interaction with secondary plant metabolites. Academic Press, NY.
Rhoades, D. F. 1983a. Herbivore population dynamics and plant chemistry. pp. 155–220. In: R. F. Denno, and M. S. McClure, eds. Variable plants and herbivores in natural and managed systems. Academic Press, NY.
Rhoades, D. F. 1983b. Responses of alder and willow to attack by tent caterpillars and webworms: evidence for pheromonal sensitivity of willows. pp. 55–68. In: P. A. Hedin, ed. Plant resistance to insects. Am. Chem. Soc., Washington, DC.
Rhoades, D. F. 1985. Pheromonal communication between plants. pp. 195–218. In: G. A. Cooper-Driver, T. Swain, and E. E. Conn, eds. Chemically mediated interactions between plants and other organisms. Plenum Press, NY.

Rhoades, D. F., and J. C. Bergdahl. 1981. Adaptive significance of toxic nectar. *Am. Natur.* 117:798–803.
Rhoades, D. F., and R. G. Cates. 1976. Toward a general theory of plant herbivore chemistry. pp. 168–213. In: J. W. Wallace, and R. L. Mansell, eds. Biochemical interaction between plants and insects. Plenum Press, NY.
Rice, B., and M. Westoby. 1986. Evidence against the hypothesis that ant-dispersed seeds reach nutrient-enriched microsites. *Ecology* 67:1270–1274.
Richards, A. J. 1986. Plant breeding systems. Allen and Unwin, London.
Richards, O. W., and R. G. Davies. 1977. Imm's general textbook of entomology, vol. II. Chapman and Hall, London.
Richards, P. W. 1936a. Ecological observations on the rain forest of Mount Dulit, Sarawak. Part I. *J. Ecol.* 24:1–37.
Richards, P. W. 1936b. Ecological observations on the rain forest of Mount Dulit, Sarawak. Part II. *J. Ecol.* 24:340–360.
Rick, C. M., and R. I. Bowman. 1961. Galapagos tomatoes and tortoises. *Evolution* 15:407–417.
Rickson, F. R. 1969. Development aspects of the shoot apex leaf, and Beltian bodies of *Acacia cornigera*. *Am. J. Bot.* 56:195–200.
Rickson, F. R. 1971. Glycogen plastids in Müllerian cells of *Cecropia peltata*—a higher green plant. *Science* 173:344–347.
Rickson, F. R. 1976. Anatomical development of leaf trichilium and Müllerian bodies of *Cecropia peltata*. L. *Am. J. Bot.* 63:1266–1271.
Rickson, F. R. 1977. Progressive loss of ant-related traits of *Cecropia peltata* on selected Caribbean Islands. *Am. J. Bot.* 64:585–592.
Rickson, F. R. 1979. Absorption of animal tissue breakdown products into a plant stem—the feeding of a plant by ants. *Am. J. Bot.* 66:87–90.
Rickson, F. R. 1980. Developmental anatomy and ultrastructure of the ant food bodies (Beccarian bodies) of *Macaranga triloba* and *M. hypoleuca* (Euphorbiaceae). *Am. J. Bot.* 67:285–292.
Ridley, H. N. 1930. The dispersal of plants throughout the world. L. Reeve, Ashford, Kent, England.
Risch, S. J., D. Andow, and M. H. Altirei. 1983. Agroecosystem diversities and pest control: data, tentative conclusions, and new research directions. *Environ. Entomol.* 12:625–629.
Robbins, C. T. 1983. Wildlife feeding and nutrition. Academic Press, NY.
Roberts, E. P., and P. D. Weigel. 1984. Habitat preference in the dark-eyed junco *(Junco hyemalis)*: the role of photoperiod and dominance. *Anim. Behav.* 32:709–714.
Roberts, P. R., and H. J. Oosting. 1958. Responses of Venus fly trap *(Dionaea muscipula)* to factors involved in its endemism. *Ecol. Monogr.* 28:193–218.
Robertson, R. J. 1973. Optimal niche space of the red-winged blackbird: spatial and temporal patterns of nesting activity and success. *Ecology* 54:1085–1093.
Robins, R. J. 1976. The nature of the stimuli causing digestive juice secretion in *Dionaea muscipula* Ellis (Venus's fly trap). *Planta* 128:263–265.
Robinson, J. G. 1986. Seasonal variation in use of time and space by the wedge-capped capuchin monkey, *Cebus olivaceus:* implications for foraging theory. Smithsonian Contributions to Zoology No. 431, 60 pp.
Robinson, S. K. 1985. Coloniality in the yellow-rumped cacique as a defence against nest predators. *Auk* 102:506–519.
Robinson, S. K. 1986a. Competitive and mutualistic interaction among females of a neotropical oriole. *Anim. Behav.* 34:113–122.
Robinson, S. K. 1986b. The evolution of social behavior and mating systems in the blackbirds (Icterinae). pp. 175–200. In: D. I. Rubenstein, and R. W. Wrangham, eds. Ecological aspects of social evolution: birds and animals. Princeton University Press, Princeton, NJ.
Robinson, S. K., and R. T. Holmes. 1982. Forging behavior of forest birds: the relationships among search tactics, diet and habitat structure. *Ecology* 63:1918–1931.
Rockwood, L. L. 1973. The effect of defoliation on seed production in six Costa Rican tree species. *Ecology* 54:1363–1369.

Rockwood, L. L. 1976. Plant selection and foraging patterns in two species of leaf cutting ants *(Atta)*. *Ecology* 57:48–61.

Rodgers, A. R., and M. C. Lewis. 1985. Diet selection in arctic lemmings *(Lemmus sibericus* and *Dicrostonyx groenlandicus)*: food preferences. *Can. J. Zool.* 63:1161–1173.

Rodman, P. S. 1984. Feeding and social systems of orangutans and chimpanzees. pp. 134–160. In: P. S. Rodman, and J. G. H. Cant, eds. Adaptations for foraging in nonhuman primates. Columbia University Press, NY.

Roessler, E. S. 1936. Viability of weed seeds after ingestion by California linnets. *Condor* 38:62–65.

Rohfritsch, O., and J. D. Shorthouse. 1982. Insect galls. pp. 131–152. In: G. Kahl, and J. Schell, eds. Molecular biology of plant tumors. Academic Press, NY.

Rohwer, S. 1977. Status signalling in Harris' sparrows: some experiments in deception. *Behaviour* 61:107–129.

Romoser, W. S. 1973. The science of entomology. Macmillan, NY.

Root, R. B. 1973. Organization of a plant-arthropod association in simple and diverse habitats: the fauna of collards *(Brassica oleracea)*. *Ecol. Monogr.* 43:95–124.

Rosenthal, G. A., and D. H. Janzen, eds. 1979. Herbivores: their interaction with secondary plant metabolites. Academic Press, NY.

Rosenthal, G. A., C. Hughes, and D. Janzen. 1982. L-Canavanine, a dietary nitrogen source for the seed predator *Caryedes brasiliensis* (Bruchidae). *Science* 217:353–355.

Rosenzweig, M. L. 1973. Habitat selection experiments with a pair of coexisting heteromyid rodent species. *Ecology* 62:327–335.

Rosenzweig, M. L. 1981. Theory of habitat selection. *Ecology* 62:327–335.

Rosenzweig, M. L. 1985. Some theoretical aspects of habitat selection. pp. 517–547. In: M. L. Cody, ed. Habitat selection in birds. Academic Press, NY.

Ross, B. A., J. R. Bray, and W. H. Marshall. 1970. Effects of long-term deer exclusion on a *Pinus resinosa* forest in north-central Minnesota. *Ecology* 51:1088–1093.

Ross, H. H. 1957. Principles of natural coexistence indicated by leafhopper populations. *Evolution* 11:113–129.

Ross, H. H. 1965. A textbook of entomology, 3rd ed. John Wiley and Sons, NY.

Rotenberry, J. T., and J. A. Wiens. 1980. Habitat structure, patchiness and avian communities in North American steppe vegetation: a multivariate approach. *Ecology* 61:1228–1250.

Rothschild, M. 1972. Secondary plant substances and warning coloration in insects. pp. 59–83. In: H. F. van Einden, ed. Insect/plant relationships. *Symp. Roy. Eng. Soc. Lond.* no. 6.

Roubik, D. W., N. M. Holbrook, and G. Parra V. 1985. Roles of nectar robbers in reproduction of the tropical treelet *Quassia amara* (Simaroubaceae). *Oecologia (Berl.)* 66:161–167.

Roughgarden, J. 1972. Evolution of niche width. *Am. Natur.* 106:683–718.

Roughgarden, J. 1979. Theory of population genetics and evolutionary ecology: an introduction. Macmillan, NY.

Roughgarden, J. 1983. The theory of coevolution. pp. 33–64. In: D. J. Futuyma, and M. Slatkin, eds. Coevolution. Sinauer Associates, Sunderland, MA.

Rowell, T. 1972. Social behavior of monkeys. Penguin Books, Middlesex, England.

Rubenstein, D. I., and R. W. Wrangham. 1986. Social evolution in birds and mammals. pp. 452–470. In: D. I. Rubenstein, and R. W. Wrangham, eds. Ecological aspects of social evolution: birds and mammals. Princeton University Press, Princeton, NJ.

Ruess, R. W., and S. J. McNaughton. 1984. Urea as a promotive coupler of plant-herbivore interactions. *Oecologia (Berl.)* 63:331–337.

Rufener, G. K. II, R. B. Hammond, R. L. Cooper, and S. K. St. Martin. 1986. Mexican bean beetle (Coleoptera: Coccinellidae) development on resistant and susceptible soybean lines in the laboratory and relation to field selection.

Ruffner, G. A., and W. D. Clark. 1986. Extrafloral nectar of *Ferocactus acanthodes* (Cactaceae): composition and its importance to ants. *Am. J. Bot.* 73:185–189.

Sanders, E. H., P. D. Gardner, P. J. Berger, and N. C. Negus. 1981. 6-

Methoxybenzoxazolinone: a plant derivative that stimulates reproduction in *Microtus montanus*. *Science* 214:67–69.
Sauer, C. O. 1969. Agricultural origins and dispersals. M.I.T. Press, Cambridge, MA.
Scala, J., K. Iott, D. W. Schwab, and F. E. Semersky. 1969. Digestive secretion of *Dionaea muscipula* (Venus's fly trap). *Plant Physiol.* 44:367–371.
Scala, J., D. W. Schwab, and E. Simmons. 1968. The fine structure of the digestive gland of Venus's fly trap. *Am. J. Bot.* 55:649–657.
Schaller, G. B. 1963. The mountain gorilla: ecology and behavior. University of Chicago Press, Chicago, IL.
Schaller, G. B., Hu Jinchu, Pan Wenshi, and Zhu Jing. 1985. The giant panda of Wolong. University of Chicago Press, Chicago, IL.
Scheline, R. R. 1978. Mammalian metabolism of plant xenobiotics. Academic Press, NY.
Schemske, D. W. 1978. Evolution of reproductive characteristics in *Impatiens* (Balsaminaceae): the significance of cleistogamy and chasmogamy. *Ecology* 59:596–613.
Schemske, D. W. 1980a. Floral ecology and hummingbird pollination of *Combretum farinosum* in Costa Rica. *Biotropica* 12:169–181.
Schemske, D. W. 1980b. The evolutionary significance of extrafloral nectar production by *Costus woodsonii* (Zingiberaceae): an experimental analysis of ant protection. *J. Ecol.* 68:959–967.
Schemske, D. W. 1983. Limits to specialization and coevolution in plant–animal interactions. pp. 67–109. In: M. H. Nitecki, ed. Coevolution. University of Chicago Press, Chicago, IL.
Schemske, D. W., and N. Brokaw. 1981. Treefalls and the distribution of understory birds in a tropical forest. *Ecology* 62:938–945.
Schemske, D. W., and C. C. Horvitz. 1984. Variation among floral visitors in pollination ability: a precondition for mutualism specialization. *Science* 225:519–521.
Schery, R. W. 1972. Plants for man. Prentis-Hall, Englewood Cliffs, NJ.
Schimper, A. F. W. 1882. Notizen uber insectenfressenden Pflanzen. *Bot. Z.* 40:225–233.
Schmucker, T., and G. Linnemann. 1959. Carnivorie. pp. 198–283. In: K. Mothes, ed. Handbuch der Pflanzenphysiologie, bd. XI. Springer-Verlag, Heidelberg.
Schneider, J. C. 1980. The role of parthenogenesis and female aptery in microgeographic ecological adaptation in the fall cankerworm, *Alsophila pometaria* Hari (Lepidoptera: Geometridae). *Ecology* 61:1082–1090.
Schnell, D. E. 1976. Carnivorous plants of the USA and Canada. John E. Blair, Winston-Salem, NC.
Schoener, T. W. 1968. The *Anolis* lizards of Binimi: resource partitioning in a complex fauna. *Ecology* 49:704–726.
Schoener, T. W. 1971. Large-billed insectivorous birds: a precipitous diversity gradient. *Condor* 73:154–161.
Schoener, T. W. 1975. Presence and absence of habitat shift in some widespread lizard species. *Ecol. Monogr.* 45:233–258.
Schoener, T. W. 1982. The controversy over interspecific competition. *Am. Sci.* 70:586–595.
Schoener, T. W. 1983. Field experiments on interspecific competition. *Am. Natur.* 122:240–285.
Scholz, A. T., R. M. Horral, J. C. Cooper, and A. D. Hasler. 1976. Imprinting to chemical cues: the basis for home stream selection in salmon. *Science* 192:1247–1249.
Schorger, A. W. 1960. The crushing of *Carya* nuts in the gizzard of the turkey. *Auk* 77:337–340.
Schowalter, T. D., J. W. Webb, and Crossley, Jr. 1981. Community structure and nutrient content of canopy arthropods in clearcut and uncut forest ecosystems. *Ecology* 62:1010–1019.
Schowalter, T. D., W. W. Hargrove, and D. A. Crossley, Jr. 1986. Herbivory in forested ecosystems. *Ann. Rev. Entomol.* 31:177–196.
Schultz, A. M. 1964. The nutrient recovery hypothesis for arctic microtine cycles. II. Ecosystem variables in relation to arctic microtine cycles. pp. 57–68. In: D. J. Crisp,

ed. Grazing in terrestrial and marine environments. Blackwell Scientific Publications, Oxford.
Schultz, A. M. 1969. A study of an ecosystem: the arctic tundra. pp. 77–93. In: G. M. Van Dyne, ed. The ecosystem concept in natural resource management. Academic Press, NY.
Schultz, J. C., and I. Baldwin. 1982. Oak leaf quality declines in response to defoliation by gypsy moth larvae. *Science* 217:149–151.
Scott, G. E., and W. D. Guthrie. 1966. Survival of European corn borer larvae on resistant corn treated with nutritional substances. *J. Econ. Entomol.* 59:1265–1267.
Scott, J. A., N. R. French, and L. W. Leetham. 1979. Patterns of consumption in grasslands. In: N. R. French, ed. Perspectives in grassland ecology. Ecol. Studies No. 32, Springer-Verlag, NY.
Scriber, J. M. 1977. Limiting effects of low leaf-water content on the nitrogen utilization, energy budget and larval growth of *Halophora cecropia* (Lepidoptera: Saturniidae). *Oecologia (Berl.)* 28:269–287.
Scriber, J. M. 1979. Effects of leaf-water supplementation upon post-ingestive nutritional indices of forb-, shrub-, and tree-feeding lepidoptera. *Entomol. Exp. Appl.* 25:240–252.
Scriber, J. M. 1984. Host plant suitability. pp. 160–202. In: W. J. Bell and R. T. Corde, eds. Chemical ecology of insects. Suncer and CMA: Sinauer Associates, Sunderland, MA.
Scriber, J. M., and P. Feeny. 1979. Growth of herbivorous caterpillars in relation to feeding specialization and to growth form of their food plants. *Ecology* 60:829–850.
Scriber, J. M., and F. Slansky. 1981. The nutritional ecology of immature insects. *Ann. Rev. Entomol.* 26:183–211.
Sculthorpe, C. D. 1967. The biology of aquatic vascular plants. Edward Arnold, NY.
Searcy, W. A. 1979. Male characteristics and pairing success in red-winged blackbirds. *Am. Sci.* 71:166–174.
Seastedt, T. R., D. A. Crossley, Jr., and W. W. Hargrove. 1983. The effects of low level consumption by anopy arthropods on the growth and nutrient dynamics of black locust and red maple trees in the Southern Appalachians. *Ecology* 64:1040–1048.
Seavey, S. R., and K. S. Bawa. 1985. Late-acting self-incompatibility in angiosperms. *Bot. Rev.* 52:234–258.
Seeley, T. D. 1977. Measurement of nest cavity volume by the honey bee *(Apis mellifera)*. *Behav. Ecol. Sociobiol.* 2:201–227.
Seeley, T. D. 1985. Honeybee ecology. Princeton University Press, Princeton, NJ.
Seif el Din, A., and M. Obeid. 1971. Ecological studies of the vegetation of the Sudan. IV. The effect of simulated grazing on the growth of *Acacia senegal* (L.) Willd. seedlings. *J. Appl. Ecol.* 8:211–216.
Seigler, D. S. 1977. Primary roles for secondary compounds. *Biochem. Syst. Ecol.* 5:195–199.
Selander, R. K. 1966. Sexual dimorphism and differential niche utilization in birds. *Condor* 68:113–151.
Self, L. S., F. Guthrie, and E. Hodgson. 1964. Adaptation of tobacco hornworms to the ingestion of nicotine. *J. Insect Physiol.* 12:224–230.
Selles, F., R. E. Karamanos, and K. E. Bouren. 1984. Changes in natural N^{15} abundance of soils associated with tillage practices. *Can. J. Soil Sci.* 64:345–354 pp.
Sernander, R. 1927. Zur. Morphologie und Biologie die Diasporen. *N. Acta Reg.* Soc. sc. Uppsaliensis, Uppsala.
Sherry, T. W., and R. T. Holmes. 1985. Dispersion patterns and habitat responses of birds in northern hardwoods forests. pp. 283–310. In: M. L. Cody, ed. Habitat selection in birds. Academic Press, NY.
Shetler, S. G. 1974. Nepenthales. pp. 958–965. In: Encyclopaedia Brittanica, 15th ed. Helen Hemingway Benton, NY.
Showers, W. B., H. C. Chiang, A. J. Keuster, R. E. Hill, G. L. Reed, A. N. Sparks, and G. L. Musick. 1975. Ecotypes of the European corn borer in North America. *Environ. Entomol.* 4:753–760.
Sikes, S. K. 1971. The natural history of the African elephant. Weidenfeld and Nicolson, London.

Silvertown, J. W. 1980. The evolutionary ecology of mast seeding in trees. *Biol. J. Linn. Soc.* 14:235–250.
Silvertown, J. W. 1982. No evolved mutualism between grasses and grazers. *Oikos* 38:253–254.
Simpson, B. B., and J. L. Neff. 1981. Floral rewards: alternatives to pollen and nectar. *Ann. Mo. Bot. Gard.* 68:301–322.
Simpson, G. G. 1964. Species density of North American recent mammals. *Syst. Zool.* 13:57–73.
Sinclair, A. R. E. 1974. The natural regulation of buffalo populations in East Africa. IV. The food supply as a regulating factor, and competition. *E. Afr. Wildl. J.* 12:291–311.
Sinclair, A. R. E., H. Dublin, and M. Borner. 1985. Population regulation of Serengeti wildebeest: a test of the food hypothesis. *Oecologia (Berl.)* 65:266–268.
Slack, A. 1979. Carnivorous plants. Ebury Press, London.
Slansky, F., Jr. 1974. Energetic and nutritional interactions between larvae of the imported cabbage butterfly, *Pieris rapae* L. and cruciferous food-plants. Ph.D. dissertation, Cornell University, Ithaca, NY.
Slansky, F., and P. Feeny. 1977. Stabilization of the rate of nitrogen accumulation by larvae of the cabbage butterfly on wild cultivated food plants. *Ecol. Monogr.* 47:209–228.
Small, J.G. C., A. Onraet, D. S. Grierson, and G. Reynolds. 1977. Studies on insect-free growth, development and nitrate-assimilating enzymes of *Drosera aliciae* Hamet. *New Phytol.* 79:127–133.
Smiley, J. 1986. Ant constancy at *Passiflora* extrafloral nectaries: effects on caterpillar survival. *Ecology* 67:516–521.
Smith, C. C. 1970. The coevolution of pine squirrels *(Tamiasciurus)* and conifers. *Ecol. Monogr.* 40:349–371.
Smith, D. R. 1979. Symphyta. pp. 3–7. In: K. V. Krambein, P. D. Hurd, Jr., D. R. Smith, and B. D. Burks, eds. Catalog of Hymenoptera in America north of Mexico. Smithsonian Institution Press, Washington, DC.
Smith, J. N. M., P. R. Grant, B. R. Grant, I. Abbott, and L. K. Abbott. 1978. Seasonal variation in feeding habits of Darwin's ground finches. *Ecology* 59:1137–1150.
Smith, K. G. 1982. Drought-induced changes in avian community structure along a moisture sere. *Ecology* 63:952–961.
Smith, L. B., and R. J. Downs. 1974. Bromeliaceae (Pitcairnioideae). *Flora Neotropica Monogr.* 14. Hafner, NY.
Smith, L. B., and R. J. Downs. 1977. Bromeliaceae (Tillandsioideae). *Flora Neotropica Monogr.* 14, pt. 2. Hafner, NY.
Smith, N. G. 1980. Some evolutionary, ecological and behavioural correlates of communal nesting by birds with wasps or bees. Acta XVII Int. Ornithol. Congr., 1199–1205.
Smith, N. J. 1981. Man, fishes and the Amazon. Columbia University Press, NY.
Smith, R. F. 1978. History and complexity integrated pest management. pp. 41–53. In: E. H. Smith, and D. Pimentel, eds. Pest control strategies. Academic Press, NY.
Smithies, B. E. 1964. The distribution and ecology of pitcher plants *(Nepenthes)* in Sarawak. In: UNESCO Humid Tropics Symposium. UNESCO, Paris.
Smythe, N. 1986. Competition and resource partitioning in the guild of neotropical terrestrial frugivorous mammals. *Ann. Rev. Ecol. Syst.* 17:169–188.
Snodgrass, R. E. 1935. Principles of insect morphology. McGraw-Hill, NY.
Snow, B. K. 1970. A field study of the bearded bellbird in Trinidad. *Ibis* 112:299–329.
Snow, B. K., and D. W. Snow. 1984. Long-term defense of a fruit tree by mistle thrushes. *Ibis* 126:300–310.
Snow, D. W. 1953. Systematics and comparative ecology of the genus *Parus* in the palearctic region. Ph.D. thesis, Oxford University.
Snow, D. W. 1962. The natural history of the oilbird, *Steatornis caripensis,* in Trinidad, W. I. II. Population, breeding ecology, and food. *Zoologica* 47:199–221.
Snow, D. W. 1965. A possible selective factor in the evolution of fruiting seasons in tropical forests. *Oikos* 15:274–281.

Snow, D. W. 1971. Evolutionary aspects of fruit-eating by birds. *Ibis* 113:194–202.
Snow, D. W. 1981. Tropical frugivorous birds and their food plants: a world survey. *Biotropica* 13:1–14.
Sohmer, S. H., and D. F. Sefton. 1978. The reproductive biology of *Nelumbo pentapetala* (Nelumbonaceae) on the Upper Mississippi River. II. The insects associated with the transfer of pollen. *Brittonia* 30:355–364.
Sorensen, A. E. 1981. Interactions between birds and fruit in a temperate woodland. *Oecologia (Berl.)* 50:242–249.
Sorensen, A. E. 1984. Nutrition, energy, and passage time: experiments with fruit preference in European blackbirds *(Turdus merula)*. *J. Anim. Ecol.* 53:545–557.
Sorenson, A. E. 1986. Seed dispersal by adhesion. *Ann. Rev. Ecol. Syst.* 17:443–463.
Sorenson, D., and W. T. Jackson. 1968. Utilization of paramecium by *Utricularia gibba*. *Planta* 83:166–170.
Soule, M. E. 1985. What is conservation biology? *BioScience* 35:727–734.
Southgate, B. J. 1979. Biology of the Bruchidae. *Ann. Rev. Entomol.* 24:449–473.
Southwood, T. R. E. 1972. The insect/plant relationship—an evolutionary perspective. pp. 3–30. In: van Emden, ed. Insect/plant relationships. Symp. Roy. Entomol. Soc. Lond. No. 6, John Wiley and Sons, NY.
Southwood, T. R. E. 1977. Habitat, the template for ecological strategies. *J. Anim. Ecol.* 46:337–365.
Southwood, T. R. E. 1985. Interactions of plants and animals: patterns and processes. *Oikos* 44:5–11.
Spedding, C. R. W. 1971. Grassland ecology. Oxford University Press, Oxford.
Spencer, S. R., and G. N. Cameron. 1983. Behavioral dominance and its relationship to habitat patch utilization by the hispid cotton rat *(Sigmodon hispidus)*. *Behav. Ecol. Sociobiol.* 13:27–36.
Stacey, P. B. 1979. Habitat saturation and communal breeding in the acorn woodpecker. *Anim. Behav.* 27:1153–1166.
Stanley, R. G., and H. F. Linskens. 1974. Pollen: biology, biochemistry, management. Springer-Verlag, Berlin.
Stanton, M. L., A. A. Snow, and S. N. Handel. 1986. Floral evolution: attractiveness to pollinators increases male fitness. *Science* 232:1625–1627.
Stapanian, M. A., and C. C. Smith. 1984. Density-dependent survival of scatterhoarded nuts: an experimental approach. *Ecology* 65:1387–1396.
Stebbins, G. L. 1971. Processes of organic evolution. Prentice-Hall, Englewood Cliffs, NJ.
Stehli, H. G., R. G. Douglas, and N. D. Newell. 1969. Generation and maintenance of gradients in taxonomic diversity. *Science* 164:947–949.
Stehr, F. W. 1982. Parasitoids and predators in pest management. pp. 135–137. In: R. L. Metcalf, and W. H. Luckmann, eds. John Wiley and Sons, NY.
Steiner, K. E. 1985. The role of nectar and oil in the pollination of *Drymonia serrulata* (Gesneriaceae) by *Epicharis* bees (Anthophoridae) in Panama. *Biotropica* 17:217–229.
Steinly, B., and M. Berenbaum. 1985. Histological effects of tannics on the mudgut epithelium of *Papilio polyseres* and *Papilio glaucus*. *Entomol. Exp. Appl.* 11:1349–1358.
Steneck, R. S. 1982. A limpet–coralline alga association: adaptations and defenses between a selective herbivore and its prey. *Ecology* 63:507–522.
Stephens, D. W., and J. R. Krebs. 1986. Foraging theory. Princeton Monographs in Behavior and Ecology, Princeton University Press, Princeton, NJ.
Stephenson, A. G. 1981a. Toxic nectar deters nectar thieves of *Catalpa speciosa*. *Am. Midl. Natur.* 105:381–383.
Stephenson, A. G. 1981b. Flower and fruit abortion: proximate causes and ultimate functions. *Ann. Rev. Ecol. Syst.* 12:253–279.
Stephenson, A. G. 1982a. Iridoid glycosides in the nectar of *Catalpa speciosa* are unpalatable to nectar thieves. *J. Chem. Ecol.* 8:1025–1034.
Stephenson, A. G. 1982b. The role of the extrafloral nectaries of *Catalpa speciosa* in limiting herbivory and increasing fruit production. *Ecology* 63:663–669.

Stephenson, A. G., and R. I. Bertin. 1983. Male competition, female choice, and sexual selection in plants. pp. 109–149. In: L. Real, ed. Pollination biology. Academic Press, NY.

Stephenson, A. G., and J. A. Winsor. 1986. *Lotus corniculatus* regulates offspring quality through selective fruit abortion. *Evolution* 40:453–458.

Stevens, L. M., A. L. Steinhaver, and J. R. Coulson. 1975. Suppression of the Mexican bean beetle on soybeans with annual innoculative releases of *Pediobuis foveolatus*. *Environ. Entomol.* 47:947–952.

Steyermark, J. A. 1961. *Brocchinia*. *Bromeliad Soc. Bull.* 1:35–41.

Steyermark, J. A. 1966. Flora del Ptari-tepui. *Acta Bot. Venez.* 1:30–104.

Steyermark, J. A. 1984. Flora of the Venezuelan Guayan. I. Sarraceniacear. *Ann. Mo. Bot. Gard.* 71:302–312.

Stiles, E. W. 1980. Patterns of fruit presentation and seed dispersal in bird-disseminated woody plants in the eastern deciduous forest. *Am. Natur.* 116:670–688.

Stiles, E. W. 1982a. Expansions of mockingbird and multiflora rose in the northeastern United States and Canada. *Am. Birds* 36:358–364.

Stiles, E. W. 1982b. Fruit flags: two hypotheses. *Am. Natur.* 120:500–509.

Stiles, E. W., and D. W. White. 1982. Additional information on bird-disseminated fruits: response to Herrera's comments. *Am. Natur.* 120:823–827.

Stiles, E. W., and D. W. White. 1986. Seed deposition patterns: influence of season, nutrients, and vegetation structure. In: A. Estrada, T. H. Fleming, C. Vazquez-Yanes, and R. Dirzo, eds. Frugivores and seed dispersal. Dr. W. Junk, Publisher, The Hague.

Stiles, F. G. 1979. Reply to Poole and Rathcke. *Science* 203:471.

Stiles, F. G., and L. L. Wolf. 1970. Hummingbird territoriality at a tropical flowering tree. *Auk* 87:467–491.

Stinner, B. R., and G. J. House. 1987. Agricultural ecosystems, conservation tillage and multiple cropping. *J. Soil Water Conserv.*, in press.

Stinner, B. R., D. A. McCartney, and W. Rubink. 1984a. Some observations on ecology of the stalk borer (*Papaipema nebris* (GN): Noctuidae) in no-tillage corn agroecosystems. *J. Ga. Entomol. Soc.* 19:229–234.

Stinner, B. R., D. A. Crossely, Jr., E. P. Odum, and R. L. Todd. 1984b. Nutrient budgets and internal cycling of N, P, K, Ca and Mg in conventional tillage, no-tillage and old-field ecosystems on the Georgia Piedmont. *Ecology* 65:354–369.

Stoner, A. W. 1979. Species-specific predation of amphipod crustacea by the pinfish *Lagodon rhomboides:* mediation by macrophyte standing crop. *Mar. Biol.* 55:201–207.

Stoutamire, W. P. 1974. Australian terrestrial orchids, thynnid wasps, and pseudocopulation. *Bull. Am. Orchid Soc.* 43:13–18.

Straw, R. M. 1956. Floral isolation in *Penstemon*. *Am. Natur.* 90:47–53.

Strickler, K. 1979. Specialization and foraging efficiency of solitary bees. *Ecology* 60:998–1009.

Strong, D. R. 1982. Harmonious coexistence of hispine beetles in experimental and natural communities. *Ecology* 63:1039–1049.

Strong, D. R. 1984. Exorcising the ghost of competition past: phytophagous insects. p. 241. In: D. R. Strong, D. Simberloff, L. G. Abele, and A. B. Thistle, eds. Ecological communities, conceptual issues and the evidence. Princeton University Press, Princeton, NJ.

Strong, D. R., J. H. Lawton, and R. Southwood. 1984a. Insects on plants. Harvard University Press, Cambridge, MA.

Strong, D. R., Jr., D. Simberloff, L. G. Abele, and A. B. Thistle. 1984b. Ecological communities—conceptual issues and the evidence. Princeton University Press, Princeton, NJ.

Struhsaker, T. T., and J. F. Oates. 1975. Comparison of the behavior and ecology of red colobus and black-and-white colobus monkeys in Uganda: a summary. pp. 103–123. In: R. H. Tuttle, ed. Socio-ecology and psychology of primates. Mouton, The Hague.

Sussman, R. W. 1977. Feeding behaviour of *Lemur catta* and *Lemur fulvus*. pp. 1–37. In: T. H. Clutton-Brock, ed. Primate ecology: feeding and ranging behaviour in lemurs, monkeys and apes. Academic Press, London.

Sussman, R. W., and P. H. Raven. 1978. Pollination by lemurs and marsupials: an archaic coevolutionary system. *Science* 200:731–736.

Svardson, G. 1949. Competition and habitat selection in birds. *Oikos* 1:157–174.

Tahvanainen, J. O., and R. B. Root. 1977. The influence of vegetation diversity on the population ecology of a specialized herbivore, *Phyllotera cruciferae* (Coleoptera: Chrysomelidae). *Oecologia (Berl.)* 10:321–346.

Talbot, L. M., and M. H. Talbot. 1963. The wildebeest in western Masailand, East Africa. *Wildl. Monogr.* 12:1–88.

Tallamy, D. W. 1985. Squash beetle feeding behavior: an adaptation against induced cucurbit defenses. *Ecology* 66:1574–1579.

Taylor, P. 1964. The genus *Utricularia* L. (Lentibulariaceae) in Africa (south of the Sahara) and Madagascar. *Kew Bull.* 18:1–245.

Tempel, A. S. 1983. Bracken fern *(Pteridium aquilinum)* and nectar-feeding ants: a nonmutualistic interaction. *Ecology* 64:1411–1422.

Temple, S. A. 1977. Plant–animal mutualism: coevolution with dodo leads to near extinction of plant. *Science* 197:885–886.

Terborgh, J. 1971. Distribution on environmental gradients: theory and a preliminary interpretation of distributional patterns in the avifauna of the Cordillera Vilcabamba, Peru. *Ecology* 52:23–40.

Terborgh, J. 1977. Bird species diversity on an Andean elevational gradient. *Ecology* 58:1007–1019.

Terborgh, J. 1980a. Vertical stratification of a neotropical forest bird community. *Proc. XVII Int. Ornithol. Cong.*, pp. 1005–1012.

Terborgh, J. 1980b. Causes of tropical species diversity. Symposium. Tropical Ecology. *Proc. XVII Int. Ornithol. Congr. II (West Berlin)*, pp. 955–961.

Terborgh, J. 1983. Five new world primates—a study in comparative ecology. Behav. Ecol. Monogr., Princeton University Press, Princeton, NJ.

Terborgh, J. 1985. Habitat selection in Amazonian birds. pp. 311–338. In: M. L. Cody, ed. Habitat selection in birds. Academic Press, NY.

Terborgh, J. 1986a. Keystone plant resources in the tropical forest. pp. 330–344. In: M. E. Soule, ed. Conservation biology. Sinauer Associates, Sunderland, MA.

Terborgh, J. 1986b. Community aspects of frugivory in tropical forests. pp. 371–384. In: A. Estrada, and T. H. Fleming, eds. Frugivores and seed dispersal. Dr. W. Junk, Publisher, Dordrecht, Netherlands.

Terborgh, J., and C. H. Janson. 1986. The socioecology of primate groups. *Ann. Rev. Ecol. Syst.* 17:111–136.

Terborgh, J., and S. Robinson. 1987. Guilds and their utility in ecology. In: J. Kikkawa (ed.) Community ecology: pattern and process. Blackwell Scientific Publications, Oxford (in press).

Terborgh, J., J. W. Fitzpatrick, and L. H. Emmons. 1984. Annoted checklist of bird and mammal species of Cocha Cashu Biological Station, Mani National Park, Peru. *Fieldiana Zool.* no. 21, 29 pp.

Tevis, L. 1958. Interrelationships between the harvester ant *Veromessor pergandei* and some desert ephemerals. *Ecology* 39:695–704.

Thien, L. B. 1980. Patterns of pollination in the primitive angiosperms. *Biotropica* 12:1–13.

Thirakhupt, K., (In Press) Foraging by sympatric parids: individual and population responses to winter food scarcity. Unpublished Ph.D. thesis.

Thompson, J. N. 1981. Reversed animal-plant interactions: the evolution of insectivorous and ant-fed plants. *Biol. J. Linn. Soc.* 16:147–155.

Thompson, J. N. 1982. Interaction and coevolution. John Wiley and Sons, NY.

Thompson, J. N., and M. F. Willson. 1979. Evolution of temperate fruit/bird interactions: phenological strategies. *Evolution* 33:973–982.

Thomson, J. D. 1978. Effect of stand composition on insect visitation in two-species mixtures of *Hieracium. Am. Midl. Natur.* 100:431–440.

Thomson, J. D. 1981. Spatial and temporal components of resource assessment by flower-feeding insects. *J. Anim. Ecol.* 50:49–59.

Thomson, J. D., and R. C. Plowright. 1980. Pollen carryover, nectar rewards, and

pollinator behavior with special reference to *Diervilla lonicera. Oecologia (Berl.)* 46:68–74.
Thorpe, W. H. 1945. The evolutionary significance of habitat selection. *J. Anim. Ecol.* 14:67–70.
Thurston, R., and R. M. Fox. 1972. Inhibition by nicotine of emergence of *Apenteles congregatus* from its host, the tobacco hornworm. *Ann. Entomol. Soc. Amer.* 65:547–550.
Tiffney, B. H. 1984. Seed size, dispersal syndromes and the rise of the angiosperms: evidence and hypothesis. *Ann. M. Bot. Gard.* 71:551–576.
Tilman, D. 1978. Cherries, ants and tent caterpillars: timing of nectar production in relation to susceptibility of caterpillars to ant predation. *Ecology* 59:686–692.
Tilman, D. 1982. Resource competition and community structure. Princeton University Press, Princeton, NJ.
Tomoff, C. S. 1974. Avian species diversity in desert scrub. *Ecology* 55:396–403.
Tonn, W. M., and J. J. Magnuson. 1982. Patterns in the species composition and richness of fish assemblages in northern Wisconsin lakes. *Ecology* 63:1149–1166.
Tordoff, H. B., and W. R. Dawson. 1965. The influence of daylength on reproductive timing in the red crossbill. *Condor* 67:416–422.
Torssell, K. B. G. 1983. Natural product chemistry. A mechanistic and biosynthetic approach to secondary metabolism. John Wiley and Sons, NY.
Trail, P. W. 1980. Ecological correlates of social organization in a communally breeding bird, the acorn woodpecker, *Melanerpes formicivorus. Behav. Ecol. Sociobiol.* 7:83–92.
Trelease, S. F., and H. M. Trelease. 1937. Toxicity to insects and mammals of foods containing selenium. *Am. J. Bot.* 24:448–451.
Tripathi, R. S. 1977. Weed problem—an ecological perspective. *Tropical Ecol.* 18:138–148.
Turcek, F. J. 1963. Color preference in fruit and seed-eating birds. *Proc. Int. Ornithol. Congr.* 13:285–292.
Turnbull, C. L., and D. C. Culver. 1983. The timing of seed dispersal in *Viola nuttallii:* attraction of dispersers and avoidance of predators. *Oecologia (Berl.)* 59:360–365.
Ulfstrand, S. 1963. Ecological aspects of irruptive bird migration in northeastern Europe. *Proc. XIII Int. Ornithol. Congr.* pp. 780–794.
USDA (U.S. Department of Agriculture). 1985. Losses in agriculture. Agric. Handbook No. 291. Agric. Res. Serv., Washington, DC.
USDA (U.S. Department of Agriculture). 1986. Research for tomorrow. 1986. Yearbook of Agriculture, U.S. Dep. of Agric., Washington, DC.
Valdeyron, G., and D. G. Lloyd. 1979. Sex differences and flowering phenology in the common fig, *Ficus carica* L. *Evolution* 33:673–685.
Van de Merendonk, S., and J. van Lenteren. 1978. Determination of mortality of greenhouse whitefly *Tiraleurodes vaporarium* (Westwood) (Homoptera: Aleyrodidae) eggs, larvae and pupae on four host-plant species: eggplant (*Solanum melongena* L.), cucumber (*Cucumis sativa* L.), tomato (*Lycopersicum esculentum* L.), and paprika (*Capsicum annuum* L.). *Med. Fac. Landbouww. Rijksuniv. Gent.* 43:421–429.
Van der Wall, S. B., and R. P. Balda. 1977. Coadaptations of the Clark's nutcracker and the pinon pine for efficient seed harvest and dispersal. *Ecol. Monogr.* 47:89–111.
Van Hook, R. I. 1971. Energy and nutrient dynamics of spider and orthopteran populations in a grassland ecosystem. *Ecol. Monogr.* 41:1–26.
Van Hook, R. R., M. G. Nielson, and H. H. Shugart. 1980. Energy and nitrogen relations for a *Macrosiphun liriodendri* (Homoptera: Aphididae) population on an east Tennessee *Liriodendron tulipfera* stand. *Ecology* 61:960–975.
Van Schaik, C. P., and J. A. R. A. M. van Hooff. 1983. On the ultimate causes of primate social systems. *Behaviour* 85:91–117.
Vaughan, T. A. 1972. Mammalogy. W. B. Saunders, Philadelphia, PA.
Verner, J., and M. F. Willson. 1966. The influence of habitats on mating systems of North American passerine birds. *Ecology* 47:143–147.
Vickery, M. L., and B. Vickery. 1981. Secondary plant metabolism. University Park Press, Baltimore, MD.

Vickery, P. J. 1972. Grazing and net primary production of a temperate grassland. *J. Appl. Ecol.* 9:307–314.
Vogel, W. G. 1966. Stem growth and apical meristem elevation related to grazing resistance of three prairie grasses. Proc. 9th Int. Grassland Congr., Sao Paulo, 1965. Secretary of Agriculture, Brazil. pp. 345–348.
Waddington, K. D. 1983. Foraging behavior of pollinators. pp. 213–329. In: L. Real, ed. Pollination biology. Academic Press, NY.
Waddington, K. D., and B. Heinrich. 1979. The foraging movements of bumblebees on vertical inflorescences: an experimental analysis. *J. Comp. Physiol.* 134:113–117.
Waddington, K. D., and L. R. Holden. 1979. Optimal foraging: on flower selection by bees. *Am. Natur.* 114:179–196.
Wagner, M. R., and P. D. Evans. 1985. Defoliation increases nutritional quality and allelochemics of pine seedlings. *Oecologia (Berl.)* 67:235–237.
Wahua, T. A. T. 1984. Rhyzosphere bacterial counts for intercropped maize (*Zea mays* L.), cowpea (*Vigna uniguiculata* L.), and 'egusi' melon (*Colosynthis vulgaris* L.). *Field Crops Res.* 8:371–380.
Waldbauer, G. P. 1968. The consumption and utilization of food by insects. *Advan. Insect. Physiol.* 5:229–289.
Wallace, A. R. 1878. Tropical nature and other essays. Macmillan, NY.
Waller, D. M. 1984. Differences in fitness between seedlings derived from cleistogamous and chasmogamous flowers in *Impatiens capensis. Evolution* 38:427–440.
Wallwork, J. A. 1970. Ecology of soil animals. McGraw-Hill, London. 283 p.
Ward, D. B., and D. Fish. 1979. Powdery *Catopsis.* pp. 74–75. In D. B. Ward, ed. Rare and endangered biota of Florida, vol. 5. University Press of Florida, Gainesville, FL.
Waser, N. M. 1978a. Competition for hummingbird pollination and sequential flowering in two Colorado wildflowers. *Ecology* 59:934–944.
Waser, N. M. 1978b. Interspecific pollen transfer and competition between co-occurring plant species. *Oecologia (Berl.)* 36:223–236.
Waser, N. M. 1979. Pollinator availability as a determinant of flowering time in ocotillo *(Fouquieria splendens). Oecologia (Berl.)* 39:107–121.
Waser, N. M. 1983a. The adaptive nature of floral traits: ideas and evidence. pp. 241–285. In: L. Real, ed. Pollination biology. Academic Press, NY.
Waser, N. M. 1983b. Competition for pollination and floral character differences among sympatric plant species: a review of evidence. pp. 277–293. In: C. E. Jones, and R. J. Little, eds. Handbook of experimental pollination biology. Van Nostrand Reinhold, NY.
Waser, N. M., and M. V. Price. 1983. Optimal and actual outcrossing in plants, and the nature of plant-pollinator interaction. pp. 341–359. In: C. E. Jones, and R. J. Little, eds. Handbook of experimental pollination biology. Van Nostrand Reinhold, NY.
Waser, N. M., and L. A. Real. 1979. Effective mutualism between sequentially flowering plant species. *Nature* 281:670–672.
Waser, P. M. 1977. Feeding, ranging and group size in the mangabey *Cercocebus albigena.* pp. 183–222. In: T. H. Clutton-Brock, ed. Primate ecology: feeding and ranging behaviour in lemurs, monkeys and apes. Academic Press, London.
Waser, P. M. 1987. Interactions among primate species. In: B. Smuts, D. L. Cheney, R. M. Wrangham, and T. T. Struhsaker, eds. Primate societies. University of Chicago Press, Chicago, IL (in press).
Waser, P. M., and R. H. Wiley. 1980. Mechanisms and evolution of spacing in animals. pp. 159–223. In: P. Armler, and J. G. Vandenbergh, eds. Handbook of behavioral neurobiology. Plenum Press, NY.
Waser, P. M., and T. J. Case. 1981. Monkeys and matrices: on the coexistence of "omnivorous" forest primates. *Oecologia (Berl.)* 49:102–108.
Watson, A. 1977. Population limitation and the adaptive value of territorial behavior in Scottish red grouse *Lagopus 1. scoticus.* pp. 19–26. In: B. Stonehouse, and C. Perrins, eds. Evolutionary ecology. University Park Press, Baltimore, MD.
Watson, A., and G. R. Miller. 1971. Territory size and aggression in fluctuating red grouse population. *J. Anim. Ecol.* 40:367–383.
Watson, A., and R. Moss. 1970. Dominance, spacing behaviour and aggression in relation to population limitation in vertebrates. pp. 167–218. In: A. Watson, ed. Animal

populations in relation to their food resources. Blackwell Scientific Publications, Oxford.

Watson, A., and R. Moss. 1972. A current model of population dynamics in red grouse. *Proc. XV Int. Ornith. Congr.* pp. 134–149.

Watson, A. P., J. N. Matthiessen, and B. P. Springett. 1982. Arthropod associates and macronutrient status of the red-ink sundew *(Drosera erythrorhiza). Aust. J. Ecol.* 7:13–22.

Watt, A. S. 1981a. A comparison of grazed and ungrazed Grassland A in East Anglian Breckland. *J. Ecol.* 69:499–508.

Watt, A. S. 1981b. Further observations on the effects of excluding rabbits from Grassland A in East Anglian Breckland: the pattern of change and factors affecting it (1936–73). *J. Ecol.* 69:509–536.

Watts, D. P. 1985. Relations between group size and composition and feeding competition in mountain gorilla groups. *Anim. Behav.* 20:1–8.

Webb, C. J., and D. G. Lloyd. 1986. The avoidance of interference between the presentation of pollen and stigmas in angiosperms II. Herkogamy. *N. Z. J. Bot.* 24:163–178.

Weber, N. A. 1972. Gardening ants: the Attines. Amer. Philosophical Soc., Philadelphia, PA.

Webster, C. C., and P. N. Wilson. 1980. Agriculture in the tropics. Longman, NY.

Wecker, S. C. 1963. The role of early experience in habitat selection by the prairie deer mouse, *Peromyscus maniculatus bairdi. Ecol. Monogr.* 33:307–325.

Wecker, S. C. 1964. Habitat selection. *Sci. Am.* 211:109–116.

Wegner, J. F., and G. Merriam. 1979. Movement by birds and small mammals between a wood and adjoining farmland habitats. *J. Appl. Ecol.* 16:349–358.

Wehner, R., R. D. Harkness, and P. Schmid-Hempel. 1983. Foraging strategies in individually searching ants *Cataglyphis bicolor* (Hymenoptera: Formicidae). Gustav Fisher Verlag, Stuttgart.

Weir, J. S., and R. Kiew. 1986. A reassessment of the relations in Malaysia between ants *(Crematogaster)* on trees *(Leptospermum* and *Dacrydium)* and epiphytes of the genus *Dischidia* (Asclepiadaceae) including "ant-plants". *Biol. J. Linn. Soc.* 27:113–132.

Weis, A. E., and W. G. Abrahamson. 1986. Evolution of host-plant manipulation by gall makers. Ecological and genetic factors in the *Solidago-Eurosta* system. *Am. Natur.* 127:681–695.

Weis, A. E., W. G. Abrahamson, and K. D. McCrea. 1985. Host gall size and oviposition success by the parasitaoid *Eurytoma gigantea. Ecol. Entomol.* 10:341–348.

Weiss, T. E., Jr. 1980. The effects of fire and nutrient availability on the pitcher plant *Sarracenia flava* L. Ph.D. thesis, University of Georgia, Athens, GA.

Weller, S. G., and R. Ornduff. 1977. Cryptic self-incompatibility in *Amsinckia grandiflora. Evolution* 31:47–51.

Werner, E. E. 1977. Species packing and niche complementarity in three sunfishes. *Am. Natur.* 111:553–578.

Westoby, M. 1974. An analysis of diet selection by large herbivores. *Am. Natur.* 108:290–301.

Westoby, M., B. Rice, J. M. Shelley, D. Haig, and J. L. Kohen. 1982. Plant's use of ants for dispersal at West Head, New South Wales. pp. 75–88. In: R. C. Buckley, ed. Ant–plant interactions in Australia. Dr. W. Junk, Publisher, The Hague.

Wheeler, W. M. 1910. Ants, their structure, development and behavior. Columbia University Press, NY.

Wheeler, W. M. 1942. Studies of neotropical ant-plants and their ants. *Bull. Mus. Comp. Zool. Harv. Univ.* 90:1–262.

Wheelwright, N. T. 1983. Fruits and the ecology of the Resplendent Quetzal. *Auk* 100:286–301.

Wheelwright, N. T. 1985. Fruit size, gape width, and the diets of fruit-eating birds. *Ecology* 66:808–818.

Wheelwright, N. T., and C. H. Janson. 1985. Colors of fruit displays of bird-dispersed plants in two tropical forests. *Am. Natur.* 126:777–799.

Wheelwright, N. T., and G. H. Orians. 1982. Seed dispersal by animals: contrasts with pollen dispersal, problems of terminology, and constraints on coevolution. *Am. Natur.* 119:402–413.
Wherry, E. T. 1935. Distribution of the North American pitcherplants. pp. 1–23. In: M. V. Walcott, ed. Illustrations of North American pitcherplants. Smithsonian Institution Press, Washington, DC.
White, R. G. 1983. Foraging patterns and their multiplier effects on productivity of northern ungulates. *Oikos* 40:377–384.
White, S. C. 1974. Ecological aspects of growth and nutrition in tropical fruit-eating birds. Ph.D. thesis, University of Pennsylvania, Philadelphia, PA.
Whitehead, D. R. 1983. Wind pollination: some ecological and evolutionary perspectives. pp. 97–108. In: L. Real, ed. Pollination biology. Academic Press, NY.
Whitford, W. G. 1978. Foraging in seed-harvester ants *Pogonomyrmex* spp. Ecology 59:185–189.
Whitham, T. G. 1978. Habitat selection by pemphigus aphids in response to resource limitation and competition. *Ecology* 59:1164–1176.
Whitham, T. G. 1980. The theory of habitat selection: examined and extended using *Pemphigus* aphids. *Am. Natur.* 115:449–466.
Whiting, M. G. 1963. Toxicity of cycads. *J. Econ. Bot.* 17:271–302.
Whittaker, R. H. 1975. Communities and ecosystems. Macmillan, NY.
Whitten, W. M. 1981. Pollination ecology of *Monarda didyma, M. clinopodia,* and hybrids (Lamiaceae) in the southern Appalachian mountains. *Am. J. Bot.* 68:435–442.
Wickler, W. 1968. Mimicry in plants and animals. Macmillan, NY.
Wiebes, J. T. 1979. Co-evolution of figs and their pollinators. *Ann. Rev. Ecol. Syst.* 10:1–12.
Wiegert, R. G. 1964. The ingestion of xylem sap by the meadow spittle bug *Philaenus spumarius* (L.) *Am. Midl. Natur.* 71:422–428.
Wiegert, R. G., and F. C. Evans. 1967. Investigations of secondary productivity in grasslands. pp. 499–518. In: E. Petrusewicz, ed. Secondary productivity of terrestrial ecosystems, vol. 2. Polish Acad. Sci., Warsaw.
Wiens, J. A. 1972. Anuran habitat selection: early experience and substrate selection in *Rana cascadae* tadpoles. *Anim. Behav.* 20:218–220.
Wiens, J. A. 1977. On competition and variable environments. *Am. Sci.* 65:590–597.
Wigglesworth, V. B. 1972. The principles of insect physiology. Chapman and Hall, London.
Wilbur, H. M., P. J. Morin, and R. N. Harris. 1983. Salamander predation and the structure of experimental communities: anuran responses. *Ecology* 64:1423–1429.
Wiley, R. H., and M. S. Wiley. 1980. Spacing and timing in the nesting ecology of a tropical blackbird: comparison of populations in different environments. *Ecol. Monogr.* 50:153–178.
Williams, S. E., and A. B. Bennett. 1982. Leaf closure in the Venus flytrap *(Dionaea muscipula):* an acid growth response. *Science* 218:1120–1122.
Williamson, G. W. 1982. Plant mimicry: evolutionary constraints. *Biol. J. Linn. Soc.* 18:49–58.
Willis, E. O. 1966. Interspecific competition and the foraging of plain-brown woodcreepers. *Ecology* 47:667–672.
Willis, E. O. 1980. Ecological roles of migratory and resident birds on Barro Colorado Island, Panama. pp. 205–226. In: A. Keast, and E. S. Morton, eds. Migrant birds in the neotropics. Smithsonian Institution Press, Washington, DC.
Willson, M. F. 1966. The breeding ecology of the yellow-headed blackbird. *Ecol. Monogr.* 36:51–77.
Willson, M. F. 1979. Sexual selection in plants. *Am. Natur.* 113:777–790.
Willson, M. F. 1983. Plant reproductive ecology. John Wiley and Sons, NY.
Willson, M. F., and R. I. Bertin. 1979. Flower-visitors, nectar production, and inflorescence size of *Asclepias syriaca. Can. J. Bot.* 57:1380–1388.
Willson, M. F., and N. Burley. 1983. Mate choice in plants: tactics, mechanisms and consequences. Monographs in Population Biology 19. Princeton University Press, Princeton, NJ.

Willson, M. F., and B. J. Rathcke. 1974. Adaptive design of the floral display in *Asclepias syriaca* L. *Am. Midl. Natur.* 92:47–57.
Willson, M. F., and J. N. Thompson. 1982. Phenology and ecology of color in bird-dispersed fruits, or why some fruits are red when they are "green". *Can. J. Bot.* 60:701–713.
Willson, M. F., J. R. Karr, and R. R. Roth. 1975. Ecological aspects of avian bill-size variation. *Wilson Bull.* 87:32–44.
Wilson, E. O. 1971. The insect societies. Belknap, Cambridge, MA.
Wilson, E. O. 1980. Caste and division of labor in leaf-cutter ants (Hymenoptera: Formicidae: *Atta*). II. The ergonomic optimization of leaf cutting. *Behav. Ecol. Sociobiol.* 7:157–165.
Wilson, E. O., and G. L. Hunt. 1966. Habitat selection by the queens of two field-dwelling species of ants. *Ecology* 47:485–487.
Wilson, E. O., and R. W. Taylor. 1967. The ants of Polynesia. *Pac. Insect Monogr.* 14:1–109.
Wilson, S. D. 1985. The growth of *Drosera intermedia* in nutrient-rich habitats: the role of insectivory and interspecific competition. *Can. J. Bot.* 63:2468–2469.
Wittenberger, J. F. 1980. Vegetation structure, food supply, and polygyny in bobolinks *(Dolichonyx oryxzivorus)*. *Ecology* 61:140–150.
Woodbury, A. M., and R. Hardy. 1948. Studies of the desert tortoise, *Gophurus agassizii*. *Ecol. Monogr.* 18:145–200.
Woolfenden, G. E., and J. W. Fitzpatrick. 1984. The Florida scrub jay: demography of a cooperative-breeding bird. Princeton University Press, Princeton, NJ.
Wooten, R. J. 1981. Paleozoic insects. *Ann. Rev. Entomol.* 26:319–344.
Wrangham, R. W. 1977. Feeding behaviour of chimpanzees in Gombe National Park, Tanzania. pp. 504–538. In: T. H. Clutton-Brock, ed. Primate ecology. Academic Press, NY.
Wrangham, R. W. 1980. An ecological model of female-bonded primate groups. *Behaviour* 75:262–300.
Wrangham, R. W. 1986. Ecology and social relationships in two species of chimpanzee. pp. 352–378. In: D. I. Rubenstein, and R. W. Wrangham, eds. Ecological aspects of social evolution-birds and mammals. Princeton University Press, Princeton, NJ.
Wyatt, R. 1980. The impact of nectar-robbing ants on the pollination system of *Asclepias curassavica*. *Bull. Torrey Bot. Club* 107:24–28.
Wyatt, R. 1981. Ant-pollination of the granite outcrop endemic *Diamorpha smallii* (Crassulaceae). *Am. J. Bot.* 68:1212–1217.
Wyatt, R. 1983. Pollinator-plant interactions and the evolution of breeding systems. pp. 51–95. In: L. Real, ed. Pollination biology. Academic Press, NY.
Zavada, M. S., and T. N. Taylor. 1986. The role of self-incompatibility and sexual selection in the gymnosperm-angiosperm transition: a hypothesis. *Am. Natur.* 128:538–550.
Zimmerman, J. L. 1982. Nesting success of dickcissels *(Spiza americana)* in preferred and non-preferred habitats. *Auk* 99:292–298.
Zimmerman, M. 1979. Optimal foraging: a case for random movement. *Oecologia (Berl.)* 43:261–267.
Zimmermann, J. 1932. Uber die extraflorale Nectarien der Angiospermen. *Bot. Centr. Beih.* 49:99–196.
Zucker, W. V. 1982. How aphids choose leaves: the roles of phenolics in host selection by a galling aphid. *Ecology* 63:972–981.

INDEX

(Page numbers in *italic* indicate definitions)

About the Editor

Dr. Warren G. Abrahamson is the David Burpee Professor of Plant Genetics at Bucknell University in Lewisburg, Pennsylvania. He teaches courses in the ecology of plant–animal interations, organic evolution, and population and community biology and is the recipient of numerous awards including a Lindback Award in 1975 and the Bucknell Class of 1956 Lectureship in 1982. Dr. Abrahamson specializes in three important areas of research: the ecological and evolutionary interations of plants and animals; fire ecology in the Southeast United States; and clonal plant population biology.

First published in the United States in 2006
by Stone Arch Books,
151 Good Counsel Drive, P.O. Box 669,
Mankato, Minnesota 56002.
www.stonearchbooks.com

Originally published in Great Britain in 2003
by A & C Black Publishers Ltd.

Library of Congress Cataloging-in-Publication Data
Wallace, Karen.
Aargh, It's an Alien! / by K. Wallace; illustrated by Michael Reid.
p. cm. — (Graphic Trax)
ISBN-13: 978-1-59889-023-5 (hardcover)
ISBN-10: 1-59889-023-9 (hardcover)
1. Graphic novels. I. Title: Aargh, it is an alien!. II. Reid, Michael, Illustrator.
III. Title. IV. Series.
PN6727.W2747A27 2006
741.5—dc22 2005026691

Summary: More than anything, Albert Twiddle wants to spend time with his parents, but Mr. and Mrs. Twiddle are always working. When Albert starts spending time with his alien friends, something interesting happens!

1 2 3 4 5 6 11 10 09 08 07 06

Printed in the United States of America.

BY K. WALLACE

illustrated by Michael Reid
colors by Jessica Fuchs

Librarian Reviewer
Laurie K. Holland
Media Specialist (National Board Certified), Edina, MN
MA in Elementary Education, Minnesota State University, Mankato, MN

Reading Consultant
Sherry Klehr
Elementary/Middle School Educator, Edina Public Schools, MN
MA in Education, University of Minnesota, MN

STONE ARCH BOOKS
Minneapolis San Diego